Lecture Notes in Mathematics

Edited by A. Dold and B. Eckmann

867

Séminaire d'Algèbre Paul Dubreil et Marie-Paule Malliavin

Proceedings, Paris 1980
(33ème Année)

Edité par M.P. Malliavin

Springer-Verlag
Berlin Heidelberg New York 1981

Editeur
Marie-Paule Malliavin
Université Pierre et Marie Curie – Mathématiques
10, rue Saint Louis en l'Ile, 75004 Paris, France

AMS Subject Classifications (1980): 05A15, 12A85, 13F15, 13G05, 13H15, 13N05, 14D99, 14F10, 14H99, 14L30, 16A08, 16A15, 16A27, 16A33, 16A38, 16A39, 16A62, 16A72, 16A74, 17B35, 20B25, 20C10, 20C20, 20G05, 22E35, 58A10, 58B20, 58C40

ISBN 3-540-10841-6 Springer-Verlag Berlin Heidelberg New York
ISBN 0-387-10841-6 Springer-Verlag New York Heidelberg Berlin

CIP-Kurztitelaufnahme der Deutschen Bibliothek
Séminaire d' Algèbre Paul Dubreil et Marie-Paule Malliavin:
Proceedings/ Séminaire d'Algèbre Paul Dubreil et Marie-Paule Malliavin. –
Berlin; Heidelberg; New York: Springer
33. 1980. Paris 1980: (33.année). – 1981.
(Lecture notes in mathematics; Vol. 867)
ISBN 3-540-10841-6 (Berlin, Heidelberg, New York);
ISBN 0-387-10841-6 (New York, Heidelberg, Berlin)
NE: GT

Printing and binding: Beltz Offsetdruck, Hemsbach/Bergstr.
2141/3140-543210

Liste des auteurs

J. Alev p. 351 - M. Bayart p. 174 - G. Besson p. 130 - J.E. Björk p.148 -
F. Couchot p. 380 - J.P. Van Deuren p. 295 - D. Eisenbud p. 141 -
J.W. Fischer p. 365 - R.M. Fossum p. 1 - J. Van Geel p. 295 - A.W. Goldie p. 396
L. Gruson p. 234 - W.H. Hesselink p. 55 - C.U. Jensen p. 234 - T. Kimura p. 38 -
A. Lévy-Bruhl-Laperrière p. 98 - M. Lorenz p. 406 - I.G. Mac Donald p. 90 -
G. Musili p. 441 - F. Van Oystaeyen p. 295 - K.W. Roggenkamp p. 421 -
C.S. Seshadri p. 441 - C. Schoeller p. 214 - A. Verschoren p. 319.

*

TABLE DES MATIERES

Robert M. FOSSUM - Invariant Theory, Representation theory, Commutative Algebra - ménage à trois 1

Tatsuo KIMURA - On the construction of some relative invariants for GL(n) (n=6,7,8) by the decomposition of the Young diagrams 38

Wim H. HESSELINK - Concentration under actions of algebraic groups 55

*

I.G. Mac DONALD - Some Conjectures for root systems and finite Coxeter groups 90

Anne LEVY-BRUHL-LAPERRIERE - Spectre du de Rham-Hodge sur l'espace projectif quaternionique 98

G. BESSON - Groupe de Lie p-adique, Immeuble et Cohomologie 130

*

David EISENBUD - Report on normal bundles of curves in P_3 141

*

Jan-Erik BJÖRK - The Bernstein class of modules on algebraic manifolds 148

Thierry LEVASSEUR - Anneaux d'opérateurs différentiels 157

*

Marc BAYART - Factorialité et séries formelles irréductibles II 174

Colette SCHOELLER - Homologie d'anneaux locaux de dimension d'immersion 3 214

L. GRUSON et C.U. JENSEN - Dimensions cohomologiques reliées aux foncteurs $\varprojlim^{(i)}$ 234

*

J.P. Van DEUREN, J. Van GEEL et F. Van OYSTAEYEN - Genus and a Riemann-Roch Theorem for non-commutative function fields in one variable 295

A.VERSCHOREN-Pour une géométrie algébrique non commutative 319

*

Jak ALEV - Sur la formule de Molien dans certaines algèbres enveloppantes 351

Joe W. FISHER - Semi-prime ideals in Rings with finite group actions revisited 365

*

François COUCHOT - Les modules artiniens et leurs enveloppes quasi-injectives 380

Alfred W. GOLDIE - The reduced rank in noetherian rings 396

*

Martin LORENZ - Prime ideals in Group Algebras of Polycyclic-by-finite Groups : Vertices and Sources 406

K.W. ROGGENKAMP - Structure of integral group rings 421

*

C. MUSILI et C.S. SESHADRI - Standard Monomial theory 441

publié avec le concours de :

Université Pierre et Marie Curie,
Première Section de l'Ecole Pratique des Hautes Etudes,
Centre National de la Recherche Scientifique.

PREVIOUS VOLUMES OF THE "SÉMINAIRE PAUL DUBREIL" WERE PUBLISHED IN THE LECTURES NOTES, VOLUMES 536 (1976), 641 (1977), 740 (1978) AND 795 (1979).

Invariant Theory, Representation Theory, Commutative algebra-ménage à trois

Robert M. Fossum

The purpose of this lecture is to give a survey of some of the interesting relationships between selected topics involved in the area mentioned in the title. Cognoscenti are aware of the fundamental interactions and know that several lecture years can be devoted to each area. Thus it is not possible to delve deeply into each subject.

One can consider these topics to live in the house of algebraic geometry-and it is in this sense that they form a ménage à trois. I have picked three problems to discuss and have tried to show how these are related. In keeping with the anology, this would be similar to the examination of the more superficial daily life of a human ménage à trois while ignoring or, at best, alluding to the more important relationships that bind the threesome together.

The discussion will proceed in historical order. There are five sections:

I. Elliptic curves - my favorite example.

II. N-ary R-forms

III. Symmetric groups

IV. Invariants in positive characteristic

V. Problems.

This report should be considered preliminary to a longer survey which will examine more deeply the topics involved. The effort has been supported by the United States National Science Foundation.

Dedicated to Barbara

I. Elliptic Curves - My favorite example

I begin with a discussion of elliptic curves, as found in [Mumford and Suominen (1972)], because it is classical, illustrates one method of obtaining examples and shows the relationship between the topics at a level many can understand.

Let k be an algebraically closed field of characteristic $\neq 2, 3$. An elliptic curve E over k is, by definition, a complete nonsingular curve of genus 1. A study of the cohomology theory of E leads to the discovery of a double covering $\pi : E \to \mathbb{P}^1_k$. If $\pi' : E \to \mathbb{P}^1_k$ is another double covering, there is an automorphism γ of $\mathbb{P}^1$ such that $\pi' = \gamma \circ \pi$. Conversely, any automorphism γ yields another (isomorphic) double covering. The branch locus of π is denoted by B_π. Then p in $\mathbb{P}^1_k$ is in B if and only if $\pi^{-1}(p)$ consists of exactly one point. Again it can be shown that B_π consists of 4 distinct points

$$\left\{\begin{pmatrix} a_1 \\ b_1 \end{pmatrix}, \begin{pmatrix} a_2 \\ b_2 \end{pmatrix}, \begin{pmatrix} a_3 \\ b_3 \end{pmatrix}, \begin{pmatrix} a_4 \\ b_4 \end{pmatrix}\right\} = B_\pi .$$

Conversely, given 4 distinct points B as above, there is an elliptic curve E and a double covering $\pi : E \to \mathbb{P}^1$ with branch locus B. So the branch loci determine elliptic curves, up to isomorphism. Thus to study elliptic curves $/k$, we can study sets of 4 distinct points in $\mathbb{P}^1_k$. If we want functions on the space of elliptic curves that are constant on isomorphism classes and distinguish isomorphism classes, it should be enough to consider functions on $\mathbb{P}^1_k$ that somehow depend on sets of 4 distinct points, no matter how they are ordered, but are also invariant when those points are moved about by $\mathrm{Aut}(\mathbb{P}^1)$.

We can begin to normalize by finding $\gamma \in \mathrm{Aut}(\mathbb{P}^1) = PGL_2(k)$ such that

$$\gamma\begin{pmatrix} a_1 \\ b_1 \end{pmatrix} = \begin{pmatrix} 0 \\ 1 \end{pmatrix} = 0$$

$$\gamma\begin{pmatrix} a_2 \\ b_2 \end{pmatrix} = \begin{pmatrix} 1 \\ 1 \end{pmatrix} = 1$$

$$\gamma\begin{pmatrix} a_3 \\ b_3 \end{pmatrix} = \begin{pmatrix} 1 \\ 0 \end{pmatrix} = \infty$$

Then $$\gamma\begin{pmatrix} a_4 \\ b_4 \end{pmatrix} = \begin{pmatrix} \lambda \\ 1 \end{pmatrix} = \lambda$$

where $\lambda \neq 0,1,\infty$. Let $B_\lambda = \{0,1,\infty,\lambda\}$.

(Such a γ is the matrix

$$\begin{pmatrix} -b_1(a_3b_2-a_2b_1), & a_1(a_3b_2-a_2b_3) \\ -b_3(a_1b_2-a_2b_1), & a_3(a_1b_2-a_2b_1) \end{pmatrix} .$$ Then

$$\gamma\begin{pmatrix} a_4 \\ b_4 \end{pmatrix} = \begin{pmatrix} (a_1b_4-a_4b_1)(a_3b_2-a_2b_3) \\ (a_3b_4-a_4b_3)(a_1b_2-a_2b_1) \end{pmatrix} ,$$ so

$$\lambda = \frac{(a_1b_4-a_4b_1)(a_3n_2-a_2b_3)}{(a_1b_2-a_2b_1)(a_3b_4-a_4b_5)} ,$$

the cross-ratio between the four points.)

The extension $k(x) \to k(x,y)$ corresponding to this normalized double covering is $y^2 = x(x-1)(x-\lambda)$.

The group of permutations on four letters S_4 acts on the branch locus just permuting the points. If S_4 is generated by its neighbor transpositions s_1, s_2, s_3, then we get

$$s_1B_\lambda = \{1,0,\infty,\lambda\}$$
$$s_2B_\lambda = \{0,\infty,1,\lambda\}$$
$$s_3B_\lambda = \{0,1,-1,\infty\}.$$

For any $w \in S_4$, we can find $\gamma_w \in PGL_2(k)$ so that

$$\gamma_w(w(0)) = 0$$

$$\gamma_w(w(1)) = 1$$

$$\gamma_w(w(\infty)) = \infty$$

Then

$$\gamma_w(w(\lambda)) =: \lambda^w.$$

A calculation shows

$$\lambda^{s_1} = 1 - \lambda$$

$$\lambda^{s_2} = \frac{\lambda}{\lambda-1}$$

$$\lambda^{s_3} = 1 - \lambda.$$

In particular we get an action

$$S_4 \times U \to U$$

where $U = \mathbb{P}^1 - \{0,1,\infty\}$ defined as above. This action extends to all of $\mathbb{P}^1_k$ by

$$0^{s_1} = 1 \qquad 1^{s_1} = 0 \qquad \infty^{s_1} = \quad ,$$

$$0^{s_2} = 0 \qquad 1^{s_2} = \infty \qquad \infty^{s_2} = 1, \quad \text{etc.}$$

Let $\mathbb{P}^1 = \mathrm{Proj}(k[X,Y])$ and consider how S_4 acts on $\mathrm{Proj}(k[X,Y])$. In general if $f(X,Y)$ is homogeneous and $w \in S_4$, the rule is

$$(w^* f(X,Y))\binom{a}{b} = f(X,Y)(w^{-1}\binom{a}{b})$$

for $\binom{a}{b} \in \mathbb{P}^1$. So we calculate:

$$(s_1^*X)\binom{a}{b} = X(s_1\binom{a}{b}) = X\binom{b-a}{b} = b - a$$

$(s_1^*Y)\binom{a}{b} = b$. Hence

$$s_1^*X = Y - X, \quad s_1^*Y = Y.$$

Likewise

$$s_3^*X = Y - X, \quad s_3^*Y = Y,$$

and

$$s_2^*X = X \quad s_2^*Y = X - Y.$$

The general action then is

$$w^*f(X,Y) = f(w^*X, w^*Y).$$

In particular, we have the 1-forms

$$V = kX + kY$$

and a (linear) action

$$S_4 \times V \to V,$$

which extends to an action

$$S_4 \times S^\bullet(V) \to S^\bullet(V)$$

on the symmetric algebra $S^\bullet(V)$ of V.

It is clear that $\dim_k V = 2$ and that V is an irreducible representation of S_4 over k. (The representation is not faithful. It is clear that s_1s_3 fixes each element, and therefore the normal subgroup V generated by s_1s_3 is in the kernel of $S_4 \to GL_k(V)$. The quotient group is S_3 and $S_3 \to GL(V)$ is an injection.)

There are now two exercises. The first is to determine the quotient

$$\mathbb{P}^1 \xrightarrow{\sigma} \mathbb{P}^1/S_4 =: E$$

(or if we want the faithful action

$$\mathbb{P}^1 \xrightarrow{\sigma} \mathbb{P}^1/S_3)_e$$

if it exists.

The second is to determine the ring of invariants

$$k[X,Y]^{S_3}$$

(where we have used the faithful action).

Then we must compare the two. The second problem is an easy exercise. Let $P(X,Y) = \frac{1}{6}\,\mathrm{Tr}(X^2) = X^2 - XY + Y^2$ and $Q(X,Y) = \frac{1}{6}\,\mathrm{Tr}(X^3) = XY(X-Y)$. Since then char $(k) \neq 2,3$, the homomorphism

$$E : k[X,Y] \to k(X,Y]^{S_3}$$

given by $E(g) = \frac{1}{6} \sum_{w \in S_3} w(g)$ is surjective. Since

$$X^3 - XP(X,Y) - Q(X,Y) = 0$$

and

$$Y^2 - XY + (X^2 - P(X,Y)) = 0,$$

it follows that

$$(k[X,Y] : k[P,Q]) \leq 6.$$

Hence

$$k[P,Q] = k[X,Y]^{S_3}.$$

We can use this to help in the solution of the first exercise. For we have a map

$$\mathrm{Proj}(k[X,Y]) \to \mathrm{Proj}\ k[P,Q].$$

Since $\deg P = 2$ and $\deg Q = 3$, we know that

$$\mathrm{Proj}\ k[P,Q] = \mathrm{Proj}\ k[P^3,Q^2] = \mathbb{P}^1_k.$$

Consider the map

$$\mathbb{P}^1 \xrightarrow{\sigma} \mathbb{P}^1$$

gotten by $\sigma\binom{a}{b} = \begin{bmatrix} 2^8 P(a,b)^3 \\ Q(a,b)^2 \end{bmatrix}$. This is well defined since $P(a,b) = 0 = Q(a,b)$ implies that $a = 0 = b$. It is also clear that

$$\sigma\, w\binom{a}{b} = \sigma\binom{a}{b}$$

for all $w \in S_4$.

Let us examine the fibres. It is clear that

$$\sigma^{-1}(\infty) = \{\binom{0}{1}, \binom{1}{1}, \binom{1}{0}\} = \{0,1,\infty\}$$

$$\sigma^{-1}(0) = \{\binom{w}{1}, \binom{\bar{w}}{1}\} \quad \text{where} \quad w^3 + 1 = 0$$

$$\sigma^{-1}\begin{pmatrix} (12)^3 \\ 1 \end{pmatrix} = \{\binom{-1}{1}, \binom{2}{1}, \binom{1}{2}\}.$$

The remaining points have complete fibres, namely

$$\sigma^{-1}\binom{j}{1} = \{\lambda, 1-\lambda, \frac{\lambda}{\lambda-1}, \frac{\lambda-1}{\lambda}, \frac{1}{\lambda}, \frac{\lambda}{1-\lambda}\}$$

where

$$j(\lambda) = 2^8\, \frac{(\lambda^2-\lambda+1)^3}{\lambda^2(\lambda-1)^2} \quad .$$

Consider $\mathbb{P}^1 - \{\infty\}$ and its preimage under σ:

$$\mathbb{P}^1 - \{0,1,\infty\} \xrightarrow{\sigma} \mathbb{P}^1 - \{\infty\}.$$

Since $\mathbb{P}^1 - \{\infty\} = \mathbb{A}^1$ with

$$\mathbb{A}^1 = \text{Spec } k[j]$$

considered as the complement of $Q^2 = 0$, and since $\mathbb{P}^1 - \{0,1,\infty\} = \mathbb{A}^1 - \{0,1\} =$ Spec $k[x,(x(x-1))^{-1}]$, we get an induced action of S_3 on

$$k[x,(x(x-1))^{-1}]$$

given by

$$s_1(x) = 1 - x$$

$$s_2(x) = \frac{x}{x-1} = + \frac{x^2}{x(x-1)} .$$

Let $p(x) = x(x-1) + 1$ and $q(x) = x(x-1)$. Again a calculation shows that

$$k[x,(x(x-1))^{-1}] = k[2^8 \frac{p(x)^3}{q(x)^2}].$$

The map $\mathbb{A}^1 - \{0,1\} \xrightarrow{\sigma} \mathbb{A}^1$ defined by

$$\sigma(\lambda) = j(\lambda) = 2^8 \frac{p(\lambda)^3}{q(\lambda)^2}$$

is that induced by $k[j] \uparrow k[x,q(x)^{-1}]$ is given by $j \to j(x)$.

So the elliptic curve E determines a branch locus $\{0,1,\infty,\lambda\} \subset \mathbb{P}^1$. Then $j(E) = 2^8 \frac{p(\lambda)^3}{q(\lambda)^2}$ in $\mathbb{A}^1$ is an invariant of E. If $j(E) = j(E')$, then $E \cong E'$.

<u>Geometry</u>: The family of elliptic curves over k is isomorphic to $\mathbb{A}^1 = \mathrm{Spec}(k[j])$. (A coarse moduli space for elliptic curves.) This j is the classical j-invariant.

<u>Invariant Theory</u>: We have seen several actions of S_3 (and S_4) on commutative rings, computed the invariants and seen how this helps to determine the geometric quotients.

<u>Representation Theory</u>: The action of S_3 (and S_4) on the linear terms of $k[X,Y]$ arises from a 2 dimensional representation of S_3. This group has three irreducible representations (when char $(k) \neq 2,3$), namely the trivial and alternating 1-dimensional representations, and this two dimensional representation.

<u>Commutative Algebra</u>: The ring of invariants of S_3 acting on $k[X,Y]$ is again a polynomial ring, whose properties are quite well understood.

II. N-ary R-forms

Let k be a field and $x_1,\dots,x_N$ variables. An N-ary R-form is a homogeneous polynomial of degree R in $k[x_1,\dots,x_N]$. A general form can be written

$$f(x_1,\dots,x_N) =: \sum_{\substack{I=(i_1,\dots,i_N)\\ I = i_1+\dots+i_N=R}} A_I x_1^{i_1} \cdots x_N^{i_N},$$

where the A_I are considered to be variables. Let $g = (g_{ij})$ be an element in $GL_N(k)$ and let it act on the variables $\{x_i\}$ by

$$gx_i := \sum_{j=1}^{N} g_{ji}x_j.$$

We define an action of g on the A_I, denoted $g^* A_I$ by

$$\sum_I (g^* A_I)(gx_1)^{i_1} \cdots (gx_N)^{i_N} = \sum_I A_I x^{i_1} \cdots x_N^{i_N},$$

and then extend this action to all polynomials in $k[\{A_I\}_{|I|=R}]$. Let $S_{N,R} = k[\{A_I\}]$. If $\chi : GL_N(k) \to k^\times$ is a linear character, an element F in $S_{N,R}$ in a χ-relative invariant if

$$g^* F = \chi(g)F$$

for all g in $GL_N(k)$ and it is an absolute invariant if

$$g^* F = F$$

for all g.

Consider the exact sequence

$$1 \to SL_N(k) \to GL_N(k) \xrightarrow{\det} k^\times \to 1.$$

since $\chi(g) = (\det g)^b$ for some $n \in \mathbb{Z}$, it follows that a relative invariant for $GL_N(k)$ is an invariant for $SL_N(k)$.

(If $k = \bar{k}$, or if k has N^{th} roots, then relative invariants can be recovered from $SL_N(k)$-invariants. Notice that for $\lambda \in k^x$, the action of $\lambda \cdot Id_N$ on A_I is given by

$$(\lambda \cdot Id_N)^* A_I = \lambda^{-R} A_I.$$

If F is homogeneous of degree g in $S_{N,R}$, then $(\lambda \cdot Id_N)^* F = \lambda^{-Rq} F$. So if F is a relative invariant, then

$$\lambda^{-Rq} F = \lambda^{Nm} F$$

and, provided $F \not\equiv 0$, then $Nm = -Rq$. Let $g \in GL_N(k)$ and write

$$g = \lambda g_1 \qquad \lambda^N = \det g$$

where $g_1 \in SL_N(k)$. Also let $p = \gcd(N,R)$ with $N = N_1 p$, $R = R_1 p$, so $q = q_1 N_1$ and $m = -q_1 R_1$. Then

$$g^* F = \lambda^* g_1^* F.$$

If F is an $SL_N(k)$-invariant, then $g^* F = \lambda^* F = \det g)^m = \lambda^{Nm} F = {}^{-Rq} F$ and F is a relative invariant for $\chi(g) = \det(g) M$.)

Cayley first asked to find all relative invariants - he wanted an algorithm, and many authors have studied the problem since Cayley and Sylvester began their investigations over 100 years ago. Faa de Bruno [1876] lists those who made contributions up to that time and Schur [1968], Popp [1977] and Springer [1977] have more recent bibliographies.

An algorithm to find these invariants led to the symbolic method for writing the invariants, but the computations became much too difficult to carry out. Not much progress has been made in explicitly describing these invariants. In particular, let $P_{N,R} := S_{N,R}^{SL_N(k)}$ denote the ring of invariants.

The following list is known

$$P_{N,1} = k.$$

$$P_{N,2} = k[\det(A_{ij})]$$

where

$$\sum_{i,j} x_i A_{ij} x_j = f(x_1,\ldots,x_N)$$

is the 2-form, $A_{ij} = A_{ji}$ and then (A_{ij}) is the general symmetric bilinear form. These calculations, as well as the next two can be found in Schur [1968]. Explicit generators and relations (syzygies) are known for

$$P_{2,R}; \quad R \leq 6$$

$$P_{3,R}; \quad R \leq 3$$

Shioda [1967] has found generators for $P_{2,7}$, $P_{2,8}$ and $P_{3,4}$. (In case $N = 2$, write $f(x_1,x_2) = \sum_{i=0}^{R} A_i x_1^{R-i} x_2^i$ so that $A_{r-i,i} = A_i$.)

In fact in case $N = 2$, $R = 3$, just as in the case $N = 2$, $R = 2$, the discriminant of the form is the only invariant.

In case $N = 2$, $R = 4$, there are 2 algebraically independent invariants:

The <u>Apolare</u> $\frac{1}{6}P$, where

$$P = 12A_0A_4 - 3A_1A_3 + A_2^2$$

and the <u>Hankel</u> determinant:

$$\det\begin{pmatrix} A_0 & \frac{A_1}{4} & \frac{A_2}{6} \\ \frac{A_1}{4} & \frac{A_2}{6} & \frac{A_3}{4} \\ \frac{A_2}{6} & \frac{A_3}{4} & A_4 \end{pmatrix} = \frac{1}{(12)^3}\det\begin{pmatrix} 12A_0 & 3A_1 & 2A_2 \\ 3A_1 & 2A_2 & 3A_3 \\ 2A_2 & 3A_3 & 12A_4 \end{pmatrix}$$

$$= \frac{1}{(12)^3}\{2\cdot(12)^2 A_0A_2A_4 - 12\cdot 9(A_0A_3^2 + A_1^2A_4) + 4\cdot 9 A_1A_2A_3 - 8A_2^3\}$$

$$= \frac{4}{(12)^3}\{72 A_0A_2A_4 + 9 A_1A_2A_3 - 27(A_0A_3^2 + A_1^2A_4) - 2A_2^3\}$$

$$=: \frac{4}{(12)^2} Q_1(A_0,\dots,A_4).$$

It follows that

$$\operatorname{disc}(A_0x_1^4 + A_1x_1^3x_2 + \dots + A_4x_2^4) = 3^{-3}(4P^3 - Q_1^2),$$

which is also invariant.

Consider two special cases:

Legendre: $x_1x_2(x_1-x_2)(x_1-\lambda x_2) = f(x_1,x_2)$.

Then

$$P(0,1,-(1+\lambda),\lambda,0) = \lambda^2 - \lambda + 1 \quad \text{and}$$

$$Q_1(0,1,-(1+\lambda),\lambda,0) = (\lambda+1)(2\lambda-1)(\lambda-2)$$

Weierstrass: $4x_1^3x_2 - g_2x_1x_2^3 - g_3x_2^4$.

Then

$$P(0,4,0,-g_2,-g_3) = 12g_2,$$

$$Q_1(0,4,0,-g_2,-g_3) = \frac{(12)^3}{4} g_3.$$

It is clear that there is a relation with elliptic curves.

In particular

$P_{2,3} = k[D]$ where

$$D = 3^3(A_0A_3)^2 = (A_1A_2)^2 + 4(A_0A_2^3 + A_1^3A_3) - 18\ A_0A_1A_2A_3$$

$P_{2,4} = k[P,Q_1]$

where P, Q_1 are as above. (From the previous section

$$3^3Q^2 = 4P^3 - Q_1^2$$

or

$$Q_1^2 = 4P^3 - 27Q^2.)$$

The problems stated here can be framed in a representation theoretic picture. Let V be a finite dimensional vector space over k. (We write $N = \dim_k V$.) Let $G = SL_k(V)$, the space of k-automorphisms of V of determinant 1. Then the Rth symmetric power, $S^R(V)$, affords a representation of G as does its dual space $S^R(V)^*$ (the <u>contragredient representation</u>). Then consider the ring of symmetric powers of $S^R(V)^*$. This is just $S_{N,R}$ when a basis $x_1,\dots,x_N$ of V is chosen. Then $P_{N,R}$ is the ring of invariants of $S_{N,R}$ by the G-action. The qth homogeneous component of $S_{N,R}$ is

$$S^q(S^R(V)^*).$$

So we can ask for the decomposition of this space into its irreducible G-subspaces. There is an invariant if there is an irreducible subspace of dimension 1 (over k). (Some restrictions on k are necessary. We assume $k = \overline{k}$ and char $(k) = 0$, for example.)

The irreducible G-representations are parameterized by sequences $m_1 \geq m_2 \geq \cdots \geq m_{N-1} \geq 0$ (of weights) cf. [Hartshorne (1966)]. So it would be sufficient to know the decompositions of these symmetric spaces in order to calculate the invariants.

For $G = SL_k(V)$ with $\dim_k V = 2$, the algebra of the irreducibles is well known. (Suppose char (k) = 0 and that we have then identified $S^q(W^*) = S^q(W)^*$. For problems that arise in case char $k = p > 0$ see IV and [Almkvist-Fossum (1978)].) The irreducible G-modules are $S^q(V)$ for $0 \le q$. Furthermore we know that for $q \ge p$

$$S^q(V) \underset{k}{\otimes} S^p(V) = \bigoplus_{v=0}^{p} S^{q+p-2v}(V),$$

a decomposition of G-modules.

(This is the "formula of Clebsch-Gordan" [Springer (1977), p. 50]. It has a rather simple proof, that we give in the appendix. It goes by induction from the decomposition

$$S^q(V) \underset{k}{\otimes} V = S^{q+1}(V) \oplus S^{q-1}(V).)$$

From this formula, it follows that the "binomial coefficient"

$$\left\{ \begin{matrix} S^q(V) \\ S^p(V) \end{matrix} \right\} := \frac{S^q(V) \otimes \cdots \otimes S^{q-p+1}(V)}{S^p(V) \otimes \cdots \otimes S'(V) \quad S^o(V)}$$

is a well defined representation of G and that

$$S^r(S^q(V)) \cong \left\{ \begin{matrix} S^{q+v-1}(V)) \\ S^{q-1}(V) \end{matrix} \right\}$$

and

$$\Lambda^r(S^q(V)) \cong \left\{ \begin{matrix} S^{q-1}(V) \\ S^{r-1}(V) \end{matrix} \right\} .$$

Apply this to the case $R = 2$.

$$S^1(S^2(V)) = S^2(V)$$

$$S^2(S^2(V)) = \frac{S^2(V) \otimes S^3(V)}{S^0(V) \otimes S^1(V)} = \frac{S^5(V) \oplus S^3(V) \oplus S^1(V)}{S^1(V)}$$

$$= S^4(V) \oplus S^0(V) \qquad \text{(1-invariant) the discriminant}$$

$$S^3(S^2(V)) = \frac{S^2(V) \otimes S^3(V) \otimes S^4(V)}{S^0(V) \otimes S^1(V) \otimes S^2(V)}$$

$$= \frac{S^3(V) \otimes S^4(V)}{S^1(V)} = \frac{S^7(V) \oplus S^5(V) \oplus S^3(V) \oplus S^1(V)}{S^1(V)}$$

$$= S^6(V) \oplus S^2(V).$$

In general

$$S^r(S^2(V)) = \bigoplus_{v=0}^{[\frac{r}{2}]} S^{2r-4v}(V)$$

and there is an invariant only in case r is even, and then only one of them. So it must be the power of the invariant from $S^2(S^2(V))$.

What about $R = 3$.

$$S^1(S^3(V)) = S^3(V).$$

$$S^2(S^3(V)) = S^3(V) \otimes S^4(V)/S^1(V) = S^6(V) \oplus S^2(V).$$

$$S^3(S^3(V)) = S^9(V) \oplus S^7(V) \oplus S^5(V).$$

$$S^4(S^3(V)) = S^{12}(V) \oplus S^8(V) \oplus S^6(V) \oplus S^4(V) \oplus S^0(V).$$

There is an invariant, the discriminant again.

In case $R = 4$

$$S^2(S^4(V)) = S^8(V) \oplus S^0(V) \oplus S^0(V),$$ yielding an invariant, proportional to P.

$$S^3(S^4(V)) = S^{12}(V) \oplus S^8(V) \oplus S^6(V) \oplus S^4(V) \oplus S^0(V),$$

yielding an invariant proportional to Q_1.

$$S^4(S^4(V)) = S^{17}(V) \oplus S^{13}(V) \oplus S^{11}(V) \oplus S^9(V) \oplus S^9(V) \oplus S^5(V) \oplus S^5(V) \oplus S^0(V),$$

the $S^0(V)$ component being an invariant corresponding to P^2.

The general problem can be stated: Determine the decomposition of the representations $S^q(S^R(V))$ into irreducible $SL_k(V)$ representations.

The ideal theoretic properties of the invariant rings are now better understood, due largely to the following result by [Hochster and Roberts (1974)].

<u>Theorem.</u> <u>Let</u> G <u>be a linearly reductive affine linear algebraic group over a field</u> K <u>(of arbitrary characteristic) acting</u> K <u>rationally on a regular noetherian</u> K-<u>algebra</u> S. <u>Then the ring of invariants</u> S^G <u>is Cohen-Macaulay</u>.

In case char $(k) = 0$, the group $SL_k(V)$ is semi-simple provided $\dim V \geq 2$, and linearly reductive. By [Fogarty (1969) V Ex. 5] the ring of invariants in the theorem above will be factorial if S is factorial. Then by [Murthy (1964)], the ring S^G will be Gorenstein.

<u>Theorem.</u> <u>The ring of invariants</u>

$$P_{N,R}$$

<u>is a factorial Gorenstein ring</u> (in case char $(k) = 0$).

(See also Geyer's paper in [Popp (1974)] where these problems are considered in case char $(k) = p > 0$.)

We conclude this section with a brief mention of the geometry involved. For more details see [Mumford-Suominen].

The group $SL_k(V)$ acts on V and hence on $\mathbb{P}(V)$, the space of lines through 0 in V. Hence there is a diagonal action of $SL_k(V)$ on a product

$$\mathbb{P}(V) \times \cdots \times \mathbb{P}(V) = \mathbb{P}(V)^{\times R},$$

and this action commutes with the natural action of the symmetric group S_R on this product.

Let $W = S^R(V)$ and define a map $\sigma : \mathbb{P}(V)^{\times R} \to \mathbb{P}(W)$ by

$$\sigma \begin{pmatrix} a_{11} & a_{12} & \cdots & a_{1R} \\ a_{21} & a_{22} & & a_{2R} \\ a_{N1} & a_{N2} & & a_{NR} \end{pmatrix} = (b_{i_1 \ldots i_N}) \quad \text{where}$$

$$\prod_{j=1}^{R} (a_{ij} X_1 + \cdots + a_{Nj} X_N) = \sum_{i_1 + \cdots + i_N = R} b_{i_1 \ldots i_N} X_1^{i_1} \cdots X_N^{i_N}.$$

Note that this action is $SL_k(V)$ equivariant, when $W = S^R(V)$ gets its $SL_K(V)$ structure through the action on V.

(On the affine level, this is just

$$(v_1, \ldots, v_R) \to v_1 \ldots v_R.)$$

Let $\mathbb{P}(V)^{(R)} = \mathbb{P}(V)^{\times R}/S_R$. So there is induced a map

$$\mathbb{P}(V)^{(R)} \to \mathbb{P}(W).$$

Now we want to study the projective space $\mathbb{P}(W)$ with this $SL_k(V)$-action. If $2 = \dim V$, then

$$\mathbb{P}(V)^{(R)} \cong \mathbb{P}_k^R$$

with an action of $SL_2(k)$.

For $R = 4$ we are "close" to the study of elliptic curves. For even $R > 4$ this is related to the study of hyperelliptic curves.

Summary and Conclusion

Geometry: There is an action of $SL_2(k)$ on $\mathbb{P}_k^R$ and in general an action of $SL_N(k)$ on $\mathbb{P}^{\binom{N+R-1}{N}-1}$ whose orbit spaces, although not always well defined have subsets that are interesting algebraic sets. In case $R = 4$ and $N = 2$, one can get a coarse moduli space (see [Mumford-Suominen, 4, Prop. 2]) for elliptic curves.

Invariant Theory: The action of $GL_k(V)$ and $SL_k(V)$ on $S^R(V)$ induces actions of these groups on the symmetric algebras $S^{\cdot}(S^R(V))$. Those invariants have been studied for many years.

Representation Theory: The irreducible representations of $SL(V)$ and $GL(V)$ are well known. The "Algebra" of these representations can aid in finding the dimensions of the space of invariants.

Commutative Algebra: The $SL_k(V)$ invariants of $S^{\cdot}(S^R(V))$ form a Gorenstein unique factorization ring, and sometimes a regular ring.

Appendix: Proof of the Clebsch-Gordan Formula.

Let char $k = 0$. Suppose V is a vector space over k. Let (r,s) be a pair of integers and define

$$M^{r,s} = \Lambda^r(V) \underset{k}{\otimes} S^s(V).$$

Then define

$$d^r_s : M^{r,s} \to M^{r-1,s+1}$$

by

$$d^r_s(v_1 \wedge v_2 \wedge \cdots \wedge v_r \otimes w) = \sum_{i=1}^{r} (-1)^{r-1} v_1 \wedge \cdots \wedge \hat{v}_i \wedge \cdots \wedge v_r \otimes v_i w$$

for $v_i \in V$ and $w \in S^s(V)$. It is clear that each d^r_s is $GL_k(V)$-equivariant.

Define also

$$e^s_r : M^{r,s} \to M^{r+1,s-1}$$

by $e^s_r(w \otimes v_1 \dots v_s) = \sum_{i=1}^{s} (w \wedge v_i) \otimes v_1 \dots \hat{v}_i \dots v_s$.

It is clear that e^s_r is also $GL_k(V)$ equivariant. A calculation shows that

$$d^{r+1}_{s-1} \; e^s_r + e^{s+1}_{r-1} \; d^r_s = (r+s)\,\mathrm{Id}.$$

Hence the complex

$$0 \to M^{r+s,0} \to M^{r+s-1,1} \to \dots \to M^{0,r+s} \to 0$$

is exact and splits as a $GL_k(V)$-sequence. Apply this to the case $\dim V = 2$ to get the complex which is $GL_k(V)$-split.

$$0 \to M^{2,r+s-2} \to M^{1,r+s-1} \to M^{0,r+s} \to 0,$$

since $M^{r,s} = 0$ for $r > 2$. Hence

$$V \otimes S^{n-1}(V) \cong S^{n-2}(V) \oplus S^{n}(V)$$

as $GL_k(V)$-modules.

<u>Theorem</u> (Clebsch-Gordan Formula): <u>There is a decomposition</u> $(m \leq n)$

$$S^n(V) \underset{k}{\otimes} S^m(V) = \bigoplus_{v=0}^{m} S^{n+m-2v}(V)$$

<u>as</u> $GL_k(V)$<u>-modules</u>.

Proof. Go by induction on m. The case $m = 1$ is the formula above.

If general, suppose the formula holds for m. Then

$$S^n(V) \otimes S^m(V) \otimes V = \bigoplus_{v=0}^{m} S^{n+m-2v}(V) \otimes V.$$

Hence we get

$$S^n(V) \otimes (S^{m+1}(V) \oplus S^{m-1}(V))$$

$$= (\bigoplus_{v=0}^{m} S^{n+m-2v+1}(V)) \oplus (\bigoplus_{v=0}^{m} S^{n+m-2v-1}(V)).$$

But

$$S^n(V) \otimes S^{m-1}(V) = \bigoplus_{v=0}^{m-1} S^{n+m-2v-1}(V)$$

by the induction hypothesis.

As the modules $S^r(V)$ are irreducible $GL_k(V)$-modules, it follows that

$$S^n(V) \otimes S^{m+1}(V) = \bigoplus_{v=0}^{m+1} S^{n+m+1-2v}(V)$$

(unless $m + 1 > n$, in which case the argument is left for the reader to complete. //

III Symmetric Groups

In this section we turn to the study of representations of finite groups, in particular the symmetric groups S_n.

Let V be a vector space over k. (In general we could consider a k-module in case k is not a field.) Suppose G is a group and

$$\rho : G \to GL_k(V)$$

is a group homomorphism yielding a representation of G over k. Suppose $n \in N$ and set

$$T^n(V) = V_k^{\otimes n} = \overbrace{V \otimes_k \cdots \otimes_k V}^{n \text{ times}}.$$

As before G acts on $T^n(V)$ by the diagonal action, viz.

$$g(v_1 \otimes \cdots \otimes v_n) = gv_1 \otimes gv_2 \otimes \cdots \otimes gv_n$$

for g in G and $v_i \in V$. Identify S_n with the permutations of $\{1,2,\ldots,n\}$ (acting on the right of elements). Then V^{n} becomes an S_n-module by

$$(v_1 \otimes \cdots \otimes v_n)w := v_{1w} \otimes \cdots \otimes v_{nw}$$

for $w \in S_n$. Note that the G and S_n actions commute.

Suppose W is a right representation of S_n over k, that is W is a right kS_n-module. Then the set

$$\mathrm{Hom}_{kS_n}(W, T^n(V)) =: W(V)$$

becomes a G representation over k. If W is a left representation, then

$$T^n(V) \underset{kS_n}{\otimes} W =: W[V]$$

is also a representation of G over k. These are called the Schur functors of V.

Examples: The Nth symmetric power of V. Let $W = ku_1$ be the free k-module of rank 1 with $wu_1 = u_1$ for all $w \in S_n$ giving the left action of S_n on W (for all n). This is the trivial kS_n-module. Then the Nth symmetric power of V is the Schur functor

$$S_k^N(V) := T^n(V) \underset{kS_n}{\otimes} ku_1 .$$

The Nth exterior power of V. This module is usually defined as the quotient of V^n modulo the submodule generated by all tensors of the form $v_1 \otimes v_2 \otimes \cdots \otimes w \otimes w \otimes \cdots \otimes v_n$. In case $\mathbb{Q} \subset k$, we can get the Nth exterior power

$$\Lambda_k^N(V) := \mathrm{Hom}_{kS_n}(ku_{-1}, T^n(V))$$

where $u_{-1}w = (-1)^{\ell(w)} u_{-1}$ is the alternating representation of S_n.

Set $S_2(V) := \mathrm{Hom}_{S_2}(ku_1, V^{\otimes 2})$.

In general, let $i_V\colon \mathrm{Hom}_{kS_2}(ku_1, V^{\otimes 2}) \to V^{\otimes 2}$ denote the inclusion defined by

$$i_V(f) := f(u_1) \quad \text{for } f \in \mathrm{Hom}_{kS_2}(ku_1, V^{\otimes 2}).$$

This map is $GL_k(V)$ equivariant. For general $N \in \mathbb{N}$, and for $1 \leq i < N$, let

$$W_i := V^{\otimes(i-1)} \otimes S_2(V) \otimes V^{\otimes(N-i-1)} .$$

Define $e_i\colon W_i \to V^{\otimes N}$ by

$$e_i := Id_V^{\otimes(i-1)} \otimes i_V \otimes Id_V^{\otimes(N-i-1)},$$

and then $e: \bigoplus_{i=1}^{N-1} W_i \quad V^{\otimes N}$ as the sum of the e_i. Then $\mathrm{Im}(e)$ is the submodule of $V^{\otimes N}$ generated by the tensors mentioned above. So

$$\Lambda_k^N(V) := \mathrm{Coker}\ e.$$

This defines the exterior powers in terms of Schur functors, independently of the characteristic. (Note: Let $S_3(V) := \mathrm{Hom}_{kS_3}(ku_1, V^{\otimes 3})$. The restriction maps along the inclusions

$$S_2 \times S_1 \to S_3 \leftarrow S_1 \times S_2$$

induce homomorphisms

$$f_1 : S_3(V) \to S_2(V) \otimes V$$

and

$$f_2 : S_3(V) \to V \otimes S_2(V).$$

The following sequence of modules and maps is a complex that is acyclic at least when $\mathbb{Q} \subset k$.

$$\text{(III.1)} \quad 0 \to S_3(V) \xrightarrow{\begin{pmatrix} -f_1 \\ f_2 \end{pmatrix}} (S_2(V) \otimes V) \oplus (V \otimes S_2(V)) \xrightarrow{(e_1, e_2)} V^{\otimes 3} \to \Lambda^3(V) \to 0 .$$

All the maps are $GL_k(V)$ equivariant.

Now it follows that $(W_1 \oplus W_2)(V) = W_1(V) \oplus W_2(V)$ as G-modules. So to know all schur functors, it is sufficient to know the indecomposable Schur functors. In case $\mathbb{Q} \subseteq k$, these are obtained from the irreducible representations of S_n. (See [Macdonald (1980)].)

Once we have a representation of G on V and then on the $W(V)$, we can form the symmetric algebra over k on $W(V)$ and ask for the invariants.

In Case $G = SL_k(V)$, and $W = ku_1$, we are back to Cayley's problem, to determine the $SL_k(V)$ invariants of $S^{\cdot}(S^R(V))$. Or we can consider representations $S_m \to GL_k(V)$ and then the associated representations of S_m on the Schur functors of V. These lead back to other topics in classical invariant theory.

Examples: Let m be an integer, $m > 1$, and suppose $\dim_k V = m$. (We want $\mathbb{Q} \subseteq k$ for some of the calculations.) Then S_m acts on V by permuting a basis of V, the <u>regular</u> representation of V over k. Then $S^\cdot(V) = k[X_1,\ldots,X_m]$, a polynomial ring. It is well known that

$$S^\cdot(V)^{S_m} = k[E_1,\ldots,E_m],$$

where

$$\prod_{i=1}^{m} (1 + X_i t) =: \sum_{r=0}^{m} E_r(X_1,\ldots,X_m)t^r.$$

The $E_1,\ldots,E_m$ are the elementary symmetric functions in the $X_1,\ldots,X_m$.

Let $\Delta = \prod_{i<j} (X_i - X_j)$. Then it is also classical that

$$S^\cdot(V)^{A_m} = k[E_1,\ldots,E_m,\Delta]$$

(where A_m is the alternating subgroup of S_m). The vector space V is decomposable as an S_m-module, in characteristic zero, and in any case contains nontrivial S_m-submodules.

Let V have basis $e_1,\ldots,e_m$ and let $a_1 = e_1 +\cdots+ e_m$. Then ka_1 is the trivial one dimensional representation of V. Let $f_{i-1} = e_i - e_1$ for $i = 2,3,\ldots,m$. Let s_i be the transposition in S_m that interchanges i and i + 1. If we suppose that $m^{-1} \in k$, then V decomposes into $V = ka_1 \oplus S_{m-1,1}$, where $E_{m-1,1}$ is the submodule spanned by $f_1,\ldots,f_{m-1}$. Note that

$$s_1 f_1 = -f_1$$

$$s_1 f_i = f_i - f_1 \quad \text{for } i > 1$$

$$s_i f_j = f_j \quad \text{for } j \notin \{i, i+1\}$$

$$s_i f_i = f_{i+1}.$$

Hence indeed $A_{m-1,1}$ is an S_m-submodule. Also $f_1 + \cdots + f_{m-1} + me_1 = a_1$ and $f_i + e_1 = e_{i+1}$.

Thus it can be seen that

$$ka_1 + S_{m-1,1} = V$$

$$ka_1 \cap S_{m-1,1} = 0$$

in case $m^{-1} \in k$.

It is now an easy task to compute $S^{\cdot}(S_{m-1,1})^{S_m}$. For it follows that

$$S^{\cdot}(V) = S^{\cdot}(ka_1) \otimes_k S^{\cdot}(S_{m-1,1}).$$

As $S^{\cdot}(ka_1) = k[E_1]$ is S_m-invariant, it follows that

$$S^{\cdot}(V)^{S_m} = k[E_1] \otimes_k S^{\cdot}(S_{m-1,1})^{S_m}.$$

Define $Y_j = X_j - \frac{1}{m} E_1$ in $S^{\cdot}(V)$ for $j = 1,\ldots,m$. Then $Y_1 + \cdots + Y_m = 0$. Let $Z_i = X_{i+1} - X_1 = Y_{i+1} - Y_1$ for $i = 1,2,\ldots,m-1$. Then the $Z_1,\ldots,Z_{m-1}$ generate $S^{\cdot}(S_{m-1,1})$ as an algebra. The map $S^{\cdot}(V) \to S^{\cdot}(S_{m-1,1})$ given by

$$X_1 \to -\frac{1}{m}(Z_1 + \cdots + Z_{m-1})$$

$$X_2 \to Z_1 - \frac{1}{m}(Z_1 + \cdots + Z_{m-1})$$

$$\vdots$$

$$X_m \to Z_{m-1} - \frac{1}{m}(Z_1 + \cdots + Z_{m-1})$$

is S_m equivariant with kernel generated by E_1. Hence

$$S^{\cdot}(S_{m-1,1})^{S_m} = k[Z_1,\dots,Z_{m-1}]^{S_m} = k[E_2(Z),\dots,E_m(Z)],$$

a polynomial ring, where, by $E_r(Z)$, we mean

$$E_r(Z) := E_r(-\frac{1}{m}(Z_1 + \cdots + Z_{m-1}),\ Z_2 - \frac{1}{m}(Z_1 + \cdots + Z_{m-1}),\dots).$$

In the particular cases $m = 2$, we get

$$E_2(-\frac{1}{2}Z_1, Z_1 - \frac{1}{2}Z_1) = \frac{1}{4}Z_1^2 E_2(-1,1) = -\frac{1}{4}Z_1^2 E_2(-1,1) = -\frac{1}{4}Z_1^2.$$

($S^{\cdot}(S_{1,1}) = k[Z_1]$ with action $s_1 Z_1 = -Z_1$.)

If $m = 3$, we get

$$E_2(-\frac{1}{3}(Z_1+Z_2), Z_1 - \frac{1}{3}(Z_1+Z_2),\ Z_2 - \frac{1}{3}(Z_1+Z_2)) = -\frac{1}{3}(Z_1^2 - Z_1Z_2 + Z_2^2)$$

and

$$E_3(Z) = \frac{1}{27}(Z_1+Z_2((2Z_1-Z_2)(Z_1-2Z_2).$$

Let $X = \frac{Z_1+Z_2}{3}$, $Y = \frac{2Z_1-Z_2}{3}$. Then $X - Y = \frac{2Z_2-Z_1}{3}$ and we get

$$E_2(-X,Y,X-Y) = -P(X,Y)$$

(See §I).

$$E_3(-X,Y,X-Y) = -Q(X,Y)$$

Also taking $Z_1 = \lambda$, $Z_2 = 1$ gives

$$E_3(Z) := \frac{1}{27}(\lambda+1)(2\lambda-1)\ -2) = \frac{1}{27}Q_1(\lambda)$$

from §II.

In general, let

$$\begin{aligned} Y_1 &= -\frac{1}{m}(Z_1 + \cdots + Z_{m-1}) \\ Y_2 &= Z_1 - \frac{1}{m}(Z_1 + \cdots + Z_{m-1}) \\ &\ \vdots \\ Y_{m-1} &= Z_{m-2} - \frac{1}{m}(Z_1 + \cdots + Z_{m-1}) \end{aligned}$$

Then $Z_{m-1} - \frac{1}{m}(Z_1 + \cdots + Z_{m-1}) = -(Y_1 + \cdots + Y_{m-1})$.

It follows that

$$E_2(Y_1, Y_2, \ldots, -(Y_1 + \cdots + Y_{m-1})) = E_2(Y_1, \ldots, Y_{m-1}) - E_1^2$$

$$= E_2(Y_1, \ldots, Y_{m-1}) - E_1(Y_1, \ldots, Y_{m-1})^2,$$

$$\vdots$$

$$E_j(Y_1, \ldots, Y_{m-1}, -(Y_1 + \cdots + Y_{m-1})) = E_j(Y_1, \ldots, Y_{m-1}) - E_{j-1}(Y_1, \ldots, Y_{m-1})E_1$$

for $j \leq m - 1$ and

$$E_m(Y_1, \ldots, Y_{m-1}, -(Y_1 + \cdots + Y_{m-1})) = -E_{m-1}E_1.$$

Our representation of S_m on $k[Y_1, \ldots, Y_{m-1}]$ is given by

$$s_i Y_j = Y_j \quad \text{if } j \notin \{i, i+1\} \text{ and } i \leq m - 1.$$

$$\left.\begin{array}{l} s_i Y_i = Y_{i+1} \\ s_i Y_{i+1} = Y_i \end{array}\right. \quad \text{if } i < m - 1$$

and

$$s_{m-1} Y_{m-1} = -(Y_1 + \cdots + Y_{m-1}).$$

The ring of invariants

$$k[Y_1, \ldots, Y_{m-1}]^{S_m} = k[P_2, P_3, \ldots, P_m]$$

where $P_r = E_r(Y_1, \ldots, Y_{m-1}) - E_{r-1}(Y_1) \cdots Y_{m-1})E_1(Y_1, \ldots, Y_{m-1})$

for $r < m$ and

$$P_m = -E_{m-1}(Y_1, \ldots, Y_{m-1})E_1(Y_1, \ldots, Y_{m-1}).$$

(The E_i are the elementary symmetric functions.)

It is probably time to state a result due to Chevalley [1955] found in Bourbaki (1968).

Say $g \in GL_k(V)$ is a pseudo-reflexion if

$$rk(Id_V - g) = 1.$$

A reflection is a pseudo-reflexion.

Theorem (Chevalley): Suppose the finite subgroup $G \subset GL_k(V)$ is generated by pseudo-reflections. Then the subring of invariants

$$S.(V)^G$$

is a polynomial ring generated by $\dim_k V$ elements (provided char $k = 0$).

So in case char $k = 0$, the calculations above explicitly find these invariants. The exercises in the same section of Bourbaki yield an example in positive characteristic where the above theorem is not true.

We mention several other results that are of interest in the case of finite groups. Again assume char $(k) = 0$ and $G \subset GL_k(V)$, with $\dim_k V < \infty$. Then G acts on the homogeneous components of the symmetric algebra $S.(V)$, so the ring of invariants is a graded ring.

For a graded ring $A = \bigoplus_{n \geq 0} A_n$, let the Hilbert-Poincaré series be given by

$$P_A(t) = \sum_{r=0} rk_k(A_r)t^r,$$

an element in $\mathbb{Z}[[t]]$.

Theorem (Molien (1897)): Let $A = S.(V)^G$. Then

$$P_A(t) = \frac{1}{(G:1)} \sum_{g \in G} \det(1-gt)^{-1}.$$

When $G = S_m$ and V is the regular representation, then

$$P_A(t) = (1-t)^{-1} \cdots (1-t^m)^{-1}$$

for $A = S.(V)^G$. When we consider the representation $S_{m-1,1}$ and consider $A = S.(S_{m-1,1})^{S_m}$, then

$$P_A(t) = (1-t^2)^{-1} \cdots (1-t^m)^{-1}.$$

On the other hand, consider the irreducible representation of S_5 of dimension 5 corresponding to the partition (3.2), in characteristic zero. The Molien series is

$$\frac{1+t^6+t^7+t^8+t^9+t^{15}}{(1-t^2((1-t^3)(1-t^4)(1-t^5)(1-t^6)}$$

which is not of the form $\Pi(1-t^{d_i})^{-1}$, and hence the ring of invariants is not a polynomial ring.

<u>Summary</u>:

<u>Geometry</u>. The irreducible representations of the symmetric groups induced actions of these groups on projective spaces. The orbit spaces sometimes classify families of varieties, eg. elliptic curves.

<u>Invariant Theory</u>. The classical symmetric functions are invariants for the regular representation of S_n. In characteristic zero, the invariants form a Cohen-Macaulay ring, but not necessarily a Gorenstein ring. And the ring need not be factorial.

<u>Representation Theory</u>. The representations of the symmetric groups are well understood [James].

Commutative Algebra. The invariant rings form a wealth of examples that can be used to substantiate or defeat conjectures. The algebra of the representations of the symmetric groups is used to find resolutions of determinantal ideals [Lascoux (1978), Nielsen (1978), (1979)].

IV. Invariants in positive characteristic.

The problems encountered in invariant theory when the ground field has positive characteristic are manifold. In the first case, there are few linearly reductive groups, so the Hochster-Roberts result does not hold. In the following discussion, the field k is assumed to have characteristic $p > 0$. The problems arise already when one considers cyclic p-groups. So another restriction will be to consider unipotent actions of order a power of p.

Let V be a finite dimensional vector space over k, say $\dim_k V = r$, and suppose $u: V \to V$ satisfies $(u-\mathrm{Id})^r = 0$, and r is the minimal such number, Then there is a basis $e_{r-1},\ldots,e_0$ of V such that

$$ue_j = e_j + e_{j-1} \qquad \text{for } 1 \leq j \leq r-1$$

and

$$ue_0 = e_0.$$

If $p^n < r \leq p^{n+1}$, then a quick calculation shows that

$$u^{p^{n+1}} = \mathrm{Id}.$$

Let $\mu_{p^{n+1}}$ denote the multiplicative cyclic group of order p^{n+1}. Then the group ring

$$k\mu_{p^{n+1}} \cong k[T]/(T^{p^{n+1}}-1).$$

Since $(T^{p^{n+1}}-1) = (T-1)^{p^{n+1}}$, this group ring is a local artin ring whose maximal ideal is generated by $T - 1$, modulo $(T-1)^{p^{n+1}}$. Hence all of the indecomposable modules are known; namely they are indexed by dimension and

$$V_r = k[T]/(T-1)^r$$

for $1 \leq r \leq p^{n+1}$.

Let $S.(V_r)$ denote the symmetric algebra on V_r over k. There is a $\mu_{p^{n+1}}$-action.

<u>Problem</u>: What is $S.(V_r)^{\mu_{p^{n+1}}}$?

In fact there are several ancillary problems.

<u>Geometry</u>. Suppose X is a smooth variety over k and G is a finite group of automorphisms of X. Then X/G need not be smooth. Classify all the singularities. (These are known in case char $(k) = 0$.)

<u>Representation Theory</u>. Let W be a representation of S_m. What the decomposition into indecomposables of

$$W \otimes_{kS_m} V^{\otimes m}?$$

<u>Commutative Algebra</u>. What are the properties of the algebra of invariants

$$S.(W \otimes_{S_m} V^{\otimes m})^{\mu_{p^{n+1}}}?$$

(The same question can be asked for the completion of this algebra at the irrelevant maximal ideal.)

This last question has some answers.

<u>Theorem.</u> <u>The algebras</u> $S.(W \underset{S_m}{\otimes} V^{m})^{\mu_{p^{n+1}}}$ <u>are factorial.</u>

The proof of this theorem depends upon the fact that

$$\mathrm{Hom}_{\mathbb{Z}}(\mu_{p^{n+1}}, k^{x}) = 0,$$

and a computation of the ideal class group due to Samuel [1964].

<u>Theorem.</u> <u>Let</u> M <u>denote the irrelevant maximal ideal of the algebra.</u> <u>Then</u>

$$\mathrm{depth}_M S.(W \underset{S_m}{\otimes} V^{\otimes m})^{\mu_{p^{n+1}}}$$

$$= \inf(\dim_k (W \underset{S_m}{\otimes} V^{\otimes m}),\ 2 + \dim_k (W \underset{S_m}{\otimes} V^{\otimes m})^{\mu})$$

This is a result of Ellingsrud and Skjelbred [1978].

Since $\dim S.(W \underset{S_m}{\otimes} V^{\otimes m})^{\mu_{p^{n+1}}} = \dim_k (W \underset{S_m}{\otimes} V^{\otimes m})$,

this result shows that there are many examples, in characteristic $p > 0$, of factorial rings that are not Cohen-Macaulay.

Almkvist and Fossum (1978) have made many calculations of the Hilbert-Poincare sereis of the ring of invariants.

$$S.(V_{r+1})^{\mu_p}$$

and

$$S.(V_{p^n})^{\mu_{p^n}}.$$

And they have completely described the decompositions of the symmetric powers of the indecomposables V_r for $1 \leq r \leq p$.

The problem in general is very difficult, partly because we cannot find a satisfactory generalization of the Clebsch-Gordan formula.

We conclude this section by referring the interested reader to Almkvist (1980).

V. Problems

In this section some of the remaining problems are discussed.

Classical Invariant Theory. Find the invariants of the N-ary R-forms.

Geometry and Representation Theory. Let S_I be an irreducible complex representation of S_n. Then S_n acts on the projective space

$$\mathbb{P}(S_I)$$

with orbit space $\mathbb{P}(S_I)/S_n$. What is the geometric meaning of this orbit space? Does it classify some nice family of varieties?

Representation Theory and Commutative Algebra. Again let S_I be an irreducible representation of S_n. Find the invariants of the symmetric algebra $S^{\cdot}(S_I)$. When is this ring Gorenstein, factorial or a complete interaction.

Problems in Characteristic $p > 0$. Let V be an indecomposable representation of $\mathbb{Z}/p^n\mathbb{Z}$ in characteristic $p > 0$. Find the decomposition of $S^n(V)$ and $\Lambda^n(V)$ for all n. Are the completions of the invariant rings factorial?

Let characteristic $(k) = p > 0$ and suppose $f(X,Y)$ is a formal group on $k[[t]]$. Then the $V_n := k[[t]]/(t^n)$ can be multiplied, viz. $V_n \otimes V_m$, and these become $k[[t]]$-modules of finite length over $k[[t]]$ through $f(X,Y)$. What are the deocmpositions of the symmetric and exterior powers?

The case above is $f(X,Y) = X + Y + XY$, giving the cyclic groups. Can one speak of invariants?

References

Almkvist, G. (1980): Invariants, mostly old ones. Pacific J. Math. 86, 1-14.

Almkvist. G. and R. Fossum (1978): Decomposition of exterior and symmetric powers of indecomposable $\mathbb{Z}/p\mathbb{Z}$-modules in characteristic p and relations to invariants. This Seminar 1976/77. Lecture Notes in Math. Berlin, Heidelberg, New York: Springer Verlag.

Bourbaki, N. (1968): Groupes et Algèbras de Lie. Chap. 4,5 et 6. Paris, Hermann.

Chevalley, C. (1955): Invariants of finite groups generated by reflections. Amer. J. Math. 67, 778-782.

Ellingsrud, G. et T. Skjelbred (1978): Profondeur d'anneaux d'invariants de polynômes en caracteristique p. C. R. Acad. Sc. Paris. 286, Ser. A., 321-322; Compositio Math. 41, 233-44 (1980).

Faa de Bruno (1876): Théorie des formes binares. Torino 1876.

Fogarty, J. (1969): Invariant Theory. New York, Benjamin.

Hartshorne, R. (1966): Ample Vector Bundles. Publ. Math. IHES, 29, 63-94.

Hochster, M. and J. Roberts (1974): Rings of invariants of reductive groups acting on regular rings are Cohen-Macaulay. Adv. in Math. 13, 115-175.

James, D. G. (1978): The Representation Theory of the Symmetric Groups. Lecture Notes in Math. No. 682, Berlin, Heidelberg, New York: Springer Verlag.

Lascoux, A. (1978): Syzygies des varietes determinantales. Adv. in Math. 30, 202-237.

Macdonald, I. G. (1980): Polynomial functors and wreath products. J. Pure and Applied Alg. 18, 173-204.

Molien, T. (1897): Uber die Invarianten der linearen substitutionsgruppen. Setz. Konig Preuss. Akad. Wiss. 1152-1156.

Mumford, D. and K. Suominen (1972): Introduction to the theory of moduli. Algebraic Geometry, Oslo 1970, Groningen: Wolters-Noordhoff.

Murthy, P. (1964): A note on factorial rings. Arch. Math. 15, 418-420.

Nielson, H. A. (1978): Tensor functors of complexes. Preprint Series 1977/78, no. 15. Aarhus Univ.

_____________ (1979): Free Resolutions of tensor forms. Preprint Series 1978/79, no. 24. Aarhus Univ.

Popp H. (1979): Classification of Algebraic Varieties and Compact Complex Manifolds. Lectures Notes in Math. no. 412. Berlin, Heidelberg, New York: Springer Verlag.

_______ (1977): Moduli Theory and Classification Theory of Algebraic Varieties. Lecture Notes in Math. No. 620. Berlin, Heidelbert, New York: Springer Verlag.

Samuel, P. (1964): Lectures on Unique Factorization Domains. Tata Institute Bombay.

Schur, I. (1968): Vorlesungen uber Invarianten-theorie. Grundl. der Math. Wiss. Bd 143. Berlin, Heidelberg, New York: Springer Verlag.

Shioda, T. (1967): On the graded ring of invariants of binary octavics. Amer. J. Math. 89, 1022-1046.

Springer, T. A. (1977): Invariant Theory. Lecture Notes in Math. No. 585. Berlin, Heidelberg, New York: Springer Verlag.

Stanley, R. P. (1979): Invariants of finite groups and their applications to combinatorics. Bull. (New Series) Amer. Math. Soc. 1, 475-571.

On the construction of some relative invariants for GL(n) (n=6,7,8) by the decomposition of the Young diagrams

Tatsuo KIMURA

Introduction.

Let V_n be a vector space spanned by skew-tensors $u_i \wedge u_j \wedge u_k$ $(1 \leq i < j < k \leq n)$ over a field k on which $G_n = GL(n, k)$ acts by $\rho(g)(u_i \wedge u_j \wedge u_k) = gu_i \wedge gu_j \wedge gu_k$ where $gu_i = \sum_r u_r g_{ri}$ for $g = (g_{ij}) \in G_n$. For $n = 6,7,8$ and $k = \mathbb{C}$, the triplet (G_n, ρ, V_n) is a reduced irreducible regular prehomogeneous vector space (See Sato-Kimura [1]). In particular, any relative invariant of (G_n, ρ, V_n) is uniquely written as the form $cf_n(x)^m$ with $c \in k^{\times}$ and $m \in \mathbb{Z}$ where $f_n(x)$ is an irreducible homogeneous polynomial of degree 4 (resp. 7,16) for $n = 6$ (resp. 7,8). In this paper, first we shall construct an $n \times n$ matrix $\varphi(x)$ whose entries are homogeneous polynomials of degree 2 (resp. 3,6) for $n = 6$ (resp. 7,8). Next, we shall construct another $n \times n$ matrix $\varphi^*(x)$ whose entries are also homogeneous polynomials of degree 2 (resp. 4,10) for $n = 6$ (resp. 7,8). Finally, we shall show that $\Phi(x) = \varphi(x)\varphi^*(x)$ is a non-zero scalar matrix and $\Phi(x) = f_n(x)\cdot I_n$ $(n = 6,7,8)$. All constructions are based on the decomposition of the Young diagrams in the following sense. First, all Young diagrams in this paper appear in the symmetric tensor of ⊟ (a column of three boxes), and hence they correspond to some polynomials on V_n. Now, for three Young diagrams A, B, C, we say that $B \times C$ is a decomposition of A (notation : $A \sim B \times C$) when A appears in the symmetric tensor of B and C. This implies that polynomials for A can be obtained from those for B and C. For $n = 6$, this was first done by M. Sato. We shall review his results in §1 in the form convenient for later use. For $n = 7$, the construction of $\varphi(x)$ was first done also by M. Sato, and that of $\varphi^*(x)$ was done by the author (See T. Kimura [8]). However, in this paper, we shall construct $\varphi^*(x)$ in the different but simpler way than in [8]. The main result for $n = 7$ is described in §2. In §3, we shall investigate the relative invariant for $n = 8$.

The author would like to express his hearty thanks to Professor Mikio SATO who kindly explained his work for $n = 6$, and to Professor Herbert POPP for his encouragement and stimulation.

§1. We shall review the M. Sato's results for $n = 6$ (See [1]). The existence of the relative invariant of degree four corresponds to the fact that the symmetric tensor S^4 ( ) contains  }6. On the other hand, S^2 ( ) contains 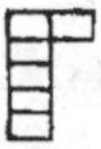 , which corresponds to the fact that there exists a 6×6 matrix $\varphi(x)$ whose entries are quadratic forms in $x \in V_6$ satisfying $\varphi(\rho(g)x) = \det g \cdot g\varphi(x)g^{-1}$ for $g \in GL(6)$. Since there exists no relatively invariant quadratic form, we have $\mathrm{tr}\varphi(x) = 0$. The decomposition  $\sim$ 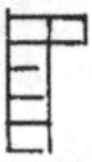 $\times$ 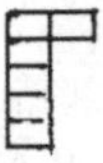 corresponds to the fact that $\varphi(x)^2 = f_6(x)\cdot I_6$ for $x \in V_6$. The explicit construction of $\varphi(x)$ is given as follows. For $x = \sum_{i<j<k} x_{ijk}\, u_i \wedge u_j \wedge u_k$, we shall define the partial derivations $\frac{\partial}{\partial u_\ell}$ by

$$(1.1) \qquad \frac{\partial}{\partial u_\ell}(u_i \wedge u_j \wedge u_k) = \delta_{i\ell}\, u_j \wedge u_k \quad \text{for } j,k \neq \ell.$$

Then, by setting $\hat{x} = \sum_i y_i \frac{\partial x}{\partial u_i}$, we have

(1.2) $\quad x \wedge \hat{x} = \sum_{i,j} \varphi_{ij}(x)\, y_i v_j \quad$ where $\; v_j \wedge u_j = \omega \; (= u_1 \wedge \cdots \wedge u_n)$.

Now $\varphi(x)$ is given by $\varphi(x) = (\varphi_{ij}(x))$. For $x_0 = u_1 \wedge u_2 \wedge u_3 + u_4 \wedge u_5 \wedge u_6$, we have

$$\text{(1.3)} \qquad \varphi(x_0) = \begin{pmatrix} 1 & & & & & \\ & 1 & & & 0 & \\ & & 1 & & & \\ & & & -1 & & \\ & 0 & & & -1 & \\ & & & & & -1 \end{pmatrix}, \text{ and } \; f_6(x_0) = 1.$$

§2. Since $\deg f_7 = 7$, the symmetric tensor S^7 ([Young diagram]) contains 7{ [Young diagram] . It has a decomposition [Young diagram] }7 ~ [Young diagram] × [Young diagram] }6. This implies there exists a 7×7 symmetric matrix $\varphi(x)$ (resp. $\varphi^*(x)$) satiafying $\varphi(\rho(g)x) = \det g \cdot g\varphi(x)\,{}^t g$ (resp. $\varphi^*(\rho(g)x) = (\det g)^2 \cdot {}^t g^{-1} \varphi^*(x) g^{-1}$ for $g \in GL(7)$, whose entries are homogeneous polynomials in x of degree three (resp. four). The matrix $\varphi(x) = (\varphi_{ij}(x))$ is first obtained by M. Sato as follows.

(2.1) $\quad x \wedge \hat{x} \wedge \hat{x} = \sum_{i,j} \varphi_{ij}(x) y_i y_j \omega \quad$ where $\; \omega = u_1 \wedge \cdots \wedge u_7$.

We shall prove that the matrix $(\varphi_{ij}(x))$ is a desired one.

Proposition 2.1. $\varphi(\rho(g)x) = \det g \cdot g\varphi(x)\,{}^t g$ for $g \in GL(7)$.

<u>Proof</u>. Note that $\varphi_{ij}(x)\omega = x \wedge \frac{\partial x}{\partial u_i} \wedge \frac{\partial x}{\partial u_j}$. Hence we have

$$(2.2)\qquad \varphi_{ij}(\rho(g)x)\omega = x(gu_1,\cdots,gu_7) \wedge \sum_{s=1}^{7}(\frac{\partial x}{\partial u_s})(gu_1,\cdots,gu_7)g_{is}$$

$$\wedge \sum_{t=1}^{7}(\frac{\partial x}{\partial u_t})(gu_1,\cdots,gu_7)g_{jt}.$$

Let E_{mn} be the matrix unit with (m, n)-entry 1, all remaining entries zero. For $m \neq n$, and $g = \exp t E_{mn}$, we have $\varphi_{ij}(d\rho(E_{mn})x)\omega = \frac{d}{dt}\varphi_{ij}(\rho(g)x)\omega|_{t=0}$

$$= (u_m \wedge \frac{\partial x}{\partial u_n}) \wedge \frac{\partial x}{\partial u_i} \wedge \frac{\partial x}{\partial u_j} + x \wedge (\delta_{im}\frac{\partial x}{\partial u_n} + u_m \wedge \frac{\partial^2 x}{\partial u_n \partial u_i}) \wedge \frac{\partial x}{\partial u_j}$$

$$+ x \wedge \frac{\partial x}{\partial u_i} \wedge (\delta_{jm}\frac{\partial x}{\partial u_n} + u_m \wedge \frac{\partial^2 x}{\partial u_n \partial u_j}) = \delta_{im}\varphi_{nj}(x) + \delta_{jm}\varphi_{in}(x).$$

Note that, for $m \neq n$, $u_m \wedge \frac{\partial x}{\partial u_n} \wedge \frac{\partial x}{\partial u_i} \wedge \frac{\partial x}{\partial u_j} + x \wedge u_m \wedge \frac{\partial^2 x}{\partial u_n \partial u_i} \wedge \frac{\partial x}{\partial u_j}$ $+ x \wedge \frac{\partial x}{\partial u_i} \wedge u_m \wedge \frac{\partial^2 x}{\partial u_n \partial u_j} = 0$. In fact, denoting by $(\frac{\partial x}{\partial u_i})_n$ the part of $(\frac{\partial x}{\partial u_i})$ which does not contain u_n, i.e., $\frac{\partial x}{\partial u_i} = (\frac{\partial x}{\partial u_i})_n + u_n \wedge \frac{\partial^2 x}{\partial u_n \partial u_i}$,

the first term $= u_m \wedge \frac{\partial x}{\partial u_n} \wedge (\frac{\partial x}{\partial u_i})_n \wedge u_n \wedge \frac{\partial^2 x}{\partial u_n \partial u_j} + u_m \wedge \frac{\partial x}{\partial u_n} \wedge u_n \wedge \frac{\partial^2 x}{\partial u_n \partial u_i} \wedge (\frac{\partial x}{\partial u_j})_n$,

the second term $= -u_m \wedge \frac{\partial x}{\partial u_n} \wedge u_n \wedge \frac{\partial^2 x}{\partial u_n \partial u_i} \wedge (\frac{\partial x}{\partial u_j})_n - x \wedge u_m \wedge u_n \wedge \frac{\partial^2 x}{\partial u_m \partial u_i}$ $\wedge \frac{\partial^2 x}{\partial u_n \partial u_j}$, and the third term $= -u_m \wedge \frac{\partial x}{\partial u_n} \wedge (\frac{\partial x}{\partial u_i})_n \wedge u_n \wedge \frac{\partial^2 x}{\partial u_n \partial u_j}$

$+ x \wedge u_m \wedge u_n \wedge \frac{\partial^2 x}{\partial u_n \partial u_i} \wedge \frac{\partial^2 x}{\partial u_n \partial u_j}$. Therefore, for $g = \exp t E_{mn}$ $(m \neq n)$,

the proposition holds. For a diagonal matrix $c = \begin{pmatrix} c_1 & & \\ & \ddots & \\ & & c_7 \end{pmatrix} \in SL(7)$, one can easily check that $\varphi_{ij}(\rho(c)x) = c_i c_j \varphi_{ij}(x)$. Q.E.D.

Example 2.2. For $x_0 = u_2 \wedge u_3 \wedge u_4 + u_5 \wedge u_6 \wedge u_7 + u_1 \wedge (u_2 \wedge u_5 + u_3 \wedge u_6 + u_4 \wedge u_7)$, we have, by simple calculation,

$$(2.3) \qquad \varphi(x_0) = \left(\begin{array}{c|c|c} -6 & 0 & 0 \\ \hline 0 & 0 & 3I_3 \\ \hline 0 & 3I_3 & 0 \end{array}\right) \quad \text{where} \quad I_3 = \begin{pmatrix} 1 & & \\ & 1 & \\ & & 1 \end{pmatrix}.$$

Remark 2.3. For $x = \sum_{i<j<k} x_{ijk} u_i \wedge u_j \wedge u_k \in V_7$, let X_i be the 6×6 skew-symmetric matrix whose entries are x_{ijk} $(j,k \neq i)$, and $Pf(x_i)$ the Pfaffian of X_i. Then we have $\varphi_{ii}(x) = \pm 6Pf(X_i)$ for $i = 1,\cdots,7$.

Now we shall construct $\varphi^*(x)$ according to the following decomposition ;

(2.4) [Young diagram] $\sim$ [Young diagram] $\times$ [Young diagram] .

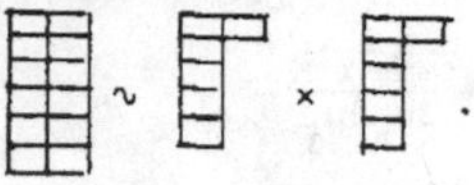

The Young diagram [Young diagram]  for $n = 7$ corresponds to the following quadratic forms $f^i_{jk}(x)$ $(i, j, k = i,\cdots,7)$.

$$(2.5) \qquad x \wedge \hat{x} = \sum_{i,j,k} f^i_{jk}(x) y_i v_{jk}, \quad \text{where} \quad v_{jk} \wedge u_j \wedge u_k = \omega.$$

For a diagonal matrix $c = \begin{pmatrix} c_1 & & 0 \\ & \ddots & \\ 0 & & c_7 \end{pmatrix}$ in $SL(7)$, we have

$$(2.6) \qquad f^i_{jk}(\rho(c)x) = \frac{c_i}{c_j c_k} f^i_{jk}(x) \quad \text{for} \quad i, j, k = 1,\cdots,7.$$

Lemma 2.4. For $m \neq n$, we have

(2.7) $$f^i_{jk}(d\rho(E_{mn})x) = \delta_{im}\, f^n_{jk}(x) - \delta_{jn}\, f^i_{mk}(x) - \delta_{kn}\, f^i_{jm}(x).$$

<u>Proof</u>. Since $f^i_{jk}(x)\omega = x \wedge \frac{\partial x}{\partial u_i} \wedge u_j \wedge u_k$, we have

$$f^i_{jk}(d\rho(E_{mn})x)\omega = (u_m \wedge \frac{\partial x}{\partial u_n}) \wedge \frac{\partial x}{\partial u_i} \wedge u_j \wedge u_k$$

$$+ x \wedge [\delta_{im}\, (\frac{\partial x}{\partial u_n}) + u_m \wedge \frac{\partial^2 x}{\partial u_n \partial u_i}] \wedge u_j \wedge u_k$$

$$= \delta_{im}\, f^n_{jk}(x)\omega - [(u_j \wedge \frac{\partial x}{\partial u_n}) \wedge \frac{\partial x}{\partial u_i} + x \wedge (u_j \wedge \frac{\partial^2 x}{\partial u_n \partial u_i})] \wedge u_m \wedge u_k.$$

For $j = n$, the last term $= -f^i_{mk}(x)\omega$. If $j, k \neq n$, then the last term

$$= -[u_j \wedge \frac{\partial x}{\partial u_n} \wedge u_n \wedge \frac{\partial^2 x}{\partial u_n \partial u_i} + u_n \wedge \frac{\partial x}{\partial u_n} \wedge u_j \wedge \frac{\partial^2 x}{\partial u_n \partial u_i}] \wedge u_m \wedge u_k = 0.$$

Q.E.D.

Now we define the 7×7 symmetric matrix $\varphi^*(x) = (\varphi^*_{st}(x))$ by

(2.8) $$\varphi^*_{st}(x) = \sum_{i,j=1}^{7} f^i_{sj}(x) f^j_{it}(x).$$

Proposition 2.5. $\varphi^*(\rho(g)x) = (\det g)^2 \cdot {}^t g^{-1}\, \varphi^*(x) g^{-1}$ for $g \in GL(7)$.

<u>Proof</u>. For a scalar matrix cI_7, it is clear. For a diagonal matrix

$c = \begin{pmatrix} c_1 & & 0 \\ & \ddots & \\ 0 & & c_7 \end{pmatrix}$ in $SL(7)$, we have $\varphi^*_{st}(\rho(c)x) = \frac{1}{c_s c_t} \varphi^*{}_{st}(x)$ by (2.6).

For $m \neq n$, by Lemma 2.4, we have $\varphi^*_{st}(d\rho(E_{mn})x)$

$$= \sum_{i,j} f^i_{sj}(d\rho(E_{mn})x)f^j_{it} + \sum_{i,j} f^i_{sj}\cdot f^j_{it}(d\rho(E_{mn})x)$$

$$= \sum_j f^n_{sj}f^j_{mt} - \delta_{sn}\sum_{i,j} f^i_{mj}f^j_{it} - \sum_i f^i_{sm}f^n_{it} + \sum_i f^i_{sm}f^n_{it}$$

$$- \sum_j f^n_{sj}f^j_{mt} - \delta_{tn}\sum_{i,j} f^i_{sj}f^j_{im} = -\delta_{sn}\varphi^*_{mt} - \delta_{tn}\varphi^*_{sm},$$ i.e., the proposition

holds for $g = \exp tE_{mn}$. Q.E.D.

Example 2.6. For x_0 in Example 2.2, the values $f^i_{jk}(x_0)$ are zero except

$f^1_{25} = f^1_{36} = f^1_{47} = f^3_{57} = f^6_{24} = -2$, $f^2_{67} = f^4_{56} = f^5_{34} = f^7_{23} = 2$, $f^2_{12} = f^3_{13} = f^4_{14} = 1$,

$f^5_{15} = f^6_{16} = f^7_{17} = -1$. Hence we have

$$(2.9) \qquad \varphi^*(x_0) = \left(\begin{array}{c|c|c} -6 & 0 & 0 \\ \hline 0 & 0 & 12I_3 \\ \hline 0 & 12I_3 & 0 \end{array}\right).$$

Remark 2.7. Note that $\varphi^*_{ss}(x)$ is a relative invariant in $u_i\ (\forall i \neq s)$ for $GL(6)$.

<u>Main Theorem</u> (for $n = 7$) (1) There exists a polynomial map φ of degree three from V_7 to 7×7 symmetric matrices satisfying $\varphi(\rho(g)x) = (\det g)\cdot g\varphi(x)\,{}^tg$ for $g \in GL(7)$.

This $\varphi(x) = (\varphi_{ij}(x))$ is given by $x \wedge \frac{\partial x}{\partial u_i} \wedge \frac{\partial x}{\partial u_j} = \varphi_{ij}(x)\omega$ $(i, j = 1,\cdots,7)$.

(2) There exists a polynomial map φ^* of degree four from V_7 to 7×7 symmetric matrices satisfying $\varphi^*(\rho(g)x) = (\det g)^2 \cdot {}^t g^{-1} \varphi^*(x) g^{-1}$ for $g \in GL(7)$. This $\varphi^*(x) = (\varphi^*_{ij}(x))$ is given by $\varphi^*_{ij}(x) = \sum_{s,t=1}^{7} f^s_{it}(x) f^t_{sj}(x)$

where $f^i_{jk}(x)\omega = x \wedge \frac{\partial x}{\partial u_i} \wedge u_j \wedge u_k$ $(i, j, k = 1,\cdots,7)$.

(3) $\Phi(x) = \varphi(x)\,\varphi^*(x)$ satisfies that $\Phi(\rho(g)x) = (\det g)^3 \cdot g\Phi(x)g^{-1}$ for $g \in GL(7)$. Moreover, $\Phi(x)$ is a scalar matrix and the relative invariant $f_7(x)$ is given by $\Phi(x) = f_7(x) \cdot I_7$. For $x_0 = u_2 \wedge u_3 \wedge u_4 + u_5 \wedge u_6 \wedge u_7 + u_1 \wedge (u_2 \wedge u_5 + u_3 \wedge u_6 + u_4 \wedge u_7)$, $f_7(x_0) = 36$.

Proof. (1) and (2) have been proved. By (2.3) and (2.9), we have $\Phi(x_0) = 36I_7$ and hence $\Phi(\rho(g)x_0) = (\det g)^3 \cdot 36I_7$. Since the orbit $\rho(G)x_0$ is Zariski-dence (See [1]), $\Phi(x)$ is a scalar matrix. The remaining part is obvious. Q.E.D.

§3. Since $\deg f_8 = 16$, the symmetric tensor S^{16} ( ) contains

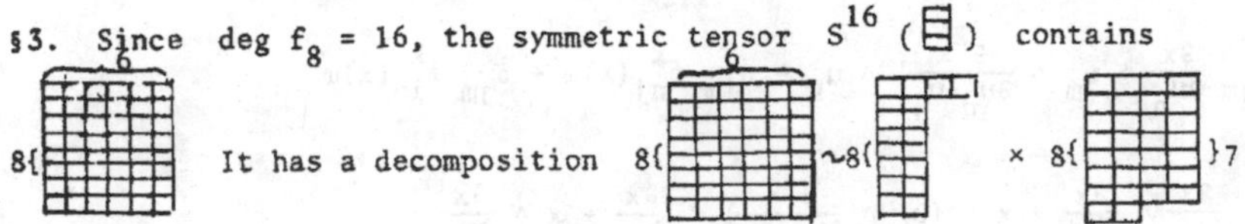

8{ It has a decomposition 8{ ~8{ × 8{ }7.

This implies that there exists a 8×8 symmetric matrix $\varphi(x)$ (resp. $\varphi^*(x)$) satisfying $\varphi(\rho(g)x) = (\det g)^2 \cdot g\varphi(x)\,{}^t g$ for $g \in GL(8)$ (resp. $\varphi^*(\rho(g)x) = (\det g)^4 \cdot {}^t g^{-1} \varphi^*(x) g^{-1}$) whose entries are homogeneous polynomials in x of degree six (resp. ten). We shall first construct $\varphi(x)$ according to the decomposition 8{ $\sim$ 7{ × 7{ . The Young diagram corresponds to the following polynomials $f^k_{ij}(x)$ $(i, j, k = 1,\cdots,8)$ of degree three.

$$x \wedge \hat{x} \wedge \hat{x} = \sum_{i,j,k} f^k_{ij}(x)\, y_i y_j v_k \quad \text{where} \quad v_k \wedge u_k = \omega \ (=u_1 \wedge \cdots \wedge u_8). \tag{3.1}$$

It is easy to check that for $c = \begin{pmatrix} c_1 & & 0 \\ & \ddots & \\ 0 & & c_8 \end{pmatrix} \in SL(8)$, we have

$$f^k_{ij}(cx) = \frac{c_i c_j}{c_k} f^k_{ij}(x). \tag{3.2}$$

Lemma 3.1. For $m \neq n$, we have

$$f^k_{ij}(d\rho(E_{mn})x) = \delta_{im} f^k_{nj} + \delta_{jm} f^k_{in} - \delta_{nk} f^m_{ij}. \tag{3.3}$$

<u>Proof</u>. Since $f^k_{ij}(x)\omega = x \wedge \frac{\partial x}{\partial u_i} \wedge \frac{\partial x}{\partial u_j} \wedge u_k$, we have $f^k_{ij}(d\rho(E_{mn})x)\omega$

$= (u_m \wedge \frac{\partial x}{\partial u_n}) \wedge \frac{\partial x}{\partial u_i} \wedge \frac{\partial x}{\partial u_j} \wedge u_k + x \wedge [\delta_{im} \frac{\partial x}{\partial u_n} + u_m \wedge \frac{\partial^2 x}{\partial u_n \partial u_i}] \wedge \frac{\partial x}{\partial u_j} \wedge u_k$

$+ x \wedge \frac{\partial x}{\partial u_i} \wedge [\delta_{jm} \frac{\partial x}{\partial u_n} + u_m \wedge \frac{\partial^2 x}{\partial u_n \partial u_j}] \wedge u_k = \delta_{im} f^k_{nj}(x)\omega + \delta_{jm} f^k_{in}(x)\omega$

$- [(u_k \wedge \frac{\partial x}{\partial u_n}) \wedge \frac{\partial x}{\partial u_i} \wedge \frac{\partial x}{\partial u_j} + x \wedge (u_k \wedge \frac{\partial^2 x}{\partial u_n \partial u_i}) \wedge \frac{\partial x}{\partial u_j} + x \wedge \frac{\partial x}{\partial u_i}$

$\wedge (u_k \wedge \frac{\partial^2 x}{\partial u_n \partial u_j})] \wedge u_m$. Since the last term is $\delta_{kn} f^m_{ij}(x)\omega$, we obtain our assertion. Q.E.D.

Now we shall define the 8×8 symmetric matrix $\varphi(x) = (\varphi_{ij}(x))$ by

$$\varphi_{ij}(x) = \sum_{s,t=1}^{8} f^s_{it}(x) f^t_{sj}(x). \tag{3.4}$$

Proposition 3.2. $\varphi(\rho(g)x) = (\det g)^2 . g\varphi(x)\, {}^t g$ for $g \in GL(8)$.

Proof. For a diagonal matrix, it is clear from (3.2). Therefore it is enough to show infinitesimally for $g = \exp t E_{mn}$ $(m \neq n)$, i.e., $\varphi_{ij}(d\rho(E_{mn})x) = \delta_{im}\varphi_{nj}(x) + \delta_{jm}\varphi_{in}$. By Lemma 3.1, $\varphi_{ij}(d\rho(E_{mn})x) = \sum_{s,t} f^s_{it}[\delta_{sm} f^t_{nj} + \delta_{jm} f^t_{sn} - \delta_{nt} f^m_{sj}] + \sum_{s,t} [\delta_{im} f^s_{nt} + \delta_{tm} f^s_{in} - \delta_{ns} f^m_{it}] f^t_{sj} = \delta_{im}\varphi_{nj} + \delta_{jm}\varphi_{in}$. Q.E.D.

Example 3.3. For $x_0 = u_1 \wedge u_2 \wedge u_3 + u_1 \wedge u_4 \wedge u_7 + u_1 \wedge u_4 \wedge u_8 + u_2 \wedge u_5 \wedge u_7 + u_3 \wedge u_6 \wedge u_8 + u_4 \wedge u_5 \wedge u_6$, the values $f^k_{ij} = f^k_{ij}(x_0)$ $(i \leq j)$ are zero except $f^1_{17} = 1$, $f^1_{18} = -1$, $f^1_{56} = 3$; $f^2_{27} = 1$, $f^2_{28} = 2$, $f^2_{46} = -3$; $f^3_{37} = -2$, $f^3_{38} = -1$, $f^3_{45} = -3$; $f^4_{47} = 1$, $f^4_{48} = -1$, $f^4_{23} = -3$; $f^5_{57} = 1$, $f^5_{58} = 2$, $f^5_{13} = 3$; $f^6_{67} = -2$, $f^6_{68} = -1$, $f^6_{12} = 3$; $f^7_{77} = f^7_{78} = 2$, $f^7_{14} = f^7_{36} = -3$; $f^8_{87} = f^8_{88} = -2$, $f^8_{14} = f^8_{25} = 3$. Then, by (3.4), we have

$$\varphi(x_0) = \left(\begin{array}{ccc|ccc|cc} & & & -30 & & & & \\ & 0 & & & 30 & & & \\ & & & & & 30 & & 0 \\ \hline -30 & & & & & & & \\ & 30 & & & 0 & & & \\ & & 30 & & & & & \\ \hline & & & & & & 20 & 10 \\ & & & 0 & & & 10 & 20 \end{array}\right). \tag{3.5}$$

Now we shall construct $\varphi^*(x)$ according to the following decomposition.

(3.6) $8\Big\{$ [Young diagram] $\sim$ [Young diagram] $\Big\}7$ $\times$ [Young diagram] .

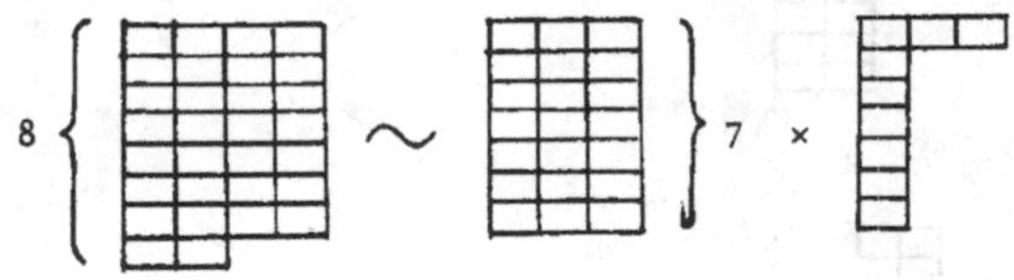

The first Young diagram of the right-hand side is related with the relative invariant for GL(7). This Young diagram for $n = 8$ implies that there exists a polynomial map Θ of degree seven from $x = \sum_{i<j<k} x_{ijk} u_i \wedge u_j \wedge u_k$ to $\Theta(x) = \sum_{i,j,k} F_{ijk}(x) u_i \cdot u_j \cdot u_k$ satisfying $\Theta(\rho(g)x) = (\det g)^3 \sum_{i,j,k} F_{ijk}(x)(g^* u_i)\cdot(g^* u_j)(g^* u_k)$ where $g^* = {}^t g^{-1}$ for $g = GL(8)$. In particular, for $m \neq n$, we have

(3.7) $F_{ijk}(d\rho(E_{mn})x) = -\delta_{in} F_{mjk}(x) - \delta_{jn} F_{imk}(x) - \delta_{kn} F_{ijm}(x)$.

Note that $F_{ttt}(x)$ is a relative invariant in u_j's ($j \neq t$) for $GL(7)$.

The last Young diagram in (3.6) corresponds to $f^k_{ij}(x)$ in (3.1). We shall define the 8×8 symmetric matrix $\varphi^*(x) = (\varphi^*_{ij}(x))$ by

(3.8) $\varphi^*_{ij}(x) = \sum_{s,t} F_{ist}(x) f^j_{st}(x) + \sum_{s,t} F_{jst}(x) f^i_{st}(x)$.

Proposition 3.4. $\varphi^*(\rho(g)x) = (\det g)^4 \cdot {}^t g^{-1} \varphi^*(x) g^{-1}$ for $g \in GL(7)$.

<u>Proof.</u> For a diagonal matrix, it is clear. For $m \neq n$, we have

$\sum_{s,t} F_{ist}(d\rho(E_{mn})x) f^j_{st}(x) + \sum_{s,t} F_{ist}(x) f^j_{st}(d\rho(E_{mn})x) = \sum_{s,t} (-\delta_{in} F_{mst} - \delta_{sn} F_{imt} - \delta_{tn} F_{ism}) f^j_{st} + \sum_{s,t} F_{ist}(\delta_{sm} f^j_{nt} + \delta_{tm} f^j_{sn} - \delta_{nj} f^m_{st}) = -\delta_{in} \sum_{s,t} F_{mst} f^j_{st} - \delta_{nj} \sum_{s,t} F_{ist} f^m_{st}$,

we obtain our assertion. Q.E.D.

Therefore, the problem has been reduced to the construction of the polynomials $F_{ijk}(x)$ of degree seven satisfying (3.7). We do this similarly as the previous sections, i.e., we use the following decompositions.

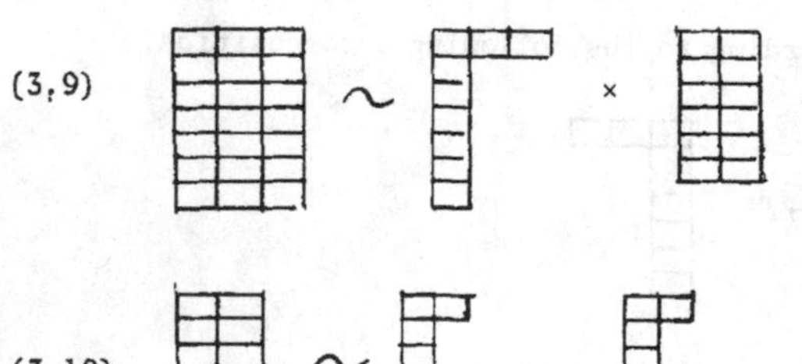

Note that the Young diagram in (3.10) corresponds to the relative invariant for GL(6). The Young diagram [Young diagram] for $n = 8$ corresponds to the following quadratic forms $f^i_{jk\ell}(x)$.

(3.11) $x \wedge \hat{x} = \sum_{i,j,k,\ell} f^i_{jk\ell}(x) y_i v_{jk\ell}$, where $v_{jk\ell} \wedge u_j \wedge u_k \wedge u_\ell = \omega$.

For a diagonal matrix $c = \begin{pmatrix} c_1 & & 0 \\ & \ddots & \\ 0 & & c_8 \end{pmatrix}$ in SL(8), we have

(3.12) $f^i_{jk\ell}(cx) = \dfrac{c_i}{c_j c_k c_\ell} f^i_{jk\ell}(x)$.

Lemma 3.5. $f^i_{jk\ell}(d\rho(E_{mn})x) = \delta_{im} f^n_{jk\ell}(x) - \delta_{nj} f^i_{mk\ell}(x) - \delta_{nk} f^i_{jm\ell}(x) - \delta_{n\ell} f^i_{jkm}(x)$ for $m \neq n$.

<u>Proof.</u> Note that $f^i_{jk\ell}(x)\omega = x \wedge \dfrac{\partial x}{\partial u_i} \wedge u_j \wedge u_k \wedge u_\ell$. Hence we have $f^i_{jk\ell}(d\rho(E_{mn})x)\omega = u_m \wedge \dfrac{\partial x}{\partial u_n} \wedge \dfrac{\partial x}{\partial u_i} \wedge u_j \wedge u_k \wedge u_\ell + x \wedge [\delta_{im} \dfrac{\partial x}{\partial u_n} + u_m \wedge \dfrac{\partial^2 x}{\partial u_n \partial u_i}] \wedge u_j \wedge u_k \wedge u_\ell = \delta_{im} f^n_{jk\ell}(x)\omega + [\dfrac{\partial x}{\partial u_n} \wedge \dfrac{\partial x}{\partial u_i} - x \wedge \dfrac{\partial^2 x}{\partial u_n \partial u_i}] \wedge u_m \wedge u_j \wedge u_k \wedge u_\ell$. If $n \neq j,k,\ell$, then the last term $= [\dfrac{\partial x}{\partial u_n} \wedge u_n \wedge \dfrac{\partial^2 x}{\partial u_m \partial u_i} - u_n \wedge \dfrac{\partial x}{\partial u_n} \wedge \dfrac{\partial^2 x}{\partial u_n \partial u_i}] \wedge u_m \wedge u_j \wedge u_k \wedge u_\ell = 0$. If $j = n$, then the last term $= [u_n \wedge \dfrac{\partial x}{\partial u_n} \wedge \dfrac{\partial x}{\partial u_i} + x \wedge u_n \wedge \dfrac{\partial^2 x}{\partial u_n \partial u_i}] \wedge u_m \wedge u_k \wedge u_\ell = -x \wedge \dfrac{\partial x}{\partial u_i} \wedge u_m \wedge u_k \wedge u_\ell = -f^i_{mk\ell}(x)\omega$. Since $f^i_{jk\ell}(x)$ is alternating with respect to j,k,ℓ, the remaining part is obvious.

Q.E.D.

The Young diagram [Young diagram] 6 for $n = 8$ corresponds to the polynomials $F_{ii',jj'}$ of degree four given by

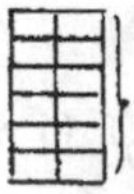

(3.13) $F_{ii',jj'}(x) = \sum_{s,t} f^s_{ii't}(x) f^t_{jj's}(x)$ (See (3.10)).

Then by (3.12) and Lemma 3.5, we have

(3.14) $F_{ii',jj'}(cx) = \frac{1}{c_i c_{i'} c_j c_{j'}} F_{ii',jj'}(x)$ for $c = \begin{pmatrix} c_1 & & 0 \\ & \cdot & \\ 0 & & \cdot\, c_8 \end{pmatrix} \in SL(8)$.

(3.15) $F_{ii',jj'}(d\rho(E_{mn})x) = -\delta_{ni}F_{mi',jj'} - \delta_{ni'}F_{im,jj'} - \delta_{nj}F_{ii',mj'} - \delta_{nj'}F_{ii',jm}$.

In particular, $F_{st,ts}(x)$ is a relative invariant in $u_i (\forall i \neq s,t)$ for $GL(6)$.

Now we shall construct $F_{ijk}(x)$ according to (3.9), i.e.,

(3.16) $\tilde{F}_{i,jk}(x) = \sum_{i',j'} f^i_{i'j'}(x)\, F_{ji',kj'}(x)$,

(3.17) $F_{ijk}(x) = \tilde{F}_{i,jk}(x) + \tilde{F}_{j,ki}(x) + \tilde{F}_{k,ij}(x)$.

Clearly $\tilde{F}_{i,jk} = \tilde{F}_{i,kj}$ and hence $F_{ijk}(x)$ in (3.17) is symmetric with respect to i,j,k.

Lemma 3.6. $F_{ijk}(x)$ in (3.17) satisfies (3.7).

<u>Proof.</u> By Lemma 3.1 and (3.15), we have $\tilde{F}_{i,jk}(d\rho(E_{mn})x) = \sum_{i',j'} [\delta_{i'm}f^i_{nj'} + \delta_{j'm}f^i_{i'n} - \delta_{ni}f^m_{i'j'}]\cdot F_{ji',kj'} + \sum_{i'j'} f^i_{i'j'}[-\delta_{nj}F_{mi',kj'} - \delta_{ni'}F_{jm,kj'} - \delta_{nj}F_{ji',mj'} - \delta_{nj'}F_{ji',km}] = \sum_{j'} f^i_{nj'}F_{jm,kj'} + \sum_{i'} f^i_{i'n}F_{ji',km} - \delta_{ni}\tilde{F}_{m,jk} - \delta_{nj}\tilde{F}_{i,mk} - \sum_{j'} f^i_{nj'}F_{jm,kj'} - \delta_{nj}\tilde{F}_{i,jm} - \sum_{j'} f^i_{i'n}F_{ji',km} = -\delta_{ni}\tilde{F}_{m,jk} - \delta_{nj}\tilde{F}_{i,mk} - \delta_{nj}\tilde{F}_{i,jm}$. Therefore by (3.17), our assertion is obvious. Q.E.D.

Example 3.7. For x_0 in Example 3.3, the values $f^i_{jk\ell}(x_0)$ $(j < k < \ell)$ are zero except $f^1_{568} = -f^1_{567} = 2$, $f^1_{125} = -f^1_{136} = -f^1_{178} = 1$; $f^2_{468} = -2$, $f^2_{214} = f^2_{236} = -f^2_{278} = 1$; $f^3_{457} = -2$, $f^3_{314} = f^3_{325} = f^3_{378} = -1$; $f^4_{238} = -f^4_{237} = 2$, $f^4_{425} = -f^4_{436} = f^4_{478} = 1$; $f^5_{138} = -2$, $f^5_{514} = f^5_{536} = f^5_{578} = 1$; $f^6_{127} = -2$, $f^6_{678} = -f^6_{614} = -f^6_{625} = 1$; $f^7_{368} = -2$, $f^7_{714} = f^7_{725} = f^7_{736} = 1$; $f^8_{257} = 2$,

$f^8_{814} = f^8_{825} = f^8_{836} = -1$. Hence the values $F_{ii',jj'}(x_0)$ $(i < i',\ j < j',\ i \le j)$ are zero except $F_{12,45} = F_{13,46} = F_{15,24} = F_{16,34} = -F_{23,56} = -F_{26,35} = 2$; $F_{12,37} = -F_{13,28} = F_{14,25} = F_{14,36} = F_{17,23} = F_{18,23} = -F_{25,36} = -F_{45,67} = F_{46,58} = -F_{47,56} = -F_{48,56} = 4$; $F_{14,14} = F_{17,48} = F_{18,47} = F_{25,25} = F_{27,58} = F_{28,57} = F_{36,36} = F_{37,68} = F_{38,68} = F_{78,78} = 6$; $-F_{12,38} = F_{13,27} = F_{45,68} = -F_{46,57} = 8$; $F_{27,57} = F_{38,68} = -12$. Hence, together with $f^k_{ij}(x_0)$ in Example 3.3, we obtain that $\tilde{F}_{i,jk}(x_0) = 0$ except $\tilde{F}_{1,23} = \tilde{F}_{2,13} = \tilde{F}_{3,12} = -42$, $\tilde{F}_{1,47} = \tilde{F}_{4,17} = \tilde{F}_{7,14} = 42$, $\tilde{F}_{1,48} = \tilde{F}_{4,18} = \tilde{F}_{8,14} = -42$, $\tilde{F}_{2,58} = \tilde{F}_{5,28} = \tilde{F}_{8,25} = -42$, $\tilde{F}_{3,67} = \tilde{F}_{6,37} = \tilde{F}_{7,36} = 42$, $\tilde{F}_{4,56} = \tilde{F}_{5,46} = \tilde{F}_{6,45} = 42$, $\tilde{F}_{7,78} = \tilde{F}_{8,77} = -84$, $\tilde{F}_{7,88} = \tilde{F}_{8,78} = 84$. Therefore, we have $F_{147} = F_{367} = F_{456} = -F_{123} = -F_{148} = -F_{258} = 126$ and $F_{788} = -F_{778} = 252$. Therefore, by (3.8), we obtain that

$$\varphi^*(x_0) = 2520 \times \left(\begin{array}{ccc|ccc|cc} 0&0&0&1&0&0&0&0\\ 0&0&0&0&-1&0&0&0\\ 0&0&0&0&0&-1&0&0\\ \hline 1&0&0&0&0&0&0&0\\ 0&-1&0&0&0&0&0&0\\ 0&0&-1&0&0&0&0&0\\ \hline 0&0&0&0&0&0&-2&1\\ 0&0&0&0&0&0&1&-2 \end{array}\right) \tag{3.18}$$

Together with (3.5), we have

$$\varphi(x_0)\varphi^*(x_0) = -(75600)I_8 = -2^4 \cdot 3^3 \cdot 5^2 \cdot 7 \cdot I_8. \tag{3.19}$$

<u>Main Theorem</u> (for $n = 8$). (1) There exists a polynomial map φ of degree six from V_8 to 8×8 symmetric matrices satisfying $\varphi(\rho(g)x) = (\det g)^2 . g\varphi(x)^t g$ for $g \in GL(8)$. This $\varphi(x) = (\varphi_{ij}(x))$ is given by $\varphi_{ij}(x) = \sum_{s,t} f^s_{it}(x) f^t_{sj}(x)$ where $f^k_{ij}(x)\omega = x \wedge \frac{\partial x}{\partial u_i} \wedge \frac{\partial x}{\partial u_j} \wedge u_k$.

(2) There exists a polynomial map φ^* of degree ten from V_8 to 8×8 symmetric matrices satisfying $\varphi^*(\rho(g)x) = (\det g)^4 \cdot {}^t g^{-1} \varphi^*(x) g^{-1}$ for $g \in GL(8)$. This $\varphi^*(x) = (\varphi^*_{ij}(x))$ is given by $\varphi^*_{ij} = \sum_{p,\gamma} (F_{ip\gamma} f^j_{p\gamma} + F_{jp\gamma} f^i_{p\gamma})$, where $\tilde{F}_{i,jk} = \sum_{i',j'} f^i_{i'j'} \sum_{s,t} f^s_{ji't} f^t_{kj's}$ and $F_{ijk} = \tilde{F}_{i,jk} + \tilde{F}_{j,ki} + \tilde{F}_{k,ij}$. Here $f^i_{jk\ell}(x)\omega = x \wedge \frac{\partial x}{\partial u_i} \wedge u_j \wedge u_k \wedge u_\ell$.

(3) $\Phi(x) = \varphi(x)\varphi^*(x)$ satisfies $\Phi(\rho(g)x) = (\det g)^6 \cdot g\phi(x)g^{-1}$ for $g \in GL(8)$. Moreover, $\Phi(x)$ is a scalar matrix and the relative invariant $f_8(x)$ is given by $\Phi(x) = f_8(x) \cdot I_8$, with $f_8(x_0) = -2^4 \cdot 3^3 \cdot 5^2 \cdot 7$ for $x_0 = u_1 \wedge u_2 \wedge u_3 + u_1 \wedge u_4 \wedge u_7 + u_1 \wedge u_4 \wedge u_8 + u_2 \wedge u_5 \wedge u_7 + u_3 \wedge u_6 \wedge u_8 + u_4 \wedge u_5 \wedge u_6$.

Proof. It is clear from (1) and (2) that $\Phi(\rho(g)x) = (\det g)^6 \cdot g\Phi(x)g^{-1}$ for $g \in GL(8)$. By (3.19), we have $\Phi(\rho(g)x_0) = -75600 \cdot (\det g)^6 \cdot I_8$. Since (G_8, ρ_8, V_8) is a prehomogeneous vector space with x_0 as its generic point, $\Phi(x)$ is a scalar matrix on the Zariski-dense orbit, i.e., $\Phi(x)$ is a scalar matrix everywhere. The remaining parts are obvious from the previous arguments. Q.E.D.

Remark 3.9. If one can say that $\tilde{F}_{i,jk}$ is symmetric with respect to i,j,k, one can take $F_{ijk} = \tilde{F}_{i,jk}$ and also $\varphi^*_{ij} = \sum_{s,t} F_{ist} f^j_{st}$.

Remark 3.10. In general, the relative invariants of prehomogeneous vector spaces are, up to a constant multiple, uniquely determined by their characters. Thanks to this fact, the b-functions (See [3], [4]) and the Fourier transforms (See [5], [6]) of $f_n(x)$ have been already calculated without knowing its explicit form. They describe the functional equations of zeta-functions obtained from $f_n(x)$. The number of orbits of (G_n, ρ, V_n) is 5 (resp. 10, 23) for $n = 6$ (resp. 7, 8) (See [7]).

References

[1] M. Sato and T. Kimura, "A classification of irreducible prehomogeneous vector spaces and their relative invariants" Nagoya Math. J. Vol. 65, (1977), 1-155.

[2] H. Weyl, Classical Groups, Princeton University Press, 1964.

[3] T. Kimura, "The b-functions and holonomy diagrams of irreducible regular prehomogeneous vector spaces", to appear.

[4] I. Ozeki, "On the Microlocal Structure of a Regular Prehomogeneous Vector Space Associated with GL(8)", to appear.

[5] M. Muro, "Some prehomogeneous vector spaces with relative invariants of degree four and the formula of the Fourier transforms, Preprint.

[6] M. Muro, "On the prehomogeneous vector spaces (GL(7), Λ_3) and (Spin(14) × GL(1),(half-spin rep.) × Λ_1) and the formulas of the Fourier transforms of the relative invariants", Preprint at RIMS-291, Kyoto University, June 1979.

[7] G. B. Gurevich, Foundation of the Theory of Algebraic Invariants, P. Noordhoff-LTD, Groningen, 1964.

[8] T. Kimura, On the relative invariant for GL(7) on the skew-tensors of rank three, Preprint.

Nagoya University, JAPAN

and

Grenoble University

FRANCE

Present address: Tsukuba University

CONCENTRATION UNDER ACTIONS OF ALGEBRAIC GROUPS

by Wim H. HESSELINK

Table of contents

0. Introduction

Part one : Concentration in affine varieties

1. The Hilbert-Mumford criterion
2. The theorem of Kempf and Rousseau
3. The stratification of N(V)

Part two : Concentration in schemes

4. Concentration in centered $G\ell(1)$-schemes
5. Regularity of the concentrator
6. Concentration under the action of a group scheme
7. The sheaf of buildings of a separated group scheme
8. Concentration under the action of a split torus
9. Optimal concentration over fields
10. A construction of Grothendieck

References

0 - Introduction

0.1 - This paper consists of two formally almost independent parts. In part one we consider actions of linear algebraic groups on affine varieties over an algebraically closed field. This part is mainly a survey of [13,15]. It concludes with a generalization of the main result of [13], which is fully proved. In part two the main ingredients of the theory are generalized to more or less arbitrary actions of group schemes on schemes. Here the reader may need good acquaintance with the scheme theory of [7,9].

0.2 - Part one : Concentration in affine varieties. By concentration we mean the following phenomenon. Let G be an algebraic group acting on a variety V with an invariant subvariety C. If $m \in \mathbb{N}$ and $\mu : G\ell(1) \to G$ is a one-parameter subgroup, the concentrator $V(\mu,m)$ consists of the points $v \in V$ such that $\mu(0)v$ is well defined and that $\mu(t)v \in C$ modulo powers t^n with $n \geqslant m$. The union of the concentrators is the concentrated cone $N(V)$. The elements of $N(V)$ are said to be concentrated or C-unstable. If G is reductive, V is affine and C is closed, then the Hilbert-Mumford criterion gives equivalent conditions for a point $v \in V$ to be concentrated. See section 1 and [15].

Optimal concentration is introduced in section 2. Given $v \in N(V)$ the problem is to find a concentrator which contains v and minimizes a certain cost function q. After an elementary exemple the Kempf-Rousseau theorem is stated, cf. [15,19].

Our stratification of the concentrated cone $N(V)$ is described in section 3. The exposition follows [13] closely. Rather unexpectedly to me, some cheap but possibly far-reaching generalizations of [13] are obtained. If both V and C are smooth we get desingularizations of the closures of the strata.

0.3 - Part two : Concentration in schemes. Formally, this part is independent of part one. Here the scheme-theoretic foundations of concentration are layed. If μ is an action of the group scheme $G\ell(1)_{\mathbb{Z}}$ on a scheme V with an invariant subscheme C, the concentrator $V(\mu,m)$ is defined as a contra-variant set-valued functor on the category of the schemes. Under suitable conditions $V(\mu,m)$ turns out to be representable. If the scheme V is separated then $V(\mu,m)$ is a subfunctor of V. Usually however, the representing scheme is not a subscheme of V. In section 5 we obtain conditions on V, μ and C which imply that $V(\mu,m)$ is (representable by) a regular scheme. This generalizes a theorem of Bialynicki-Birula, cf. [2]. Some commutative algebra is used here.

Let $\mu : G\ell(1)_S \to G$ be a morphism of group schemes over a base scheme S.

If G acts on an S-scheme V with an invariant subscheme C, we get a concentrator $V(\mu,m)$ as above. The concentrator $G(\mu,m)$ of the interior action of G on itself with $C = G$, is a group functor in a natural way. Moreover the group functor $G(\mu,m)$ acts on the concentrator $V(\mu,m)$. In section 7 the vector building $Vb(G,S)$ is constructed. It may be considered as a partition of the set of the pairs (μ,m) such that $V(\mu,m) = V(\nu,n)$ holds whenever (μ,m) and (ν,n) belong to the same class $\delta \in Vb(G,S)$. The cost function q mentioned above is a morphism from the associated sheaf $\underline{Vb}(G)$ to the sheaf of the locally constant rational functions.

If G is a split torus over a connected affine scheme S the determination of the concentrators in V reduces to linear inequalities in the space of the weights. The exposition in section 8 is original, but the ideas go back to [2,15,17]. The Kempf-Rousseau theorem on optimal concentration mentioned above is generalized and proved in section 9. We are forced to work with a reductive group G over a field k acting on a separated k-scheme V. The condition that V is affine, can be weakened slightly. Section 10 serves as an appendix. It contains a construction of Grothendieck with a discussion of some aspects not mentioned in the available reference.

0.4 - As usual we refer to [7,9,10,11] by the symbols E G A or S G A followed by the appropriate sequence of numbers. As we use the new edition [9] of E G A I, our schemes are not necessarily separated. In part one the varieties are meant to be separated schemes of finite type over the given field. So they are not necessarily reduced or irreducible.

0.5 - A conversation with Bialynicki-Birula greatly inspired me in the research which led to part two. Part one grew out of a lecture delivered in Paris on invitation of Mme Malliavin. I am thankful for the opportunity she offered me, to publish the two parts together in the Seminaire Dubreil. I dedicate this paper to the memory of my son Mark Hessel who lived and died only four weeks old during the preparation of this manuscript.

<u>Part one</u> : <u>Concentration in affine varieties</u>

1 - The Hilbert-Mumford criterion

1.1 Let G be a linear algebraic group over an algebraically closed field k, cf. [3] or [14]. A triple (V,ρ,C) consisting of an affine variety V, an action $\rho: G \times V \to V$ and a G-invariant closed subvariety C is called an <u>affine centered</u> G-<u>variety</u>. The subvariety C is called the <u>center</u>. If V is a finite dimensional vector space, ρ is a linear action $G \to G\ell(V)$, and C is an invariant subspace then (V,ρ,C) is called a centered G-space.

Let (V,ρ,C) be an affine centered G-variety. We shall write $g\,v = \rho(g,v)$. Consider a morphism of algebraic groups $\lambda : G\ell(1) \to G$ and a point $v \in V$. We write $\lim \lambda(t)\,v = w$ if there is a morphism of varieties $h : \mathbb{A}^1 \to V$ with $h(0) = w$ and $h(t) = \lambda(t)\,v$ for all $t \neq 0$. Here $G\ell(1)$ is embedded in the affine line $\mathbb{A}^1$ in the obvious way. A point $v \in V$ is called concentrated if there exists a morphism of algebraic groups $\lambda : G\ell(1) \to G$ and a point $c \in C$ with $\lim \lambda(t)\,v = c$. The set of the concentrated points is called the concentrated cone $N(V)$.

1.2 Theorem (Hilbert-Mumford). Let G be reductive. Let (V,ρ,C) be an affine centered G-variety. The following conditions on $v \in V$ are equivalent :

a) $v \in N(V)$.

b) the closure of the orbit $G\,v$ meets the center C.

c) $f(v) = 0$ for every G-invariant function f which vanishes on C.

The implications $a \Rightarrow b$ and $b \Rightarrow c$ are trivial. The implication $c \Rightarrow b$ is a consequence of the theorem of Haboush, cf. [6] 4(b). The implication $b \Rightarrow a$ is proved in [15] 1.4. Under the assumptions of the theorem it follows that $N(V)$ is a G-invariant closed subvariety of V.

1.3 Example Let $p \in \mathbb{A}^2$ (the affine plane) be a fixed non-zero vector. Let G be the stabilizer of p in the group $G\ell(2)$. Let $C = \{0\}$. Then the cone $N(\mathbb{A}^2)$ is not closed.

1.4 Example Consider $G = S\ell(2)$ acting by substitution on the space V of the cubic forms v in two variables x and y. So we have $(g\,v)\,(x,y) = v\,((x\,y).g)$ where the dot means a matrix multiplication. We use the co-ordinates $\alpha,\beta,\gamma,\delta$ on V given by :

$$v = \alpha x^3 + 3\,\beta x^2\,y + 3\,\gamma x\,y^2 + \delta y^3 .$$

It is known that the ring of the invariant functions on V is generated by the function

$$f(v) = (\alpha\,\delta - \beta\gamma)^2 - 4\,(\alpha\gamma - \beta^2)\,(\delta\,\beta - \gamma^2).$$

If we put $C = \{0\}$ it follows from 1.2 that $N(V)$ consists of the points v with $f(v) = 0$. Now the variety $N(V)$ has singularities. In fact, it is not normal.

<u>Remark</u> (D. Bartels). Now let C be the orbit of x^3 in V, which is not closed. The closure of the orbit of $v = x^2y$ meets the set C but there is no morphism of algebraic groups $\lambda : G\ell(1) \to G$ with $\lim \lambda(t)v \in C$. So in theorem 1.2 the condition that C is closed, cannot be dropped.

2 - The theorem of Kempf and Rousseau [15,19].

2.1 Let G be a reductive group and let v be a concentrated point of a G-module V (here $C = \{0\}$). On a conference in Les Plans sur Bex, Switzerland, March 1977, Kempf formulated an optimality criterion which defines a not too large class of morphisms $\lambda : G\ell(1) \to G$ with $\lim \lambda(t)v = 0$. An elementary exposition is given in [18]. We need a slightly different procedure. Our optimal class $\Lambda(v)'$ consists of "fractional" one-parameter subgroups, or co-weights. Let us first give an example.

2.2 <u>Example</u> <u>Ternary quartic forms</u> Let $G = S\ell(3,\mathbb{C})$ be the group of the complex 3×3 matrices with determinant one. Let T be the subgroup of the diagonal matrices in G. A one-parameter subgroup λ of T is of the form

$$\lambda(t) = [a,b,c](t) := \operatorname{diag}(t^a, t^b, t^c)$$

with $a+b+c = 0$ and $a,b,c \in \mathbb{Z}$. If we also admit $a,b,c \in \mathbb{Q}$ with $a+b+c=0$, then it is called a <u>fractional one-parameter subgroup</u> of T. Let $M(G)$ be the set of all fractional one-parameter subgroups of G. We use a <u>norm</u> $q : M(G) \to \mathbb{Q}$ satisfying:

(i) If $\lambda = [a,b,c]$ then $q(\lambda) = a^2 + b^2 + c^2$.

(ii) If $\lambda(t) = g\,\mu(t)g^{-1}$ then $q(\lambda) = q(\mu)$.

Let V be the G-module of the homogeneous forms v of degree 4 in the indeterminates x,y,z with the action:

$$(gv)(x,y,z) = v((x,y,z).g).$$

The weight vectors of the space V with respect to T are the 15 monomials $x^i y^j z^k$ with $i,j,k \geq 0$ and $i+j+k = 4$. Using the tripels (i,j,k) as barycentric co-ordinates in the plane $X(T) \otimes \mathbb{Q}$ we get an equilateral triangle of weights:

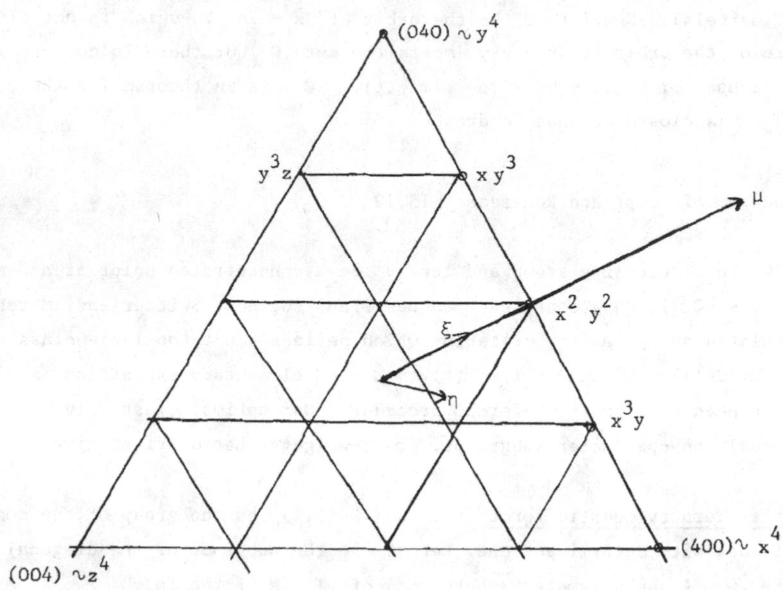

Now consider $v = x^4 + 2x^3y - 2xy^3 - y^4$. This form only uses the four encircled weights. The one-parameter subgroup $\mu = [1,1,-2]$ sends v to zero

$$\mu(t)\, v = t^4 v \to 0 \quad (t \to 0) .$$

Now we want to minimize $q(\lambda)$ under the constraint that $\lambda(t)\, v = O(t)$, where O is the Landau symbol. So the fractional one-parameter subgroup $\xi = [1/4, 1/4, -1/2]$ is better. However there still is a better one. As $v = (x+y)^3\,(x-y)$, there is a substitution $g \in G$ with $gv = x^3y$. Now $\eta = [5/14, -1/14, -4/14]$ satisfies $\eta(t)\, gv = t\, gv = O(t)$ and $q(\eta) < q(\xi)$. Therefore $\lambda \in M(G)$ given by $\lambda(t) = g^{-1}\, \eta(t)\, g$ satisfies $\lambda(t)\, v = O(t)$ and $q(\lambda) < q(\xi)$. It turns out that λ is optimal in the sense of 2.6 below.

The plane $M(T)$ of the fractional one-parameter subgroups of T is the dual of $X(T) \otimes \mathbb{Q}$. The norm q induces an inner product on $M(T)$ and hence an identification of $M(T)$ with $X(T) \otimes \mathbb{Q}$. This identification has been used to draw μ, ξ and η as vectors in the above diagram.

2.3 Let G be a linear algebraic group. The set of morphisms of algebraic groups $\lambda : G\ell(1) \to G$ is denoted $Y(G)$. The set $M(G)$ is obtained by a localization procedure. In fact, if $\lambda \in Y(G)$ and $m \in \mathbb{N}$, let $m\lambda \in Y(G)$ be given by $m\lambda(t) = \lambda(t^m)$. On $Y(G) \times \mathbb{N}$ the equivalence relation $\sim$ is defined by

$$(\mu, m) \sim (\nu, n) \iff n\mu = m\nu$$

The equivalence class of (μ,m) is called the <u>fractional one-parameter subgroup</u> or <u>co-weight</u> μ/m. Let $M(G)$ denote the set of the co-weights. The interior action of G on itself given by $\mathrm{int}(g)h = ghg^{-1}$, extends to an action of G on $M(G)$ also denoted by int.

Choose a maximal torus T in G. The set $Y(T)$ is considered as a $\mathbb{Z}$-module and $M(T)$ is identified with the vector space $Y(T)\otimes\mathbb{Q}$. A map $q : M(G) \to \mathbb{Q}$ is called a <u>norm</u> if we have :

(i) The restriction of q is a positive definite quadratic form on the vector space $M(T)$.

(ii) If $\lambda \in M(G)$ and $g \in G$ then $q(\mathrm{int}(g)\lambda) = q(\lambda)$.

Let V be a G-module. Every co-weight $\lambda = \mu/m$ induces a grading $V = \Sigma_r V_{r,\lambda}$, where $r \in \mathbb{Q}$ and :

$$V_{r,\lambda} = \{v \in V \mid \mu(t)\, v = t^{mr} v\}.$$

If V is faithful, the map $q_V : M(G) \to \mathbb{Q}$ given by :

$$q_V(\lambda) := \Sigma\, r^2 \dim(V_{r,\lambda})$$

is a norm on $M(G)$. It follows that a norm on $M(G)$ exists.

2.4 Let (V,ρ,C) be an affine centered G-variety. Let $\lambda = \mu/m$ be a co-weight. The <u>concentrator</u> $V(\lambda)$ is the set of the points $v \in V$ such that $\mu(t)v$ belongs to C, modulo powers t^n with $n \geqslant m$, and that $\lim \mu(t)v$ exists. The formal definition is postponed to section 4. If (V,ρ,C) is a centered G-space, the grading introduced in 2.3 enables us to write :

$$V(\lambda) = \Sigma_{r\geqslant 1} V_{r,\lambda} + \Sigma_{r\geqslant 0} C_{r,\lambda}.$$

If (V,ρ,C) is an affine centered G-variety, we may construct a centered G-space (V',ρ',C') and a G-equivariant closed immersion $j : V \to V'$ such that C is the schematic inverse image $j^{-1}(C')$. Then the concentrator $V(\lambda)$ is the schematic inverse image $j^{-1}(V'(\lambda))$. See 4.3(b) and 8.5 below, or [15].

2.5 An important special case is the G-variety (G,int,G). Now the concentrator $G(\lambda)$ is a closed subgroup of G. It consists of the elements $g \in G$ such that $\lim \lambda(t)\, g\, \lambda(t)^{-1}$ exists. If G is reductive, $G(\lambda)$ is a parabolic subgroup of G, cf. [17] p.55.

There is an equivalence relation $\sim$ on $M(G)$ such that $\lambda \sim \mu$ holds if and only if there is $g \in G(\lambda)$ with $\mu = \mathrm{int}(g)\lambda$. The quotient set is called the <u>vector building</u> $Vb(G)$. If $\lambda \in M(G)$ its equivalence class is denoted $[\lambda] \in Vb(G)$. If $V = (V,\rho,C)$ is an affine centered G-variety then $G(\lambda)$ stabilizes $V(\lambda)$. In particular, if $\lambda \sim \mu$ then $V(\lambda) = V(\mu)$ and $G(\lambda) = G(\mu)$. See 7.1 below.

2.6 Now we fix a norm $q : M(G) \to \mathbb{Q}$, cf. 2.3 . Let X be a subset of an affine centered G-variety (V,ρ,C) . The set X is said to be <u>concentrated</u> if there is $\lambda \in M(G)$ with $X \subset V(\lambda)$. We define

$$q^{\star}(X) := \inf \{ q(\lambda) \mid \lambda \in M(G) \quad , \quad X \subset V(\lambda) \}$$
$$\Lambda(X) := \{\lambda \in M(G) \mid X \subset V(\lambda) \quad , \quad q(\lambda) = q^{\star}(X)\}$$

A co-weight λ is called <u>optimal</u> if $\lambda \in \Lambda(X)$.

<u>Theorem</u> (Kempf-Rousseau [15,19]). Let X be a concentrated subset of an affine centered G-variety (V,ρ,C).

a) $\Lambda(X)$ is a non-empty union of equivalence classes in $M(G)$.

b) If T is a torus in G , then $M(T) \cap \Lambda(X)$ contains at most one element.

c) If G is reductive then $\Lambda(X) \in Vb(G)$.

Below in 8.6,8.9 and 9.4 we shall prove generalized versions of this theorem.

2.7 <u>Example</u> Let $V = \mathbb{A}^3$ and $C = \{0\}$. Let $G \subset G\ell(V)$ be the solvable group of the matrices

$$x(t,a) = \begin{pmatrix} t & 0 & 0 \\ 0 & t & 0 \\ 0 & a & t^{-2} \end{pmatrix} \qquad t \in k \setminus \{0\} \text{ , } a \in k$$

We use the norm $q = q_V$ cf. 2.3 . Let $v \in V$ be the point $(1,0,0)$. Clearly $\lambda \in M(G)$ given by $\lambda(t) = x(t,0)$ is optimal. The class $[\lambda]$ only contains λ itself. Choose $g = x(1,a)$ with $a \neq 0$. Since $gv = v$ the co-weight $\mu = \mathrm{int}(g)\lambda$ is also optimal. Since $\mu \neq \lambda$, this shows that in 2.6(c) the reductivity assumption cannot be dropped.

3 - The stratification of N(V)

3.1 In this section we assume that G is reductive and equipped with a norm $q : M(G) \to \mathbb{Q}$, and that (V,ρ,C) is an affine centered G-variety. Recall that $N(V)$ is a G-invariant closed subvariety of V , cf. 1.2 . Let X be a concentrated subset of V . By 2.5 and 2.6 we may define $P(X) := G(\lambda)$ and $S(X) := V(\lambda)$ where $\lambda \in \Lambda(X)$ is chosen arbitrarily. It is clear that $S(X)$ is concentrated with $\Lambda(S(X)) = \Lambda(X)$. Therefore we have $X \subset S(X) = S(S(X))$. The set X is called <u>saturated</u> if $X = S(X)$. Clearly $S(X)$ is saturated, it is called the <u>saturation</u> of X . The parabolic subgroup $P(X)$ is called the <u>Kempf group</u> of X . It satisfies

$$P(X) = \{g \in G \mid g\, X \subset S(X)\} ,$$

see [13] 2.8 . If v is a concentrated point of V we write $q^{\star}(v) = q^{\star}(\{v\})$,etc.

3.2 Lemma a) The number of conjugacy classes of saturated subsets of V is finite.
b) If X is saturated then X and GX are closed in V.
c) If $s \in \mathbb{Q}$ then $\{ v \in V \mid q^{\star}(v) < s \}$ is closed in V.

The proof of [13] 2.9 is easily adapted to this situation. Some facts related to (a) and (b) are contained in 8.9 and 10.2 below.

3.3 In the cone $N(V)$ we define two equivalence relations

$$x \approx y \iff \Lambda(x) = \Lambda(y) .$$
$$x \sim y \iff \text{there is } g \in G \text{ with } \Lambda(g\,x) = \Lambda(y)$$

The set $[x] := \{y \in N(V) \mid y \approx x\}$ is called the blade of x. The set $G[x] = \{y \in N(V) \mid y \sim x\}$ is called the stratum of x. Using lemma 3.2 we obtain the following proposition cf. [13] 4.2.

3.4 Proposition Let $v \in N(V)$
a) $[v] = \{x \in S(v) \mid q^{\star}(x) = q^{\star}(v)\}$.
b) The blade $[v]$ is open in the closed set $S(v)$.
c) $G[v] = \{x \in G\,S(v) \mid q^{\star}(x) = q^{\star}(v)\}$.
d) The stratum $G[v]$ is open in the closed set $G\,S(v)$.
e) $N(V)$ is a finite disjoint union of the strata.

3.5 Let $v \in N(V)$. Put $Y = G[v]$ and $Z = G\,S(v)$. Consider the map

$$\tau\colon G \times^{P(v)} S(v) \longrightarrow Z .$$

The lefthand side is the quotient of $G \times S(v)$ under the right $P(v)$-action $(g,x)p = (gp,p^{-1}x)$. Let $[g,x]$ represent the $P(v)$-class of (g,x). The morphism is given by $\tau[g,x] = gx$. Since $G/P(v)$ is a projective variety, the morphism is proper. See action 10 below. The following result is an immediate generalization of [13] 4.5.

Proposition The inverse image $\tau^{-1}Y$ is equal to $G \times^{P(v)} [v]$. The restriction $\tau\colon \tau^{-1}(Y) \longrightarrow Y$ is a proper bijective morphism. Therefore it is a finite morphism and a universal homeomorphism, cf. E G A III 4.4.2 and IV 2.4.5.

3.6 Theorem Assume $\mathrm{char}(k) = 0$. Then the restriction $\tau\colon \tau^{-1}Y \longrightarrow Y$ is an isomorphism.

Proof By 3.5 and E G A IV 15.2.3 and 17.9 it suffices to prove that the restriction of τ to $\tau^{-1}Y$ is an unramified morphism. The morphism τ is G-equivariant. If $x \in [v]$ then $[v] = [x]$. Therefore it suffices to prove that the morphism τ is unramified at the point $[1,v]$.

We may choose a centered G-space (V', ρ', C') and a G-equivariant closed immersion $j : V \to V'$ such that C is the schematic inverse image $j^{-1}(C')$, see the remark in 8.5 below. If $\lambda \in M(G)$ then $V(\lambda) = j^{-1}(V'(\lambda))$, see 4.3(b) below. It follows that $\Lambda(x) = \Lambda(j(x))$ for every $x \in V$. So we have $P(v) = P(j(v))$ and $S(v) = j^{-1}S(j(v))$. Therefore $G \times^{P(v)} S(v)$ is a closed subvariety of $G \times^{P(j(v))} S(j(v))$. So it suffices to show that the morphism :

$$\tau' : G \times^{P(j(v))} S(j(v)) \to V'$$

is unramified at the point $[1, j(v)]$. This means that the tangent mapping $d\tau'$ at this point should be injective. So it suffices to prove the following lemma.

3.7 <u>Lemma</u> Assume $\mathrm{char}(k) = 0$. Let $V = (V, \rho, C)$ be a centered G-space. Let $v \in N(V)$. Let $\underline{g}$ and $\underline{p}(v)$ denote the Lie algebras of G and $P(v)$, respectively. Then we have

$$\underline{p}(v) = \{X \in \underline{g} \mid X v \in S(v)\} .$$

Proof Choose $\lambda \in \Lambda(v)$. This co-weight λ induces gradings $V = \Sigma V_r$ and $\underline{g} = \Sigma \underline{g}_r$ with $r \in \mathbb{Q}$, see 2.3 . It is clear that $g_r V_s \subset V_{r+s}$ and that $\underline{p}(v) = \Sigma_{r \geq 0} \underline{g}_r$ and that :

$$v \in S(v) = \Sigma_{r \geq 1} V_r + \Sigma_{r \geq 0} C_r .$$

If $X \in \underline{p}(v)$ then it is obvious that $X v \in S(v)$. Suppose $X \in \underline{g} \setminus \underline{p}(v)$ satisfies $X v \in S(v)$. Write $X = \Sigma_{r \geq m} X_r$ with $X_r \in \underline{g}_r$ and $X_m \neq 0$, so that $m < 0$.

Now X_m is a nilpotent element of the semi-simple Lie algebra $[\underline{g}, \underline{g}]$. By the theory of Jacobson-Morozov, cf. [5] § 11, we can choose $H \in \underline{g}_o$ and $Y \in \underline{g}_{-m}$ such that $[H, X_m] = 2X_m$, $[H, Y] = -2Y$ and $[X_m, Y] = -H$. Let K be the connected subgroup of G such that its Lie algebra $\underline{k}$ is spanned by X_m, Y and H . Let $S \subset K$ be the centralizer of H , that is the torus with Lie algebra kH . The torus $\mathrm{Im}(\lambda)$ in G normalizes K and $\underline{k}$, and it centralizes H and S . So we have a reductive group $L = \mathrm{Im}(\lambda) K$ of semi-simple rank 1 , with a maximal torus $T = \mathrm{Im}(\lambda) S$. Let Π be the character group of T , let $\{1, w\}$ be the Weyl group of L with respect to T and let $\{\alpha, -\alpha\}$ be the root system. We may assume that $X_m \in \underline{\ell}_{\alpha}$ and $Y \in \underline{\ell}_{-\alpha}$. It follows that :

$$(\alpha, \lambda) = m < 0 .$$

Let (,) be the inner product on $M(T)$ such that $(\mu, \mu) = q(\mu)$ for all $\mu \in M(T)$. We use this inner product to identify

$$M(T) = M(T)^{\star} = \Pi \otimes_{Z} \mathbb{Q} .$$

A finite subset of Π is called a <u>diagram</u>. If E is a T-module and $e \in E$, the diagram $R(e)$ is defined as the smallest subset R of Π with $e \in \Sigma_{\pi \in R} E_{\pi}$.

If R is a diagram and $\mu \in M(T)$ we define :

$$(R,\mu) := \inf \{(\pi,\mu) \mid \pi \in R\} .$$

Since V and C are L-modules, we have a quotient module $U = V/C$. Let $u \in U$ denote the image of the point $v \in V$. Since $v \in V(\lambda)$ we have $(R(v),\lambda) \geq 0$ and $(R(u),\lambda) \geq 1$. Since λ is optimal, we have $q(\mu) \geq q(\lambda)$ for every $\mu \in M(T)$ with $(R(v),\mu) \geq 0$ and $(R(u),\mu) \geq 1$.

Coming back to the $\mathbb{Q}$-grading of V induced by λ , we write $v = \Sigma v_r$ with $v_r \in V_r$ and $u = \Sigma u_r$ with $u_r \in U_r$. Since $X\, v \in S(v)$ we have $x_m v_o = 0$ in V and $X_m u_1 = 0$ in U . By the representation theory of $\underline{k} = \underline{s\ell}(2)$ in characteristic zero, it follows that $(R(v_o),\alpha) \geq 0$ and $(R(u_1),\alpha) \geq 0$. On the other hand it is obvious that :

$$(R(v-v_o),\lambda) > 0 \qquad \text{and} \qquad (R(u-u_1),\lambda) > 1 .$$

Now we obtain a positive number ε such that for every rational number t with $0 \leq t < \varepsilon$ the co-weight $\mu_t = \lambda + t\,\alpha$ satisfies $(R(v),\mu_t) \geq 0$ and $(R(u),\mu_t) \geq 1$. The optimality of λ implies that :

$$q(\lambda) \leq q(\mu_t) = q(\lambda) + 2(\alpha,\lambda)t + q(\alpha)t^2 .$$

It follows that $(\alpha,\lambda) \geq 0$. This contradicts the inequality $(\alpha,\lambda) = m < 0$ obtained earlier.

3.8 <u>A variation of the definitions</u> If $v \in N(V)$, let the <u>c-blade</u> $[v]^o$ be defined as the connected component of the blade $[v]$ which contains the point v . The closure of $[v]^o$ is called the <u>c-saturation</u> $S_o(v)$. The variety $G[v]^o$ is called the <u>c-stratum</u> of v .

By 3.5 the c-stratum is a connected component of the stratum. It is easy to see that $G\,S_o(v)$ is the closure of the c-stratum $G[v]^o$, and that $G \times^{P(v)} [v]^o$ is the inverse image $\tau^{-1}(G[v]^o)$. It seems possible that $\tau^{-1}(G\,S_o(v))$ is larger than $G \times^{P(v)} S_o(v)$. The cone $N(V)$ is a disjoint union of the c-blades. It is a finite disjoint union of the c-strata.

3.9 <u>The regular case</u> Assume that C is regular and contained in the regular locus of V . By a generalization of a theorem of Bialynicki-Birula, see 5.8 below, all concentrators are regular. So the saturation $S(V)$ and the bundle $G \times^{P(v)} S(v)$ are regular. In particular the connected components are irreductible. Therefore the c-saturation $S_o(v)$ is a connected component of $S(v)$, so it is regular as well. Both the c-blade $[v]^o$ and the c-stratum $G\,[v]^o$ are irreductible.

Assume moreover that char(k) = 0 . By 3.6 the morphism τ induces an isomorphism between $G[v]^o$ and $G \times^{P(v)} [v]^o$. So the c-stratum $G[v]^o$ is a regular and connected subvariety of $N(V)$ with closure $G S_o(v)$. The morphism

$$\tau : G \times^{P(v)} S_o(v) \to G S_o(v)$$

is a desingularization : it is proper and birational, and the variety on the lefthand side is regular.

Remark In [13] section 4, we considered the regular case where moreover the center C consists of one point. Then the distinction between blades and c-blades, etc..., vanishes. This distinction also vanishes if (V,ρ,C) is a centered G-space. For then concentrators are linear and hence connected.

Part two : Concentration in schemes

4 - Concentration in centered $G\ell(1)$-schemes

4.1 In this section G is the multiplicative group scheme $G\ell(1)$ over $\mathbb{Z}$, with co-ordinate ring $\mathbb{Z}[T, T^{-1}]$. It is an open subscheme of the affine line $\mathbb{A}$ = Spec ($\mathbb{Z}[T]$). We consider $\mathbb{A}$ as a monoid scheme with multiplication $\nu : \mathbb{A}\times\mathbb{A} \to \mathbb{A}$ given by the co-morphism $\nu^\star$ with $\nu^\star(T) = T \otimes T$. Let 0 and ε be the morphisms Spec $(\mathbb{Z}) \to \mathbb{A}$ given by the co-morphism with $0^\star(T) = 0$ and $\varepsilon^\star(T) = 1$. If $m \in \mathbb{N}$ we define the closed subscheme

$$\mathbb{A}(m) = \mathrm{Spec}(Z[T] / (T^m)) .$$

A centered G-scheme V is a triple (V,μ,C) where V is a scheme, $\mu : G\times V \to V$ is an action of G on V , and C is a G-invariant subscheme of V , to be called the center . The concentrator functor Φ of a centered G-scheme V at speed $m \in \mathbb{N}$ is defined such that for every scheme X the set $\Phi(X)$ consists of the morphisms $f : \mathbb{A}\times X \to V$ satisfying the conditions :

a) The restriction $f\,|\,G\times X$ equals $\mu(-,f(\varepsilon,-))$.

b) The restriction $f\,|\,\mathbb{A}(m)\times X$ factorizes over the center C .

It is clear Φ is a contravariant functor from the category (Sch) of the schemes to the category of the sets. If X is a scheme, one easily verifies that the presheaf on X given by $U \to \Phi(U)$ is a sheaf. So Φ is a sheaf on (Sch) in the sense of E G A I 2.4.3.

If X is a scheme let $h(X)$ be the functor Hom(?,X). In many cases the functor Φ is representable by a scheme Y with an isomorphism $\Phi \cong h(Y)$, see E G A 0_I1 .

Representability of Φ by Y means the existence of $\varphi \in \Phi(Y)$ such that for every scheme X and every $f \in \Phi(X)$ there is a unique morphism $y : X \to Y$ with $f = \varphi(?,y)$. Then the scheme Y is called the concentrator scheme.

The realization morphisms $i,p : \Phi \to h(V)$ are defined by $i(f) = f(\varepsilon,-)$ and $p(f) = f(0,-)$. By condition (b) the morphism p factorizes over the subfunctor $h(C)$

4.2 Proposition Let (V,μ,C) be a centered G-scheme. Assume that V is affine and that C is closed in V. Then Φ is representable by an affine scheme Y and the realization morphism $i : Y \to V$ is a closed immersion.

Proof. Let B be the co-ordinate ring of V. Let $p : B \to B[T,T^{-1}]$ be the co-morphism of the multiplication $\mu : G \times V \to V$. We can write

$$p(b) = \sum_{n \in \mathbb{Z}} p_n(b)\, T^n \quad , \quad b \in B \ .$$

Since μ is an action the mappings p_n are the projections corresponding to a $\mathbb{Z}$-grading $\sum B_n$ of the ring B. Let J be the ideal of the closed subscheme C. Since C is G-invariant, J is homogeneous, say $J = \sum J_n$.

Since Φ is a sheaf on the category (Sch) of the schemes, we may restrict Φ to the category (Aff) of the affine schemes without loosing information, cf. EGA I.2.3.6 . so let $X = \mathrm{Spec}(R)$ be an affine scheme and let $f : \mathbb{A} \times X \to V$ be a morphism. The co-morphism $u : B \to R[T]$ of f can be written :

$$u(b) = \sum_{n \geqslant 0} u_n(b)\, T^n \quad , \quad b \in B \ .$$

The morphism $f(\varepsilon,-)$ has co-morphism $u' : B \to R$ with $u' = \sum u_n$. So the morphism $\mu(-,f(\varepsilon,-))$ has the co-morphism $u'' : B \to R[T,T^{-1}]$ given by

$$u''(b) = \sum_{n \in \mathbb{Z}} u'(p_n(b))\, T^n \quad , \quad b \in B \ .$$

Now condition 4.1(a) is equivalent to the condition that $u = u''$. So it follows from 4.1(a) that $u'(\sum_{n<0} B_n) = 0$. The restriction $f|\mathbb{A}(m) \times X$ has co-morphism $\overline{u} : B \to R[T]/(T^m)$ given by $\overline{u} = u(\mathrm{mod}(T^m))$. This restriction factorizes over C if and only if $\overline{u}(J) = 0$. So we have $f \in \Phi(X)$ if and only if

$$u(b) = \sum_{n \geqslant 0} u'(p_n(b))\, T^n \quad , \quad b \in B$$

$$u'(\sum_{n<0} B_n + \sum_{n<m} J_n) = 0 \ .$$

Since u is determined by u' the realization morphism $i : \Phi \to h(V)$ is injective. It is clear that the subfunctor $i\Phi$ of $h(V)$ is represented by the closed subscheme of V given by the ideal generated by :

$$\sum_{n<0} B_n + \sum_{n<m} J_n \ .$$

4.3. Let Φ and Φ' be the concentrator functors at speed m of centered G-schemes (V,μ,C) and (V',μ',C'), respectively. Let $j : V \to V'$ be a G-equivariant morphism with a restriction $j_1 : C \to C'$. Let $j_\star : \Phi \to \Phi'$ be the morphism of functors with $j_\star(f) = jf$. Let (i,p) and (i',p') be the realization morphisms of Φ and Φ', respectively. We clearly have

$$(i',p')j_\star = h(j \times j_1)(i,p) : \Phi \to h(V') \times h(C') .$$

<u>Proposition</u> Assume that $j : V \to V'$ is an immersion. The morphism $j_\star$ is injective, so that Φ is isomorphic to $j_\star \Phi \subset \Phi'$.

a) If j_1 induces an open immersion of C into $j^{-1}(C')$, then we have $j_\star \Phi = (i',p')^{-1} h(V \times C)$.

b) If j is a closed immersion and $C = j^{-1}(C')$, then $j_\star \Phi = (i')^{-1} h(V)$.

c) If j and j_1 are open immersions, then $j_\star \Phi = (p')^{-1} h(C)$.

Proof a) and b). Let X be a scheme and $f \in \Phi'(X)$. Assume $i'(f) \in h(V)(X)$. Assume either that $p'(f) \in h(C)(X)$ or that j is a closed immersion and $C = j^{-1}(C')$. Since $i'(f) : X \to V'$ factorizes over V and V is G-invariant the restriction $f \mid G \times X$ factorizes over V. So the subscheme $f^{-1}(V)$ of $\mathbb{A} \times X$ contains $G \times X$ and is therefore schematically dense and open, cf. EGA I 5 4. In the first case $f^{-1}(V)$ also contains the set $0 \times X$. In the second case $f^{-1}(V)$ is closed. Any how we have $f^{-1}(V) = \mathbb{A} \times X$. So there is a unique morphism $g : \mathbb{A} \times X \to V$ with $f = jg$. It is clear that the restriction $g \mid \mathbb{A}(m) \times X$ factorizes over $j^{-1}(C')$. In the first case we know that $g(0,-)$ factorizes over C. In both cases it follows that $g \in \Phi(X)$ and $j_\star(g) = f$.

c) Now let $f \in \Phi'(X)$ and $p'(f) \in h(C)(X)$. The complement E of V in V' is a G-invariant closed subset of V'. Since $p'(f) : X \to C'$ factorizes over C and $E \cap C = \emptyset$, the image of f does not meet E. So there is a unique $g : \mathbb{A} \times X \to V$ with $f = jg$. Since $g(0,?)$ factorizes over C and $f \mid \mathbb{A}(m) \times X$ factorizes over C', we obtain $g \in \Phi(X)$ and $f = j_\star(g)$.

4.4. <u>Lemma</u> Let (V,ρ,C) be a centered G-scheme. Let $(V_\alpha)_{\alpha \in I}$ and $(C_\alpha)_{\alpha \in I}$ be families of G-invariant open subschemes of V and C, respectively, with $C_\alpha \subset V_\alpha$ for every $\alpha \in I$, and $\bigcup C_\alpha = C$. Assume that the concentrator functors Φ_α of (V_α,ρ,C_α) at speed m are representable by schemes Y_α. Then the concentrator functor Φ of (V,ρ,C) at speed m is representable by a scheme Y and the canonical morphisms $Y_\alpha \to Y$ form a family of open immersions which covers Y.

Proof. By 4.3(c) the inclusions $j^\alpha : V_\alpha \to V$ induce inclusions $j^\alpha_\star : \Phi_\alpha \to \Phi$ which identify Φ_α with $p^{-1}(h(C_\alpha))$. If X is a scheme and $f : h(X) \to \Phi$ is a morphism of functors, then $f^{-1}\Phi_\alpha$ is representable by the open subscheme $X_\alpha = p(f)^{-1}(C_\alpha)$ of X. Here f is identified with an element of $\Phi(X)$ so that $p(f)$ is a morphism $X \to C$. The family $(X_\alpha)_\alpha$ is a covering of X. So the morphisms j^α are representable by open immersions, cf. EGA 0_I.1.7.7, and they form a covering of Φ. As Φ is a sheaf on the category (Sch) of the schemes, the assertions follows from EGA I.2.4.3.

4.5. A G-scheme (V,μ) is called <u>locally affine</u> if every G-invariant open subset U of V is covered by G-invariant affine open subsets of U. In 8.7 below it will be proved that it suffices to cover V itself by G-invariant affine open subsets.

<u>Theorem</u> Let (V,μ,C) be a locally affine centered G-scheme. The concentrator functor Φ at speed m is representable by a scheme Y. The realization morphism $(i,p) : Y \to V\times C$ is an immersion. The morphism $i : Y \to V$ is locally immersive. The morphism $p : Y \to C$ is affine.

<u>Proof</u> Let U be the largest open subset of V such that C is closed in U. Since U is G-invariant it has a covering by G-invariant affine open subsets V_α. We use the notation of 4.4 with $C_\alpha = C \cap V_\alpha$. By 4.2 the concentrators Φ_α are representable by schemes Y_α. By 4.4 the concentrator Φ is representable by a scheme Y. The image $(i,p)\,Y$ in $V\times C$ is covered by the family $(V_\alpha \times C_\alpha)_\alpha$. For every α the morphism $Y_\alpha \to V_\alpha \times C_\alpha$ is a closed immersion and $Y_\alpha = (i,p)^{-1}(V_\alpha \times C_\alpha)$. So $(i,p) : Y \to V\times C$ is an immersion, cf. EGA I 4.2.4. Since the restrictions $i : Y_\alpha \to V_\alpha$ are immersions, the morphism $i : Y \to V$ is locally immersive. For every α the restriction $p_\alpha : Y_\alpha \to C_\alpha$ is affine and $Y_\alpha = p^{-1}(C_\alpha)$. Since $C = \bigcup_\alpha C_\alpha$ this proves that $p : Y \to C$ is affine.

4.6 <u>Examples with curves</u> over an algebraically closed field k. In the examples (a) and (b) the action is locally affine. The third curve is the easiest example of a not locally affine action. In all cases we choose the center C equal to the curve itself. As condition 4.1(b) is trivially satisfied, the speed m is irrelevant.

a) Let $\mathbb{P}$ be the projective line over k with homogeneous co-ordinates $[r,s]$. Let $\mu : G\times\mathbb{P} \to \mathbb{P}$ be the action with $t\,[r,s] = [r,ts]$, so that $0 = [1,0]$ and $\infty = [0,1]$ are the invariant points. The concentrator scheme Y is the sum of the affine line $\mathbb{A}_k$ and the singleton $\{\infty\}$. The realization $(i,p) = Y \to \mathbb{P}^2$ is not a closed immersion. The morphism $i : Y \to \mathbb{P}$ is bijective and not an immersion.

b) Let Z be the non-separated line over k with a double origin. It is obtained as the quotient $Y/\sim$ of the sum $Y = \{0,1\} \times \mathbb{A}_k$ of two lines, under the equivalence relation $\sim$ with $(0,s) \sim (1,s)$ whenever $s \neq 0$. The action $t(r,s) = (r,ts)$ of G on Y induces an action μ of G on Z. It turns out that Y is the concentrator scheme of (Z,μ,Z) with $i : Y \to Z$ as the quotient morphism.

c) Independently, Bialynicki-Birula and Luna showed me the following example. In the G-scheme $\mathbb{P}$ of example a) we identify the points 0 and ∞ to obtain a cubic curve V over k with a double-point c. The quotient morphism $\pi : \mathbb{P} \to V$ induces an isomorphism between $\mathbb{P} \setminus \{0,\infty\}$ and $V \setminus \{c\}$. It satisfies $\pi^{-1}(c) = \{0,\infty\}$. Since the scheme V is separated there is at most one action $\mu : G \times V \to V$ such that $\pi : \mathbb{P} \to V$ is equivariant. Assume for the moment that the action μ on V exists. Then (V,μ) is not locally affine, since the only invariant open neighbourhood of the double-point c is V itself.

The existence of μ can be proved as follows. We may assume that V is the closed subvariety given by the equation $x^3 = (x+y)yz$ in the projective plane with homogeneous co-ordinates $[x,y,z]$. The point c has co-ordinates $[0,0,1]$. Now consider the affine G-variety

$$U_1 = \{ (a,b) \in \mathbb{A}_k^2 \mid ab = 0 \}$$

with the action $t(a,b) = (ta, t^{-1}b)$. Let U be the invariant open subset $U_1 \setminus \{(0,0)\}$. Let U_2 be a copy of U_1. Let $\sigma : U \to U_2$ be the equivariant open immersion given by $\sigma(a,b) = (b^{-1}, a^{-1})$. Let S be the union of U_1 and U_2 with U and $\sigma(U)$ identified through σ. Then S is a union of two projective lines which intersect in the two double-points p_1 and p_2 of U_1 and U_2, respectively. Let $\varphi_1 : U_1 \to V$ be the morphism given by

$$\varphi_1(a,b) = [\,(a+b)(a+b-1),\ a-b^2,\ (a+b-1)^3\,] .$$

Since $\varphi_1(b^{-1},a^{-1}) = \varphi(a,b)$ on U, the morphism φ_1 extends uniquely to a morphism $\varphi : S \to V$. One verifies that φ is a twofold étale covering. Using SGA 1 VIII one proves that the action of G on S descends to V.

The concentrator scheme of the G-action on S is a sum of two affine lines. Again using descent theory one verifies that the concentrator scheme of V is an affine line $\mathbb{A}_k$. The realization morphism $i : \mathbb{A}_k \to V$ is bijective but not locally immersive.

5 - Regularity of the concentrator scheme

5.1 In this section we use the conventions of 4.1. In particular, G is $G\ell(1)$ over $\mathbb{Z}$. We define an <u>$\mathbb{A}$-scheme</u> to be a pair (V,μ) such that V is a scheme and $\mu : \mathbb{A} \times V \to V$ is an action of the monoid $\mathbb{A}$. So we have :

a) $\mu(\nu,?) = \mu(?,\mu) : \mathbb{A}\times\mathbb{A}\times V \to V$

b) $\mu(\varepsilon,?) = 1_V : V \to V$

The <u>basis</u> V_o of an $\mathbb{A}$-scheme (V,μ) is defined as the equalizer $\mathrm{Ker}\ (0,1_V)$ where $0 : V \to V$ is the morphism $\mu(0,?)$. Clearly $0 : V \to V$ factorizes over a <u>projection</u> $\mu_o : V \to V_o$.

The $\mathbb{A}$-scheme (V,μ) is called an <u>$\mathbb{A}$-bundle</u> if every point $v \in V_o$ has an open neighbourhood U in V_o such that the $\mathbb{A}$-scheme $\mu_o^{-1} U$ is isomorphic to an $\mathbb{A}$-scheme $\mathbb{A}^r \times U$ with an action μ' given by :

$$\mu'(t,(a_1,\dots,a_r\,;u)) = (t^{w(1)}\ a_1,\dots,t^{w(r)}\ a_r\,;u)$$

for every triple of morphisms $t : X \to \mathbb{A}$, $a : X \to \mathbb{A}^r$, $u : X \to U$. The weights $w(1),\dots,w(r)$ are positive integers which may depend on the point v in a locally constant way.

5.2 Let Φ be the concentrator functor of a centered G-scheme (V,μ,C) at speed m. If X is a scheme and $f \in \Phi(X)$ and $t \in h(\mathbb{A})(X)$, let $tf \in \Phi(X)$ be given by :

$$tf = f\ (\nu(t,-),-) : \mathbb{A}\times X \to V$$

This defines a mophism of functors $h(\mathbb{A})\times\Phi$, which is an action of the monoid functor $h(\mathbb{A})$ on the set functor Φ . As in 5.1 the <u>basis</u> Φ_o is defined as the equalizer of the two morphisms of functors $0,1 : \Phi \to \Phi$. So Φ_o is a subfunctor of Φ with a <u>projection</u> $\mu_o : \Phi \to \Phi_o$.

5.3 <u>Lemma</u> a) The realization morphism $p : \Phi \to h(C)$ factorizes over the subfunctor $h(C)^G$ of the G-invariant sections of $h(C)$.

b) The basis Φ_o is the equalizer of the morphisms $i,p : \Phi \to h(V)$.

c) The morphism p induces an isomorphism $\Phi_o \cong h(C)^G$.

Proof a) Let X be a scheme and $f \in \Phi(X)$. Let ν' be the restriction $\nu\,|\,G\times\mathbb{A}$ of the multiplication ν of $\mathbb{A}$. The equalizer E of the morphisms $f(\nu',?)$ and $\mu(?,f)$ from $G\times\mathbb{A}\times X$ to V is a subscheme of $G\times\mathbb{A}\times X$. Since it contains $G\times G\times X$, the subscheme E is schematically dense and open. Let $j : G\times X \to G\times A\times X$ be the closed immersion with $j(g,x) = (g,0,x)$. Now $j^{-1}E$ is an open subscheme of the group scheme $G\times X$ over X . One verifies that $j^{-1}E$ is the stabilizer of the section $p(f) \in h(C)(X)$. So it is an open subgroup scheme of $G\times X$ over X, and therefore it is equal to $G\times X$. This proves that $p(f)$ is G-invariant (and that $E = G\times\mathbb{A}\times X$).

b) and c) Let Ψ be the equalizer $\mathrm{Ker}(i,p)$. It is clear that $\Phi_o \subset \Psi$. By part(a) we have a morphism $p : \Psi \to h(C)^G$. If X is a scheme and $c \in h(C)^G(X)$ we may define $q(c) \in \Phi_o(X)$ by $q(c) = cp_2$ where $p_2 : \mathbb{A} \times X \to X$ is the projection. Clearly q is a morphism of functors $h(C)^G \to \Phi_o$ with $pq(c) = c$ for every $c \in h(C)^G(X)$. Consider $f \in \Psi(X)$. Then $i(f) = p(f)$. So by 4.1(a) the restriction $f \mid G \times X$ equals $\mu(-,p(f))$. By part (a) it follows that $f \mid G \times X$ equals the restriction of $qp(f) = p(f)p_2$ to $G \times X$. Now it follows from lemma 5.4 below that $qp(f) = f$. This proves that $\Phi_o = \Psi$ and that $p : \Psi \to h(C)^G$ is bijective.

5.4 Lemma Let $\sigma : X \to S$ and $\pi : \mathbb{A} \times X \to S$ be morphisms of schemes. Let $p_2 : \mathbb{A} \times X \to X$ denote the projection. Assume that the restrictions of τ and σp_2 to $G \times X$ are equal. Then $\tau = p_2$.

Proof. Let E be the equalizer $\mathrm{Ker}(\tau, \sigma p_2)$. It is a subscheme of $\mathbb{A} \times X$ and it contains $G \times X$. So it is a schematically dense open subscheme of $\mathbb{A} \times X$, cf. EGA I 5.4. Now it suffices to prove that E contains all points of $\mathbb{A} \times X$. So we may assume that $X = \mathrm{Spec}(k)$ where k is a field. We can choose an affine open neighbourhood S' of $\tau(0)$ in S. The morphisms τ and σ factorize over S'. So we may assume that $S = S'$. Since S is now separated, E is closed and hence equal to $\mathbb{A} \times X$.

5.5. Assume that Φ is representable by a concentrator scheme Y. The action of $h(\mathbb{A})$ on Φ induces an action of $\mathbb{A}$ on Y, so that Y is an $\mathbb{A}$-scheme. The basis Y_o represents Φ_o. By 5.3 it is equal to the equalizer $\mathrm{Ker}(i,p)$. It also represents the functor $h(C)^G$, so we may identify $Y_o = C^G$.

Assume that the scheme C is separated, or that the action of G on C is locally affine. Then the functor $h(C)^G$ is representable by a closed subscheme C^G of C. In the first case we may refer to SGA 3 VIII 6.5 (e). The second case is proved with the methods of 4.5.

5.6 Lemma Let B a $\mathbb{Z}$-graded noetherian ring. Assume that the subring B_o is a local ring with maximal ideal M_o and that $M = M_o + \sum_{n \neq o} B_n$ is a maximal ideal of B.

a) If $P' \subset P$ are finitely generated $\mathbb{Z}$-graded B-modules, we have

$$P = P' + MP \iff P_M = P'_M \iff P = P'$$

b) Let J be a homogeneous ideal in B. Assume that the local rings B_M and $(B/J)_M$ are regular. Fix $m \in \mathbb{N}$ and put $L = \sum_{n<o} B_n + \sum_{n<m} J_n$. Then $(B/LB)_M$ is also regular.

c) Assume that B_M is regular and that $B_n = 0$ for all $n < 0$. Then B_o is noetherian and regular. B is isomorphic to a graded polynomial B_o-algebra $B_o[T_1,\dots,T_r]$ with $T_1,\dots,T_r$ homogeneous of positive degree.

Proof a) Replacing P by P/P' we may assume that $P' = 0$. If $P = MP$ there is $d \in M$ such that $(1-d)P = 0$, cf. [1] p.20, so that $P_M = 0$. Assume $P_M = 0$. Let x be a homogeneous element of P. There is $b \in B$ with $bx = 0$ and $b \notin M$. Write $b = \Sigma b_n$ with $b_n \in B_n$. Then we have $b_n x = 0$ for all n. Since b_o is invertible it follows that $x = 0$. This proves $P = 0$. If $P = 0$ then $P = MP$, trivially.

b) Since J is homogeneous it is contained in M. We may choose homogeneous elements $x_1,\ldots,x_r$ in J and $x_{r+1},\ldots,x_s$ in M such that the images $x'_1,\ldots,x'_r$ form a basis of $(J+M^2)/M^2$ and that $x'_1,\ldots,x'_s$ form a basis of M/M^2, in both cases over the field B/M. Since the local rings B_M and $(B/J)_M$ are regular, we have

$$J_M = \sum_{i=1}^{r} B_M x_i \quad \text{and} \quad M_M = \sum_{i=1}^{s} B_M x_i$$

see the proof of EGA 0_{IV} 17.1.9. By part (a) it follows that $x_1,\ldots,x_r$ generate J and that $x_1,\ldots,x_s$ generate M. Let L' be the ideal in B generated by the elements x_i which belong to the subspace L. So, if the degree of a homogeneous element $b \in B$ is denoted by $d(b)$, the ideal L' is generated by the set

$$\{ x_i \mid d(x_i) < 0 \ , \quad \text{or} \quad i \leqslant r \quad \text{and} \quad d(x_i) < m \}$$

Since the ideal L'_M is generated by regular parameters, the local ring $(B/L')_M$ is regular. So it suffices to prove that $L' = LB$. By part (a) it suffices to prove $L \subset L' + ML$. Let $b \in B_n$ with $n < 0$. Then $b \in M$, so we may write $b = \sum_{i=1}^{s} x_i b_i$ with $d(x_i) + d(b_i) = n$. For every i we have $d(x_i) < 0$ or $d(b_i) < 0$, so that $x_i b_i \in L' \cup ML$. This proves $b \in L' + ML$. Now let $b' \in J_n$ with $n < m$. We may write $b' = \sum_{i=1}^{r} x_i b'_i$ with $d(x_i) + d(b'_i) = n$. For every $i \leqslant r$ we have $d(x_i) < m$ or $d(b'_i) < 0$, so that $x_i b'_i \in L' \cup ML$. This proves $b' \in L' + ML$. Therefore $L \subset L' + ML$.

c) This part is well known but we give a proof for the lack of a reference. Let $x_1,\ldots,x_s$ be homogeneous elements of M such that the images form a basis of M/M^2 over B/M. Since B_M is regular it follows from part (a) that $(x_1,\ldots,x_s)$ is a B-regular sequence generating M. We may assume that $x_1,\ldots,x_t \in M_o$ and that $d(x_i) \geqslant 1$ whenever $i \geqslant t+1$. Then the sequence $(x_1,\ldots,x_t)$ is B_o-regular and it generates M_o over B_o. So $B_o \cong B/\Sigma_{n>o} B_n$ is a noetherian regular local ring of dimension t. Put $r = s-t$. We clearly have a surjective morphism $\varphi : B_o[T_1,\ldots,T_r] \to B$ with $\varphi(T_i) = x_{t+i}$. Since $B_o[T_1,\ldots,T_r]$ is a domain of dimension s, the map φ is bijective. If we give T_i the degree of x_{t+i}, then φ is an isomorphism of graded B_o-algebras.

5.7 <u>Proposition</u> Let (V,μ) be an $\mathbf{A}$-scheme. Assume that the projection $\mu_o : V \to V_o$ is an affine morphism, that V is locally noetherian and that V_o is contained in the regular locus of V. Then (V,μ) is an $\mathbf{A}$-bundle and the basis V_o is regular.

Proof We may assume that V_o is affine. Then V is affine as well, say with coordinate ring B. The action μ corresponds to a positive grading $B = \sum_{n \geq o} B_n$. The projection $\mu_o : V \to V_o$ corresponds to the injection $B_o \to B$. Let $v \in V_o$ correspond to a prime ideal p of B_o. Put $S = B_o \setminus p$. The localization $S^{-1}B$ is a graded ring which satisfies the assumptions of 5.6(c). Therefore $S^{-1}B_o$ is a regular local ring and there are homogeneous elements $x_1,\ldots,x_r$ of positive degree in $S^{-1}B$ such that the morphism from $S^{-1}B_o[T_1,\ldots,T_r]$ to $S^{-1}B$ which sends T_i to x_i, is an isomorphism. Choose $s_o \in S$ and homogeneous elements $t_1,\ldots,t_r$ in B with $x_i = t_i/s_o$. The morphism

$$\varphi : B_o[T_1,\ldots,T_r] \to B \;,\; \varphi(T_i) = t_i$$

is such that $S^{-1}\varphi$ is an isomorphism. By EGA I 6.6.4 there is $s \in S$ such that the localization $s^{-1}\varphi$ is an isomorphism. Thus we have a trivialization of a neighbourhood of V.

5.8 <u>Theorem</u> Let (V,μ,C) be a locally affine centered G-scheme.Assume that V is locally noetherian and that C^G is contained in the regular locus of both V and C. Fix $m \in \mathbf{N}$. The concentrator scheme Y at speed m is an $\mathbf{A}$-bundle and its basis $Y_o = C^G$ is regular.

Proof By the proof of 4.5 we may assume that V is affine, and hence noetherian, and that C is closed in V. By 4.2 the concentrator is represented by a closed subscheme Y of V. By 5.5, Y is a noetherian affine $\mathbf{A}$-scheme. By 5.7 it remains to prove that Y_o is contained in the regular locus of Y.

In the notations of 4.2 the co-ordinate ring of V is a $\mathbf{Z}$-graded ring $B = \Sigma B_n$, the center C is given by a homogeneous ideal J and the concentrator Y corresponds to the ideal LB where $L = \sum_{n<o} B_n + \sum_{n<m} J_n$. A point $v \in Y_o$ corresponds to a prime ideal $P = P_o + \sum_{n \neq o} B_n$ with $J_o \subset P_o$. The local rings $\mathcal{O}_{V,v} = B_P$ and $\mathcal{O}_{C,v} = (B/J)_P$ are regular. Put $S = B_o \setminus P_o$. Application of lemma 5.6(b) on the graded ring $S^{-1}B$ yields that the local ring $\mathcal{O}_{Y,v} = (B/LB)_P$ is regular. So v is a regular point of Y.

5.9 <u>Remarks</u> Theorem 5.8 is a generalization of a theorem of Bialynicki-Birula, cf. [2] 4.1. He considered the case that V is of finite type over an algebraically closed field and that $C = V$. The regularity of the invariant subscheme C^G holds more generally, cf. [8].

6. Concentration under the action of a group scheme

6.1 Let G be a group scheme over a base scheme S. Clearly, $G\ell(1)_S$ is a group scheme over S. Let $Y(G,S)$ denote the set of the morphisms of group schemes $\mu : G\ell(1)_S \to G$. If $\lambda,\mu \in Y(G,S)$ commute, we define $\lambda + \mu \in Y(G,S)$ by :

$$(\lambda + \mu)(t) = \lambda(t)\mu(t) = \mu(t)\lambda(t)$$

for every section $t : X \to G\ell(1)_S$. If $\mu \in Y(G,S)$ and $m \in \mathbb{Z}$ then $m\mu \in Y(G,S)$ is defined by $m\mu(t) = \mu(t)^m$. As multiplication with $m \in \mathbb{N}$ turns out to be injective, we get an equivalence relation $\sim$ on $Y(G,S) \times \mathbb{N}$ with $(\mu,m) \sim (\nu,n)$ if and only if $n\mu = m\nu$. The equivalence class of (μ,m) is called the <u>co-weight</u> μ/m. The set of the co-weights is denoted $M(G,S)$. We consider $Y(G,S)$ as a subset of $M(G,S)$ by the identification $\mu = \mu/1$. The partially defined addition on $Y(G,S)$ extends to a partially defined addition on $M(G,S)$. The interior action $\mathrm{int}(g)h = ghg^{-1}$ of G on itself induces an action of the group of sections $G(S)$ on $M(G,S)$ If G is commutative, $Y(G,S)$ is a $\mathbb{Z}$-module and $M(G,S) = Y(G,S) \otimes \mathbb{Q}$ is a vector space over $\mathbb{Q}$.

6.2 A <u>centered</u> <u>G-scheme</u> (V,ρ,C) is a triple such that V is a scheme over S, that $\rho : G \times_S V \to V$ is an action over S and that C is a G-invariant subscheme of V, to be called the center. If $\mu \in Y(G,S)$ and $m \in \mathbb{N}$, the <u>concentrator functor</u> $V(\mu,m)$ is defined as the concentrator with respect to the center C and the speed m of the action μ given as the composition :

$$G\ell(1) \times V = G\ell(1)_S \times_S V \xrightarrow{\ \mu\ } G \times_S V \xrightarrow{\ \rho\ } V .$$

We give $V(\mu,m)$ the structure of an S-functor through the morphism $h(\sigma)\, i : V(\mu,m) \to h(S)$, where $\sigma : V \to S$ is the structural morphism of V. So if X is an S-scheme, $V(\mu,m)(X/S)$ consists of the elements $f \in V(\mu,m)(X)$ such that $i(f) : X \to V$ is an S-morphism. By lemma 5.4 this condition is equivalent with the condition that $f : \mathbb{A} \times X \to V$ is an S-morphism where $\mathbb{A} \times X$ is identified with the S-scheme $\mathbb{A}_S \times_S X$.

It follows that $p : V(\mu,m) \to h(C)$ is a morphism of S-functors. For if $f \in V(\mu,m)(X/S)$ then $p(f) : X \to C$ is an S-morphism. Since the Yoneda functor h commutes with fibered products we get a morphism $(i,p): V(\mu,m) \to h(V \times_S C)$.

6.3 <u>Proposition</u> a) The morphism $(i,p) : V(\mu,m) \to h(V \times_S C)$ is injective.
b) If V is separated over S, then $i : V(\mu,m) \to h(V)$ is injective.

Proof. Let $f_1, f_2 \in V(\mu,m)(X)$. Assume $i(f_1) = i(f_2)$. Assume either that $p(f_1) = p(f_2)$ or that V is separated over S. Let E be the equalizer $\mathrm{Ker}(f_1,f_2)$. It follows from 4.1(a) that E contains $G\ell(1) \times X$, so it is a schematically dense open subscheme of $\mathbf{A} \times X$. If $p(f_1) = p(f_2)$ then E contains all points of $\mathbf{A} \times X$. If V is separated over S then E is closed in $\mathbf{A} \times X$, cf. EGA I 5.2.5. So in both cases we have $E = \mathbf{A} \times X$ and hence $f_1 = f_2$.

6.4 Definition We use the realization morphism (i,p) to identify the concentrator $V(\mu,m)$ with a subfunctor of $h(V \times_S C)$. Now we have $V(\mu,m) = V(\nu,n)$ whenever $(\mu,m) \sim (\nu,n)$, cf. 6.1. In the locally affine case this follows by reduction to 4.2. The general proof requires descent theory cf. SGA 1 VIII 5. Now we may define $V(\lambda) := V(\mu,m)$ if λ is a co-weight represented by (μ,m).

The group scheme G itself is considered as the centered G-scheme (G, int, G). Since the interior action int works by automorphisms the concentrator $G(\lambda)$ is a subgroup functor of the group functor $h(G \times_S G)$ over S. Let $h(G \times_S G)$ act co-ordinate-wise on $h(V \times_S C)$. Then we have the following lemma, which goes back to [17].

6.5 Lemma $V(\lambda)$ is $G(\lambda)$-invariant.

Proof Write $\lambda = \mu/m$ with $\mu \in Y(G,S)$ and $m \in \mathbf{N}$. We consider $V(\lambda) = V(\mu,m)$ and $G(\lambda) = G(\mu,m)$ in the abstract sense of 6.2. Let X be an S-scheme. Consider $f \in V(\lambda)(X/S)$ and $k \in G(\lambda)(X/S)$. So we have S-morphisms $f : \mathbf{A} \times X \to V$ and $k : \mathbf{A} \times X \to G$. Condition 4.1(a) may be expressed by :

$$k(t,x) = \mu(t)\, k(\varepsilon,x)\, \mu(t)^{-1} \quad , \quad f(t,x) = \mu(t)\, f(\varepsilon,x)$$

for every scheme X' and every pair of morphisms $t : X' \to G\ell(1)$, $x : X' \to X$. The product $kf : \mathbf{A} \times X \to V$ satisfies :

$$kf(t,x) = k(t,x)\, f(t,x) \, .$$

So it is an S-morphism satisfying 4.1(a). Condition 4.1(b) reads as follows. If $t : X' \to \mathbf{A}$ factorizes over $\mathbf{A}(m)$ then $f(t,x)$ factorizes over C for every $x : X' \to X$. Since C is G-invariant it follows that $kf(t,x)$ factorizes over C as well. This proves that $kf \in V(\lambda)(X/S)$. In this way we get an action of $G(\lambda)$ on $V(\lambda)$ over S. It is obvious that the morphisms of the actions commute with the realization morphisms of 6.4.

6.6 Remark on base change Let $U \to S$ be a morphism of schemes. Then G_U is a group scheme over U and (V_U, ρ_U, C_U) is a centered G_U-scheme. If $\lambda \in M(G,S)$ then $\lambda_U \in M(G_U,U)$ is defined in the obvious way. One verifies that $V(\lambda)_U = V_U(\lambda_U)$.

6.7 Problem Let V be affine over S and let C be closed in V. Then $V(\lambda)$ is representable by an S-scheme Y. Assume that V and C are smooth over S. Does it follows that Y is smooth over S ?

7 The sheaf of buildings of a separated group scheme

7.1 In this section G is a separated group scheme over a base scheme S. Let (V,ρ,C) be a centered G-scheme which is separated over S. If $\lambda \in M(G,S)$ we use the injective morphism $i : V(\lambda) \to h(V)$ to identify the concentrator $V(\lambda)$ with a subfunctor of $h(V)$, see 6.4(b). This identification is called the <u>separated realization</u> of $V(\lambda)$.

Similarly, $G(\lambda)$ is identified with a subgroup functor of $h(G)$. It follows from 6.5 that the subfunctor $V(\lambda)$ of $h(V)$ is invariant under the action of the subgroup functor $G(\lambda)$ of $h(G)$. Using 5.3 we get a morphism of group functors $p : G(\lambda) \to G^{\lambda}$, where G^{λ} is the centralizer of λ. Let $G(\lambda)_+$ be the kernel of p. Clearly $G(\lambda)$ is the semi-direct product of $G(\lambda)_+$ and G^{λ}.

If $\lambda, \mu \in M(G,S)$ and $g \in G(\lambda)(S)$ are such that $\mu = \mathrm{int}(g)\lambda$, then $V(\mu) = g\, V(\lambda) = V(\lambda)$ and in particular $G(\mu) = G(\lambda)$. So we have an equivalence relation $\sim$ on $M(G,S)$ such that $\lambda \sim \mu$ if and only if $\mu = \mathrm{int}(g)\lambda$ with $g \in G(\lambda)(S)$. The set of the equivalence classes is called the <u>vector building</u> $Vb(G,S)$. The class of λ is denoted $[\lambda]$. The concentrators $V([\lambda]) := V(\lambda)$ and $G([\lambda]) := G(\lambda)$ are clearly well-defined subfunctors of $h(V)$ and $h(G)$, respectively. It is easy to see that $G([\lambda])_+ := G(\lambda)_+$ is also well-defined.

7.2 If X is an S-scheme, G_X is a separated group scheme over X. So we may define $Y(G,X) := Y(G_X,X)$, $M(G,X) := M(G_X,X)$ and $Vb(G,X) := Vb(G_X,X)$. The functors $Y(G,?)$, $M(G,?)$ and $Vb(G,?)$ are considered as pre-sheaves on the category (Sch)/S of the S-schemes. Let $\underline{Y}(G)$, $\underline{M}(G)$ and $\underline{Vb}(G)$ be the associated (f p q c)-sheaves on (Sch)/S, cf. S G A 3 IV 4.3 and 6.3. As $Y(G,?)$ is already a sheaf we have $\underline{Y}(G) = Y(G,?)$. If X is quasi-compact then $\underline{M}(G)(X) = M(G,X)$. If not then $M(G,X)$ is contained in $\underline{M}(G)(X)$ and the other elements of $\underline{M}(G)(X)$ need unbounded denominators.

Consider $\lambda \sim \mu$ in $M(G,X)$. Since $G_X(\lambda)(X)$ is the semi-direct product of $G_X(\lambda)_+(X)$ and $G_X^{\lambda}(X)$, there is a unique $g \in G_X(\lambda)_+(X)$ with $\mu = \mathrm{int}(g)\lambda$. It follows that $Vb(G,X)$ is contained in $\underline{Vb}(G)(X)$.

<u>Remark</u> If (V,ρ,C) is a separated centered G-scheme and $\delta \in \underline{Vb}(G)(S)$, one may use 6.6. to define a concentrator-sheaf $V(\delta)$. We shall not use this here.

7.3 <u>Lemma</u> Assume that G is reductive, cf. S G A 3 XIX 2.7, and that S is affine. Then $\underline{Vb}(G)(S) = Vb(G,S)$.

Proof If $\delta \in Vb(G,X)$ then $G_X(\delta)$ is a parabolic subgroup of G_X with unipotent radical $G_X(\delta)_+$, see S G A 3 XXVI 6.1. Let $Q = Par(G,?)$ be the (fpqc)-sheaf such that $Q(X)$ is the set of the parabolic subgroups of G_X, see S G A 3 XXVI 3. The morphism of pre-sheaves $Vb(G,?) \to Q$ mapping $\delta \in Vb(G,X)$ to $G_X(\delta) \in Q(X)$ induces a morphism of sheaves $P : \underline{Vb}(G) \to Q$.

Consider $\delta \in \underline{Vb}(G)(S)$. Let F be the subsheaf of $\underline{M}(G)$ such that $F(X)$ consists of the co-weights $\lambda \in \underline{M}(G)(X)$ with $[\lambda] = \delta_X$. Let R be the unipotent radical of $P(\delta)$. If $\lambda, \mu \in F(X)$ then $G(\lambda)_+ = R_X$ and there is a unique $g \in R(X)$ with $\mu = int(g)\lambda$. So F is a principal homogeneous R-sheaf . By SGA 3 XXVI 2.2 , it is trivial. So there is a co-weight $\lambda \in M(G,S)$ with $\delta = [\lambda]$. This proves that $\delta \in Vb(G,S)$.

Remark I have the impression that $\underline{Vb}(G)(S) = Vb(G,S)$ holds already if G is a smooth affine group scheme over an affine scheme S . I need a positive answer to 6.7.

7.4 Lemma Let $j : T \to G$ be a monomorphism of group schemes over S and assume that T is commutative. Then $\underline{Vb}(T) = \underline{M}(T)$ is a sheaf of vector spaces over $\mathbb{Q}$ and the induced morphism $j_\star : \underline{Vb}(T) \to \underline{Vb}(G)$ is injective.

Proof The first assertion is trivial. It suffices to prove that the map from $Y(T,S)$ to $Vb(G,S)$ is injective. Let $\lambda, \mu \in Y(T,S)$ have the same image in $Vb(G,S)$. Choose $g \in G(\lambda)(S)$ with $\mu = int(g)\lambda$. Put $\nu = \lambda - \mu$ in $Y(T,S)$, cf. 6.1 . We have :

$$\nu(t) = \lambda(t)\, g\, \lambda(t)^{-1} g^{-1} = (int(\lambda(t))g)g^{-1}$$

for every section t of $G\ell(1)_S$. Since $g \in G(\lambda)(S)$ there is an S-morphism $f : \mathbb{A}_S \to G$ entending $\nu : G\ell(1)_S \to G$. Since ν is a morphism of separated group schemes over S , it follows that ν is trivial, so that $\lambda = \mu$.

Remark So we may consider $\underline{Vb}(T)$ as a subsheaf of $\underline{Vb}(G)$. If T is a maximal locally-split subtorus of G the space $Vb(T,S)$ is called an apartment of $Vb(G,S)$. If G is of finite presentation over S , the building $Vb(G,S)$ is the union of the apartments, see SGA 3 IX 2.11 and 6.8.

7.5 The interior action of G on itself induces an action of the sheaf of groups $h(G)$ on the sheaf of buildings $\underline{Vb}(G)$. Let $\mathbb{Q}_S$ be the (fpqc)-sheaf such that $\mathbb{Q}_S(X)$ is the ring of the locally constant functions on X with values in $\mathbb{Q}$. A morphism of sheaves $q : \underline{Vb}(G) \to \mathbb{Q}_S$ is called a norm if for every S-scheme X the following conditions hold :

a) Let $\delta \in \underline{Vb}(G)(X)$. If $g \in h(G)(X)$ then $q(int(g)\delta) = q(\delta)$. The function $q(\delta)$ is non-negative. If $q(\delta) = 0$ on X , then $\delta = 0$.

b) If $j : T \to G_X$ is a morphism of group schemes over X and T is commutative, the expression

$$\beta(\delta,\delta') = q\, j_\star(\delta + \delta') - q\, j_\star(\delta) - q\, j_\star(\delta')$$

defines a bilinear form β on the $\mathbb{Q}_S(X)$-module $\underline{Vb}(T)(X)$.

7.6 <u>Proposition</u> Let $U \to S$ be a finite morphism, faithfully flat and of finite presentation. Let F be a locally free $\mathcal{O}_U$-module and let $\rho : G_U \to G\ell(F)$ be a morphism of group schemes over U. Assume that the kernel of ρ is quasi-finite over U. Then there is an associated norm $q = q_F$ on $\underline{Vb}(G)$.

<u>Construction</u> Let X be an S-scheme. Put $Z = X \times_S U$. The projection $p : Z \to X$ is faithfully flat, finite and of finite presentation. If F' is a locally free $\mathcal{O}_Z$-module, the direct image $p_* F'$ is a locally free $\mathcal{O}_X$-module and $\mathrm{rank}(p_* F')$ is a locally constant rational function on X, cf. [1] p. 123. Now let $\delta \in Vb(G)(X)$. Assume that X is small enough, so that $\delta = [\mu/m]$ with $\mu \in Y(G,X)$ and $m \in \mathbb{N}$. The action $\rho\mu_Z$ of $G\ell(1)_Z$ on the $\mathcal{O}_Z$-module F_Z induces a direct sum decomposition $F_Z = \sum (F_Z)_n$ such that $\rho\mu_Z(t)$ acts on $(F_Z)_n$ by multiplication with t^n. Now we define :

$$q(\delta) := m^{-2} \sum n^2 \ \mathrm{rank}(p_*(F_Z)_n) \in \mathbb{Q}_S(X) .$$

One verifies that $q(\delta)$ is independent of the choice of the pair (μ,m). If X is arbitrary, the function $q(\delta)$ is well-defined at least locally on X, so it is well-defined on X. The verification that q is a norm is left to the reader. Compare 2.3.

7.7 <u>Applications</u> a) Assume that G has a Lie algebra L which is a locally free $\mathcal{O}_S$-module, and that the adjoint action $G \to G\ell(L)$ has a quasi-finite kernel. The associated norm $q = q_L$ might be called the <u>Killing-norm</u>. In this case $U = S$.
b) If G is semi-simple then case a) applies, cf. SGA 3 II 4.11 and XXII 5.7.14.
c) Assume that S is locally noetherian and normal (or geometrically unibranche), and that G is reductive. Then $\underline{Vb}(G)$ admits a norm. In fact, let T be the coradical of G, cf. SGA 3 XXII 6.2. The torus T splits over some finite étale covering U of S, cf. SGA 3 X 5.15. Now there is a free $\mathcal{O}_U$-module F' with an injective morphism $T_U \to G\ell(F')$. Put $F = L_U \oplus F'$ where L is the Lie algebra of G. Then q_F is a norm on $\underline{Vb}(G)$.

7.8 <u>Example</u> The two tori of SGA 3 X 1.6 do not admit norms. Let me briefly describe the second one. Let S be a curve over an algebraically closed field with two irreducible components S_1 and S_2 which meet in two points p_1 and p_2. Clearly S is covered by the two open subsets $U_i = S \setminus \{p_i\}$. The intersection $U_1 \cap U_2$ is the disjoint union of the two open subsets $V_i = S \setminus S_i$. Now the torus T over S is defined as follows. The restriction $T|U_i$ is $G\ell(1)^2 \times U_i$ with multiplicative co-ordinates (v,w_i). The restriction $T|U_1 \cap U_2$ has co-ordinates (v,w). The restriction maps are given by $v|V_j = v$, and $w_i|V_j = w$ if $i \leqslant j$, and $w_2|V_1 = vw$. The sheaf $F = \underline{Vb}(T)$ satisfies $F(U_i) = F(V_j) = \mathbb{Q}^2$. For the sets

U_i, V_j are connected and T splits over each. If $i \leqslant j$ the restriction $F(U_i) \to F(V_j)$ is the identity. The map $F(U_2) \to F(V_1)$ has the matrix $a = \begin{pmatrix} 1 & 0 \\ 1 & 1 \end{pmatrix}$. Therefore a norm q on $\underline{Vb}(T)$ would induce a positive definite quadratic form $q : \mathbb{Q}^2 \to \mathbb{Q}$ with $q(a(x)) = q(x)$ for every $x \in \mathbb{Q}^2$. This is impossible. So T does not admit a norm.

8 Concentration under the action of a split torus

8.1 In this section T is a split torus over a connected affine scheme S with co-ordinate ring k. Let Π be the group of the characters $\pi : T \to G\ell(1)_k$. It is a free $\mathbb{Z}$-module of finite rank. The co-ordinate ring of T is identified with the group algebra $k[\Pi]$. Let Y be the dual $\mathbb{Z}$-module of Π through the duality $(\, , \,)$. We may identify $Y = Y(T,k)$. If $\pi \in \Pi$, and $\mu \in Y$, and t is a section of $G\ell(1)_k$ then we have $\pi(\mu(t)) = t^{(\pi,\mu)}$. See S G A 3 VIII. Now let M be the vector space $Y \otimes \mathbb{Q}$. Then we have :

$$M = M(T,k) = Vb(T,k) .$$

8.2 A locally trivial vector bundle V over S with a linear action $\rho : T \to \mathrm{Aut}(V/S)$ is called a T-<u>bundle</u>. The co-ordinate ring of a T-bundle V is the symmetric algebra $S_k(F)$ on the module F of the linear functions on V. The k-module F is projective and of finite type. If $\pi \in \Pi$ let F_π be the set of the linear functions $f \in F$ with $f(\rho(t)v) = \pi(t)\, f(v)$ for all sections t of T and v of V. Now F is a direct sum ΣF_π. So the k-modules F_π are projective and of finite type, and the set $R(V) = \{\pi \in \Pi \mid F_\pi \neq 0\}$ is finite. If $\pi \in \Pi$ let F^π be the quotient $F/\Sigma_{\chi \neq \pi} F_\chi$. The symmetric algebra $S_k(F^\pi)$ is the co-ordinate ring of the sub-bundle V_π of V which contains the sections v of V with $\rho(t)\, v = \pi(t)\, v$ for every section t of T. Clearly $V_\pi \neq 0$ if and only if $\pi \in R(V)$. The bundle V is the direct sum $\sum V_\pi$ with $\pi \in R(V)$.

If $R \subset \Pi$, we define $V[R] := \Sigma_{\pi \in R} V_\pi$. If $v : X \to V$ is a morphism of schemes the <u>diagram</u> $R(v)$ is defined as the smallest subset R of Π such that v factorizes over the sub-bundle $V[R]$. Clearly, $R(v)$ is contained in $R(V)$, so it a finite set .

8.3 We define a <u>centered</u> T-<u>bundle</u> to be a triple (V,ρ,C) such that (V,ρ) is a T-bundle and C is a T-invariant sub-bundle. If $\pi \in \Pi$ then clearly (V_π, π, C_π) is a centered T-sub-bundle with action $\rho|V_\pi = \pi$.

Consider $\lambda \in M$. We use the realization morphism (i,p) to identify the concentor scheme $V(\lambda)$ with a closed subscheme of the T-bundle $V \oplus C = V \times_S C$, see 4.2 and 6.4 . First we determine the concentrator schemes $V_\pi(\lambda)$ of the centered

T-bundles (V_π, π, C_π) . If $(\tau,\lambda) \geqslant 1$ then $V_\pi(\lambda) = V_\pi \oplus O$. If $0 < (\pi,\lambda) < 1$ then $V_\pi(\lambda) = C_\pi \oplus O$. If $(\pi,\lambda) = 0$ then $V_\pi(\lambda)$ is the image, say D_π , of the diagonal morphism $\Delta: C_\pi \to C_\pi \oplus C_\pi$. If $(\pi,\lambda) < 0$ then $V_\pi(\lambda) = 0$. It follows that :

$$V(\pi) = \sum_{(\pi,\lambda)\geqslant 1} (V_\pi \oplus O) + \sum_{0<(\pi,\lambda)<1} C_\pi \oplus O) + \sum_{(\pi,\lambda)=0} D_\pi$$

$$iV(\lambda) = \sum_{(\pi,\lambda)\geqslant 1} V_\pi + \sum_{(\pi,\lambda)\geqslant 0} C_\pi$$

8.4 A finite subset of Π is called a <u>diagram</u>. A pair $\sigma = (\sigma_1, \sigma_2)$ of diagrams with $\sigma_1 \subset \sigma_2$ is called a <u>two-diagram</u>. The set of the two-diagrams is ordered by co-ordinate wise inclusion.

Let (V,ρ,C) be a centered T-bundle. If σ is a two-diagram we define the sub-bundle $V[\sigma] := V[\sigma_1] + C[\sigma_2]$, see 8.2 . If $v : X \to V$ is a morphism of schemes there is a unique smallest two-diagram $\sigma = \sigma(v)$ such that v factorizes over $V[\sigma]$. We have $\sigma(v) = (R(pv), R(v))$ where $p : V \to V/C$ is the quotient morphism.

8.5 <u>Lemma</u> Let (V,ρ,C) be a centered T-scheme. Assume that V is affine and of finite type over k . Assume that C is closed in V and of finite presentation over V . Then there is a centered T-bundle (V',ρ',C') and a T-equivariant closed immersion $j : V \to V'$ such that $C = j^{-1}(C')$.

Proof By S G A I 4.7.3 the coordinate ring B of V is multi-graded, say $B = \sum B_\pi$ with $\pi \in \Pi$, such that B_π consists of the functions $b \in B$ with $b(\rho(t)v) = \pi(t)\, b(v)$ for every section t of T and every section v of V . As V is of finite type over k we may choose a finite set of homogeneous generators $x_1,\dots,x_m$ of the k-algebra B . Let J be the ideal of C . As C is of finite presentation over V the ideal J is finitely generated. Since C is T-invariant we may choose a finite set of homogeneous generators $x_{m+1},\dots,x_n$ of the ideal J . Assume that $x_i \in B_{\pi(i)}$. Then we define the centered T-bundle (V',ρ',C') by :

$$V' = \mathbb{A}^n_k \;;\; C' = \{ w \mid w_{m+1} = \dots = w_n = 0 \} .$$

$$\rho' = \operatorname{diag}(\pi(1),\pi(2),\dots,\pi(n)) .$$

We clearly have a T-equivariant closed immersion $j : V \to V'$ with $C = j^{-1}(C')$ given by :

$$j(v) = (x_1(v),\dots,x_n(v))$$

<u>Remark</u> One may use Cartier's lemma on dual actions, cf. [17] p.25, to give a similar construction in the case of an arbitrary affine group scheme over a field

instead of a torus over a ring.

8.6 Let (V,ρ,C) be a centered T-scheme over S. Assume that V is affine and of finite type over S and that C is closed in V and of finite presentation over V. Let $q : M \to \mathbb{Q}$ be a positive definite quadratic form. If $v : X \to V$ is a morphism of schemes let $M(v)$ be the set of the elements $\lambda \in M$ such that v factorizes over the subscheme $iV(\lambda)$ of V, and let

$$q^{\star}(v) = \inf \{q(\lambda) \mid \lambda \in M(v)\}$$

An element $\lambda \in M(v)$ with $q(\lambda) = q^{\star}(v)$ is called <u>optimal</u> for v. If $\lambda \in M$ is optimal for some morphism v, then λ is called <u>balanced</u> for (V,ρ,C).

<u>Proposition</u> a) Let $v : X \to V$ be a morphism such that $M(v)$ is non-empty. Then there is a unique co-weight $\lambda \in M$ which is optimal for v.
b) The number of co-weights λ which are balanced for (V,ρ,C), is finite.

Proof. Choose a T-equivariant closed immersion j of V into a centered T-bundle (V',ρ',C') such that $C = j^{-1}(C)$. By 4.3 (b) we have $M(v) = M(jv)$ for every morphism $v : X \to V$. So we may assume that (V,ρ,C) itself is a centered T-bundle. If R is a diagram and $\lambda \in M$ we define

$$(R,\lambda) = \inf \{(\pi,\lambda) \mid \pi \in R\}$$

Let $\sigma = \sigma(v)$ be the two-diagram of a morphism $v : X \to V$. It follows from 8.3 that we have

$$M(v) = \{\lambda \in M \mid (\sigma_1,\lambda) \geqslant 1 \text{ and } (\sigma_2,\lambda) \geqslant 0\}$$

If λ_1 and λ_2 are different elements of $M(v)$ with $q(\lambda_1) = q(\lambda_2)$, then $\lambda = \frac{1}{2}(\lambda_1 + \lambda_2)$ belongs to $M(v)$ and $q(\lambda) < q(\lambda_1)$. This proves the uniqueness of an optimal co-weight λ. The existence follows from the finiteness of the sets σ_1 and σ_2, see [12] section 3.

The number of co-weights which are balanced for (V,ρ,C), is bounded by the number of two-diagrams σ such that σ_2 is contained in the finite set $R(V)$.

8.7 A T-scheme (V,ρ) is called <u>locally affine</u> if every T-invariant open subset U of V is covered by T-invariant affine open subsets of U.

<u>Lemma</u> Let (V,ρ) be a T-scheme which is covered by T-invariant affine open subsets. Then (V,ρ) is locally affine.

Proof We may assume that V is affine. Let U be a T-invariant open subset of V. Its complement E is considered as a reduced closed subscheme of V. By EGA I 4.6.2. the subscheme E is T-invariant. As in 8.5 the coordinate ring B of V is multi-graded, say $B = \Sigma B_{\pi}$, and the ideal J of the subscheme E is homogeneous. So we may choose a set L of homogeneous generators of the ideal J. Now U is covered by the T-invariant affine open subsets $D(f) = \{p \in V \mid f \notin p\}$ with $f \in L$.

Remarks a) It is essential that the group scheme T is diagonalizable. For example, the usal action of $S\ell(2)_k$ on the affine plane $\mathbb{A}_k^2$ over a field k, is not locally affine.

b) Sumihiro has obtained normality conditions on the k-scheme V which imply that any torus action ρ on V is locally affine, cf. [20] 3.11. An example of a not locally affine torus action is given in 4.6(c).

8.8 A triple of diagrams $\tau = (\tau_o, \tau_1, \tau_2)$ is called a three-diagram if it satisfies $\tau_o \cup \tau_1 \subset \tau_2$ and $\tau_o \cap \tau_1 = \emptyset$. If τ is a three-diagram and (V,ρ,C) is a centered T-bundle, let $V[\tau]$ be the sub-bundle of $V \oplus C$ given by :

$$V[\tau] = (V[\tau_1] \oplus 0) + (C[\tau_2 \setminus (\tau_o \cup \tau_1)] \oplus 0) + D[\tau_o]$$

where D is the image of the diagonal morphism $C \to C \oplus C$. If R is a diagram and $\lambda \in M$, let $\tau(\lambda,R)$ be the three-diagram defined by :

$$\begin{aligned} \tau_o(\lambda,R) &= \{\pi \in R \quad (\pi,\lambda) = 0\} \\ \tau_1(\lambda,R) &= \{\pi \in R \quad (\pi,\lambda) \geq 1\} \\ \tau_2(\lambda,R) &= \{\pi \in R \quad (\pi,\lambda) \geq 0\} \end{aligned}$$

If R contains $R(V)$ then $V[\tau(\lambda,R)]$ is equal to the realization of $V(\lambda)$ in $V \oplus C$, cf. 8.3.

8.9 Let (V,ρ,C) be a locally affine centered T-scheme. If $\lambda \in M$ the concentrator functor $V(\lambda)$ is representable by a subscheme $V(\lambda)$ of $V \times_k C$, cf. 4.5.

Proposition Assume that V is locally of finite type over k, that C is locally of finite presentation over V, and that C is quasi-compact. Then the number of different concentrator schemes $V(\lambda)$ in $V \times_k C$ is finite.

Proof Just as in the proof of 4.5 we choose a T-invariant open subset U of V such that C is a closed subscheme of U. We choose a covering $(V_\alpha)_{\alpha \in I}$ of U by T-invariant affine open subsets. Now $C_\alpha := C \cap V_\alpha$ is a closed subscheme of V_α. The family $(C_\alpha)_{\alpha \in I}$ is an affine open covering of C. Since C is quasi-compact we may assume that the index set I is finite. By 8.5 we may choose centered T-bundles $(V'_\alpha, \rho'_\alpha, C')$ and T-equivariant closed immersions $j_\alpha : V_\alpha \to V'_\alpha$ with $C_\alpha = j_\alpha^{-1}(C'_\alpha)$.

If $\lambda \in M$ the concentrator $V(\lambda)$ is a closed subscheme of the open subscheme $\cup (V_\alpha \times_k C_\alpha)$ of $V \times_k C$. By 4.3. we have

$$V(\lambda) \cap (V_\alpha \times_k C_\alpha) = V_\alpha(\lambda) = V'_\alpha(\lambda) \cap (V_\alpha \times_k C_\alpha) .$$

Put $R = \cup R(V'_\alpha)$, cf. 8.2 . By 8.8 the concentrator scheme $V'_\alpha(\lambda)$ is equal to $V'_\alpha[\tau(\lambda,R)]$. This proves that $V(\lambda) = V(\mu)$ holds whenever $\tau(\lambda,R) = \tau(\mu,R)$. So the number of different concentrator schemes is bounded by the number of three-diagrams τ with $\tau_2 \subset R$.

9 Optimal concentration over fields

9.1 In this section G is a reductive algebraic group over a field k . We choose a fixed norm $q : \underline{Vb}(G) \to \mathbb{Q}_k$, cf. 7.7(c) . We identify $\mathbb{Q}_k(k) = \mathbb{Q}$. Let (V,ρ,C) be a separated centered G-scheme. A morphism of k-schemes $v : X \to V$ is said to be <u>concentrated</u> if there is $\delta \in Vb(G,k)$ with $v \in V(\delta)(X)$, see 7.1. We define :

$$q^\star(v,k) = \inf \{ q(\delta) \mid \delta \in Vb(G,k) , v \in V(\delta)(X)\}$$

If $v \in V(\delta)(X)$ and $q(\delta) = q^\star(v,k)$ then δ is called <u>optimal</u>.

9.2 <u>Lemma</u> Let T be a maximal split k-torus in G .

a) Every split k-torus in G is $G(k)$-conjugate to a subtorus of T .

b) Let $\delta \in Vb(G,k)$. Then $\delta \in Vb(T,k)$ if and only if $T \subset G(\delta)$.

c) Let $\delta_1,\delta_2 \in Vb(G,k)$. Then there is a split k-torus T' in G such that $\delta_1,\delta_2 \in Vb(T',k)$.

Proof (a) See [4] 4.21 (b) Assume $T \subset G(\delta)$. Write $\delta = [\mu/m]$ with $\mu \in Y(G,k)$ and $m \in \mathbb{N}$. By [4] 11.6 there is $g \in G(\delta)(k)$ such that $int(g)\mu \in Y(T,k)$. Therefore $\delta \in Vb(T,k)$, see remark 7.4. The other implication follows from the commutativy of T . (c) By [4] 4.13 the intersection $G(\delta_1) \cap G(\delta_2)$ contains a maximal split k-torus T' of G . Now (b) implies $\delta_1,\delta_2 \in Vb(T',k)$.

<u>Remark</u> This lemma is well known. It seems to me that the parts (a) and (b) remain valid if the field k is replaced by a ring k such that every projective k-module of finite type is free, compare SGA 3 XXVI 6 and [16] . Part (c) however fails already if the field k is replaced by an artinian local ring, cf. example 8.6 (d) below.

9.3 <u>Proposition</u> Let T be a maximal split k-torus in G . Let U be a T-invariant affine open subset of V . Assume that U is of finite type over k and that C is a closed subscheme of U . Let $v : X \to V$ be a morphism of schemes with $q^\star(v,k) < \infty$. Then there is a unique optimal $\delta \in Vb(G,k)$.

Proof If $v' : X \to V$ is a morphism and T' is a k-torus in G , let :

$$f(v',T') := \inf \{q(\delta) \mid \delta \in Vb(T',k) , v' \in V(\delta)(X)\}$$

By 9.2 we have :

$$q^\star(v,k) = \inf \{f(v,\mathrm{int}(g^{-1})T) \mid g \in G(k)\}$$

If $g \in G(k)$ then $f(v,\mathrm{int}(g^{-1})T) = f(gv,T)$. Let H be the subset of $G(k)$ of the elements g with $f(gv,T) < \infty$. Then we have :

$$q^\star(v,k) = \inf \{f(gv,T) \mid g \in H\}$$

Now use the notations of 7.1 . So Π is the character group of T , and $Y = Y(T,k)$ and $M = Vb(T,k)$. Clearly (U,ρ,C) is a centered T-scheme satisfying the assumptions of 8.6 . By 4.3(c) we have $V(\delta) = U(\delta)$ for every $\delta \in M$. Therefore, if $g \in H$, the morphism $gv : X \to V$ factorizes over U and :

$$f(gv,T) = \inf \{q(\delta) \mid \delta \in M , gv \in U(\delta)(X)\} .$$

By 8.6(b) the number of values $f(gv,T)$ with $g \in H$ is finite. So there is $h \in H$ with $q^\star(v,k) = f(hv,T)$. By 8.6(a) there is $\delta' \in M$ with $hv \in U(\delta')(X)$ and $f(hv,T) = q(\delta')$. Now $\delta = \mathrm{int}(h^{-1})\delta'$ is an optimal element for v .

As for the uniqueness of the optimal δ , by 9.2(c) we may assume that $G = T$. Then clearly we may assume that $V = U$. Now we are reduced to the case of 8.6(a).

9.4 Let K be the separable closure of the field k . It follows from 7.3 that the building $Vb(G,k)$ may be identified with the invariant subset of $Vb(G,K)$ under the action of the Galois group Γ of K over k . Since q is a morphism of sheaves the map $q = q_K$ is Γ-invariant and the map q_k is the restriction $q \mid Vb(G,k)$. If $\delta \in Vb(G,k)$ then $V(\delta)_K = V_K(\delta)$ by 6.6. If $v : X \to V$ is a morphism of k-schemes we define $q^\star(v,K) := q^\star(v_K,K)$. It is obvious that $q^\star(v,K) \leqslant q^\star(v,k)$.

<u>Theorem</u> Let T be a maximal K-torus in G_K . Let U be a T-invariant affine open subset of V_K . Assume that U is of finite type over K and that C_K is a closed subscheme of U . Let $v : X \to V$ be a k-morphism with $q^\star(v,K) < \infty$. Then there is a unique $\delta \in Vb(G,k)$ with $v \in V(\delta)(X)$ and $q(\delta) = q^\star(v,K)$. In particular $q^\star(v,k) = q^\star(v,K)$.

Proof Since K is separably closed the torus T is split. By 9.3 we have a unique $\delta \in Vb(G,K)$ with $v_K \in V_K(\delta)(X_K)$ and $q(\delta) = q^\star(v,K)$. Clearly δ is Γ-invariant, so $\delta \in Vb(G,k)$. It follows that $v \in V(\delta)(X)$. The uniqueness of δ is trivial.

9.5 Remarks In respect to Kempf's theorem, cf. [15] , the new features are : (1) the field K is only separably closed, (2) we are able to concentrate a morphism $v : X \to V$, not just a point $v \in V(k)$, (3) the scheme is not necessarily affine. All three points are essentially trivial. The points (1) and (2) were introduced in [12] . Point (3) depends on the rather artificial introduction of an intermediate affine scheme U . This generalization is non-empty and in some sense best-possible, but as yet I have no applications.

9.6 Examples a) Let $G = G\ell(1)_k$ with the obvious norm. Let (P,μ,C) be the centered G-scheme consisting of the projective line $\mathbb{P}$, the action μ of 4.6(a), and the center $C = \{0,\infty\}$. Fix a point $v \in \mathbb{P}(k)$ outside of the center. There are two unequal optimal elements in Vb(G,k). For the element v does not know how to choose between 0 and ∞ as a possible target. Now C is closed and affine, though not contained in an invariant affine open subset of $\mathbb{P}$.

b) If $k \subset K$ is an inseparable field extension, we may have $q^{\star}(v,K) \neq q^{\star}(v,k)$, see [12] 5.6 .

c) As for the dependence of the choice of the norm, we refer to [12] section 7 and [13] 4.10 .

d) Definition 9.1 makes good sense if the field k is replaced by a ring k with a connected spectrum. So let k be a local ring with a maximal ideal $m \neq 0$. Let $G = S\ell(2)_k$ act on $V = \mathbb{A}^2_k$ with the zero section as a center. Let $V \in V(k/m)$ be the point with co-ordinates $v_1 = 1 (\mathrm{mod}\ m)$ and $v_2 = 0\ (\mathrm{mod}\ m)$. The co-weight λ with $\lambda(t) = \mathrm{diag}(t,t^{-1})$ is optimal. Choose $a \in m$ with $a \neq 0$ and put $g = \begin{pmatrix} 1 & 0 \\ a & 1 \end{pmatrix}$. We have $gv = v$, so that $\mu = \mathrm{int}(g)\lambda$ is optimal as well. However $[\mu] \neq [\lambda]$ in Vb(G,k). In fact $V(\mu) = g\,V(\lambda) \neq V(\lambda)$ as subschemes of V .

10 A construction of Grothendieck, cf. SGA 3 XIII 1 .

10.1 Let G be a group scheme over a scheme S , acting on an S-scheme V . Let P be a subgroup scheme of G , acting on an S-scheme X . Let $j : X \to V$ be a P-equivariant morphism of S-schemes. We consider the right-action of P on the product $G \times_S X$ given by $(g,x)\,p = (gp,p^{-1}x)$. All S-schemes are identified with (fpqc)-scheaves on the category (Sch)/S of the S-schemes. Let $G \times^P X$ be the (fpqc)-quotient sheaf $(G \times_S X)/P$. We clearly have a commutative diagram of (fpqc)-sheaves :

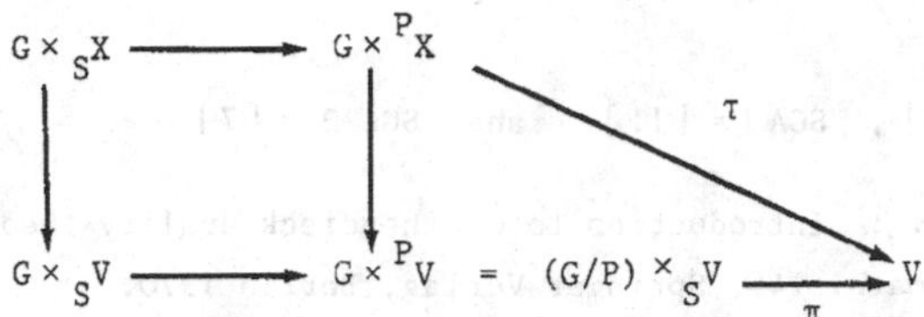

The lefthand square is a pull-back diagram. The morphism τ is given by $\tau(g,x)P = g\ j(x)$. The identification is given by $(g,v)P = (gP,gv)$. The morphism π is the projection.

Assume that G/P is representable by an S-scheme and that the morphism $G \to G/P$ is faithfully flat and quasi-compact. If the morphism $j : X \to V$ is quasi-affine it follows from SGA 1 VIII 7.9 that $G \times^P X$ is representable. If moreover j is a closed immersion then $G \times^P X$ is a closed subscheme of $G \times^P V$

10.2 Now assume that G is reductive and that P is a parabolic subgroup, cf. SGA 3 XIX 2.7 and XXVI 1 . Then G/P is a projective S-scheme, so that π is a proper morphism. Therefore, if $j : X \to V$ is a closed immersion, then τ is proper. In that case the schematical image of τ is a well-defined closed subscheme of V , cf. EGA I 6.10.5 . This image is independent of the choice of the parabolic subgroup P normalizing X . So it may be denoted GX .

REFERENCES

Recall that EGA = [9,10], SGA 1 = [11] and SGA 3 = [7]

1 - Altman, A., Kleiman, S. : Introduction to Grothendieck duality theory. Lecture notes in math. 146, Springer Verlag, Berlin 1970.

2 - Bialynicki-Birula, A. : Some theorems on actions of algebraic groups. Annals of Math. 98 (1973) 480 - 497.

3 - Borel, A. : Linear algebraic groups. Benjamin, New York, 1969.

4 - Borel, A., Tits, J. : Groupes réductifs. Publ. Math. de l'IHES 27 (1965) 55-152.

5 - Bourbaki, N. : Groupes et algèbres de Lie, chapitres 7 et 8. Hermann, Paris 1975.

6 - Demazure, M. : Démonstration de la conjecture de Mumford [d'après W. Haboush] Séminaire Bourbaki, exp. 462. Springer Verlag, Berlin 1976.

7 - Demazure, M., Grothendieck, A : Schémas en Groupes (SGA 3). Lecture notes in math. 151, 152, 153, Springer Verlag, Berlin 1970.

8 - Fogarty, J., Norman, P. : A fixed-point characterization of linearly reductive groups. In contributions to algebra. Academic Press, New York 1977, p. 151 - 155.

9 - Grothendieck, A., Dieudonné, J.A. : Eléments de Géométrie Algébrique I. Springer Verlag, Berlin 1971.

10 - Grothendieck, A., Dieudonné, J.A. : Eléments de Géométrie Algébrique. Publ. Math. de l'I.H.E.S. 11,20,24,28,32 (1961 - 1967).

11 - Grothendieck, A : Revêtements étales et groupe fondamental (SGA 1). Lecture notes in math. 224. Springer Verlag, Berlin 1971.

12 - Hesselink, W.H. : Uniform instability in reductive groups. J. reine u. angewandte Math. 303/304 (1978) 74-96.

13 - Hesselink, W.H. : Desingularizations of Varieties of nullforms. Inventiones math. 55 (1979) 141 - 163.

14 - Humphreys, J.E. : Linear algebraic groups. Springer Verlag, New York 1975.

15 - Kempf, G.R. : Instability in invariant theory. Annals of Math. 108(1978) 299-316.

16 - Lam, T.Y. : Serre's conjoncture. Lecture notes in math. 635. Springer Verlag Berlin 1978.

17 - Mumford, D. : Géométric Invariant Theory. Ergebnisse Bd 34. Springer Verlag, Berlin 1965.

18 - Ness, L. : Mumford's numerical function and stable projective hypersurfaces. In : Algebraic geometry. Copenhagen 1978. Lecture notes in math. 732. Springer Verlag, Berlin 1979, p. 417-453.

19 - Rousseau,G : Immeubles sphériques et théorie des invariants. CR. Acad. Sc. Paris 286, A 247 - 250 (1978).

20 - Sumihiro,H. : Equivariant completion II. J. of Math. Kyoto Univ. 15 (1975) 573 - 605.

Groningen, february 1980.

SOME CONJECTURES FOR ROOT SYSTEMS AND FINITE COXETER GROUPS

I.G. MACDONALD

A. "Dyson's conjecture" and generalizations.

In 1962 F.J. Dyson [2] conjectured that the constant term in

$$(1) \qquad \prod_{1 \leqslant i \neq j \leqslant n} (1 - u_i u_j^{-1})^k$$

(where $u_1, \ldots, u_n$ are independent variables and k is a positive integer) should be

$$(nk)!/(k!)^n .$$

This conjecture was soon proved by various people [3] [4] [10], and more generally that the constant term in

$$(2) \qquad \prod_{i \neq j} (1 - u_i u_j^{-1})^{a_i}$$

(where the a_i are arbitrary non-negative integers) is equal to

$$(a_1 + \ldots + a_n)!/a_1! \ldots a_n! .$$

When $a_1 = \ldots = a_n = k$, (2) reduces to (1).

I shall discuss some possible analogues and generalizations of (1). I do not know whether there are corresponding analogues of (2). For a more detailed account of these and similar conjectures, see [7].

The fractions $u_i u_j^{-1}$ in (1) remind one of the roots of a root system of type A_{n-1}. So let R be any reduced root system, spanning a finite-dimensional real Hilbert space V of dimension ℓ, and let W be the Weyl group of R, which acts on V and hence on the symmetric algebra $S(V)$. It is well-known that the subalgebra $S(V)^W$ of W-invariants is generated by ℓ algebraically independent homogeneous elements φ_i of degrees $d_1,\dots,d_\ell$, and that the order of W is equal to the product of the d_i.

<u>Example</u>. - The set of vectors $v_i - v_j (i \neq j)$ in $\mathbb{R}^n$ (where $v_1,\dots,v_n$ is the standard basis of $\mathbb{R}^n$) is a root system of type A_{n-1}. In this example V is the hyperplane $x_1 + \dots + x_n = 0$ in $\mathbb{R}^n$, W is the symmetric group S_n acting on V by permuting the coordinates, and we may take $\varphi_i = x_1^i + \dots + x_n^i$ for $i = 2,\dots,n$, so that the degrees d_i are $2,3,\dots n$.

For each root $\alpha \in R$ we introduce a formal exponential e^α, which may be regarded as the element corresponding to α in the group sing $\mathbb{Z}[Q]$ of the lattice $Q(\cong \mathbb{Z}^\ell)$ generated by R in V.

<u>Conjecture</u> (C1). <u>For each integer</u> $k \geqslant 0$, <u>the constant term in</u>

$$\prod_{\alpha \in R} (1 - e^\alpha)^k$$

<u>should be equal to</u>

$$\prod_{i=1}^{\ell} \binom{kd_i}{k}.$$

In the example above, the e^α are $e^{v_i - v_j} = u_i u_j^{-1}$ where $u_i = e^{v_i}$, and

$$\prod \binom{kd_i}{k} = \binom{2k}{k}\binom{3k}{k}\cdots\binom{nk}{k} = \frac{(nk)!}{(k!)^n}$$

in agreement with (1).

It is enough to prove (C1) for the irreducible root systems. I cannot do this in general : however, (C1) is true

(a) for all R and $k = 1$ or 2 ;

(b) for R of classical type (A,B,C,D) and any k .

Before coming to these verifications, we shall write (C1) in an equivalent form. Let G be a compact connected Lie group, T a maximal torus in G , and R the root system of (G,T). We may regard the e^{α} as characters of the torus T . Choose a system R^{+} of positive roots and write

$$\Delta(t) = \prod_{\alpha \in R^{+}} (e^{\alpha/2}(t) - e^{-\alpha/2}(t))$$

for $t \in T$, so that

$$|\Delta(t)|^{2} = \prod_{\alpha \in R} (1 - e^{\alpha}(t))$$

is a positive real-valued fonction on T , which occurs in Weyl's integration formula

$$\int_{G} f(x)\, dx = \frac{1}{|W|} \int_{T} f(t)\, |\Delta(t)|^{2}\, dt$$

in which f is any continuous class-function on G , and dx , dt are normalized Haar measures.

The conjecture (C1) is then equivalent to :

(C1') $$\int_{T} |\Delta(t)|^{2k}\, dt = \prod_{i=1}^{\ell} \binom{kd_i}{k}$$

<u>for all integers</u> $k \geqslant 0$.

The equivalence of (C1) with (C1') follows form the fact that integration over T kills all but the trivial character, hence picks out the constant term. We may remark that (C1') makes sense if the integer k is replaced by a complex number s with $\mathrm{Re}(s) > 0$, and the binomial coefficients on the right are replaced by the appropriate combination of gamma functions.

In order to verify (C1') for types B,C and D we shall use an integral formula due to Selberg [9], which generalizes Euler's beta integral. Let a,b,c be complex numbers satisfying

$$\mathrm{Re}(a) > -1 \ , \ \mathrm{Re}(b) > -1 \ , \ \mathrm{Re}(c) > -\min\left(\frac{1}{n}\ , \ \frac{\mathrm{Re}(a)+1}{n-1}\ , \ \frac{\mathrm{Re}(b)+1}{n-1}\right)$$

and let $J_n(a,b,c)$ denote the integral

$$\int_0^1 \cdots \int_0^1 \prod_{i=1}^n (x_i^a(1-x_i)^b) \cdot \left|\prod_{i<j}(x_i-x_j)\right|^{2c} dx_1 \ldots dx_n \ .$$

Then Selberg's formula is

$$J_n(a,b,c) = \prod_{r=1}^n \frac{\Gamma(rc+1)\,\Gamma(a+(r-1)c+1)\,\Gamma(b+(r-1)c+1)}{\Gamma(c+1)\,\Gamma(a+b+2+(n+r-2)c)} .$$

If we make the substitution $x_i = \sin^2\theta_i$ in Selberg's integral, it becomes :

$$\int_0^\pi \cdots \int_0^\pi \prod_{i=1}^n (\sin^{2a+1}\theta_i \cos^{2b+1}\theta_i) \prod_{i<j} \left|\sin(\theta_i-\theta_j)\sin(\theta_i+\theta_j)\right|^{2c} d\theta_1 \ldots d\theta_n$$

When R is of type B,C or D the integral in (C1') is of this form, for suitable values of a,b,c, and hence can be evaluated by Selberg's formula.

B. "q-analogues" of Dyson's conjecture.

Many familiar functions have q-analogues : the gamma function, hypergeometric function, etc. For a simple example, a q-analogue of the binomial coefficient :

$$\binom{n}{r} = \frac{n(n-1)\ldots(n-r+1)}{1\cdot 2\ldots r}$$

is the "q-binomial coefficient" (or Gaussian polynomial) obtained by replacing each integer s in the above expression by $1+q+\ldots+q^{s-1}$:

$$\begin{bmatrix} n \\ r \end{bmatrix} = \frac{(1-q^n)(1-q^{n-1})\ldots(1-q^{n-r+1})}{(1-q)(1-q^2)\ldots(1-q^r)}$$

is in fact a polynomial in q , with integer coefficients, and reduces to the usual binomial coefficient when $q = 1$.

In this vein, a q-analogue of the conjecture (C1) is

Conjecture (C2). For each positive integer k , the constant term (i.e. involving q

<u>but not the e^{α}) in the product</u>

$$\prod_{\alpha \in R^+} \prod_{i=1}^{k} (1 - q^{i-1} e^{-\alpha})(1 - q^i e^{\alpha})$$

<u>should be equal to</u>

$$\prod_{i=1}^{\ell} \begin{bmatrix} kd_i \\ k \end{bmatrix}$$

Clearly (C2) implies (C1) by setting $q = 1$.

I have less evidence for (C2) than for (C1). It is true for types A_1 and A_2 and all values of k (Andrews [1]) ; also for all R and $k = 1,2$.

Conjecture (C2) can be reformulated as a statement about a compact Lie group G; it is equivalent to

<u>Conjecture</u> (C2'). $\displaystyle\int_G \det \prod_{j=1}^{k-1} (1 - q^j \, \mathrm{Ad}(x)) \, dx = \prod_{i=1}^{\ell} \prod_{j=1}^{k-1} (1 - q^{km_i+j})$

where as before dx is normalized Haar measure on G , and the m_i are the exponents of the Weyl group W of G (so that $m_i = d_i - 1$) .

For a proof that (C2) $\Leftrightarrow$ (C2'), we refer to [7] .

When $k = 1$, (C2') is clearly true (both sides are equal to 1). When $k = 2$, on replacing q by $-q$ the assertion is that

$$\int_G \det(1 + q \, \mathrm{Ad}(x)) \, dx = \prod_{i=1}^{\ell} (1 + q^{2m_i+1}) .$$

Here the left-hand side is the Poincaré polynomial of the graded algebra $(\Lambda \underline{\mathbf{g}})^G$, where $\underline{\mathbf{g}}$ is the Lie algebra of G ; but it is well-known (via de Rham cohomology) that $(\Lambda \underline{\mathbf{g}})^G \cong H^{\star}(G ; \mathbb{R})$, and that $H^{\star}(G ; \mathbb{R})$ is an exterior algebra generated by elements of degrees $2m_i+1 (1 \leqslant i \leqslant \ell)$. Hence (C2') and therefore (C2) and (C1) are true for any root system R and $k = 1,2$.

Finally, we way remark that (C2) and (C2') make sense and are true when $k = +\infty$. In this case (C2) asserts that the constant term in the formal infinite product

$$\prod_{\alpha \in R^+} \prod_{i=1}^{\infty} (1 - q^{i-1} e^{-\alpha})(1 - q^i e^{\alpha})$$

is equal to

$$\prod_{i=1}^{\infty} (1 - q^i)^{-\ell}$$

and this assertion is equivalent to the main theorem of [5] for the affine root system $S(R)$. Likewise , when $k = \infty$, (C2') is the equivalent assertion that :

$$\int_G \prod_{j=1}^{\infty} \det(1 - q^j \operatorname{Ad}(x))\, dx = 1 .$$

Thus the conjecture (C2) may be regarded as a truncated version of the main theorem of [5] , for an affine root system of type $S(R)$. One can formulate an analogous conjecture for the other types of affine root systems [7].

C. Mehta's conjecture and generalizations

In his book Random Matrices [8] , Mehta conjectured that

$$\int_{\mathbb{R}^n} e^{-|x|^2/2} |D(x)|^{2k}\, dx = (2\pi)^{n/2} \prod_{r=1}^{n} \frac{(rk)!}{k!} \tag{3}$$

for any positive integer k (or more generally for a complex number k with $\operatorname{Re}(k) > 0$). Here $dx = dx_1 \dots dx_n$ is Lebesgue measure, $|x|^2 = x_1^2 + \dots + x_n^2$ and $D(x) = \prod_{i<j} (x_i - x_j)$.

Bombieri observed that (3) can be established by use of Selberg's integral . Put $x_i = \frac{1}{2}(1 + (2a)^{-\frac{1}{2}} y_i)$ in $J_n(a,a\ ;\ k)$ (where $a > 0$) ; if we then let $a \to \infty$ we obtain (3) .

To generalize (3), let G be a finite Coxeter group, i.e. a finite group of isometries of $\mathbb{R}^n$ generated by reflections r in hyperplanes through the origin. The equations of these hyperplanes are of the form $h_r(x) = \sum_{i=1}^{n} a_i x_i = 0$. Normalize each h_r (up to sign) by requiring that $\sum a_i^2 = 2$, and let $P(x) = \prod_r h_r(x)$ be the product of these normalized linear forms, the product being over all reflections r in G. As before, let d_i be the degrees of the fundamental polynomial invariants of G .

<u>Conjecture</u> (C3).
$$\int_{\mathbb{R}^n} e^{-|x|^2/2} |P(x)|^{2k} dx = (2\pi)^{n/2} \prod_{i=1}^{n} \frac{(kd_i)!}{k!}.$$

When G is the symmetric group S_n, acting on R^n by permuting the factors, (C3) reduces to (3) above.

When G is a dihedral group (so that $n = 2$), the conjecture (C3) can be verified directly (e.g. by transforming to polar coordinates in the integral). When G is of type B_n or type D_n, (C3) is true for all k (A. Regev), again by use of Selberg's integral. Finally, (C3) is true for $k = 1$ and G a Weyl group [6].

REFERENCES

[1] G.E. ANDREWS.
Problems and Prospects for Basic Hypergeometric Functions, in Theory and Application of Special Functions, ed. R. Askey, Academic Press (1975).

[2] F.J. DYSON.
J. Math. Phys. 3(1962) 140-156.

[3] I.J. GOOD.
J. Math. Phys. 11 (1970) 1884.

[4] J. GUNSON.
J. Math. Phys. 3(1962) 752-753.

[5] I.G. MACDONALD.
Affine root systems and Dedekind's η-function, Inv. Math. 15 (1972) 91-143.

[6] I.G. MACDONALD.
The volume of a compact Lie group, Inv. Math. 56 (1980) 93-95.

[7] I.G. MACDONALD.
Some conjectures for root systems and finite Coxeter groups, to appear.

[8] M.L. MEHTA.
Random Matrices, Academic Press (1967).

[9] A. SELBERG.
Norsk Mat. Tidsskrift 26 (1944) 71-78.

[10] K. WILSON.
J. Math. Phys., 3(1962) 1040-1043.

Queen Mary College
University of London
Department of Pure Mathematics
Mile End Road
LONDON E 1 4 N S

SPECTRE DU DE RHAM-HODGE SUR L'ESPACE
PROJECTIF QUATERNIONIQUE

par Anne Lévy-Bruhl-Laperrière

Le but de cette étude est de déterminer par des méthodes de représentations d'algèbres de Lie, le spectre du de Rham-Hodge sur les formes différentielles de l'espace projectif quaternionique. Les théorèmes 12 et 17 le donneront pour les formes de degré 1 et 2.

0. Généralités. La méthode utilisée pour obtenir ces résultats repose sur le théorème, démontré dans [4] et [5] :

Théorème 0. *Soit* G *un groupe de Lie compact semi-simple d'algèbre de Lie* $\mathfrak{g}$, K *un sous-groupe fermé d'algèbre de Lie* $\mathfrak{k}$, $\mathfrak{p}$ *un supplémentaire de* $\mathfrak{k}$ *dans* $\mathfrak{g}$ *et* $\mathfrak{p}^*_{\mathbb{C}}$ *le dual de son complexifié. On suppose que* $[\mathfrak{k},\mathfrak{p}] \subset \mathfrak{p}$, $[\mathfrak{p},\mathfrak{p}] \subset \mathfrak{k}$. *Soit* $\mathcal{C}^{\mu}(G, F_1)$ *l'espace des fonctions* $\mathcal{C}^{\infty}$ *de* G *dans* $F_1 = \Lambda^s \mathfrak{p}^*_{\mathbb{C}} \otimes F$, s *entier* ≥ 1, *qui sont* K*-équivariantes pour la représentation* $\mu = \lambda_0^{*s} \otimes \lambda$, λ_0^* *étant la représentation adjointe de* K *dans* $\mathfrak{p}^*_{\mathbb{C}}$ *et* λ *étant une représentation de* K *dans* F. *Soit* Δ_G *l'opérateur de de Rham-Hodge.*

Dans $\mathcal{C}^{\mu}(G,F_1)$ on a :

$$\text{Spectre } \Delta_G = \bigcup_{i=1}^{q} \{(\Lambda_\rho + 2\delta_G | \Lambda_\rho), \text{ pour toutes les représentations irréductibles } \rho \text{ de } G \text{ telles que } \rho|_K \text{ contient } \mu_i\}$$

où $\mu_1,\ldots,\mu_q$ désignent les composantes irréductibles de $\mu = \lambda_o^{*s} \otimes \lambda$, Λ_ρ le poids dominant de la représentation irréductible ρ de G, δ_G la demi-somme des racines positives de $\mathfrak{g}$ et $(\ |\ )$ le produit scalaire dans le dual de la sous-algèbre de Cartan de $\mathfrak{g}$ choisie.

Pour démontrer ce théorème, le résultat suivant dû à A. Ikeda et Y. Taniguchi [2] a été utilisé :

Proposition 1. Soit (G,K) une paire symétrique (compacte) où G est un groupe de Lie semi-simple connexe compact. On munit $M = G/K$ de la métrique riemannienne G-invariante obtenue par restriction de l'opposé de la forme de Killing de $\mathfrak{g}$, soit Δ l'opérateur de Laplace sur M défini par cette métrique. Alors, en identifiant $\mathcal{C}^{\infty}(\Lambda^p M)$ à $\mathcal{C}^{\infty}(G,K\ ;\ \Lambda^p(\mathfrak{g}/\mathfrak{K})^*_{\mathbb{C}})$ on a :

$$\Delta = -C$$

C étant l'élément de Casimir de $\mathfrak{g}$.

Ce résultat peut s'étendre au cas où G est un groupe de Lie semi-simple, connexe, unimodulaire, K est un sous-groupe maximal compact, comme nous le prouverons dans la proposition 3 suivante. Soient alors $\mathfrak{g}$ l'algèbre de Lie de G, $\mathfrak{k}$ l'algèbre de Lie de K. Soit B la forme de Killing de $\mathfrak{g}$ et $\underline{p}$ l'orthogonal de $\mathfrak{k}$ dans $\mathfrak{g}$ pour la forme de Killing B. On a alors

$$\mathfrak{g} = \mathfrak{k} \oplus \underline{p} \qquad \text{avec} \qquad [\mathfrak{k},\mathfrak{k}] \subset \mathfrak{k},\ [\mathfrak{k},\underline{p}] \subset \underline{p}\ ,\ [\underline{p},\underline{p}] \subset \underline{p}$$

B est négative définie sur $\mathfrak{k}$, positive non dégénérée sur $\underline{p}$.

Soit $(X_i)_{1 \leq i \leq m}$ une base orthonormale de $\underline{p}$

$(X_a)_{m < a \leq n}$ une base pseudo-orthonormale de $\mathfrak{k}$ (i.e. $(B(X_a,X_a) = -1)$.

On a alors :

$$[X_i,X_j] = \sum_{m<a\leq n} c^a_{ij} X_a \qquad 1\leq i,j\leq m$$

$$[X_a,X_i] = \sum_{1\leq j\leq m} c^j_{a,i} X_j \qquad 1\leq i\leq m \; ; \; m< a\leq n \; ;$$

l'invariance de la forme B conduit à :

(1) $c^{\gamma}_{\alpha\beta} = -c^{\gamma}_{\beta\alpha}$

(2) $c^a_{ij} = c^i_{aj}$ car $B([X_i,X_j],X_a) + B([X_i,[X_a,X_j]]) = 0$

Soit π la projection de G sur M = G/K. $\mathcal{E}^{\infty}(\Lambda^s M)$ et $\mathcal{E}^{\mu}(G\,;\Lambda^s(\underline{p}^*_{\mathbb{C}}))$ sont identifiées par :

$$\alpha \longrightarrow \tilde{\alpha}$$

$$\tilde{\alpha}(g)(Y_1,\dots,Y_s) = (\pi^*\alpha)(Y_1,\dots,Y_s)(g)$$

pour g appartenant à G, $Y_1,\dots,Y_s$ appartenant à $\mathfrak{g}$, $\underline{p}^*$ étant considéré comme l'ensemble des formes linéaires sur $\mathfrak{g}$ nulles sur $\mathfrak{k}$.

$\eta \in \mathcal{E}^{\mu}(G, \Lambda^s(\underline{p}^*_{\mathbb{C}}))$ est déterminée par le système de fonctions $\mathcal{E}^{\infty}$: $\eta(X_{i_1},\dots,X_{i_s})$, $1\leq i_1< \dots < i_s\leq m$.

Munissons M de la métrique riemannienne G-invariante provenant de la restriction de l'opposée de la forme de Killing à $\underline{p}$ et soit dm la mesure sur M définie par la métrique riemannienne.

Si ϕ , $\psi \in \mathcal{E}^{\infty}(\Lambda^s M)\cap L^2$ posons :

$$(\phi,\psi)_M = \int_M \langle \phi,\psi\rangle \, dm$$

où $\langle\,,\rangle$ sur $\Lambda^s M$ dénote l'extension canonique de la métrique hermitienne de M.

Dans toute la suite on ne considèrera que des formes de $\mathcal{E}^{\infty}_o(\Lambda^s M)$, c'est-à-dire les formes $\mathcal{E}^{\infty}$ à support compact dans M.

Considérons sur G une mesure de Haar bi-invariante dg. On a alors $\int_K dg < +\infty$. Sur $\mathcal{E}^{\mu}(G\,;\Lambda^s(\underline{p}^*_{\mathbb{C}}))$, considérons le produit scalaire défini par

$$(\xi,\eta)^* = \int_G \langle \xi(g), \eta(g)\rangle \, dg \, .$$

Alors il existe une constante c (voir [7] page 380)

$$(\alpha,\beta)_M = c(\tilde{\alpha},\tilde{\beta})^* \, .$$

Puisque $\alpha \in \mathcal{E}^{\infty}_o(\Lambda^s M)$ on a $\tilde{\alpha} \in \mathcal{E}^{\infty}(G,\Lambda^s \,\underline{p}^*_{\mathbb{C}})$ et $(\tilde{\alpha},\tilde{\alpha})^* < +\infty$. Sur $\mathcal{E}^{\mu}(G,\Lambda^s \,\underline{p}^*_{\mathbb{C}})$ considérons l'opérateur D défini par

$$(D\eta)(X_{i_1},\dots,X_{i_{s+1}}) = \sum_{u=1}^{s+1} (-1)^{u-1} X_{i_u} \eta(X_{i_1}\dots\hat{X}_{i_u}\dots X_{i_s})$$

$$1 \leq i_1 < i_2 \quad \dots < i_{p+1} \leq m$$

On a alors $\widetilde{d\alpha} = D\tilde{\alpha}$, $\alpha \in \mathcal{E}^\infty(\Lambda^s M)$, ceci provenant du fait que $[\underline{p},\underline{p}] \subset \underline{k}$.

Soit D^* l'opérateur défini par :

$$(D^*\xi)(X_{i_1}\dots X_{i_{s-1}}) = -\sum_{k=1}^{m} X_k\, \xi(X_k, X_{i_1}\dots X_{i_{s-1}})$$

$$1 \leq i_1 < \dots < i_{p-1} \leq n .$$

D^* est l'adjoint de D ; en effet pour ξ, η tels que

$$\xi \in \mathcal{E}^\infty(G, \Lambda^s(\underline{p}^*_{\mathbb{C}})) \quad \text{supp}\,\xi \ \text{compact} ,$$

$$\eta \in \mathcal{E}^\infty(G, \Lambda^{s-1}(\underline{p}^*_{\mathbb{C}})) \quad \text{supp}\,\eta \ \text{compact} ,$$

on a : $(D\xi,\eta)^* = (\xi, D^*\eta)^*$.

$$(\xi,D\eta)^* = \frac{1}{s!}\int_G dg\left\{\sum_{1\leq i_1,\dots,i_s\leq m} \xi(X_{i_1},\dots,X_{i_s}).\left[\sum_{u=1}^{s}(-1)^{u-1}X_{i_u}\eta(X_{i_1},\dots\hat{X}_{i_u},\dots X_{i_s})\right]\right\}$$

$$= \frac{1}{s!}\sum_{1\leq i_1,\dots i_s\leq m}\int_G dg\left(\xi(X_{i_1}\dots X_{i_s}).\sum_{u=1}^{s}(-1)^{u-1}X_{i_u}\eta(X_{i_1},\dots\hat{X}_{i_u}\dots X_{i_s})\right)$$

$$= \frac{1}{s!}\sum_{1\leq i_1,\dots,i_s\leq m}\int_G\left[\left(\sum_{u=1}^{s}(-1)^u X_{i_u}.\xi(X_{i_1},\dots X_{i_s})\right)\eta(X_{i_1},\dots\hat{X}_{i_u},\dots X_{i_s})\right]dg,$$

d'après [4] page 379.

$$= -\frac{1}{(s-1)!}\sum_{1\leq j_1,\dots j_{s-1}\leq m}\int_G\left(\sum_{k=1}^{n}X_k.\xi(X_k,X_{j_1}\dots X_{j_{s-1}})\right).\eta(X_{j_1}\dots X_{j_{s-1}})dg$$

$$= (D^*\xi,\eta)^*.$$

Posons $\Delta^0 = DD^* + D^*D$. On a alors :

$$(\Delta\alpha,\beta)_M = c(\Delta^0\tilde{\alpha},\tilde{\beta})^* \quad \text{quels que soient} \quad (\alpha,\beta)\in\mathcal{E}^\infty_0(\Lambda^s M)$$

et même quels que soient $(\alpha,\beta)\in\mathcal{E}^\infty(\Lambda^s M)$ telles que :

$$\int_M \langle\Delta\alpha,\Delta\alpha\rangle\, dm < +\infty , \quad \int_M \langle\Delta\beta, \Delta\beta\rangle\, dm < +\infty ,$$

et $\int_M \langle\alpha,\alpha\rangle \, dm < +\infty$ $\int_M \langle\beta,\beta\rangle \, dm < +\infty$ on a $(\Delta\alpha,\beta)_M = c(\Delta^{\circ}\tilde{\alpha},\tilde{\beta})^{*}$

donc $\boxed{\widetilde{\Delta\alpha} = \Delta^{\circ}\tilde{\alpha}}$

Lemme 2. Pour tout $\alpha \in \mathcal{E}^{\mu}(G,\Lambda^{s} p_{\mathbb{C}}^{*})$, on a :

$$\Delta^{\circ}\alpha = -\left[-\sum_{k=m+1}^{n} X_k^2 + \sum_{k=1}^{m} X_k^2\right]\alpha = -C\alpha$$

où C est l'élément de Casimir de G.

Preuve : Soient $1 \leq i_1 < \ldots < i_s \leq m$, α à support compact

$$(DD^{*}\alpha)(X_{i_1},\ldots X_{i_s}) = -\sum_{u=1}^{s}(-1)^{u-1}\sum_{k=1}^{m} X_{i_u} X_k \alpha(X_k, X_{i_1},\ldots \hat{X}_{i_u},\ldots X_{i_s})$$

$$(D^{*}D\alpha)(X_{i_1},\ldots X_{i_s}) = -\sum_{k=1}^{m} X_k (D\alpha)(X_k, X_{i_1}\ldots X_{i_s})$$

$$= -\sum_{k=1}^{m} X_k^2 \alpha(X_{i_1},\ldots X_{i_s}) - \sum_{k=1}^{m}\sum_{u=1}^{s}(-1)^{u} X_k X_{i_u} \alpha(X_k, X_{i_1},\ldots \hat{X}_{i_u},\ldots X_{i_s})$$

$$\Delta^{\circ}\alpha = -\sum_{k=1}^{m} X_k^2 \alpha(X_{i_1},\ldots X_{i_s}) - \sum_{k=1}^{m}\sum_{u=1}^{s}(-1)^{u-1}\left[X_{i_u},X_k\right]\alpha(X_k, X_{i_1},\ldots \hat{X}_{i_u},\ldots X_{i_s})$$

Nous allons utiliser les propriétés de la base $(X_1,\ldots X_n)$ de $\mathfrak{g}$ considérée :

$$\left[X_i,X_j\right] = \sum_{m<a\leq n} C_{ij}^{a} X_a \qquad \text{si } 1 \leq i,\, j \leq m$$

$$\left[X_a,X_i\right] = \sum_{1\leq j\leq m} c_{a,i}^{j} X_j \qquad \text{si } 1 \leq i \leq m,\ m < a \leq n$$

avec
$$(*) \quad \begin{cases} c_{\alpha\beta}^{\gamma} = -c_{\beta\alpha}^{\gamma} \\ c_{ij}^{a} = c_{aj}^{i} \end{cases}$$

$$-\sum_{k=1}^{m}\sum_{u=1}^{s}(-1)^{u-1}\left[X_{i_u},X_k\right]\alpha(X_k, X_{i_1},\ldots \hat{X}_{i_u},\ldots X_{i_s})$$

$$= -\sum_{k=1}^{m}\sum_{u=1}^{s}(-1)^{u-1}\sum_{a=m+1}^{n} c_{i_u k}^{a} X_a \alpha(X_k, X_{i},\ldots \hat{X}_{i_u},\ldots X_{i_s}), \text{ car } \left[X_{i_u},X_{i_k}\right]\in \mathfrak{k}$$

pour $m+1 \leq a \leq n$

$$X_a \alpha(X_{i_1},\dots X_{i_s})(g) = \frac{d}{dt}(\alpha(g.\exp t X_a)(X_{i_1},\dots X_{i_s}))_{t=o}$$

$$= \frac{d}{dt}(\alpha(g)(\exp t X_a.X_{i_1},\dots \exp t X_a.X_{i_s}))_{t=o}$$

$$= \sum_{u=1}^{s} (-1)^{u-1} \alpha(g)([X_a,X_{i_u}],X_{i_1},\dots \hat{X}_{i_u},\dots X_{i_s})$$

$$= \sum_{u=1}^{s} (-1)^{u-1} \sum_{k=1}^{m} c^k_{a,i_u} \alpha(X_k,X_{i_1},\dots \hat{X}_{i_u},\dots X_{i_s})$$

car $[X_a,X_{i_u}] \in \underline{p}$, $m+1 \leq a \leq n$, $1 \leq i_u \leq m$; d'où :

$$- \sum_{k=1}^{m} \sum_{u=1}^{s} (-1)^{u-1} \sum_{a=m+1}^{n} c^a_{i_u} X_a \alpha(X_k,X_1,\dots,\hat{X}_{i_u},\dots,X_{i_s})$$

$$= - \sum_{a=m+1}^{n} \sum_{u=1}^{s} (-1)^{u-1} \sum_{k=1}^{m} c^a_{i_u} X_a \alpha(X_k,X_1,\dots,\hat{X}_{i_u},\dots X_{i_s})$$

$$= - \sum_{a=m+1}^{n} X_a \Big[\sum_{u=1}^{s} (-1)^{u-1} \sum_{k=1}^{m} (-c^k_{a\, i_u}) \alpha(X_k,X_1,\dots \hat{X}_{i_u},\dots X_{i_s})$$

$$= + \sum_{a=m+1}^{n} X_a X_a \alpha(X_{i_1},\dots X_{i_s}) = \sum_{a=m+1}^{n} X_a^2 \alpha(X_{i_1},\dots X_{i_s})$$

car $c^a_{i_u k} = - c^k_{a\, i_u}$ en utilisant (*) ;

$$\Delta^o \alpha = -(- \sum_{k=m+1}^{n} X_k^2 \alpha + \sum_{k=1}^{m} X_k^2 \alpha) \;;$$

or l'élément de Casimir C de $\mathfrak{g}$ vaut

$$C = \sum_{k=1}^{m} X_k^2 - \sum_{k=m+1}^{n} X_k^2 \quad \text{puisque } B|\underline{p} \text{ est définie positive}$$

et que $B|\mathfrak{k}$ est définie négalive.

Nous avons donc montré que quel que soit

$$\alpha \in \mathcal{E}^\infty(\Lambda^s M) \quad \text{on a } \widetilde{\Delta\alpha} = -C\tilde{\alpha}$$

où C est l'élément de Casimir de $\mathfrak{g}$.

Proposition 3. Soit G un groupe de Lie semi-simple, unimodulaire, K un sous-groupe maximal compact. En identifiant $\mathcal{E}^{\infty}(\Lambda^s G/K)$ et $\mathcal{E}^{\mu}(G,\Lambda^s p^*_{\mathbb{C}})$ on a :

$$\Delta = -C$$

où C est l'élément de Casimir de $\mathfrak{g}$ et $\mathfrak{g} = \mathfrak{k} \oplus p$.

1. Etude du groupe symplectique et de quelques sous-groupes.

Dans $\mathbb{C}^{2n}$ considérons la forme alternée de matrice

$$J = \begin{pmatrix} 0 & Id \\ -Id & 0 \end{pmatrix}$$

dans la base $(f_1,\dots f_{2n})$. Soit $Sp(n,\mathbb{C})$ le groupe de cette forme. L'algèbre de Lie, $sp(n,\mathbb{C})$, de $Sp(n,\mathbb{C})$ est l'algèbre des matrices complexes $(2n,2n)$ de la forme

$$\begin{pmatrix} X_1 & X_2 \\ X_3 & -{}^tX_1 \end{pmatrix} \quad X_2 \text{ et } X_3 \text{ symétriques.}$$

En désignant par $(e_{ij})_{1\leq i\leq 2n,\ 1\leq j\leq 2n}$ la base usuelle de $M_{2n}(\mathbb{C})$, l'algèbre $sp(n,\mathbb{C})$ a pour base la réunion de $\left\{e_{i,j}-e_{j+n,i+n}, 1\leq i\leq n,\ 1\leq j\leq n\right\}$, $\left\{e_{i+n,j}+e_{j+n,i},\ 1\leq i\leq j\leq n\right\}$ et de $\left\{e_{i,j+n}+e_{j,i+n},\ 1\leq i\leq j\leq n\right\}$.

Il est bien connu (cf. [5] page 140) que si on pose

$$h_{ii} = e_{ii} - e_{i+n,i+n} \qquad 1\leq i\leq n$$

$$e_{\omega_i-\omega_j} = e_{ij} - e_{j+n,i+n} \qquad i\neq j$$

$$e_{\omega_i+\omega_j} = e_{i+n,j} + e_{j+n,i} \qquad i<j$$

$$e_{-\omega_i-\omega_j} = e_{i,j+n} + e_{j,i+n} \qquad i<j$$

$$e_{-2\omega_i} = e_{i,i+n} \qquad e_{2\omega_i} = e_{i+n,i}$$

alors $\mathfrak{h} = \overset{n}{\underset{i=1}{\oplus}} \mathbb{C}\, h_{ii}$ est une sous-algèbre de Cartan de $sp(n,\mathbb{C})$, les racines relatives à $\mathfrak{h}$ sont :

$$\{\pm 2\omega_i\} \cup \{\omega_i-\omega_j,\ i\neq j\} \cup \{\omega_i+\omega_j,\ i\neq j\} \cup \{-\omega_i-\omega_j,\ i\neq j\}$$

et on peut prendre pour base des racines positives

$$\omega_1-\omega_2, \omega_2-\omega_3, \ldots, \omega_{n-1}-\omega_n,\ 2\omega_n\ .$$

1.1. <u>Forme de Killing sur</u> $sp(n,\mathbb{C})$

La forme bilinéaire symétrique dans $sp(n,\mathbb{C})$ définie par :

$$(X,Y) \longrightarrow \mathrm{Tr}\ XY$$

est invariante. En effet

$$\mathrm{Trace}\ ([A,C] \times B) + \mathrm{Trace}(A \times [B,C])$$

$$= \mathrm{Trace}\ (ACB-CAB) + \mathrm{Trace}\ (ABC-ACB) = 0$$

Elle est donc proportionnelle à la forme de Killing sur $sp(n,\mathbb{C})$ (cf. [3] exercice 3.9 page 104). On a donc pour tout $(X,Y) \in sp(n,\mathbb{C})^2$:

$$B(X,Y) = \alpha\,\mathrm{Trace}(XY).$$

Calculons α en posant $X = Y = \left(\begin{array}{c|c} \mathrm{Id} & 0 \\ \hline 0 & -\mathrm{Id} \end{array}\right)$

On a $\mathrm{Ad}X.\mathrm{Ad}X \quad \left(\begin{array}{c|c} M_1 & M_2 \\ \hline M_3 & M_4 \end{array}\right) \longrightarrow \left(\begin{array}{c|c} 0 & 4M_2 \\ \hline 4M_3 & 0 \end{array}\right)$

$\mathrm{Trace}(\mathrm{Ad}\ X \circ \mathrm{Ad}\ X) = 4(n^2+n) = \alpha 2n$ et $\alpha = 2(n+1)$

<u>Proposition</u> 4. <u>Dans</u> $sp(n,\mathbb{C})$ <u>la forme de Killing est donnée par</u> $B(X,Y) = 2(n+1)\ \mathrm{Tr}(XY)$.

1.2. <u>Poids dominants des représentations irréductibles de dimension finie de</u> $sp(n,\mathbb{C})$.

On a $(\omega_i-\omega_{i+1})(h) = B(h, \frac{1}{4(n+1)}(h_i-h_{i+1}))$ quel que soit $h \in \mathfrak{h}$

$(2\omega_n)(h) = B(h, \frac{2h_n}{4(n+1)})$ quel que soit $h \in \mathfrak{h}$

Si α et β sont deux éléments de $\mathfrak{h}^*$:

$$(\alpha|\beta) = B(h_\alpha, h_\beta).$$

Un poids de $sp(n,\mathbb{C})$ est de la forme :

$$\begin{aligned}\lambda &= m_1\omega_1+\ldots+m_n\omega_n \\ &= m_1(\omega_1-\omega_2)+\ldots+(m_1+\ldots+m_i)(\omega_i-\omega_{i+1})+\ldots+(m_1+\ldots+m_n)\omega_n\end{aligned}$$

et est dominant si et seulement si :

$$(1)\quad \begin{cases} \text{Pour tout } i\ ,\ 1\leq i\leq n-1,\ \text{on a}\quad \langle\lambda,\omega_i-\omega_{i+1}\rangle\in\mathbb{Z}^+ \\ \langle\lambda,\ 2\omega_n\rangle\in\mathbb{Z}^+ \end{cases}$$

$$\langle\lambda,\omega_i-\omega_{i+1}\rangle = 2\,\frac{(\lambda|\omega_i-\omega_{i+1})}{(\omega_i-\omega_{i+1}|\omega_i-\omega_{i+1})} = m_i-m_{i+1}$$

$$\langle\lambda,2\omega_n\rangle = m_n .$$

La condition (1) est équivalente à

$$\begin{cases} m_i-m_{i+1}\in\mathbb{Z}^+ & 1\leq i\leq n-1 \\ m_n\in\mathbb{Z}^+ \end{cases}$$

Ceci est encore équivalent à :

$$\begin{cases} m_i\in\mathbb{Z}^+ \quad \text{quel que soit } i. \\ \text{et} \\ m_1\geq m_2\geq\ldots\geq m_n\geq 0. \end{cases}$$

<u>Proposition</u> 5. <u>Les poids dominants des représentations irréductibles de dimension finie de</u> $sp(n,\mathbb{C})$ <u>sont</u> :

$$m_1\omega_1+\ldots+m_n\omega_n \quad \text{où}\quad m_i\in\mathbb{Z}\quad \text{pour}\quad i=1,\ldots,n$$

et

$$m_1\geq m_2\geq\ldots\geq m_n\geq 0.$$

1.3. <u>Inclusion de</u> $Sp(n-1,\mathbb{C})\times Sp(1,\mathbb{C})$ <u>dans</u> $Sp(n,\mathbb{C})$.

Considérons dans $\mathbb{C}^{2n}$ les sous-espaces vectoriels $\mathbb{C}f_1\oplus\mathbb{C}f_{n+1}$ et $\bigoplus_{i=2}^{n}(\mathbb{C}f_i\oplus\mathbb{C}f_{i+n})$ et considérons le sous-groupe de $Sp(n,\mathbb{C})$ conversant globalement ces deux sous-espaces vectoriels : il est isomorphe à $Sp(1,\mathbb{C})\times Sp(n-1,\mathbb{C})$, son algèbre de Lie est $sp(1,\mathbb{C})\oplus sp(n-1,\mathbb{C})$, ensemble des matrices de la forme

$$\begin{pmatrix} a_{11} & 0 & \vdots & a_{31} & 0 \\ 0 & X_{22} & \vdots & 0 & X_{42} \\ \cdots & \cdots & \cdots & \cdots & \cdots \\ a_{13} & 0 & \vdots & -a_{11} & 0 \\ 0 & X_{24} & \vdots & 0 & -{}^tX_{22} \end{pmatrix}$$ où X_{42} et X_{24} sont symétriques.

On a :

$$sp(n,\mathbb{C}) = sp(1,\mathbb{C}) \oplus sp(n-1,\mathbb{C}) \oplus \underline{p}$$

où $\underline{p}$ a pour base $\{e_{i,1}-e_{n+1,i+n}\ ;\ e_{1,i}-e_{n+i,n+1}\ ;\ e_{1,n+i}+e_{i,n+1}\ ;$ $e_{n+1,i}+e_{n+i,1}\ ;\ 2\leq i\leq n\}$.

Recherchons quelle est la représentation λ_o ; pour $2\leq i\leq n$:

$$[e_{11}-e_{n+1,n+1},\ e_{i,1}-e_{n+1,i+n}] = -e_{i,1}+e_{i+n,n+1}$$

$$[e_{11}-e_{n+1,n+1},\ e_{1,i}-e_{i+n,1+n}] = e_{1,i}-e_{n+1,i+n}$$

$$[e_{11}-e_{n+1,n+1},\ e_{1,n+i}+e_{i,n+1}] = e_{1,n+i}+e_{i,n+1}$$

$$[e_{11}-e_{n+1,n+1},\ e_{n+1,i}+e_{n+i,1}] = -(e_{1,n+i}+e_{n+i,1})$$

$$[e_{n+1,1},\ e_{1,i}-e_{i+n,n+1}] = e_{n+1,i}+e_{1,n+i}$$

$$[e_{n+1,1},\ e_{i,1}-e_{1+n,i+n}] = 0$$

$$[e_{n+1,1},\ e_{1,n+i}+e_{i,n+1}] = e_{n+1,n+i}-e_{i,1}$$

$$[e_{n+1,1},\ e_{n+1,i}+e_{n+i,1}] = 0$$

$$[e_{1,n+1},\ e_{1,i}-e_{i+n,n+1}] = 0$$

$$[e_{1,n+1},\ e_{i,1}-e_{1+n,i+n}] = -e_{1,i+n}-e_{i,n+1}$$

$$[e_{1,n+1},\ e_{1,n+i}+e_{i,n+1}] = 0$$

$$[e_{1,n+1},\ e_{n+1,i}+e_{n+i,1}] = e_{1,i}-e_{n+i,n+1}$$

On remarque que c'est la somme de $2(n-1)$ fois la représentation de $sp(1,\mathbb{C})$ de poids dominant ω_1 (cette représentation est bien de dimension 2).

<u>Action de</u> $sp(n-1,\mathbb{C})$ <u>sur</u> <u>p</u> : on suppose $2 \leq k < n$ et on note δ_i^j le symbole de Kronecker.

Pour $2 \leq i \leq n$, $2 \leq j \leq n$ on a :

$$[e_{ij}-e_{j+n,i+n},\ e_{k,1}-e_{n+1,k+n}] = \delta_j^k\, e_{i,1} - \delta_j^k\, e_{n+1,i+n}$$

$$[e_{i,j}-e_{j+n,i+n},\ e_{1,k}-e_{n+k,n+1}] = \delta_i^k\, e_{j+n,n+1} - \delta_i^k\, e_{1,j}$$

$$[e_{i,j}-e_{j+n,i+n},\ e_{1,n+k}+e_{k,n+1}] = \delta_j^k\, e_{i,n+1} + \delta_j^k\, e_{1,i+n}$$

$$[e_{i,j}-e_{j+n,i+n},\ e_{n+1,k}+e_{n+k,1}] = -\delta_i^k\, e_{j+n,1} - \delta_i^k\, e_{n+1,j}\ ;$$

pour $2 \leq i \leq j \leq n$ on a :

$$[e_{i,n+j}+e_{j,n+i},\ e_{k,1}-e_{n+1,k+n}] = 0$$

$$[e_{i,n+j}+e_{j,n+i},\ e_{1,k}-e_{n+k,n+1}] = -\delta_j^k\, e_{i,n+1} - \delta_i^k\, e_{j,n+1} - \delta_k^i\, e_{1,n+j} - \delta_k^j\, e_{1,n+i}$$

$$[e_{i,n+j}+e_{j,n+i},\ e_{1,n+k}+e_{k,n+1}] = 0$$

$$[e_{i,n+j}+e_{j,n+i},\ e_{n+1,k}+e_{n+k,1}] = \delta_j^k\, e_{i,1} + \delta_i^k\, e_{j,1} - \delta_k^i\, e_{n+1,n+j} - \delta_k^j\, e_{n+1,n+i}$$

$$[e_{n+i,j}+e_{j+n,i},\ e_{k,1}-e_{n+1,k+n}] = \delta_j^k\, e_{n+i,1} + \delta_i^k\, e_{j+n,1} + \delta_k^i\, e_{n+1,j} + \delta_k^j\, e_{n+1,i}$$

$$[e_{n+i,j}+e_{j+n,i},\ e_{1,k}-e_{n+k,n+1}] = 0$$

$$[e_{n+i,j}+e_{j+n,i},\ e_{1,n+k}+e_{k,n+1}] = \delta_j^k\, e_{n+i,n+1} + \delta_i^k\, e_{n+j,n+1} - \delta_k^i\, e_{1,j} - \delta_k^j\, e_{1,i}$$

$$[e_{n+i,j}+e_{j+n,i},\ e_{n+1,k}+e_{n+k,1}] = 0$$

On remarque en particulier que pour $2 \leq i \leq n$

$$[e_{ii}-e_{i+n,n+i},\ e_{k,1}-e_{n+1,k+n}] = \delta_i^k\, e_{i,1} - \delta_i^k\, e_{n+1,i+n}$$

$$[e_{ii}-e_{i+n,n+i},\ e_{1,k}-e_{n+k,n+1}] = \delta_i^k\, e_{n+i,n+1} - \delta_i^k\, e_{1,i}$$

$$[e_{ii}-e_{i+n,n+i},\ e_{1,n+k}+e_{k,n+1}] = \delta_i^k\, e_{i,n+1} + \delta_i^k\, e_{1,n+i}$$

$$[e_{ii}-e_{i+n,n+i},\ e_{n+1,k}+e_{n+k,1+n}] = -\delta_i^k\, e_{i+n,1+n} - \delta_i^k\, e_{n+1,i}$$

Lemme 6. La représentation irréductible de $sp(n-1,\mathbb{C})$ de dimension finie de poids dominant ω_1 est $2(n-1)$.

Preuve : La dimension de cette représentation est donnée par la formule de Weyl.

$$\frac{\prod_{\alpha>0}(\omega_1+\delta_K|\alpha)}{\prod_{\alpha>0}(\delta_K|\alpha)}$$

où $\delta_K = \sum_{i=1}^{n-1}(n-i)\omega_i$.

La dimension vaut donc :

$$\prod_{2\leq j\leq n-1}\left[\frac{(\omega_1+\delta_K|\omega_1-\omega_j)}{(\delta_K|\omega_1-\omega_j)} \times \frac{(\omega_1+\delta_K|\omega_1+\omega_j)}{(\delta_K|\omega_1+\omega_j)}\right] \frac{(\omega_1+\delta_K|2\omega_1)}{(\delta_K|2\omega_1)}$$

$$= \prod_{2\leq j\leq n-1}\left[\frac{n-(n-j)}{n-1-(n-j)} \times \frac{n+n-j}{n-1+n-j}\right] \frac{2n}{2(n-1)} = 2(n-1).$$

Proposition 7. L'action de $sp(n-1,\mathbb{C})$ sur p par ad est la somme de deux fois la représentation irréductible de $sp(n-1,\mathbb{C})$, dont le poids dominant est ω_1 et dont les poids sont $\{\pm\omega_i,\ 1\leq i\leq n-1\}$.

Preuve : Les sous-espaces $\bigoplus_{2\leq k\leq n}\left[\mathbb{C}(e_{k,1}-e_{n+1,k+n}) \oplus \mathbb{C}(e_{n+1,k}+e_{n+k,1})\right]$ et $\bigoplus_{2\leq k\leq n}\left[\mathbb{C}(e_{1,k}-e_{n+k,n+1}) \oplus \mathbb{C}(e_{k,n+1}+e_{1,n+k})\right]$ sont de dimension $2(n-1)$, sont stables sous l'action de $sp(n-1,\mathbb{C})$ et contiennent la représentation de poids dominant ω_1.

Remarquons que la représentation de poids dominant ω_1 est sa propre duale : en effet les poids de cette représentation duale sont les opposés de ceux de ω_1 c'est-à-dire $\{\pm\omega_i,\ 1\leq i\leq n-1\}$, ce sont exactement les poids de la représentation de poids dominant ω_1.

1.4. Représentation de $Sp(1,\mathbb{C}) \times Sp(n-1,\mathbb{C})$ et étude de λ_o.

Le travail de J.Y. Lepowsky ([6] pages 67 à 69) caractérise les poids dominants des représentations irréductibles de $sp(n-1,\mathbb{C}) \oplus sp(1,\mathbb{C})$. Ce sont les formes linéaires de la sous-algèbre des matrices diagonales de

$sp(1,\mathbb{C}) \oplus sp(n-1,\mathbb{C})$ appartenant à la famille :

$$\{m_1\omega_1+m_2\omega_2+\ldots+m_n\omega_n,\ m_i \in \mathbb{Z} \qquad i=1,\ldots,n\ ,\ m_2 \geq m_3 \geq \ldots \geq m_n,\ m_1 \geq 0\}$$

($(m_2,\ldots,m_n)$ correspondant à l'action de $sp(n-1,\mathbb{C})$). On peut alors énoncer le résultat suivant :

Proposition 8. La représentation adjointe λ_o de $sp(1,\mathbb{C}) \oplus sp(n-1,\mathbb{C})$ dans p est irréductible et de poids dominant $\omega_1+\omega_2$ et est sa propre duale.

Preuve : La représentation λ_o contient la représentation irréductible de $sp(1,\mathbb{C}) \oplus sp(n-1,\mathbb{C})$ de poids $\omega_1+\omega_2$. Les images du vecteur $e_{2,1}-e_{n+1,2+n}$ sous l'action de $sp(n-1,\mathbb{C})$ donnent $\bigoplus_{2\leq k\leq n}[\mathbb{C}(e_{k,1}-e_{n+1,k+n}) \oplus \mathbb{C}(e_{n+k,1}+e_{n+1,k})]$, et sous l'action de $sp(1,\mathbb{C})$, $e_{2,n+1}+e_{1,n+2}$, vecteur de poids ω_1 sous l'action de $sp(n-1,\mathbb{C})$. La représentation obtenue est donc exactement la représentation irréductible de $sp(1,\mathbb{C}) \oplus sp(n-1,\mathbb{C})$ de poids dominant $\omega_1+\omega_2$. Elle coïncide avec sa duale.

1.5. Décomposition d'une représentation de $sp(n,\mathbb{C})$ restreinte à $sp(1,\mathbb{C}) \oplus sp(n-1,\mathbb{C})$.

Le résultat suivant est dû à J.Y. Lepowsky ([6], page 71 et 72, théorème 6 et corollaire) :

Théorème 9. Soit $\lambda = \sum_{i=1}^{n} a_i\omega_i$ le poids dominant d'une représentation irréductible de $sp(n,\mathbb{C})$, $\mu = \sum_{i=1}^{n} b_i\omega_i$ le poids dominant d'une représentation irréductible de $sp(1,\mathbb{C}) \oplus sp(n-1,\mathbb{C})$. La multiplicité de la représentation de poids dominant μ de $sp(1,\mathbb{C}) \oplus sp(n-1,\mathbb{C})$ dans la restriction à $sp(1,\mathbb{C}) \oplus sp(n-1,\mathbb{C})$ de la représentation de $sp(n,\mathbb{C})$ de poids dominant λ vaut 0 sauf si $\sum_{i=1}^{n} (a_i+b_i)$ appartient à $2\mathbb{Z}$ et si

$A_1 = a_1-\max(a_2,b_2) \geq 0$, $A_i = \min(a_i,b_i)-\max(a_{i+1},b_{i+1}) \geq 0$, $2 \leq i \leq n-1$

et, alors, cette multiplicité vaut :

$$m(\lambda , \mu) = F_{n-1}(\tfrac{1}{2}(b_1-A_1+ \sum_{i=2}^{n} A_i) \; ; A_2, A_3,\dots,A_n)$$

$$-F_{n-1}(\tfrac{1}{2}(-b_1-A_1+ \sum_{i=2}^{n} A_i)-1 \; ; A_2, A_3,\dots,A_n)$$

où $F_k(s \; ; t_1, t_2,\dots,t_k)$ est le nombre de façons de mettre s balles identiques dans k boites distinctes de capacités respectives $t_1,\dots,t_k$ et vaut :

$$F_k(s \; ; t_1, t_2,\dots t_k) = \sum_{L\subset\{1,2,\dots k\}} (-1)^{|L|}\binom{k-1-|L|+s-\sum_{i\in L} t_i}{k-1}$$

où $|L|$ est le cardinal de L, $\binom{x}{y}$ est le coefficient du binôme et vaut 0 si $x < y$.

1.6. Etude de Λ^2_μ.

On a vu au paragraphe 1.4 que la représentation μ de $Sp(1,\mathbb{C}) \times Sp(n-1,\mathbb{C})$ dans l'espace vectoriel $\bigoplus_{2\leq i\leq n} [\mathbb{C}(e_{i,1}-e_{n+1,i+n}) \oplus \mathbb{C}(e_{n+1,i}+e_{n+i,1})]$ $\bigoplus_{2\leq i\leq n} [\mathbb{C}(e_{1,i}-e_{n+i,n+1}) \oplus \mathbb{C}(e_{i,n+1}+e_{1,n+i})]$ est irréductible et de poids dominant $\omega_1+\omega_2$. De plus sa restriction à $Sp(n-1,\mathbb{C})$ se décompose en la somme directe de deux représentations irréductibles de $Sp(n-1,\mathbb{C})$ de poids dominant identique : ω_2.

Proposition 10. La représentation $\mu\wedge\mu$ se décompose en la somme des représentations irréductibles de $Sp(1,\mathbb{C}) \times Sp(n-1,\mathbb{C})$ de poids dominants respectifs : $2\omega_1+\omega_2+\omega_3$, $2\omega_1$ $2\omega_2$ de multiplicité 1.

Preuve : a) Les poids de la représentation μ sont tous de multiplicité 1 et constituent l'ensemble $\mathcal{P} = \{\varepsilon_1\omega_1+\varepsilon_i\omega_i\ ,\ 2\leq i\leq n, (\varepsilon_1,\varepsilon_2)\in\{+1,-1\}^2\}$. Les poids de la représentation $\mu\wedge\mu$ sont donc les éléments de la famille :

$$\{(\varepsilon_1+\varepsilon_1')\omega_1+\varepsilon_i\omega_i+\varepsilon_j\omega_j,\ 2\leq i<j\leq n,\ (\varepsilon_1,\varepsilon_i)\in\{+1,-1\}^2,$$

$$(\varepsilon_1',\varepsilon_j)\in\{+1,-1\}^2\}\cup\{(\varepsilon_1+\varepsilon_1')\omega_1+\varepsilon_i\omega_i+\varepsilon_i'\omega_i,\ 2\leq i\leq n,\ (\varepsilon_1,\varepsilon_i)\in\{+1,-1\}^2,$$

$$(\varepsilon_1',\varepsilon_i')\in\{+1,-1\}^2,\ (\varepsilon_1,\varepsilon_i)\neq(\varepsilon_1',\varepsilon_i')\}.$$

On remarque que les seuls poids dominants de cette famille sont :

$2\omega_1+\omega_2+\omega_3$	de multiplicité 1
$\omega_2+\omega_3$	de multiplicité 2
$2\omega_1$	de multiplicité n-1
$2\omega_2$	de multiplicité 1
0	de multiplicité 2(n-1)

On voit que $2\omega_1+\omega_2+\omega_3$ et $2\omega_2$ sont les poids dominants de deux représentations irréductibles contenues dans $\mu\wedge\mu$ car $2\omega_2$ n'est pas un poids de la représentation de poids dominant $2\omega_1+\omega_2+\omega_3$ alors que $2\omega_1, \omega_2+\omega_3$ et 0 en sont.

b) Etudions la multiplicité $m(\omega_2+\omega_3)$ du poids $\omega_2+\omega_3$ dans la représentation de $Sp(1,\mathbb{C})\times Sp(n-1,\mathbb{C})$ de poids dominant $2\omega_1+\omega_2+\omega_3$ en utilisant la formule de Freudenthal ([1] page 122) :

$$[(2\omega_1+\omega_2+\omega_3+\delta,\ 2\omega_1+\omega_2+\omega_3+\delta)-(\omega_2+\omega_3+\delta,\omega_2+\omega_3+\delta)]\ m(\omega_2+\omega_3)$$

$$= 2\sum_{\alpha>0}\sum_{i=1}^{\infty} m(\omega_2+\omega_3+i\alpha)(\omega_2+\omega_3+i\alpha,\alpha)\ .$$

En fait le calcul se limite à :

$$[(2\omega_1+\omega_2+\omega_3+\delta,2\omega_1+\omega_2+\omega_3+\delta)-(\omega_2+\omega_3+\delta,\omega_2+\omega_3+\delta)]\ m(\omega_2+\omega_3)$$

$$= 2m(2\omega_1+\omega_2+\omega_3)(2\omega_1+\omega_2+\omega_3,\ 2\omega_1)$$

$$= 4(2\omega_1+\omega_2+\omega_3,\omega_1) = 8$$

et on obtient : $\delta = \omega_1+\sum_{i=2}^{n}(n+1-i)\,\omega_i$,

d'où : $[9+n^2+(n-1)^2-(1+n^2+(n-1)^2)]\ m(\omega_2+\omega_3) = 16$

$m(\omega_2+\omega_3) = 2.$

Etudions la multiplicité $m'(\omega_2+\omega_3)$ du poids $\omega_2+\omega_3$ dans la représentation de $Sp(n-1,\mathbb{C})$ de poids dominant $2\omega_2$.

$$[(2\omega_2+\delta,2\omega_2+\delta)-(\omega_2+\omega_3+\delta,\omega_2+\omega_3+\delta)]\ m'(\omega_2+\omega_3)$$

$$= 2\sum_{\alpha>0}\sum_{i=1}^{\infty} m(\omega_2+\omega_3+i\alpha)(\omega_2+\omega_3+i\alpha,\alpha)$$

$$= 2\ m(2\omega_2)(\omega_2+\omega_3+\omega_2-\omega_3,\omega_2-\omega_3)$$

$$= 2\times 2 = 4\ ;$$

d'où :

$$[(n+1)^2+(n-1)^2-(n^2+n^2)]\ m'(\omega_2+\omega_3) = 4$$

et : $$m'(\omega_2+\omega_3) = 2.$$

On remarque donc que la somme des multiplicités du poids $\omega_2+\omega_3$ dans les représentations irréductibles de $Sp(1,\mathbb{C})\times Sp(n-1,\mathbb{C})$ de poids dominants respectifs $2\omega_1+\omega_2+\omega_3$ et $2\omega_2$ est exactement égale à la multiplicité du poids $\omega_2+\omega_3$ dans $\mu\wedge\mu$. La représentation de $Sp(1,\mathbb{C})\times Sp(n-1,\mathbb{C})$ de poids dominant $\omega_2+\omega_3$ n'intervient pas dans la décomposition de $\mu\wedge\mu$.

c) Etudions maintenant la multiplicité du poids $2\omega_1$ dans la représentation irréductible de $Sp(1,\mathbb{C})\times Sp(n-1,\mathbb{C})$ de poids dominant $2\omega_1+\omega_2+\omega_3$.

Les racines positives (pour le choix donné en 1.4) sont les

$$\{2\omega_j,\ 1\leq j\leq n\}\cup\{\omega_j+\varepsilon_k\omega_k,\ 2\leq j<k\leq n,\ \varepsilon_k = \pm 1\}\ .$$

$$[(2\omega_1+\omega_2+\omega_3+\delta,\ 2\omega_1+\omega_2+\omega_3+\delta)-(2\omega_1+\delta,\ 2\omega_1+\delta)]\ m(\omega_1)$$

$$= 2\sum_{1\leq j\leq n}\sum_{i=1}^{\infty} m(2\omega_1+2i\,\omega_j)\ (2\omega_1+2i\omega_j,\ 2\omega_j)$$

$$+\ 2\sum_{\substack{2\leq j<k\leq n\\ \varepsilon_k=\pm 1}}\sum_{i=1}^{\infty} m(2\omega_1+i(\omega_j+\varepsilon_k\omega_k))(2\omega_1+i\omega_j+i\varepsilon_k\omega_k,\omega_j+\varepsilon_k\omega_k)$$

or $m(2\omega_1+2i\omega_j) = 0$ pour $i>1$ si $j\neq 2$, pour $i\geq 0$ si $j=2$,

car $(2\omega_1+\omega_2+\omega_3)-(2i\omega_j+2\omega_1) = \omega_2+\omega_3-2i\omega_j$

$$= \omega_2-\omega_j+\omega_3-\omega_j-2(i-1)\omega_j \in \Delta^+ \quad \text{pour} \quad i \leq 1.$$

$$m(2\omega_1+i(\omega_j+\omega_k)) = 0 \quad \text{pour} \quad i \geq 1 \quad (j<k)$$

car $2\omega_1+\omega_2+\omega_3-(2\omega_1+i\omega_j+i\omega_k)$

$$= \omega_2-i\omega_j+\omega_3-i\omega_k \in \Delta^+ \quad \text{pour} \quad i \leq 1$$

$$m(2\omega_1+i(\omega_j-\omega_k)) = 0 \quad \text{pour} \quad i \geq 3 \quad \text{et} \quad 2<j<k, \ \text{pour} \quad i \geq 1 \quad \text{et} \quad j=2$$

car $2\omega_1+\omega_2+\omega_3-i\omega_j+i\omega_k-2\omega_1 = \omega_2+\omega_3+i\omega_k-i\omega_j$

$$= (\omega_2-\omega_3)+3\sum_{3\leq s\leq n-1}(\omega_s-\omega_{s+1})+\frac{3}{2}\,2\omega_n-i\,[(\omega_j-\omega_{j+1})+\ldots+(\omega_{k-1}-\omega_k)]$$

donc $[(2\omega_1+\omega_2+\omega_3+\delta,2\omega_1+\omega_2+\omega_3+\delta)-(2\omega_1+\delta,2\omega_1+\delta)]\ m(2\omega_1)$

$$= 2\sum_{2<j\leq n} m(2\omega_1+2\omega_j)(2\omega_1+2\omega_j,2\omega_j)$$

$$+ 2\sum_{2\leq j<k\leq n} m(2\omega_1+\omega_j+\omega_k)(2\omega_1+\omega_j+\omega_k,\ \omega_j+\omega_k)$$

$$+ 2\sum_{2\leq j<k\leq n} m(2\omega_1+\omega_j-\omega_k)(2\omega_1+\omega_j-\omega_k,\ \omega_j-\omega_k)$$

$$+ 2\sum_{2<j<k\leq n} m(2\omega_1+2(\omega_j-\omega_k))(2\omega_1+2(\omega_j-\omega_k),\omega_j-\omega_k)$$

d'où $(4n-4)\ m(2\omega_1) = 8\sum_{2<j\leq n} m(2\omega_1+2\omega_j)+4\sum_{2\leq j<k\leq n} m(2\omega_1+\omega_j+\omega_k)$

$$+4\sum_{2\leq j<k\leq n} m(2\omega_1+\omega_j-\omega_k) + 8\sum_{2<j<k\leq n} m(2\omega_1+2(\omega_j-\omega_k)) .$$

En utilisant le fait que des poids conjugués sous l'action du groupe de Weyl ont même multiplicité, la multiplicité cherchée vaut :

$$4(n-1)m(2\omega_1) = 8(n-2)m(2\omega_1+2\omega_2)+4(n-1)(n-2)m(2\omega_1+\omega_2+\omega_3)$$
$$+ 4(n-2)(n-3)m(2\omega_1+2\omega_3-2\omega_4).$$

Il nous faut donc calculer :

$$m(2\omega_1+2\omega_2) \quad \text{et} \quad m(2\omega_1+2\omega_3-2\omega_4)$$

$$2\omega_1+2\omega_2 = 2\omega_1+\omega_2+\omega_3+(\omega_2-\omega_3)$$

donc $\left[(2\omega_1+\omega_2+\omega_3+\delta,2\omega_1+\omega_2+\omega_3+\delta)-(2\omega_1+2\omega_2+\delta,2\omega_1+2\omega_2+\delta)\right]\ m(2\omega_1+2\omega_2)$

$$= 2m(2\omega_1+\omega_2+\omega_3)(2\omega_1+2\omega_2+\omega_2-\omega_3,\omega_2-\omega_3)$$

et $m(2\omega_1+2\omega_2) = \dfrac{2 \times 0}{-4} = 0$.

Recherchons $m(2\omega_1+2\omega_3-2\omega_4)$; comme les poids $2\omega_1+2\omega_3-2\omega_4$ et $2\omega_1+2\omega_2-2\omega_3$ sont conjugués et comme la multiplicité du poids $2\omega_1+2\omega_2-2\omega_3$ est nulle, on en déduit :

$$m(2\omega_1+2\omega_3-2\omega_4) = 0$$

donc $\boxed{m(2\omega_1) = n-2}$

$2\omega_1$ n'est pas un poids de la représentation irréductible de $Sp(1,\mathbb{C})\times Sp(n-1,\mathbb{C})$ de poids dominant $2\omega_2$. On en déduit donc que la somme des multiplicités du poids $2\omega_1$ dans les représentations de poids dominants $2\omega_1+\omega_2+\omega_3$ et $2\omega_2$ est $n-2$, elle est $n-1$ dans $\mu\wedge\mu$. La représentation irréductible de poids dominant $2\omega_1$ intervient donc avec la multiplicité 1 dans $\mu\wedge\mu$.

d) Etudions la multiplicité du poids 0 dans la représentation $\mu\wedge\mu$. Un simple calcul de dimension donne :

$$\frac{4(n-1)\left[4(n-1)-1\right]}{2} = \dim \rho_1+\dim\rho_2+\dim\rho_3+k\ \dim\rho_4$$

où $\dim\rho_i$ désigne les dimensions des représentations irréductibles de $Sp(1,\mathbb{C}) \times Sp(n-1,\mathbb{C})$ de poids dominants valant, pour $i=1$, $2\omega_1+\omega_2+\omega_3$, pour $i=2$, $2\omega_2$, pour $i=2$, $2\omega_1$, pour $i=4$, 0 et où k désigne la multiplicité de la représentation irréductible de poids dominant 0 dans la décomposition de $\mu\wedge\mu$.

Les calculs de dimension donnent :

$$\dim\rho_1 = 3(n-2)(2n-1) = 6n^2-15n+6$$
$$\dim\rho_2 = 2n^2-3n+1$$
$$\dim\rho_3 = 3$$
$$\dim\rho_4 = 1$$

d'où $k + 8n^2 - 18n + 10 = 8n^2 - 18n + 10$

$k = 0$.

La multiplicité de la représentation de poids dominant 0 dans $\mu\wedge\mu$ est 0.

2. Spectre du de Rham-Hodge sur l'espace projectif quaternionique.

2.1. Formes de degré 1.

Le théorème 0 nous conduit à rechercher les représentations irréductibles de $Sp(n,\mathbb{C})$ qui, restreintes à $Sp(1,\mathbb{C})\times Sp(n-1,\mathbb{C})$ contiennent la représentation μ^* étudiée à la proposition 8. Le théorème 9 nous permet de trouver ces représentations.

2.1.1. Proposition 1 1. *Les représentations irréductibles de* $Sp(n,\mathbb{C})$ *qui lorsqu'on les restreint à* $Sp(1,\mathbb{C})\times Sp(n-1,\mathbb{C})$ *contiennent la représentation de* $Sp(1,\mathbb{C})\times Sp(n-1,\mathbb{C})$ *de poids dominant* $\omega_1+\omega_2$ *la contiennent avec multiplicité* 1 *et ont pour poids dominants respectifs.*

$$a(\omega_1+\omega_2) \qquad a\in\mathbb{N}^*$$

$$(a+2)\omega_1+a\omega_2$$

$$(a+1)\omega_1+a\omega_2+\omega_3 .$$

De plus leurs dimensions respectives sont :

$$\frac{2a+2n-1}{(2n-1)(2n-2)}\binom{a+2n-2}{2n-3}\binom{a+2n-3}{2n-3}$$

$$\frac{3(2a+2n+1)}{(2n-1)(2n-2)}\binom{a+2n}{2n-3}\binom{a+2n-3}{2n-3}$$

$$\frac{a(a+2)(a+n)(a+2n)(a+2n-2)}{(2n-1)(n-2)(2n-3)^2(n-1)}\binom{a+2n-2}{2n-5}\binom{a+2n-4}{2n-5}$$

Preuve : En appliquant le théorème 9 avec $b_1=b_2=1$, $b_3=\ldots=b_n=0$ une représentation de $sp(n,\mathbb{C})$ de poids dominant $a_1\omega_1+\ldots+a_n\omega_n$ conviendra si et seulement si :

$$a_1-\max(a_2,1)\geq 0$$
$$\min(a_2,1)-\max(a_3,0)\geq 0$$
$$\min(a_i,0)-\max(a_{i+1},0)\geq 0 \qquad i\geq 3$$

donc $a_4=\ldots=a_n=0$.

Deux cas se présentent :

a) $a_2\geq 1$:

alors on doit avoir :

$$a_1-a_2\geq 0$$
$$1-a_3\geq 0$$

d'où $a_1\geq a_2$ $\quad a_3\leq 1$

d'où les représentations de poids dominants :

$$\begin{cases} (1)\ a_1\omega_1+a_2\omega_2 \\ (2)\ a_1\omega_1+a_2\omega_2+\omega_3 \end{cases} \qquad a_1\geq a_2$$

b) $a_2 = 0$

on doit alors avoir $a_1-1\geq 0$, $0-a_3\geq 0$ d'où $a_3 = 0$;

d'où les représentations de poids dominants

$$(3)\ \{a_1\omega_1 \qquad a_1\geq 1 .$$

Il reste la condition $\sum_{i=1}^{n}(a_i+b_i)\in 2\mathbb{Z}$;

d'où pour (1) $a_1+a_2+1+1\in 2\mathbb{Z}$ donc $a_1+a_2\in 2\mathbb{Z}$ et

$$a_1=a_2+2p \text{ avec } p\in\mathbb{N}$$

(2) $a_1+a_2+1+1+1\in 2\mathbb{Z}$ donc $a_1+a_2\in 2\mathbb{Z}+1$ et

$$a_1=a_2+2p+1 \quad p\in\mathbb{N}$$

(3) $a_1+1+1\in 2\mathbb{Z}$ donc $a_1 = 2p$, $p\in\mathbb{N}^*$.

Calculons dans chacun de ces cas la multiplicité de la représentation de poids dominant $\omega_1+\omega_2$ de $sp(1,\mathbb{C})\oplus sp(n-1,\mathbb{C})$ dans la restriction de ces représentations irréductibles de $sp(n,\mathbb{C})$ à $sp(1,\mathbb{C})\oplus sp(n-1,\mathbb{C})$.

$$m((a_2+2p)\omega_1+a_2\omega_2,\omega_1+\omega_2)$$

$$= F_{n-1}(\tfrac{1}{2}(1-2p+1)\ ;\ 1,0\ldots0) - F_{n-1}(\tfrac{1}{2}(-1-2p+1)-1\ ;\ 1,0\ldots)$$

$$= F_{n-1}(1-p\ ;\ 1,\ 0\ldots0) - F_{n-1}(-p-1\ ;\ 1,0\ldots0) = \begin{cases} 1 & \text{si } p=0 \\ 1 & \text{si } p=0 \\ 0 & \text{autrement} \end{cases}$$

d'où les poids : $\begin{cases} a(\omega_1+\omega_2) \\ (a+2)\omega_1+a\omega_2 \end{cases}$ $a \in \mathbb{N}^*$

$$m((a_2+2p+1)\omega_1+a_2\omega_2+\omega_3, \omega_1+\omega_2)$$

$$= F_{n-1}(\tfrac{1}{2}(1-2p-1)\ ;\ 0,0\ldots0) - F_{n-1}(\tfrac{1}{2}(-1-2p-1+0)-1\ ;\ 0,\ldots0)$$

$$= F_{n-1}(-p\ ;\ 0,\ldots0) - F_{n-1}(-p-2\ ;\ 0,\ldots0) = \begin{cases} 1 & \text{si } p=0 \\ 0 & \text{si } p\neq 0 \end{cases}$$

d'où le poids $(a+1)\omega_1+a\omega_2+\omega_3$ $a \in \mathbb{N}^*$

$m(a\omega_1, \omega_1+\omega_2) = 0$ quel que soit a.

La dimension d'une représentation de $Sp(n,\mathbb{C})$ de poids dominant $k_1\omega_1+k_2\omega_2$, $k_1 \geq k_2$ est

$$\frac{k_1+n-(k_2+n-1)}{n-(n-1)} \times \prod_{3\leq j\leq n} \frac{k_1+n-(n+1-j)}{n-(n+1-j)} \quad \frac{k_1+n+k_2+n-1}{2n-1} \prod_{3\leq j\leq n} \frac{k_1+2n+1-j}{2n+1-j}$$

$$\frac{k_1+n}{n} \prod_{3\leq j\leq n} \frac{k_2+n-1-(n+1-j)}{n-1-(n+1-j)} \prod_{3\leq j\leq n} \frac{k_2+n-1+n+1-j}{n-1+n+1-j} \quad \frac{k_2+n-1}{n-1}$$

$$= \frac{(k_1-k_2+1)(k_1+k_2+2n-1)}{2n-1} \quad \frac{k_1+n}{n} \quad \frac{(k_1+n-1)\ldots(k_1+2)(k_1+n+1)\ldots(k_1+2n-2)}{(n-1)\ldots\ldots 2 \quad (n+1)\ \ldots\ldots(2n-2)}$$

$$\frac{k_2+n-1}{n-1} \quad \frac{(k_2+n-2)\ldots(k_2+1)(k_2+n)\ldots(k_2+2n-3)}{(n-2)\ldots\ldots 1 \quad (n)\ \ldots\ldots(2n-3)}$$

$$= \frac{(k_1-k_2+1)(k_1+k_2+2n-1)}{2n-1} \times \frac{(k_1+2n-2)\ldots(k_1+2)}{(2n-2)!} \times \frac{(k_2+1)\ldots(k_2+2n-3)}{(2n-3)!}$$

$$= \frac{(k_1-k_2+1)(k_1+k_2+2n-1)}{(2n-1)\quad(2n-2)} \binom{k_1+2n-2}{2n-3} \binom{k_2+2n-3}{2n-3}$$

La dimension de la représentation de $Sp(n,\mathbb{C})$ de poids dominant

$a(\omega_1+\omega_2)$, $a \geq 1$ est donc $\dfrac{2a+2n-1}{(2n-1)(2n-2)}\dbinom{a+2n-2}{2n-3}\dbinom{a+2n-3}{2n-3}$

$(a+2)\omega_1+\omega_2$, $a \geq 1$ est donc $\dfrac{3(2a+2n+1)}{(2n-1)(2n-2)}\dbinom{a+2n}{2n-3}\dbinom{a+2n-3}{2n-3}$

La dimension d'une représentation irréductible de $Sp(n,\mathbb{C})$ de poids dominant $k_1\omega_1+k_2\omega_2+k_3\omega_3$, $k_1 \geq k_2 \geq k_3 \geq 0$ est :

$$\prod_{1\leq i<j\leq n}\frac{(k_1\omega_1+k_2\omega_2+k_3\omega_3+\delta\,|\,\omega_i+\omega_j)(k_1\omega_1+k_2\omega_2+k_3\omega_3+\delta\,|\,\omega_i-\omega_j)}{(\delta\,|\,\omega_i+\omega_j)\qquad(\delta\,|\,\omega_i-\omega_j)}\prod_{1\leq i\leq n}\frac{(k_1\omega_1+k_2\omega_2+k_3\omega_3+\delta\,|\,2\omega_i)}{(\delta\,|\,2\omega_i)}$$

$$=\prod_{4\leq j}\frac{(k_1+n+n+1-j)\,[(k_1+n)-(n+1-j)]}{(n+n+1-j)\ (n-(n+1-j))}\quad\frac{(k_1+n+k_2+n-1)(k_1+n-k_2-n+1)}{(2n-1)}$$

$$\times\frac{(k_1+n+k_3+n-2)(k_1+n-k_3-n+2)}{(2n-2)2}\quad\frac{2(k_1+n)}{2n}\times\frac{2(k_2+n-1)}{2(n-1)}\times\frac{2(k_3+n-2)}{2(n-2)}$$

$$\times\frac{(k_2+n-1+k_3+n-2)(k_2+n-1-k_3-n+2)}{2n-3}\prod_{j\geq 4}\frac{(k_2+n-1+n+1-j)\,[(k_2+n-1)-(n+1-j)]}{(2n-j)(j-2)}$$

$$\times\prod_{j\geq 4}\frac{(k_3+n-2+n+1-j)(k_3+n-2-n-1+j)}{(n-2+n+1-j)(n-2-n-1+j)}$$

$$=\prod_{n\geq j\geq 4}\frac{(k_1+2n+1-j)(k_1+j-1)(k_2+2n-j)(k_2+j-2)(k_3+2n-j-1)(k_3+j-3)}{(2n+1-j)(j-1)(2n-j)(j-2)(2n-j-1)(j-3)}$$

$$\times\frac{(k_1+k_2+2n-1)(k_1-k_2+1)(k_1+k_3+2n-2)(k_1-k_3+2)(k_1+n)(k_2+n-1)(k_3+n-2)}{2(2n-1)(2n-2)n(n-1)(n-2)(2n-3)}$$

$$\times\ (k_2+k_3+2n-3)(k_2-k_3+1)$$

$$=\frac{(k_1+2n-3)\ldots(k_1+3)}{(2n-3)\ldots3\times 2}\times\frac{(k_2+2n-4)\ldots(k_2+2)}{(2n-4)\ldots2}\times\frac{(k_3+2n-5)\ldots(k_3+1)}{(2n-5)\ldots1}$$

$$\times\frac{(k_1+k_2+2n-1)(k_1-k_2+1)(k_1+k_3+2n-2)(k_1-k_3+2)(k_2+k_3+2n-3)(k_2-k_3+1)}{(2n-1)(2n-2)(2n-3)}$$

$$=\binom{k_1+2n-3}{2n-5}\frac{1}{(2n-3)(2n-4)}\times\binom{k_2+2n-4}{2n-5}\frac{1}{2n-4}\binom{k_3+2n-5}{2n-5}$$

$$\times \frac{(k_1+k_2+2n-1)(k_1-k_2+1)(k_1+k_3+2n-2)(k_1-k_3+2)(k_2+k_3+2n-3)(k_2-k_3+1)}{(2n-1)(2n-2)(2n-3)}$$

$$= \frac{(k_1-k_2+1)(k_1-k_3+2)(k_2-k_3+1)(k_1+k_2+2n-1)(k_1+k_3+2n-2)(k_2+k_3+2n-3)}{(2n-1)(2n-2)(2n-3)^2\ (2n-4)^2}$$

$$\times \binom{k_1+2n-3}{2n-5}\binom{k_2+2n-4}{2n-5}\binom{k_3+2n-5}{2n-5}$$

2.1.2. Théorème 1 2. Le spectre du de Rham-Hodge sur les formes de degré 1 de l'espace projectif quaternionique est formé de :

. $\frac{a(a+2n-1)}{4(n+1)}$ de multiplicité $\frac{2a+2n-1}{(2n-1)(2n-2)}\binom{a+2n-2}{2n-3}\binom{a+2n-3}{2n-3}$

. $\frac{a^2+a(2n+1)+2n+2}{4(n+1)}$ de multiplicité $\frac{3(2a+2n+1)}{(2n-1)(2n-2)}\binom{a+2n}{2n-3}\binom{a+2n-3}{2n-3}$

. $\frac{2a^2+4an+4n-3}{8(n+1)}$ de multiplicité

$\frac{a(a+2)(a+n)(a+2n)(a+2n-2)}{(2n-1)(n-2)(2n-3)^2(n-1)}\binom{a+2n-2}{2n-5}\binom{a+2n-4}{2n-5}$, où a est un entier non nul

Preuve Pour :

$$\Lambda_\rho = a\omega_1 + a\omega_2$$
$$\Lambda_\rho = (a+2)\omega_1 + a\omega_2$$
$$\Lambda_\rho = (a+1)\omega_1 + a\omega_2 + \omega_3$$

calculons

$$(\Lambda_\rho + 2\delta_G | \Lambda_\rho)$$

$$(a\omega_1 + a\omega_2 + 2\sum_{i=1}^{n}(n+1-i)\omega_i \,|\, a\omega_1 + a\omega_2)$$

$$= \frac{1}{8(n+1)}\ ((a+2n)a+(a+2n-2)a) = \frac{a(a+2n-1)}{4(n+1)}$$

$$((a+2)\omega_1 + a\omega_2 + 2\sum_{i=1}^{n}(n+1-i)\omega_i \,|\, (a+2)\omega_1 + a\omega_2)$$

$$= \frac{1}{8(n+1)}\ ((a+2+2n)(a+2)+(a+2n-2)a)$$

$= \frac{1}{8(n+1)}\ (2a^2+a(4n+2)+4n+4) = \frac{1}{4(n+1)}\ (a^2+a(2n+1)+2n+2)$

$((a+1)\omega_1+a\omega_2+\omega_3+2\sum_{i=1}^{n}(n+1-i)\omega_i\,/\,(a+1)\omega_1+a\omega_2+\omega_3)$

$= \frac{1}{8(n+1)}\ ((a+1)(a+1+2n)+a(a+2n-2)+2n-4)$

$= \frac{1}{8(n+1)}\ (2a^2+4an+4n-3)$

2.2. Formes de degré 2.

2.2.1. Proposition 1.3. Les représentations irréductibles de $Sp(n,\mathbb{C})$ qui, restreintes à $Sp(1,\mathbb{C})\times Sp(n-1,\mathbb{C})$, contiennent la représentation irréductible de poids dominant $2\omega_1+\omega_2+\omega_3$, ont pour poids dominants respectifs :

$(a+2)\omega_1+a\omega_2 \quad a\in\mathbb{N}^*$

$(a+1)\omega_1+a\omega_2+\omega_3 \quad a\in\mathbb{N}^*$

$(a+2)\omega_1+a\omega_2+\omega_3+\omega_4 \quad a\in\mathbb{N}^*$

et le poids $2\omega_1+\omega_2+\omega_3$ y a pour multiplicité 1.

Leurs dimensions sont respectivement :

$$\frac{3(2a+2n+1)}{(2n-1)(2n-2)}\binom{a+2n}{2n-3}\binom{a+2n-3}{2n-3}$$

$$\frac{a(a+2)(a+n)(a+2n)(a+2n-2)}{(2n-1)(n-2)(2n-3)^2(n-1)}\binom{a+2n-2}{2n-5}\binom{a+2n-4}{2n-5}$$

$$\frac{3a(a+1)(a+3)(a+4)(a+2n-3)(a+2n-2)(a+2n)(a+2n+1)(2a+2n+1)}{2(2n-1)(2n-2)(2n-3)(2n-4)^2(2n-5)^2(2n-6)}\binom{a+2n-2}{2n-7}\binom{a+2n-5}{2n-7}$$

Preuve : Une représentation irréductible de $Sp(n,\mathbb{C})$ de poids dominant $a_1\omega_1+a_2\omega_2+a_3\omega_3+\ldots+a_n\omega_n$ contient lorsqu'on la restreint à $Sp(1,\mathbb{C})\times Sp(n-1,\mathbb{C})$ la représentation irréductible de poids dominant $2\omega_1+\omega_2+\omega_3$ si :

$A_1 = a_1 - \max(a_2,1) \geq 0$

$A_2 = \min(a_2,1) - \max(a_3,1) \geq 0$

$A_3 = \min(a_3,1) - \max(a_4,0) \geq 0$

$A_i = \min(a_i, 0) - \max(a_{i+1}, 0) \geq 0 \qquad 4 \leq i \leq n-1$

et $\sum_{i=1}^{n} a_i + 4 \in 2\mathbb{Z}$.

a) $a_2 \geq 1 \longrightarrow a_1 \geq a_2 \geq 1$ et $1-\max(1,a_3) \geq 0$, $a_i = 0 \quad i \geq 4$ donc $a_3 > 1$ est impossible d'où ; $a_3 \leq 1$ et $a_3 \geq a_4$;

si $a_3 = 1$ on a $1-a_4 \geq 0$ donc $a_4 = 0$ ou $a_4 = 1$

si $a_3 = 0$ on a $a_4 = 0$;

la condition $\sum_{i=1}^{n} a_i \in 2\mathbb{Z}$ donne les poids de la forme

$(a+2p)\omega_1 + a\omega_2 \qquad a \in \mathbb{N}^* \qquad p \in \mathbb{N}$

$(a+2p+1)\omega_1 + a\omega_2 + \omega_3 \qquad a \in \mathbb{N}^* \qquad p \in \mathbb{N}$

$(a+2p)\omega_1 + a\omega_2 + \omega_3 + \omega_4 \qquad a \in \mathbb{N}^* \qquad p \in \mathbb{N}$

b) $a_2 = 0 \qquad a_1 - 1 \geq 0 \qquad o-\max(1,a_3) \geq 0$ est impossible.

Il nous reste à trouver la multiplicité de la représentation irréductible de poids $2\omega_1 + \omega_2 + \omega_3$ dans la restriction de ces représentations à $Sp(1,\mathbb{C}) \times Sp(n-1,\mathbb{C})$.

c) $(a+2p)\omega_1 + a\omega_2 \qquad a \in \mathbb{N}^* \qquad p \in \mathbb{N}$.

La multiplicité du poids $2\omega_1 + \omega_2 + \omega_3$ vaut

$$F_{n-1}(\tfrac{1}{2}(2-2p) \ ; \ 0,\ldots0) - F_{n-1}(\tfrac{1}{2}(-2-2p)-1 \ ; \ 0,0\ldots0)$$

$$= F_{n-1}(1-p \ ; \ 0,\ldots0) - F_{n-1}(-2-p \ ; \ 0,\ldots0)$$

Or $F_k(s \ ; \ 0,0\ldots0) = 0$ si $s \neq 0$ et 1 si $s=0$

$F_k(s \ ; \ t_1,t_2,\ldots t_k) = 0$ si $s_k < 0 \qquad F_k(s \ ; \ t_1,t_2,\ldots t_k) = 0$ si $s < \sum_{i=1}^{k} t_i$

$F_k(0 \ ; \ t_1, t_2,\ldots t_k) = 0$ si $\sum_{i=1}^{k} t_i \neq 0$;

d'où

$$F_{n-1}(1-p \ ; \ 0,\ldots0) - F_{n-1}(-2-p \ ; \ 0,\ldots0) = \begin{cases} 1 & \text{si } p=1 \\ 0 & \text{si } p\neq 1 \end{cases}$$

Seuls les poids de la forme $(a+2)\omega_1 + a\omega_2 \quad a \in \mathbb{N}^*$ interviennent et la multiplicité du poids $2\omega_1 + \omega_2 + \omega_3$ est alors 1.

Dimension de cette représentation : elle a été calculée dans la preuve de la proposition 10 et vaut

$$\frac{3(2a+2n+1)}{(2n-1)(2n-2)} \binom{a+2n}{2n-3} \binom{a+2n-3}{2n-3}$$

d) $(a+2p+1)\omega_1+a\omega_2+\omega_3 \qquad a\in\mathbb{N}^* \qquad p\in\mathbb{N}$

$$F_{n-1}(\tfrac{1}{2}(2-2p-1+1) \ ; \ 1, 0,\ldots 0) - F_{n-1}(\tfrac{1}{2}(-2p-1-2+1)-1 \ ; \ 1, 0\ldots 0)$$

$$= F_{n-1}(1-p \ ; \ 1, 0\ldots 0) - F_{n-1}(-p-2 \ ; \ 1, 0\ldots 0)$$

$$= F_{n-1}(1-p \ ; \ 1, 0\ldots 0) = \begin{cases} 1 \text{ si } p=0 \\ 0 \text{ si } p\neq 0 \end{cases}$$

Seuls les poids de la forme $(a+1)\omega_1+a\omega_2+\omega_3$, $a\in\mathbb{N}^*$ sont tels que la multiplicité de $2\omega_1+\omega_2+\omega_3$ est non nulle et vaut 1.

La dimension de ces représentations a déjà été calculée (proposition 10) et vaut :

$$\frac{a(a+2)(2a+2n)(a+2n)(a+2n-2)}{(2n-1)^2(2n-2)(2n-3)(n-2)(n+1)(n+2)} \binom{a+2n-2}{2n-5} \binom{a+2n-4}{2n-5}$$

e) $(a+2p)\omega_1+a\omega_2+\omega_3+\omega_4 \qquad a\in\mathbb{N}^* \qquad p\in\mathbb{N}$

la multiplicité du poids $2\omega_1+\omega_2+\omega_3$ vaut

$$F_{n-1}(\tfrac{1}{2}(+2-2p) \ ; \ 0,\ldots 0) - F_{n-1}(\tfrac{1}{2}(-2-2p)-1 \ ; \ 0,\ldots 0)$$

$$= F_{n-1}(1-p \ ; \ 0,\ldots 0) - F_{n-1}(-2-p \ ; \ 0,\ldots 0)$$

$$= \begin{cases} 0 \text{ si } p\neq 1 \\ 1 \text{ si } p=1 \ . \end{cases}$$

Seuls les poids de la forme $(a+2)\omega_1+a\omega_2+\omega_3+\omega_4$, $a\in\mathbb{N}^*$, $p\in\mathbb{N}$ conviennent et la multiplicité de $2\omega_1+\omega_2+\omega_3$ est alors 1. La dimension d'une telle représentation est :

$$\prod_{1\leq i<j\leq n} \frac{((a+2)\omega_1+a\omega_2+\omega_3+\omega_4+\delta|\omega_i-\omega_j)((a+2)\omega_1+a\omega_2+\omega_3+\omega_4+\delta|\omega_i+\omega_j)}{(\delta|\omega_i-\omega_j)(\delta|\omega_i+\omega_j)}$$

$$\times \prod_{1\leq i\leq n} \frac{((a+2)\omega_1+a\omega_2+\omega_3+\omega_4+\delta|2\omega_i)}{(\delta|2\omega_i)}$$

$$= \frac{[n+a+2-(n+a-1)]\,[n+a+2-(n-1)]\,[n+a+2-(n-2)]\,\prod_{5\leq j}[(n+a+2)-(n+1-j)]}{\prod_{2\leq j\leq n}(n-(n-j+1))}$$

$$\times\ \frac{[n+a+2+n+a-1]\,[n+a+2+n-1]\,[n+a+2+n-2]\,\prod_{j\geq 5}[(n+a+2+n+1-j)]}{\prod_{2\leq j}(n+n-j+1)}\ \times\ \frac{2(n+a+2)}{2n}$$

$$\times\ \frac{[(n+a-1)-(n-1]\,[(n+a-1)-(n-2)]\,\prod_{j\geq 5}[(n+a-1)-(n+1-j)]}{\prod_{3\leq j}[(n-1)-(n-j+1)]}\ \times\ \frac{2(n+a-1)}{2(n-1)}$$

$$\times\ \frac{(n+a-1+n-1)(n+a-1+n-2)\,\prod_{j\geq 5}(n+a-1+n+1-j)}{\prod_{3\leq j}(n-1+n-j+1)}\ \times\ \frac{2(n-1)}{2(n-2)}$$

$$\times\ \frac{\{(n-1)-(n-2)]\,\prod_{j\geq 5}[(n-1)-(n-1-j)]\,[(n-1)+(n-2)]\,\prod_{j\geq 5}(n-1+n-1-j)}{\prod_{j\geq 4}(n-2-(n-j+1))(n-2+(n-j+1))}$$

$$\times\ \prod_{j\geq 5}\frac{[(n-2)-(n-j+1)]\,[(n-2)+n-j-1]}{[(n-3)-(n-j+1)]\,[(n-3)+n-j-1]}\ \frac{2(n-2)}{2(n-3)}\ =$$

$$= \frac{3a(a+1)(a+3)(a+4)(a+2n-3)(a+2n-2)(a+2n+1)(2a+2n+1)}{2(n-1)(2n-1)(2n-2)(2n-3)(2n-4)^2(2n-5)^2(2n-6)}\binom{a+2n-2}{2n-7}\binom{a+2n-5}{2n-7}$$

<u>Proposition</u> 1.4. <u>Les représentations irréductibles de</u> $Sp(n,\mathbb{C})$ <u>qui, restreintes à</u> $Sp(1,\mathbb{C})\times Sp(n-1,\mathbb{C})$, <u>contiennent la représentation irréductible de poids dominant</u> $2\omega_1$ <u>ont pour poids dominants</u> :

$$(a+2)\omega_2+a\omega_2 \qquad a\in\mathbb{N}\,.$$

<u>La multiplicité de la représentation de poids dominant</u> $2\omega_1$ <u>dans leur restriction à</u> $Sp(1,\mathbb{C})\times Sp(n-1,\mathbb{C})$ <u>est</u> 1 <u>et leur dimension vaut</u> :

$$\frac{3(2a+2n-1)}{(2n-1)(2n-2)}\ \binom{a+2n}{2n-3}\binom{a+2n-3}{2n-3}$$

<u>Preuve</u> : Puisque :

$b_1 = 2 \quad b_i = 0 \quad i\geq 2$, on doit avoir :

$$A_1 = a_1-\max(a_2,0) = a_1-a_2\geq 0$$

$A_i = \min(a_i,b_i)-\max(a_{i+1},b_{i+1}) = 0-a_{i+1} \geq 0$, d'où $a_i = 0 \quad i \geq 3$;

donc des poids de la forme :

$$a_1\omega_1+a_2\omega_2 \quad \text{avec} \quad a_1-a_2 \geq 0$$

et $\quad a_1+a_2+2 \in 2\mathbb{Z}$ donc $a_1=a_2+2p \qquad p \in \mathbb{N}$.

Recherchons la multiplicité de la rerpésentation irréductible de poids dominant $2\omega_1$ dans les restrictions à $Sp(1,\mathbb{C}) \times Sp(n-1,\mathbb{C})$.

$$F_{n-1}(\tfrac{1}{2}(2-2p) \ ; 0,\ldots 0) - F_{n-1}(\tfrac{1}{2}(-2-2p)-1 \ ; 0,\ldots 0)$$

$$= F_{n-1}(1-p \ ; 0,\ldots 0) = \begin{cases} 0 & \text{si} \quad p \neq 1 \\ 1 & \text{si} \quad p=1 \end{cases} ;$$

donc les poids convenables sont de la forme :

$$(a+2)\omega_1+a\omega_2 \qquad a \in \mathbb{N}.$$

La multiplicité trouvée est 1. La dimension a déjà été calculée.

<u>Proposition</u> 15. <u>Les repésentations irréductibles de</u> $Sp(n,\mathbb{C})$ <u>qui restreintes à</u> $Sp(1,\mathbb{C}) \times Sp(n-1,\mathbb{C})$ <u>contiennent la représentation irréductible de poids dominant</u> $2\omega_2$ <u>ont pour poids dominants</u> :

$2\omega_1$

$2\omega_1+\omega_2+\omega_3$

$a\omega_1+a\omega_2+2\omega_3 \qquad a \in \mathbb{N}-\{0,1\}$.

<u>La multiplicité de la représentation de poids dominant</u> $2\omega_2$ <u>dans leurs restrictions est</u> 1 <u>et leurs dimensions respectives sont</u> :

$n(2n+1)$

$\dfrac{(n+1)(2n+1)(2n-1)(n-2)}{2}$

$\dfrac{a(a-1)(2a+2n-1)(a+2n)(a+2n-1)}{2(2n-1)(2n-2)(2n-3)(2n-4)} \dbinom{a+2n-3}{2n-5} \dbinom{a+2n-4}{2n-5}$

<u>Preuve</u> : On doit avoir pour une représentation de $Sp(n,\mathbb{C})$ de poids dominant $a_1\omega_1+a_2\omega_2+\ldots+a_n\omega_n$ répondant au problème :

$A_1 = a_1-\max(a_2,2) \geq 0$

$A_2 = \min(a_2,2)-a_3 \geq 0$

$i \geq 3$ $A_i = \min(a_i,0) - \max(a_{i+1},0) \geq 0$, d'où $a_i = 0$ pour $i \geq 4$.

Deux cas se présentent : $a_2 < 2$ et $a_2 \geq 2$

a) $a_2 < 2$ $A_1 = a_1 - 2 \geq 0$

$A_2 = a_2 - a_3 \geq 0$ et alors $a_2 \geq a_3 \geq 0$;

d'où $a_2 = 1$ et $a_3 = 1$ ou $a_3 = 0$

$a_2 = 0$ et $a_3 = 0$

Cas α) si $a_2 = 0$ on a $a_3 = 0$ donc $a_1 \geq 2$ et $\sum_{i=1}^{n} (a_i + b_i) \in 2\mathbb{Z}$

donc $a_1 = 2p$, $p \in \mathbb{N}^*$; et le poids est de la forme : $2p\omega_1$.

Calculons la multiplicité de $2\omega_2$ dans la restriction.

$$F_{n-1}(\tfrac{1}{2}(-(2p-2) ; 0,\ldots 0) - F_{n-1}(\tfrac{1}{2}(-(2p-2))-1 ; 0,\ldots 0)$$

$$= F_{n-1}(-p+1 ; 0,\ldots 0) - F_{n-1}(-p ; 0,\ldots 0)$$

or $p \in \mathbb{N}^*$; donc cette multiplicité est nulle sauf pour $p=1$ où elle vaut 1.

La représentation trouvée est de poids dominant $\boxed{2\omega_1.}$

Cas β) $a_2 = 1$ d'où $a_3 = 0$ ou 1

Si $a_3 = 0$, on doit avoir : $a_1 + 1 \in 2\mathbb{Z}$ d'où $a_1 = 2p+1$ et $a_1 \geq 2$ donc $p \in \mathbb{N}^*$. Le poids est de la forme :

$$(2p+1)\omega_1 + \omega_2$$

La multiplicité cherchée est :

$$F_{n-1}(\tfrac{1}{2}(-(2p+1-2)+1) ; 1, 0,\ldots 0) - F_{n-1}(\tfrac{1}{2}(-(2p+1-2)+1)-1 ; 1, 0\ldots 0)$$

$$= F_{n-1}(-p ; 1, 0,\ldots 0) - F_{n-1}(-p-1 ; 1, 0\ldots 0) = 0$$

Si $a_3 = 1$; on doit avoir $a_1 + 2 \in 2\mathbb{Z}$ d'où $a_1 = 2p$ avec $p \in \mathbb{N}^*$.

Le poids est de la forme :

$$2p\omega_1 + \omega_2 + \omega_3 .$$

La multiplicité cherchée est :

$$F_{n-1}(\tfrac{1}{2}(-(2p-2)) ; 0, 0\ldots 0) - F_{n-1}(\tfrac{1}{2}(-(2p-2)-1 ; 0,\ldots 0)$$

$$= F_{n-1}(-p+1 ; 0,\ldots 0) - F_{n-1}(-p-1 ; 0,\ldots 0)$$

est non nulle pour $p = 1$ et vaut alors 1.

La représentation trouvée est de poids dominant $\boxed{2\omega_1+\omega_2+\omega_3}$

b) $a_2 \geq 2$ $\quad A_1 = a_1 - a_2 \geq 0$

$A_2 = 2 - a_3 \geq 0 \qquad 0 \leq a_3 \leq 2.$

On a donc à étudier les cas possibles pour a_3 : 0,1,2.

Cas α) $a_3 = 0$; on doit avoir $a_1+a_2 \in 2\mathbb{Z}$ d'où $a_1 = a_2+2p$, $p \in \mathbb{N}$,

$A_1 = a_1 - a_2 \quad A_2 = 2 \quad A_i = 0 \quad i \geq 2.$

La multiplicité cherchée vaut :

$$F_{n-1}(\tfrac{1}{2}(-2p+2)\ ;\ 2,\ 0\ldots 0) - F_{n-1}(\tfrac{1}{2}(-2p+2)-1\ ;\ 2,\ 0\ldots 0)$$

$= F_{n-1}(-p+1\ ;\ 2,\ 0\ldots 0) - F_{n-1}(-p\ ;\ 2,\ 0\ldots 0) = 0$ car $-p+1 \leq 1 < 2$.

Cas β) $a_3 = 1$; on a alors $a_1 = a_2+2p+1$, $p \in \mathbb{N}$, $a_2 \geq 2$.

La multiplicité cherchée vaut :

$$F_{n-1}(\tfrac{1}{2}(-2p-1+1)\ ;\ 1,\ 0\ldots 0) - F_{n-1}(\tfrac{1}{2}(-2p-1+1)-1\ ;\ 1,\ 0\ldots 0)$$

$= F_{n-1}(-p\ ;\ 1,\ 0\ldots 0) - F_{n-1}(-p-1\ ;\ 1,\ 0\ldots 0)$

$= 0$

Cas γ) $a_3 = 2$; on a alors $a_1 = a_2+2p \quad p \in \mathbb{N} \quad a_2 \geq 2$.

La multiplicité cherchée vaut, car $A_1=2p$, $A_2=0$, $A_i=0 \quad i \geq 3$,

$$F_{n-1}(\tfrac{1}{2}(-2p)\ ;\ 0,\ldots 0) - F_{n-1}(-p-1\ ;\ 0,\ldots 0)$$

et est nulle sauf pour $p=0$ auquel cas elle vaut 1.

On a donc comme poids : $\boxed{a\omega_1+a\omega_2+2\omega_3}$ avec $a \in \mathbb{N}-\{0,1\}$.

Les dimensions des représentations trouvées ont été déjà toutes calculées précédemment.

2.2.2. Théorème 16. Les représentations irréductibles de $Sp(n,\mathbb{C})$ qui restreintes à $Sp(1,\mathbb{C}) \times Sp(n-1,\mathbb{C})$ contiennent l'une des composantes irréductibles de $\Lambda^2\mu$ ont pour poids respectifs :

$(a+2)\omega_1+a\omega_2 \qquad a \in \mathbb{N}$

$a\omega_1+a\omega_2+2\omega_3 \qquad a \in \mathbb{N}-\{0,1\}$

$(a+1)\omega_1+a\omega_2+\omega_3 \qquad a \in \mathbb{N}-\{0\}$

$$(a+2)\omega_1+a\omega_2+\omega_3+\omega_4 \qquad a\in\mathbb{N}-\{0\}$$

et chacune d'entre elles ne contient dans sa restriction que l'une des composantes irréductibles de $\Lambda^2\mu$ excepté les représentations de poids dominants $(a+2)\omega_1+a\omega_2$ qui en contiennent 2 pour a entier naturel.

Preuve : C'est le condensé des trois propositions de 2.2.1.

2.2.3. Théorème 17.

Le spectre du de Rham-Hodge sur les formes de degré 2 de l'espace projectif quaternionique est formé de :

(i) $\frac{1}{4(n+1)}\left[a^2+a(2n+1)+2(2n+1)\right]$ avec $a\in\mathbb{N}$,

de multiplicité $\frac{6(2a+2n+1)}{(2n-1)(2n-2)}\binom{a+2n}{2n-3}\binom{a+2n-3}{2n-3}$

(ii) $\frac{1}{4(n+1)}\left[a^2+a(2n-1)+2n-2\right]$ avec $a\in\mathbb{N}-\{0,1\}$

de multiplicité $\frac{a(a-1)(2a+2n-1)(a+2n)(a+2n-1)}{2(2n-1)(2n-2)(2n-3)(2n-4)}\binom{a+2n-3}{2n-5}\binom{a+2n-4}{2n-5}$

(iii) $\frac{1}{4(n+1)}\left[a^2+2na+2n-1\right]$ avec $a\in\mathbb{N}-\{0\}$

de multiplicité $\frac{a(a+2)(a+n)(a+2n)(a+2n-2)}{(n-1)(n-2)(2n-1)(2n-3)^2}\binom{a+2n-2}{2n-5}\binom{a+2n-4}{2n-5}$

(iv) $\frac{1}{4(n+1)}\left[a^2+2na+4n-2\right]$ avec $a\in\mathbb{N}-\{0\}$ de multiplicité

$$\frac{3a(a+1)(a+3)(a+4)(a+2n-3)(a+2n-2)(a+2n)(a+2n+1)(2a+2n+1)}{2(2n-1)(2n-2)(2n-3)(2n-4)^2(2n-5)^2(2n-6)}\binom{a+2n-2}{2n-7}\binom{a+2n-5}{2n-7}$$

Preuve : Ce sont les mêmes calculs que pour la démonstration du théorème 12 mais appliquées aux résultats du théorème 16.

Bibliographie :

[1] J.E. Humphreys. Introduction to Lie Algebras and representation theory. Springer Verlag, 1972.

[2] A. Ikeda et Y. Taniguchi. Spectra and eigenforms of the Laplacian on S^n and $P^n(\mathbb{C})$. Osaka Journal of Math 15 (1978) pages 515-546.

[3] N. Jacobson. Lie algebras. Interscience, New-York, 1962.

[4] A. Lévy-Bruhl-Laperrière. Spectre du de Rham Hodge sur l'espace projectif complexe. Séminaire d'algèbre Paul Dubreil. Lecture Notes 641, pages163-188.

[5] A. Lévy-Bruhl-Laperrière. Spectre du de Rham Hodge sur les formes de degré supérieur à 1 sur l'espace projectif complexe. Bulletin des Sciences Mathématiques, 104, 1980, pages 135-143.

[6] J.Y. Lepowsky. Representations of semi-simple Lie groups and an enveloping algebra decomposition. P.H.D. 1970, résumé dans:Bulletin of the American Math. Society, 1971, volume 77, pages 601-605.

[7] Y. Matsushima et S. Murakami. On vector bundle valued harmonic forms and automorphic forms on symmetric riemannian manifolds. Annals of Maths, volume 79,n°2, 1963, pages 365-416.

GROUPE DE LIE p-ADIQUE, IMMEUBLE ET COHOMOLOGIE

G. BESSON

Les relations qui existent entre la cohomologie d'un groupe de Lie réel et celle de son algèbre de Lie sont connues depuis longtemps. Pour un groupe de Lie réel G d'algèbre de Lie $\mathfrak{g}$, un sous-groupe compact maximal K de G d'algèbre $\mathfrak{k}$ et un sous-groupe Γ discret de G, Borel [1] a notamment défini et étudié un homomorphisme $j_\Gamma : H^*(\mathfrak{g}, \mathfrak{k} : R) \to H^*(\Gamma : R)$ (G agissant trivialement sur R).

Récemment, J.L. Dupont [4] a donné une interprétation explicite de cet homomorphisme : il associe à un cocycle relatif de $\mathfrak{g}$ modulo $\mathfrak{k}$ une forme harmonique G-invariante sur le quotient G/K et de là un cocycle sur Γ.

Lorsque l'on a cherché à obtenir dans le cas p-adique (ou plus généralement dans le cas d'un corps muni d'une valuation discrète et complet) des résultats analogues à ceux du cas réel, il s'est avéré que l'objet susceptible de remplacer l'espace symétrique G/K était l'immeuble de Bruhat-Tits T du groupe G. L'immeuble T est un complexe polysimplicial dont chaque cellule est un produit (fini) de simplexes, dont l'existence a été démontrée pour un groupe réductif G sur un corps muni d'une valuation discrète à caractéristique résiduelle différente de 2 ou à corps résiduel parfait [5].

Le but de cet article est de préciser dans ce cas une construction (semblable à celle de Dupont) qui associe à tout cocycle simplicial G-invariant sur T un cocycle sur Γ. A un complexe hypersimplicial, c'est-à-dire à un ensemble semi-simplicial dont chaque simplexe est un complexe polysimplicial (voir définition plus précise ci-dessous), on associera d'abord deux complexes doubles dont on démontra l'isomorphisme. Ce résultat sera ensuite utilisé dans la construction proprement dite.

I - LES DEUX COMPLEXES.

Définition - On appelle complexe hypersimplicial X la donnée d'une suite X_p de complexes polysimpliciaux et pour tout p la donnée de $p+1$ morphismes de complexes polysimpliciaux $\varepsilon_i : X_p \to X_{p-1}$ $(i=0, 1,\dots,p)$. Si Δ^p désigne le simplexe standard de dimension p, l'opérateur ε_i correspond à l'inclusion $\varepsilon^i : \Delta^{p-1} \to \Delta^p$ de la i-ème face.

Définition - On appelle n-cochaine φ sur X une suite de n-cochaines $\varphi^{(p)}$ sur le produit $\Delta^p \times X_p$ vérifiant pour tout p et tout $i=0,1,\dots,p$ les conditions :

$$(\varepsilon^i \times id)^*(\varphi^{(p)}) = (id \times \varepsilon_i)(\varphi^{(p-1)}) \qquad (I)$$

C'est en fait ce que l'on peut appeler par analogie avec [4] une n-cochaine de la réalisation $\|X\|$ de X en un complexe polysimplicial définie comme quotient de $\coprod_p \Delta^p \times X_p$ par les identifications : $\varepsilon^i(\tau) \times \xi \sim \tau \times \varepsilon_i(\xi)$ pour tout p, tout simplexe τ de Δ^{p-1}, toute cellule ξ de X_p et tout $i=0, 1,\dots,p$.

On notera $C^n(X)$ l'ensemble des n-cochaines sur X. On peut décomposer $C^n(X)$ en $\bigoplus_{k+\ell=n} C^{k\ell}(X)$ où $C^{k\ell}(X)$ est l'ensemble des $(k+\ell)$-cochaines nulles pour tout p en dehors des cellules obtenues comme produit d'un k-simplexe de Δ^p et d'une ℓ-cellule de X_p. Soient d_Δ et d_X les opérateurs cobords respectivement sur Δ^p et X_p ; le cobord sur $\Delta^p \times X_p$ est $d = d_\Delta \times id \pm id \times d_X$ et $(C^*(X),d)$ est donc le complexe total associé au complexe double $(C^{k\ell}(X)\ ;\ d_\Delta \times id,\ id \times d_X)$.

Considérons par ailleurs le complexe double $(\mathcal{C}^{k\ell}(X)\ ;\ \delta, d_X)$ où $\mathcal{C}^{k\ell}(X)$ est l'ensemble des ℓ-cochaines sur X_k et $\delta = \sum_i (-1)^i \varepsilon_i^*$.

Théorème - Les deux complexes doubles $(C^{k\ell}(X)\ ;\ d_\Delta \times id,\ id \times d_X)$ et $(\mathcal{C}^{k\ell}(X)\ ;\ \delta, d_X)$ sont isomorphes.

Démonstration. Définissons $\mathcal{J} : C^{k\ell}(X) \to \mathcal{C}^{k\ell}(X)$ par $\mathcal{J}(\varphi)(\xi) = \varphi^{(k)}(\Delta^k \times \xi)$ pour toute cellule ξ de dimension ℓ de X_k. On vérifie que $\mathcal{J}\circ(d_\Delta \times id) = \delta\circ\mathcal{J}$. En effet, si ξ est une ℓ-cellule de X_{k+1} et si $\varphi \in C^{k\ell}(X)$,

$$\begin{aligned}(\delta\circ\mathcal{J})(\varphi)(\xi) &= (\sum_i(-1)^i\, \varepsilon_i^*\, \mathcal{J}(\varphi))(\xi) = \sum_i(-1)^i\, \mathcal{J}(\varphi)(\varepsilon_i\xi) = \\ &= \sum_i(-1)^i\, \varphi^{(k)}(\Delta^k \times \varepsilon_i\xi)\end{aligned}$$

$$\begin{aligned}(\mathcal{J}\circ(d_\Delta \times id))(\varphi)(\xi) &= ((d_\Delta \times id)(\varphi))^{(k+1)}(\Delta^{k+1} \times \xi) = \\ &= \sum_i(-1)^i \varphi^{(k+1)}(\varepsilon_i\Delta^k \times \xi)\end{aligned}$$

L'égalité résulte alors de la condition (I).

Par ailleurs, $\mathfrak{J}$ commute clairement à d_X, c'est-à-dire $\mathfrak{J}\circ(\mathrm{id}\times d_X) = d_X\circ\mathfrak{J}$.

Avant de définir l'application inverse de $\mathfrak{J}$, introduisons quelques notations. Toute suite $I = (i_0,i_1,\dots,i_k)$, $0 \leqslant i_o \leqslant i_1 \leqslant \dots \leqslant i_k \leqslant p$, définit une inclusion naturelle que l'on note $\mu^I : \Delta^k \to \Delta^p$: si $(j_1,j_2,\dots,j_\ell)$ est la suite complémentaire de I dans $\{0,1,\dots,p\}$, $\mu^I = \varepsilon^{j_1}\varepsilon^{j_2}\dots\varepsilon^{j_\ell}$. On définit de même $\mu_I = \varepsilon_{j_1}\varepsilon_{j_2}\dots\varepsilon_{j_\ell}$. On note γ_I la k-cochaine sur Δ^p qui vaut 1 sur le simplexe $\mu^I(\Delta^k)$ et s'annule sur les autres k-simplexes. Enfin pour $I = (i_o,\dots,i_k)$ on pose $|I| = k$.

Définissons alors l'homomorphisme $\mathcal{E}: \mathcal{E}^{k\ell}(X) \to C^{k\ell}(X)$ par

$$\mathcal{E}(\omega)^{(p)} = 0 \quad \text{si} \quad p < k$$

$$\mathcal{E}(\omega)^{(p)} = \sum_{\substack{|I|=k \\ I\subset\{0,\dots,p\}}} \gamma_I \times \mu_I^{*}(\omega) \quad \text{si} \quad p \geqslant k.$$

Soient τ un k-simplexe de Δ^{p-1} et ξ une ℓ-cellule de X_p :

$$\mathcal{E}(\omega)^{(p)}(\varepsilon^i(\tau)\times\xi) = \sum_{\substack{|I|=k \\ I\subset\{0,\dots,p\}}} \gamma_I(\varepsilon^i(\tau))\omega(\mu_I(\xi))$$

$$= \sum_{\substack{|I|=k \\ I\subset\{0,\dots,p\} \\ I\not\ni i}} \gamma_I(\varepsilon^i(\tau))\omega(\mu_I(\xi))$$

$$\mathcal{E}(\omega)^{(p-1)}(\tau\times\varepsilon_i(\xi)) = \sum_{\substack{|I'|=k \\ I'\subset\{0,\dots,p\}}} \gamma_{I'}(\tau)\omega(\mu_{I'}\cdot\varepsilon_i(\xi)),$$

ce qui permet de vérifier la condition (I) pour $\mathcal{E}(\omega)$ (par exemple en choisissant la numérotation des sommets de telle sorte que $i=p$).

De plus, $\mathcal{E}(\omega)^{(k)} = \sum_{\substack{|I|=k \\ I\subset\{0,\dots,k\}}} \gamma_I \times \mu_I^{*}(\omega) = \gamma_{\{0,\dots,k\}}\times\omega$. Donc $\mathfrak{J}\circ\mathcal{E}(\omega) = \omega$.

Inversement, pour un k-simplexe τ de Δ^p et une ℓ-cellule ξ de X_p,

$$(\mathcal{E}\circ\mathfrak{J}(\varphi))^{(p)}(\tau\times\xi) = \sum_{\substack{|I|=k \\ I\subset\{0,\dots,p\}}} \gamma_I(\tau)\mathfrak{J}(\varphi)\ (\mu_I(\xi)).$$

$$= (\varphi)(\mu_{I_\tau}(\xi)) = \varphi^{(k)}(\Delta^k\times\mu_{I_\tau}(\xi))$$

où I_τ désigne l'ensemble d'indices correspondant aux sommets de τ. D'où, en utilisant la condition (I) :

$$(\mathcal{E}\circ\mathcal{J}(\varphi))^{(p)}(\tau\times\xi)=\varphi^{(p)}(\mu^{I}\tau(\Delta^k)\times\xi)=\varphi^{(p)}(\tau\times\xi).$$

II - L'HOMOMORPHISME j_Γ

Considérons maintenant un corps F muni d'une valuation discrète à corps résiduel parfait ou de caractéristique différente de 2. Soient G un groupe réductif sur F, T son immeuble de Bruhat-Tits et Γ un sous-groupe discret de G. Notons $N\Gamma$ (resp. $\overline{N\Gamma}$) l'ensemble semi-simplicial dont le p-simplexe est $N\Gamma(p)=\{(\gamma_1,\dots,\gamma_p)\in\Gamma^p\}$ (resp. $\overline{N\Gamma}(p)=\{(\bar\gamma_0,\bar\gamma_1,\dots,\bar\gamma_p)\in\Gamma^{p+1}\}$) et dont les morphismes "faces" sont définis par
$\varepsilon_i(\gamma_1,\dots,\gamma_p)=(\gamma_1,\dots,\gamma_{i-1},\gamma_i\gamma_{i+1},\gamma_{i+2},\dots,\gamma_p)$
(resp. $\varepsilon_i(\bar\gamma_0,\dots,\bar\gamma_p)=(\bar\gamma_0,\dots,\hat{\bar\gamma}_i,\dots,\bar\gamma_p)$) et notons $\pi:\overline{N\Gamma}\to N\Gamma$ la projection définie par $\pi(\bar\gamma_0,\dots,\bar\gamma_p)=(\bar\gamma_0\bar\gamma_1^{-1},\bar\gamma_1\bar\gamma_2^{-1},\dots,\bar\gamma_{p-1}\bar\gamma_p^{-1})$. C'est un fibré principal de groupe Γ qui a été étudié notamment par Segal [6]. A ce fibré principal, on associe le fibré (de fibre T), $\bar\pi:\overline{N\Gamma}\times_\Gamma T\to N\Gamma$ où $\overline{N\Gamma}\times_\Gamma T$ est le complexe hypersimplicial défini par $(\overline{N\Gamma}\times_\Gamma T)_p=\overline{N\Gamma}(p)\times_\Gamma T$.

Si φ est un q-cocycle Γ-invariant sur T à valeurs réelles, on en déduit un q-cocycle $\vec\varphi\in C^q(\overline{N\Gamma}\times_\Gamma T)$, la restriction $\vec\varphi^{(p)}$ de $\vec\varphi$ à $\Delta^p\times\overline{N\Gamma}(p)\times_\Gamma T$ étant simplement obtenue par image réciproque de φ dans $\Lambda^p\times\overline{N\Gamma}(p)\times T$ et passage au quotient en utilisant l'invariance de φ.

Dans la suite, toutes les cohomologies seront à valeurs réelles et on omettra de le préciser. Notons Z^{top} l'espace topologique sous-jacent au complexe polysimplicial Z.

Lemme 1 - On a les isomorphismes :

(i) $H(C^*(N\Gamma))\simeq H^*(\|N\Gamma\|^{top})$

(ii) $H(C^*(\overline{N\Gamma}\times_\Gamma T))\simeq H^*(\|\overline{N\Gamma}\times_\Gamma T^{top}\|)$.

Le premier point est clair car $H(C^*(N\Gamma))=H^*(\|N\Gamma\|)$ (cf. [7]). Pour (ii), il faut tout d'abord justifier l'écriture du second membre. L'action de Γ sur T se prolonge naturellement à T^{top} (chaque cellule de T étant un produit de simplexes, elle est naturellement paramétrée par les coordonnées barycentriques de ses facteurs ce qui permet de définir le prolongement par linéarité) ; cela donne un sens à $\overline{N\Gamma}\times_\Gamma T^{top}$. Par ailleurs, si $Y=\{Y_p,\varepsilon_i\}$ est un ensemble semi-simplicial dont les simplexes Y_p sont des espaces topologiques, on peut de manière analogue au cas des complexes hypersimpliciaux définir sa réalisation, notée encore $\|Y\|$, comme quotient de $\coprod(\Delta^p)^{top}\times Y_p$ par les relations $\varepsilon^i(\tau)\times\xi\sim\tau\times\varepsilon_i(\xi)$ pour tout p, tout point τ de $(\Delta^{p-1})^{top}$, tout point ξ

de Y_p et tout $i=0,\ldots,p$. Il est alors facile de voir que $\|\overline{N\Gamma} \times_\Gamma T^{top}\| = \|\overline{N\Gamma} \times_\Gamma T\|^{top}$ et le lemme en résulte puisque $H(C^*(\overline{N\Gamma} \times_\Gamma T)) = H^*(\|\overline{N\Gamma} \times_\Gamma T\|)$.

Lemme 2 - On a les isomorphismes :

$$\text{(i) } \mathfrak{J} : H(C^*(\overline{N\Gamma} \times_\Gamma T)) \xrightarrow{\sim} H(\mathcal{E}^*(\overline{N\Gamma} \times_\Gamma T))$$

$$\text{(ii) } \mathfrak{J} : H(C^*(N\Gamma)) \xrightarrow{\sim} H(\mathcal{E}^*(N\Gamma)).$$

Cela résulte de la première partie en prenant comme complexe hypersimplicial X respectivement $\overline{N\Gamma} \times_\Gamma T$ et $N\Gamma$.

Lemme 3 - On a les isomorphismes :

$$H(\mathcal{E}^*(\overline{N\Gamma} \times_\Gamma T)) \simeq H(\mathcal{E}^*(\overline{N\Gamma} \times_\Gamma T^{top})) \simeq H(\mathcal{E}^*(N\Gamma)).$$

Le premier découle à nouveau de résultats généraux (cf. [7]) et le second du fait que la fibre T^{top} du fibré $\overline{N\Gamma} \times_\Gamma T^{top} \to N\Gamma$ est contractile.

D'où le diagramme commutatif :

$$\begin{array}{ccccc}
\mathcal{E}^*(T^{top})^\Gamma & \to & H^*(\|\overline{N\Gamma} \times_\Gamma T^{top}\|) & \underset{\rho^*}{\xrightarrow{\sim}} & H^*(\|N\Gamma\|^{top}) \\
 & & \wr & & \wr \\
\mathcal{E}^*(T)^\Gamma & \to & H(C^*(\overline{N\Gamma} \times_\Gamma T)) & & H(C^*(N\Gamma)) \\
 & & \mathfrak{J}\downarrow\wr & & \mathfrak{J}\downarrow\wr \\
 & & H(\mathcal{E}^*(\overline{N\Gamma} \times_\Gamma T)) & \to & H(\mathcal{E}^*(N\Gamma)) = H^*(\Gamma) \\
 & & \wr & \nearrow & \\
 & & H(\mathcal{E}^*(\overline{N\Gamma} \times_\Gamma T^{top})) & &
\end{array}$$

et l'isomorphisme ρ^* peut être induit par une section ρ du fibré $\|\overline{N\Gamma} \times_\Gamma T^{top}\| \to \|N\Gamma\|^{top}$. Se donner la section ρ revient à se donner pour tout p une section $\rho^{(p)}$ de la projection $(\Delta^p)^{top} \times \overline{N\Gamma}(p) \times_\Gamma T^{top} \to (\Delta^p)^{top} \times N\Gamma(p)$ telle que le diagramme suivant commute pour tout $i=0,\ldots,p$:

$$\begin{array}{ccc}
(\Delta^{p-1})^{top} \times N\Gamma(p) & \xrightarrow{\rho^{(p)} \circ (\varepsilon^i \times id)} & (\Delta^p)^{top} \times \overline{N\Gamma}(p) \times_\Gamma T^{top} \\
\Big\downarrow id \times \varepsilon_i & & \Big\downarrow id \times \varepsilon_i \\
(\Delta^{p-1})^{top} \times N\Gamma(p-1) & \xrightarrow{(\varepsilon^i \times id) \circ \rho^{(p-1)}} & (\Delta^p)^{top} \times \overline{N\Gamma}(p-1) \times_\Gamma T^{top}
\end{array}$$

Il est alors équivalent, en identifiant $N\Gamma = \overline{N\Gamma}/\Gamma$ de construire une suite $r^{(p)}$ d'applications simpliciales Γ-équivariantes $r^{(p)} : (\Delta^p)^{top} \times \overline{N\Gamma}(p) \to T^{top}$ telles que

$$r^{(p)} \circ (\varepsilon^i \times id) = r^{(p-1)} \circ (id \times \varepsilon_i)$$

Si ψ est un cocycle simplicial Γ-équivariant de T^{top} et $\bar{\psi}$ son image réciproque sur $\|\overline{N\Gamma} \times_\Gamma T^{top}\|$, il est immédiat que $r^*(\psi)$ induit $\rho^*(\bar{\psi})$ par passage au quotient par Γ. Si en particulier la classe de $\bar{\psi}$ dans $H^*(\|\overline{N\Gamma} \times_\Gamma T^{top}\|)$ correspond à celle de $\bar{\varphi}$ dans $H(C^*(\overline{N\Gamma} \times_\Gamma T))$, leur image commune dans $H^*(\Gamma)$ est représentée par une q-cochaine notée $j_\Gamma(\varphi)$ et définie pour tout $\sigma = (\gamma_1,\ldots,\gamma_q) \in N\Gamma(q)$ par

$$j_\Gamma(\varphi)(\sigma) = r^*(\psi)(\Lambda^q \times \bar{\sigma}) = \psi(r(\Delta^q \times \bar{\sigma}))$$

où $\bar{\sigma}$ est un relèvement de σ dans $\overline{N\Gamma}(q)$, c'est-à-dire $\bar{\sigma} = (\bar{\gamma}_o,\bar{\gamma}_1,\ldots,\bar{\gamma}_q)$ et $\gamma_i = \bar{\gamma}_{i-1}\bar{\gamma}_i^{-1}$, $i=1,\ldots,q$.

Choisissons arbitrairement un sommet 0 de T et une contraction $g : [0,1] \times T^{top} \to T^{top}$ de T^{top} sur 0 le long des géodésiques ($g([0,1] \times \{x\})$ est la géodésique de x à 0). Si t est un point de $(\Delta^q)^{top}$ de coordonnées barycentriques $(t_o,t_1,\ldots,t_q)$, posons $s_i = t_i+t_{i+1}+\ldots+t_q$. On définit alors pour $\bar{\sigma} = (\bar{\gamma}_o,\ldots,\bar{\gamma}_q) \in \overline{N\Gamma}(q)$ et $\sigma = (\gamma_1,\ldots,\gamma_q) \in N\Gamma(q)$

$$r^{(q)}(t,\bar{\sigma}) = \bar{\gamma}_o^{-1} g_{s_1}(\gamma_1 g_{s_{2/s_1}}(\gamma_2 g_{s_{3/s_2}} \ldots g_{s_{q/s_{q-1}}}(\gamma_q)\ldots))$$

et $r(\Delta^q \times \bar{\sigma})$ est le simplexe de T^{top} dont les sommets sont $0,\gamma_1 0,\gamma_1\gamma_2 0,\ldots,\gamma_1\gamma_2\ldots\gamma_q 0$.

Il ne reste plus qu'à construire ψ . Pour cela, en adaptant une idée de Whitney [9] on va associer à toute n-cochaine φ sur T une n-forme différentielle $W\varphi = \eta$ qui sera de classe C^∞ sur le complémentaire du squelette de codimension 1 de T. Soit $\Theta = \Theta_1 \times\ldots\times \Theta_k$ une n-cellule produit des simplexes $\Theta_1,\ldots,\Theta_k$ de dimension respective $n_1,\ldots,n_k$. On définit d'abord la n-forme η_Θ dont le support est ét(Θ), c'est-à-dire la réunion des espaces topologiques fermés sous-jacents aux cellules de dimension maximale dont Θ est une face. Si $\Theta = \Theta_1\times \ldots \times \Theta_k \times \ldots \times\Theta_\ell$ est l'une d'elles, où les Θ_i sont numérotés de telle sorte que Θ_i soit une face de Θ_i pour $i=1,\ldots,k$, la cellule Θ s'identifie en fait à $\Theta_1 \times\ldots \times \Theta_k \times \ldots\times \Theta_\ell$ où, pour $i=k+1,\ldots,\ell$, les Θ_i sont des 0-simplexes. On pose alors

$$\eta_\Theta \big/ \Theta^{top} = \varphi(\Theta).\bigwedge_{i=1}^{\ell} \chi_i$$

où χ_i est la forme différentielle définie comme suit : en désignant par $\mu_{io},\ldots,\mu_{in_i}$ les coordonnées barycentriques des simplexes Θ_i, on définit sur Θ_i^{top}

$$\chi_i = n_i! \sum_{\lambda=0}^{n_i} (-1)^\lambda \mu_{i\lambda} d\mu_{io}\Lambda\ldots\Lambda\widehat{d\mu_{i\lambda}}\Lambda\ldots\Lambda d\mu_{in_i} \quad .$$

Enfin on pose $\eta = \Sigma\eta_\Theta$, la somme étant étendue à toutes les n-cellules Θ de T. C'est en tout point de T^{top} une somme finie.

Remarquons maintenant que les coordonnées $\mu_{i\lambda}$ sont des fonctions C^∞ sur l'adhérence de toute cellule de dimension maximale, ce qui donne un sens à la différentielle extérieure $d\eta_\Theta$ de η_Θ. Rappelons par ailleurs que l'on peut définir sur $\mathcal{C}^*(T)$ l'opérateur cobord de la manière suivante. Notons $\tilde{\Theta}$ la cochaine qui vaut 1 sur la cellule Θ et 0 ailleurs. Pour tout $i=1,\ldots,\ell$, soient $\Theta_i = [P_o,\ldots,P_{n_i}]$ et $\Theta_{iP} = [P,P_o,\ldots,P_{n_i}]$ une face quelconque du simplexe Θ_i contenant Θ_i : alors

$$d\tilde{\Theta} = \sum_{i=1}^{\ell} (-1)^{\varepsilon_i} \sum_{\Theta_{iP}} (\Theta_1\times\ldots\times\Theta_{i-1}\times\Theta_{iP}\times\Theta_{i+1}\times\ldots\times\Theta_\ell)^\sim$$

où $\varepsilon_i = \dim(\Theta_1\times\ldots\times\Theta_{i-1})$.

<u>Lemme</u> 4 - <u>L'opération</u> W <u>de Whitney commute avec le cobord</u> :

$$W(d\varphi) = d(W\varphi).$$

Puisque $\eta = W\varphi$ est en tout point une somme finie de η_Θ, il suffit de le montrer lorsque $\varphi = \tilde{\Theta}$ donc $\eta = \eta_\Theta$. Dans ce cas,

$$d\eta_\Theta = \sum_{i=1}^{\ell} (-1)^{\varepsilon_i} \chi_1\wedge\ldots\wedge\chi_{i-1}\wedge d\chi_i\wedge\chi_{i+1}\wedge\ldots\wedge\chi_\ell \quad \text{et}$$

$$W(d\tilde{\Theta}) = \sum_{i=1}^{\ell} (-1)^{\varepsilon_i} \chi_1\wedge\ldots\wedge\chi_{i-1}\wedge\Big(\sum_{\Theta_{iP}}\chi_{iP}\Big)\wedge\chi_{i+1}\wedge\ldots\wedge\chi_\ell .$$

Le lemme résultera donc de l'égalité $d\chi_i = \sum_{\Theta_{iP}} \chi_{iP}$. Or, $d\chi_i = (n_i+1)!\,d\mu_{io}\wedge\ldots\wedge d\mu_{in_i}$. D'autre part,

$$\frac{1}{(n_i+1)!}\sum_{\Theta_{iP}}\chi_{iP} = \sum_{\Theta_{iP}}\Big(\mu_P d\mu_{io}\wedge\ldots\wedge d\mu_{in_i} + \sum_{\lambda=0}^{n_i}(-1)^{\lambda+1}\mu_{i\lambda}d\mu_P\wedge\ldots\wedge\widehat{d\mu_{i\lambda}}\wedge\ldots\wedge d\mu_{in_i}\Big)$$

$$= \sum_{\Theta_{iP}}\mu_P d\mu_{io}\wedge\ldots\wedge d\mu_{in_i} + \sum_{\lambda=0}^{n_i}(-1)^{\lambda+1}\mu_{i\lambda}d\Big(\sum_{\Theta_{iP}}\mu_P\Big)\wedge d\mu_{io}\wedge\ldots\wedge\widehat{d\mu_{i\lambda}}\wedge\ldots\wedge d\mu_{in_i}$$

$$= \sum_{\Theta_{iP}}\mu_P d\mu_{i_o}\wedge\ldots\wedge d\mu_{in_i} + \sum_{\lambda=0}^{n_i}(-1)^{\lambda}\mu_{i\lambda}d\Big(\sum_{j=0}^{n_i}\mu_{ij}\Big)\wedge d\mu_{io}\wedge\ldots\wedge\widehat{d\mu_{i\lambda}}\wedge\ldots\wedge d\mu_{in_i}$$

$$= \sum_{\Theta_{iP}}\mu_P d\mu_{io}\wedge\ldots\wedge d\mu_{in_i} + \sum_{\lambda=0}^{n_i}\mu_{i\lambda}d\mu_{io}\wedge\ldots\wedge d\mu_{in_i}$$

$$= d\mu_{io}\wedge\ldots\wedge d\mu_{in_i}$$

(On a utilisé dans le calcul le fait que la somme des coordonnées barycentriques dans Θ_i est égale à 1, donc que la somme des différentielles est nulle).C.Q.F.D

On a vu que $\eta = W\varphi$ est une forme C^∞ sur le complémentaire du squelette de codimension 1 de T et que pour toute cellule Θ de dimension maximale, on pouvait prolonger $\eta/_{\Theta^{top}}$ à l'adhérence $\overline{\Theta^{top}}$ de Θ^{top} en une forme C^∞. Les prolongements construits à partir de deux cellules adjacentes Θ et Θ' ne coïncident pas sur $\overline{\Theta^{top}} \cap \overline{\Theta'^{top}} = F$; cependant leur restriction (en tant que forme différentielle) à la frontière commune F définit une même forme sur F. Plus précisément, si $i : R \to \overline{\Theta^{top}}$ et $i' : F \to \overline{\Theta'^{top}}$ sont les deux inclusions, on a le

Lemme 5 - $i^*(\eta/_{\overline{\Theta^{top}}}) = i'^*(\eta/_{\overline{\Theta'^{top}}})$.

Si Θ et Θ' sont deux cellules de ét(Θ), le résultat pour η_Θ vient de ce que la différentielle extérieure commute à l'image réciproque et que $i^*\mu = i'^*\mu$, d'où $i^*d\mu = i'^*d\mu$ pour toute coordonnée barycentrique μ de Θ. Si seule Θ est une cellule de ét(Θ), il faut vérifier que $i^*(\eta/_{\overline{\Theta^{top}}}) = 0$: cela résulte de la nullité sur F de l'une au moins des coordonnées barycentriques $\mu_{i\lambda}$ de Θ. Le résultat pour η s'en déduit par linéarité.

Le lemme 5 permet de définir sans ambiguité la n-cochaine simpliciale ψ déduite de φ ; en conservant toujours la notation $\eta = W\varphi$, on associe à toute n-cellule ρ de T^{top} l'intégrale de η sur ρ :

$$\varphi(\rho) = \int_\rho \eta = \int_\rho W\varphi .$$

Lemme 6 - Si φ est un cocycle de T, alors ψ est un cocycle de T^{top}.

En effet, le lemme 5 permet d'écrire $\psi(\rho)$ comme une somme d'intégrales dont les domaines sont contenus chacun dans l'adhérence d'une cellule de dimension maximale. L'application de la formule de Stokes et du lemme 4 donne le résultat cherché.

Lemme 7 - Si ρ est la cellule de T^{top} correspondant à la cellule ρ_o de T, alors $\psi(\rho) = \varphi(\rho_o)$.

Par linéarité, on peut supposer que $\varphi = \tilde{\Theta}$. Il faut alors montrer que $\int_\rho \eta_\Theta = \delta_{\rho_o \Theta}$. Si $\rho_o \neq \Theta$, ou bien $\rho \not\subset \overline{\text{ét}(\Theta)}$, ou bien ρ est contenu dans la frontière de ét(Θ) : dans le premier cas la nullité de $\int_\rho \eta_\Theta$ est évidente, et dans le second elle résulte du lemme 5.

Si $\rho_o = \Theta$, on se ramène immédiatement au cas où $\rho_o = \Theta$ est un simplexe. La démonstration (classique) se fait alors par récurrence sur la dimension de Θ.

- Si $\Theta = [P]$, on vérifie bien que $\psi([P]) = \mu_P(P) = 1$.

- Si $\dim \Theta = n \geq 1$, soit Θ' une face de Θ ; alors
$d\tilde{\Theta}' = \tilde{\Theta} + \sum_{\tau \neq \Theta} \tilde{\tau} \Rightarrow \int_\rho \eta_\Theta = \int_\rho W(d\tilde{\Theta}') = \int_\partial d(W\tilde{\Theta}') = \int_{\partial\Theta} W\tilde{\Theta}' = \int_{\Theta'} \eta_{\Theta'} = 1.$

Proposition - Le cocycle simplicial ψ de T^{top} ainsi associé au cocycle φ de T est Γ-invariant ; l'image réciproque $\bar{\psi}$ de ψ sur $\|\overline{N\Gamma} \times_\Gamma T^{top}\|$ et l'image réciproque $\bar{\varphi}$ de φ sur $\|\overline{N\Gamma} \times_\Gamma T\|$ induisent des classes de cohomologie dans $H^*(\|\overline{N\Gamma} \times_\Gamma T^{top}\|)$ et $H(C^*(\overline{N\Gamma} \times_\Gamma T))$ qui se correspondent.

La Γ-invariance de ψ résulte des définitions linéaires de l'action de Γ sur T^{top} et de la construction de ψ.

D'après le lemme 7, la restriction $\mathcal{C}^*(\|\overline{N\Gamma} \times_\Gamma T^{top}\|) \to C^*(\overline{N\Gamma} \times_\Gamma T)$ dont on sait qu'elle induit un isomorphisme en cohomologie envoie $\bar{\psi}$ sur $\bar{\varphi}$. Cela démontre la proposition.

On peut conclure la discussion précédente en énonçant le

Théorème - L'application qui à un n-cocycle Γ-invariant φ sur T associe le n-cocycle sur Γ défini par

$$(\gamma_1,\dots,\gamma_n) \mapsto \int_{\Delta(\gamma_1,\dots,\gamma_n)} W\varphi = \psi(\Delta(\gamma_1,\dots,\gamma_n))$$

(où ψ est le coclycle simplicial sur T^{top} associé à φ comme défini ci-dessus, où O est un sommet de T et $\Delta(\gamma_1,\dots,\gamma_n)$ le simplexe géodésique de T^{top} de sommets $O, \gamma_1 O, \dots, \gamma_1\dots\gamma_n O$) induit par passage aux classes de cohomologie un homomorphisme $j_\Gamma : H(\mathcal{C}^*(T)^\Gamma) \to H^*(\Gamma)$.

Le seul point restant à vérifier est en effet que si φ est un cobord sur T, alors ψ est un cobord sur T^{top}, ce qui résulte du lemme 4.

Remarque 1 - Un cas particulier intéressant est celui où l'immeuble T est un arbre (c'est-à-dire où le groupe G est de rang semi-simple 1). On peut alors préciser davantage le cocycle ψ et donc l'homomorphisme j_Γ.

Si φ est un 0-cocyle sur T, c'est une fonction constante sur les sommets de T puisque l'immeuble est connexe : par le procédé précédemment décrit, on lui associe la fonction constante ψ sur T^{top} qui prend la même valeur. Ce résultat est d'ailleurs le même quelle que soit la dimension de T.

Les 1-cocycles sur T sont des fonctions sur l'ensemble des arêtes de T. Si Θ est l'une d'elles, où les coordonnées barycentriques sont μ_o et μ_1, la forme $\eta = W\varphi$ est définie sur Θ^{top} par $\eta/_{\Theta^{top}} = \mu_o d\mu_1 - \mu_1 d\mu_o$. Quant aux 1-simplexes de T^{top} qui interviennent effectivement dans le calcul de j_Γ, ce sont les $r(\Delta^1 \times \bar{\sigma})$ dont les deux extrémités sont O et $\gamma_1 O$. La géodésique entre ces deux points est une réunion finie de simplexes ξ_ν de T et

$j_\Gamma(\varphi)(\{\gamma_l\})$ est égal à la somme des valeurs de φ sur les ξ_ν.

Remarque 2 - Dans l'article qui a inspiré celui-ci [4] Dupont considère l'homomorphisme composé

$$H^*(\mathfrak{g},\mathfrak{k}; R) \to H^*(G/K,R) \to H^*(\Gamma,R)$$

où G est un groupe de Lie réel connexe, K un sous-groupe compact maximal, $\mathfrak{g}$ et $\mathfrak{k}$ leurs algèbres de Lie respectives et Γ un sous-groupe discret de G (le premier homomorphisme, en fait un isomorphisme, est dû à Chevalley et Eilenberg [3]). De plus Van Est a démontré par ailleurs [8] que

$$H^*(\mathfrak{g},\mathfrak{k}; R) \overset{\sim}{\to} H^*_{ct}(G,R)$$

où $H_{ct}(G,R)$ est la cohomologie continue de G à valeurs dans R, d'où un homomorphisme

$$H^*_{ct}(G,R) \to H^*(\Gamma,R).$$

Or Borel et Wallach [2] ont démontré récemment dans le cas p-adique l'existence d'un isomorphisme

$$H^*_{ct}(G,R) \to H(\mathcal{E}^*(T)^G)$$

de la cohomologie continue de G dans la cohomologie du complexe des cohaines réelles G-invariantes sur T.

Si $\varkappa \in H^q_{ct}(G,R)$ et si φ est un q-cocycle G-invariant sur T représentant l'image de $\varkappa$, la construction précédente permet donc de définir un homomorphisme

$$H^*_{ct}(G,R) \to H^*(\Gamma,R)$$

analogue à celui du cas réel.

BIBLIOGRAPHIE

[1] A.Borel. Stable Real Cohomology of Arithmetic Groups. Annales E.N.S. (4), 7 (1974), p.235-272.

[2] A.Borel and N.Wallach. Continuous Cohomology, Discrete Subgroups and Representation of Reductive Groups. Annals of Math. Studies n°94. Princeton University Press. (1979).

[3] C.Chevalley and S.Eilenberg. Cohomology Theory of Lie Groups and Lie Algebras. Trans. of Am. Math. Soc. 68 (1948), p.85-124

4 J.L.Dupont. Simplicial De Rham Cohomology and Characteristic Classes of Flat Bundles. Topology 15 (1976), p.233-245.

5 G.Rousseau. Immeubles des groupes réductifs sur les corps locaux. Thèse (1977). Université de Paris-Sud. Orsay.

6 G.Segal. Classifying Spaces and Spectral Sequences Publ. Math. de l'I.H.E.S. 34 (1968), p.105-112.

7 E.H.Spanier. Algebraic Topology. Mc Graw Hill (1966). New-York.

8 W.T.van Est. Group Cohomology and Lie Algebra Cohomology in Lie Groups I.II. Proc. Kon. Neder. Akad. Wetensch. A56 - Indag. Math. 15 (1953), p.484-504.

9 H.Whitney. Geometric Integration Theory. Princeton University Press. (1957).

G.BESSON
Université Pierre et Marie Curie
couloir 45-46 - 5ème étage
4, Place Jussieu
75230 Paris - Cedex 05

REPORT ON NORMAL BUNDLES OF CURVES IN $\mathbb{P}_3$

by

David Eisenbud*

There has recently been a little flurry of activity devoted to an apparently new corner of the hoary subject of smooth curves in projective 3-space, namely the study of the normal bundles of such curves. The subject is still embryonic, but there are many concrete and interesting problems. I wish here to surrey the results obtained so far, which treat curves of degree ≤ 5 and rational curves.

Throughout curves will be non singular, and everything will take place over $\mathbb{C}$, though this last is probably inessentiel.

A good deal of the activity in this subject seems to have been triggered by a remark of Grauert, to the effect that one had (18 months ago) no single example of a curve in $\mathbb{P}_3$ whose normal bundle was indecomposable.

What would be the simplest possible such example ? Smooth curves of degrees 1, 2, or 3 are rational or complete intersections, and every vector bundle over $\mathbb{P}_1$, decomposes. The only non-rational smooth curves of degree 4 are the elliptic quartics obtained as the intersection of two quadric surfaces, which again have decomposable normal bundles because they are complete intersection So one must go to degree 5 for the first interesting examples.

---

* I am grateful to the hospitality of the SFB for theoretical mathematics in Bonn,on whose premises I learned most of what is contained in this report.

There are two sorts of non-rational smooth curves of degree 5 in $\mathbb{P}_3$ (by Castel-nuovo or by R-R directly) : the <u>elliptic curves and curves of genus</u> 2 (hyperelliptic curves). Now, h^o (5 points on an elliptic curve) = 5 - so there is a natural embedding in $\mathbb{P}_4$ - and the elliptic quintics are projections of this. On the other hand the hyperelliptic curves are not projections, and in fact they are arithmetically normal ; i.e. $\sum H_o(\mathcal{O}_C(n))$ is integrally closed and we will treat these hyperelliptic curves first.

$$R = k\,[X_0,X_1,X_2,X_3] \twoheadrightarrow \sum H_o(\mathcal{O}_C(n)) \quad \text{is an epimorphism.}$$

Now

$$h^o(\mathcal{O}_C(2)) = h^o(\mathcal{O}_C(2)) - h^1(\mathcal{O}_C(-2) \otimes K)$$

$$= 10 + 1 - 2 = 9.$$

So such a curve lies on a unique, singular or non singular, quadric,

The distinction between hyperelliptic curves an smooth and an singular conics is fundamental, as one can see in another way : let C be <u>any</u> genus 2 curve, K (of degree 2) its canonical series.If D is a divisor of degree 5, then h^o D = 4 and $|D| = \mathbb{P}_3$. Thus of the 5 points of D we may choose 3 to come from 2K ; if a 4th does too, then the image of C in $\mathbb{P}_3$ lies on a quadric cone :

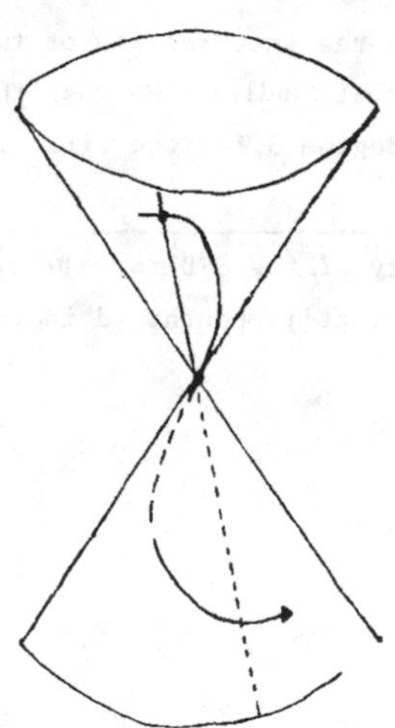

(a hyperplane section of the cone through the vertex is 2 lines and meets the curve in 5 points). If only 3 points of D are in 2K , then the image of C in $\mathbb{P}_3$ lies on a nonsingular quadric, isomorphic to $\mathbb{P}_1 \times \mathbb{P}_1$, as a divisor of class (3,2) :

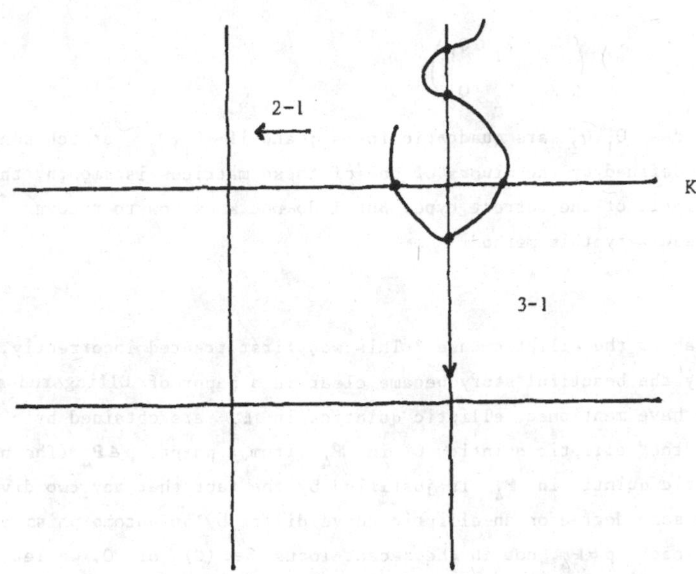

<u>Theorem</u> (Van de Ven -Comptes Rendus- Spring 1979). <u>If</u> C <u>is nonsingular curve of degree</u> 5 <u>and genus</u> 2 <u>in</u> $\mathbb{P}_3$, <u>then</u> N_C <u>decomposes if and only if</u> C <u>lies on a singular quadric</u>.

In the situation of the Theorem, it is easy to write equations for C , so one can "see" the normal bundle of C very directly as $(I_C/I_C^2)^*$; in fact, if C lies on a non singular quadric

$X_0X_3-X_1X_2$, then its equations are the 2×2 minors of a matrix of the form :

$$\begin{pmatrix} X_0 & X_1 & Q_1 \\ X_2 & X_3 & Q_2 \end{pmatrix}$$

while if C lies on a singular quadric $X_0X_2-X_1^2$ then its equations are the minors of :

$$\begin{pmatrix} X_0 & X_1 & Q_1 \\ X_1 & X_2 & Q_2 \end{pmatrix}$$

where in each case Q_1,Q_2 are quadratic forms ; and if Q_1,Q_2 are chosen such that the curve defined by the minors of one of these matrices is smooth, then it is hyperelliptic of the correct type. But I do not know how to recover Van de Ven's result by this method.

What about the elliptic case ? This was first treated incorrectly, but eventually the beautiful story became clear in a paper of Ellingsrud and Laksov ; as we have mentioned, elliptic quintics in $\mathbb{P}_3$ are obtained by projection of "the" elliptic quintic C in $\mathbb{P}_4$ from a point $p\in\mathbb{P}_4$ (The notion of "the" elliptic quintic in $\mathbb{P}_4$ is justified by the fact that any two divisor classes of the same degree on an elliptic curve differ by an automorphism of the curve). So for each $p\in\mathbb{P}_4$, not in the secant locus Sec (C) of C, we let C_p be the projection of C from p and we write N_p for its normal bundle. A Chern class computation shows that if N_p decomposes, then it does so as $M_o\otimes\mathcal{L}^2\oplus M_o^{-1}\otimes\mathcal{L}^2$, where M_o is a line bundle of degree 0 and $\mathcal{L}$ is the bundle corresponding to the hyperplane section of C in $\mathbb{P}_4$. The main result is :

Theorem. (Ellingsrud-Laksov-Comptes Rendus-to appear). For each line bundle M_o of degree 0 on C there is a quintic hypersurface $H_{M_o}\subset\mathbb{P}_4$, such that :

1) $H_{M_o} = H_{M_o^{-1}}$

2) $\bigcup_{M_o} H_{M_o} = \mathbb{P}_4$; $H_{M_o}\cap H_{M'_o}\subset \mathrm{Sec}(C)$ if $M_o \neq M'^{\pm 1}_o$

3) $H_{M_o}\subset$ Sec C $\Longleftrightarrow M_o = \mathcal{O}$, in which case H_{M_o} = Sec C .

4) For any $p\in\mathbb{P}_4$-Sec C $p\in H_{M_o}$ iff there is an exact sequence of vector bundles.

$$0 \longrightarrow M_o^{-1}\otimes\mathcal{L}^2 \longrightarrow N_p \longrightarrow M_o\otimes\mathcal{L}^2 \longrightarrow 0$$

5) <u>If</u> $M_o^2 \neq \mathcal{O}$, <u>the sequence above splits</u> ; <u>if</u> $M_o^2 = \mathcal{O}$, <u>then for suitable</u> (conjecturally : for all) $p \in H_{M_o}$ - Sec C, N_p <u>is indecomposable.</u>

A second proof of this result has been found by Van de Ven and my self ; its key point is to exploit the fact that all the equations of C in $\mathbb{P}_4$ are of the same degree (2), and all the relations on them are generated by linear relations. In fact, the homogeneous ideal of C is generated by the 4 x 4 Pfaffians of a matric of the form :

$$\begin{pmatrix} 0 & 0 & X_0 & X_1 & X_3 \\ 0 & 0 & X_1 & X_2 & X_4 \\ -X_0 & -X_1 & 0 & A & B \\ -X_1 & -X_2 & -A & 0 & C \\ -X_3 & -X_4 & -B & -C & 0 \end{pmatrix}$$

where A, B and C are linear forms depending on the embedding chosen for C in $\mathbb{P}_4$.

So far, because of our quest for indecomposable normal bundles, we have left out what ought to the first and simplest case of all, the case of rational curves. Last fall, Van de Ven and I set out to plug this gap, and I now want to describe some of the results we have obtained.

First of all, any rank 2 bundle on $\mathbb{P}_1$ is of the form $B = \mathcal{O}_{\mathbb{P}_1}(a) \oplus \mathcal{O}_{\mathbb{P}_1}(b)$ for some integers a and b. The numbers a and b are analytic but not topological invariants of the bundle, but $a + b = \deg(\det B)$ is, and it is correspondingly easier to compute.

Every rational curve of degree n in $\mathbb{P}_3$ is the projection from some $p \cong \mathbb{P}_{n-4} \subset \mathbb{P}_n$ of "the rational normal n-ic" C in $\mathbb{P}_n$ (up to ambient automorphisms), so we may parametrize every thing by projection centers p, and write C_p for a rational n-ic in $\mathbb{P}_3$, and N_p for its normal bundle. The normal bundle N_p is obtained from N_C by factoring out a sub-bundle isomorphic to $\overset{n-4}{\oplus} \mathcal{O}_{\mathbb{P}_n}(1)\big|_C$ induced by P, and one computes easily from this that if $N_p = \mathcal{O}_{\mathbb{P}_1}(a) \oplus \mathcal{O}_{\mathbb{P}_1}(b)$, then $a + b = 4n-2$.

From general principles, one now expects the general C_p to have $N_p = \mathcal{O}_{\mathbb{P}_1}(2n-1) \oplus \mathcal{O}_{\mathbb{P}_1}(2n-1)$, a "balanced bundle", and this is in fact the case. (Example : the curve parametrized by $(s^n, s^{n-1}t, st^{n-1}, t^n)$). So the first question is : how unbalanced can N_p be ?

With the above notation, if C_p is smooth and $n > 3$, then :

Proposition : $n+3 \leq a,b \leq 3n-5$.

One now can construct examples to show that every a,b subject to the above conditions actually occurs as the normal bundle of a smooth rational curve. One way to evaluate the computations is to use the following result, which gives a geometric description of a and b :

Theorem : Let $C \subset \mathbb{P}_3$ be a smooth, rational curve of degree n, $T(C)$ its tangent surface (the union of its tangent lines, which is a surface of degree $2n-2$). Let L be a line meeting $T(C)$ transversely and consider $\Gamma \subset C \times L$,

$$\Gamma = \{(C,p) \,|\, T_C \ni p\} .$$

Then $N_C = \mathcal{O}((2n-1)+a) \oplus \mathcal{O}((2n-1)-a)$ iff Γ lies on the graph of a map $C \longrightarrow L$ of degree $(n-1)-a$.

Using this we can exhibit an interesting example of a maximally unbalanced curve :

Example. Let D be the twisted cubic, $T(D)$ its tangent surface. $T(D)$ is the projection of a ruled surface S of degree 4 in $\mathbb{P}_5$ (S is $\mathbb{P}_1 \times \mathbb{P}_1$, embedded linearly in the first factor and quadratically in the second). Let H be the hyperplane section on S, R the ruling. The divisor $H + (n-4)R$ is very ample for all $n \geq 4$, and is a rational curve (since intersection with rulings induces an isomorphism with $\mathbb{P}_1$). The general divisor in this class will project to a smooth rational n-ic C in $\mathbb{P}_3$ lying on $T(D)$ and meeting each tangent line once. We claim that $N_C = \mathcal{O}_{\mathbb{P}_1}(n+3) \oplus \mathcal{O}_{\mathbb{P}_1}(3n-5)$. To check this, let L be a general line in $\mathbb{P}_3$, and consider the map $\varphi : C \to L$ obtained by carrying $z \in C$ to $T_{T(D),z} \cap L \in L$. If, for some $z \in C$, $T_{C,z} \cap L = \bar{z}$, then clearly $\varphi(z) = \bar{z}$. By the Theorem, it will be enough to show that φ is a map if degree 3 ; or equivalently, that the number of points $z \in C$ for which $T_{T(D),z}$ contains a given point, is 3. But, as a local computation shows, $T_{T(D),z}$ is just the osculating plane of D at the point of D whose tangent passes through z. Since the dual curve to D has degree 3, we are done. ||

Added in Proof : After the above was written, I have became aware of papers by Ghione and Sacchiero, and Sacchiero (Preprints, Univ. of Ferrara), giving bounds on the normal bundles of rational curves with ordinary singularities in $\mathbb{P}_n$ (known to Van de Ven and me for $n=3$), and some (singular) examples of such curves.

The Bernstein class of modules on algebraic manifolds

By Jan-Erik Björk

Introduction

In [1] and [2] a remarkable theory about the Weyl algebra $A_n(\mathbb{C})$; and its modules was established by I.N. Bernstein. In these notes we discuss how his results can be extended when the affine space $\mathbb{C}^n$ is replaced by a non-singular algebraic manifold, always defined over the complex field.

These extensions of Bernstein's results are based upon a study of filtered rings whose associated graded rings are commutative regular noetherian rings. The whole theory is developped in [3 , Chapter 2 and 3] and here we shall review the essential results, leaving out most details of the proofs. In Section 1 we discuss this general class of filtered rings, while the remainder of these notes is devoted to the study of the rings of differential operators on non-singular affine algebraic varieties.

1. - Regular filtered rings

Let A be a ring which is equipped with a filtration $\Sigma = \{ \Sigma_o, \Sigma_1, \dots \}$. This means that Σ_o is a subring of A -containing the unit element 1- and $\Sigma_o \subset \Sigma_1 \subset \Sigma_2 \subset \dots$ and $\bigcup_{v \geqslant 0} \Sigma_v = A$, and finally the inclusions $\Sigma_k \Sigma_v \subset \Sigma_{k+v}$ hold for all pairs k and $v \geqslant 0$. Of course, the inclusions mean that if $Q \in \Sigma_k$ and if $P \in \Sigma_v$ then the product $QP \in \Sigma_{k+v}$.

Starting from this filtration on A we obtain the associated graded ring :

$$gr(A) = \oplus \Sigma_v/\Sigma_{v-1} = \Sigma_o \oplus \Sigma_1/\Sigma_o \oplus \Sigma_2/\Sigma_1 \oplus \dots$$

This leads to :

1.1. Definition A is called a regular filtered ring if A is equipped with a filtration Σ such that gr(A) is a regular commutative noetherian ring.

1.2. The dimension of gr(A) . If m is a maximal ideal in the regular noetherian ring gr(A) , then the localisation $gr(A)_m$ is a regular local ring and it has some dimension which may depend on the choice of m . From now on we shall make the simplifying assumption that gr(A) has a pure dimension, which means that there exists some integer ω such that $\dim(gr(A)_m) = \omega$ for all maximal ideals m in gr(A) .

Of course, as a definition of the dimension of the local rings $gr(A)_m$ we can take the Krull dimension - or the maximal number of a system of parameters - or the dimension of the vector space m/m^2 over the residue field $gr(A)/m$.

1.3. The integer μ . When A is a regular filtered ring and gr(A) has some pure dimension ω , then it follows that the global homological dimension of gr(A) is equal to ω . It follows easily that the ring A is left and right noetherian - i.e. one - sided ideals of A are finitely generated - and that the ring A also has a finite global homological dimension which is $\leqslant \omega$. In many interesting examples a strict inequality holds here and we put :

$$\mu = \text{gl. dim}(A) \quad \text{so that} \quad \mu \leqslant \omega$$

and it is the case where strict inequality holds which will be of particular interest. Examples of regular filtered rings A for which $\mu < \omega$ are given in Section 2 .

1.4. The dimension of finitely generated A-modules. From now on A is a regular noetherian ring where gr(A) has some pure dimension ω . Let us now consider a finitely generated left A-module M . (The case with finitely generated right A-modules can be treated in the same way, so we restrict the attention to left A-modules).

A filtration on M consists of an increasing sequence of additive subgroups $\{\Gamma_v\}$, where $\Gamma_v = \{0\}$ when $v << 0$ and $\cup\Gamma_v = M$ and $\Sigma_k \Gamma_v \subset \Gamma_{k+v}$ hold for all $k \geqslant 0$ and all v .

When $\Gamma = \{\Gamma_v\}$ is a filtration on M we can introduce the associated graded gr(A)-module : $gr_\Gamma(M) = \oplus \Gamma_v/\Gamma_{v-1}$. We say that Γ is a <u>good filtration</u> if and only if $gr_\Gamma(M)$ is a finitely generated gr(A)-module.

<u>Remark</u> M can always be equipped with some good filtration. For example if $\xi_1 \dots \xi_s$ generate M as an A-module and if we put $\Gamma_v = \Sigma_v \xi_1 + \dots + \Sigma_v \xi_s$ for $v \geqslant 0$, while $\Gamma_v = 0$ when $v < 0$, then $\{\Gamma_v\}$ is a good filtration.

Suppose now that Γ is some good filtration on M . Then $gr_\Gamma(M)$ is a finitely generated gr(A)-module and we can compute its Krull dimension and define the integer :

$$d_\Gamma(M) = \text{Kr.dim}_{gr(A)}(gr_\Gamma(M)) .$$

In general, M can be equipped with many different good filtrations. But it turns out that the integers $d_\Gamma(M)$ do not depend on the choice of Γ , i.e. we have the equalities $d_\Gamma(M) = d_\Omega(M)$ for all pairs of good filtrations Γ and Ω on M.

This common integer is therefore denoted by d(M) and simly called the <u>dimension of the A-module M</u> .

<u>Remark</u> See [3, Chapter 3 Section 3] for more details. In particular Lemma 6.2. in [3, Chapter 3] proves the claim above.

1.5. <u>The integer</u> j(M) . When M is a finitely generated left A-module then the Ext-groups $Ext_A^v(M,A)$ are defined for all $v \geqslant 0$. If μ = gl. dim(A) then $Ext_A^v(M,A) = 0$ for all $v > \mu$. Besides, using the fact that gl. dim(A) is finite, it can be proved that at least one of these Ext-groups is non-zero and this leads to:

<u>Definition</u> Put : $j(M) = \inf \{v \geqslant 0 : Ext_A^v(M,A) \neq 0\}$

At this stage we can announce an important result which connects the two integers d(M) and j(M) :

1.6. <u>Theorem</u> <u>The equality</u> $d(M) + j(M) = \omega$ <u>holds for every finitely generated left (or right) A-module</u> M .

The proof of theorem 1.6. requires several steps and the details appear in [3, Chapter 3] . See in particular Theorem 7.1 in [3, Chapter 3] .

Using Theorem 1.6. and the observation that $0 \leq j(M) \leq \mu$ holds we obtain :

1.7 <u>Corollary</u> <u>If</u> M <u>is a (non-zero) finitely generated</u> A-<u>module then</u> $d(M) \geq \omega - \mu$.

<u>Remark</u> Of course, this inequality is only interesting in the case when $\mu < \omega$. The inequality in 1.7. is called <u>Bernstein's inequality</u> and it leads to :

1.8. <u>The Bernstein class</u> $\mathcal{B}(A)$ which consists of the family of all finitely generated left or right A-modules M, satisfying $d(M) = \omega - \mu$.

We can prove some facts about modules in the Bernstein class. In fact, using the proof of Theorem 1.6. we first obtain a duality between left and right A-modules in the Bernstein class. The result is :

1.9. <u>Theorem</u> <u>If</u> $M \in \mathcal{B}_{\ell}(A)$, i.e. <u>if</u> M <u>is a left</u> A-<u>module in the Bernstein class, then the right</u> A-<u>module</u> $\mathrm{Ext}_A^{\omega-\mu}(M,A) = M^{\star} \in \mathcal{B}_r(A)$ <u>and finally</u> $M \cong \mathrm{Ext}_A^{\omega-\mu}(M^{\star},A)$.

Finally, using this duality and the fact that the ring A is left and right noetherian, it is not difficult to prove that modules in the Bernstein class also satisfy the descending chain condition on submodules and this gives :

1.10. <u>Corollary</u> <u>If</u> $M \in \mathcal{B}(A)$ <u>then</u> M <u>has a finite length as an</u> A-<u>module</u>, i.e. <u>there exists a finite composition series</u> :

$0 = M_0 \subset M_1 \subset \ldots \subset M_t = M$, <u>where</u> $M_1 \ldots M_{t-1}$ <u>are</u> A-<u>submodules of</u> M <u>and the factor modules</u> M_i/M_{i-1} <u>are irreducible for all</u> i.

<u>Remark</u> See [3, Chapter 3, Section 7.12] for more details.

1.11 <u>Some open problems</u> The finiteness property of modules in $\mathcal{B}(A)$ suggests the following questions.

<u>Problem 1</u> - Let M and N be two left A-modules which both $\in \mathcal{B}(A)$. Consider now the Ext-groups $\mathrm{Ext}_A^{\nu}(M,N)$ which are defined for all $\nu \geq 0$, and they vanish when $\nu > \mu$. These Ext-groups are modules over $Z(A)$, where $Z(A)$ is the <u>center</u> of the ring A, i.e. the subring which consists of all central elements a in A, i.e elements for which $ax = xa$ for all x in A.

<u>Is it true that</u> $\mathrm{Ext}_A^v(M,N)$ <u>are finitely genrated</u> $Z(A)$-<u>modules</u> ?

<u>Remark</u> Starting form a left A-module M in $\mathcal{B}(A)$ and a right A-module N in $\mathcal{B}(A)$ we can also consider the Tor-groups : $\mathrm{Tor}_v^A(N,M)$ and ask the same question as above, i.e. whether these groups are finitely generated Z(A)-modules for all $0 \leqslant v \leqslant \mu$.

These two problems are directly related to each other. In fact, using the duality between $\mathcal{B}_\ell(A)$ and $\mathcal{B}_r(A)$ we can prove that :

$\mathrm{Ext}_A^v(M,N) \cong \mathrm{Tor}_{\mu-v}^A(M^\star,N)$ hold for all pairs M and N in $\mathcal{B}_\ell(A)$

and these isomorphisms show that the question posed in Problem 1 is equivalent to the similar question for Tor-groups.

<u>Remark</u> See [3, Chapter 3, Theorem 7.15] for the isomorphism above.

<u>A final Remark</u> Problem 1 seems to be quite difficult to settle when it is posed in such a generality. If we make the additional assumption that Z(A) is a field contained in Σ_o and that gr(A) is a finitely generated algebra over its subfield Z(A), then it is very likely that Problem 1 has a positive answer.

For example, in Section 2 below we shall study a family of regular filtered rings where this condition on Z(A) is satisfied and establish the finite-dimensionality of the Ext-groups $\mathrm{Ext}_v(M,N)$ when M and N both $\varepsilon\ \mathcal{B}(A)$.

2. - <u>The rings</u> $\mathcal{D}(V)$

Let us consider a non-singular affine algebraic variety V , defined over the complex field $\mathbb{C}$. We can assume that V is connected also and thus V is the locus of some prime ideal $\mathcal{P}$ in a polynomial ring $\mathbb{C}[z_1 \dots z_N]$. In fact, this follows from the wellknown fact that V always can be imbedded into some affine space $\mathbb{C}^N$ so that V appears as an algebraic submanifold of $\mathbb{C}^N$ and then the Nullstellen Satz shows that $V = \mathcal{P}^{-1}(0) = \{z \in \mathbb{C}^N ,\ p(z) = 0 \text{ for all } p \in \mathcal{P}\}$, where $\mathcal{P}$ is some prime ideal in $\mathbb{C}[z_1 \dots z_N]$.

The dimension of V is defined as the Krull dimension of the $\mathbb{C}[z]$ - module $\mathbb{C}[z]/\mathcal{P}$ and it is well known that this dimension equals the dimension of V where V is viewed as a complex analytic manifold.

This integer is denoted by dim(V) and usually we shall put $n = \dim(V)$.

2.1 <u>The ring</u> A(V) This is the ring of regular affine functions on V , i.e. we can put $A(V) = C[z]/\wp$ when $V = \wp^{-1}(0)$ as above. Since V is non-singular it follows that A(V) is a regular noetherian ring and A(V) has pure dimension equal to $n = \dim(V)$. The maximal ideals of A(V) correspond to points on V , i.e. if m is a maximal ideal in A(V) then there exists a unique point ξ on V such that $m = \{f \in A(V) : f(\xi) = 0\}$.

2.2 <u>The Lie algebra</u> $\mathcal{g}(V)$ This is the Lie algebra of $\mathbb{C}$-linear derivations on the ring A(V) . Since V is non-singular it can be proved that $\mathcal{g}(V)$ is a finitely generated and projective A(V)-module whose rank = dim(V) . Geometrically the elements in $\mathcal{g}(V)$ are so called affine vector fields on the algebraic manifold V and they generate the (complex) tangent spaces $T_\xi(V)$ for all points ξ in V .

We refer to [3, Chapter 3, Section 2] for more details. In particular to theorem 2.2. there.

2.3 <u>The ring</u> $\mathcal{D}(V)$ This is the ring of $\mathbb{C}$-linear differential operators on V . By definition $\mathcal{D}(V)$ is the ring of $\mathbb{C}$-linear operators on A(V) which is generated by the derivation operators from $\mathcal{g}(V)$ and the "zero-order multiplication operators" defined by elements from the ring A(V) itself.

2.4 <u>An example</u> If V is the affine space $\mathbb{C}^n$ then $A(V) = \mathbb{C}[z_1 \dots z_n]$ and here $\mathcal{g}(V)$ is the free A(V)-module of rank n generated by the usual derivation operators $\partial_j = \partial/\partial z_j, j = 1,2,..n$. Thus $\mathcal{D}(\mathbb{C}^n)$ is the $\mathbb{C}$-algebra generated by $z_1,\dots,z_n$ and $\partial_1,\dots,\partial_n$.

Leibniz's rule shows that $\partial_j(z_j f) = f + z_j \partial_j(f)$ for all polynomials p and this means that the $\mathbb{C}$-linear operator : $\partial_j z_j = z_j \partial_j + 1$ where 1 is the identity operator. So in the ring $\mathcal{D}(\mathbb{C}^n)$ we see that the elements ∂_j and z_j do not commute. In fact, the commutators $\partial_j z_j - z_j \partial_j = 1$ for each $1 \leqslant j \leqslant n$.

The non-commutative ring $\mathcal{D}(\mathbb{C}^n)$ is denoted by $A_n(\mathbb{C})$ and it is called the Weyl algebra in n variables, with coefficients in $\mathbb{C}$. We refer to [3 ,Chapter 1] for a detailed study of $A_n(\mathbb{C})$. Among those results we mention <u>Stafford's Theorem</u> which asserts that every one-sided ideal of $A_n(\mathbb{C})$ can be generated by 2 elements.

2.5. <u>The Σ-filtration on $\mathcal{D}(V)$</u> <u>and the ring</u> $gr(\mathcal{D}(V))$. By definition $\mathcal{D}(V)$ is a ring of differential operators and it has a natural filtration Σ , where $\Sigma_o = A(V)$ are the zero-order differential operators, while $\Sigma_1 = A(V) + \mathcal{J}(V)$ are the first order differential operators, and in general $\Sigma_k = \{ Q \in \mathcal{D}(V)$: the differential operator Q has order $\leqslant k \}$.

<u>Remark</u> It is easily seen that $\Sigma_k = S_k + \Sigma_{k-1}$ where S_k is the $\mathbb{C}$-subspace of $\mathcal{D}(V)$ generated by all k-fold products of elements from $\mathcal{J}(V)$.

Next, using the fact that the Lie algebra $\mathcal{J}(V)$ contains enough vector fields to generate the tangent spaces at all points in V , it follows easily that the associated graded ring : $gr(\mathcal{D}(V)) = \oplus \Sigma_v / \Sigma_{v-1}$ can be identified with $A(T^*(V))$ where $T^*(V)$ is the cotangent bundle over V .

Of course, here $T^*(V)$ is an algebraic manifold whose dimension is $2\dim(V)$ and we conclude that $gr(\mathcal{D}(V))$ is a regular noetherian ring of pure dimension equal to $2n$, where $n = \dim(V)$. In particular $\mathcal{D}(V)$ is a regular filtered ring where we can apply the results from section 1 . To do this we first need

2.6 <u>The equality</u> $gl.\dim(\mathcal{D}(V)) = \dim(V)$. In the case when $V = \mathbb{C}^n$ so that $\mathcal{D}(\mathbb{C}^n) = A_n(\mathbb{C})$ it was proved by J-E Roos in [4] that the global homological dimension of the Weyl algebra $A_n(\mathbb{C})$ is equal to n . It is not difficult to generalize the methods of his proof to the case when $A_n(\mathbb{C})$ is replaced by $\mathcal{D}(V)$ and then prove that $gl.\dim(\mathcal{D}(V)) = n = \dim(V)$, for every non-singular affine algebraic variety V defined over the complex field $\mathbb{C}$.

In fact, this follows from a quite general result in [3, p.83] .

2.7 <u>The Bernstein class</u> $\mathcal{B}(V)$. If $\dim(V) = n$ then $\dim(gr(\mathcal{D}(V)) = 2n$ so with the notations from Section 1 we have $\omega = 2n$, while the integer $\mu = gl.\dim(\mathcal{D}(V))$ is $\omega/2 = n$.

This gives the <u>Bernstein class</u> $\mathcal{B}(\mathcal{D}(V))$ which consists of finitely generated left (or right) $\mathcal{D}(V)$-modules M satisfying $d(M) = n$. To simplify the subsequent notations we put : $\mathcal{B}(V) = \mathcal{B}(\mathcal{D}(V))$.

When $V = \mathbb{C}^n$ so that $\mathcal{D}(V) = A_n(\mathbb{C})$ then I.N. Bernstein proved that if M and N belong to the Bernstein class of left $A_n(\mathbb{C})$-modules, then the Ext-groups

$\mathrm{Ext}^{v}_{A_n(\mathbb{C})}(M,N)$ are finite-dimensional complex vector spaces. Actually Bernstein did not state this explicitly, but the proof is an easy consequence of his results in [2] . See also [3, Chapter 1, Section 6] for this deduction.

Starting from this result we can generalize it to arbitrary non-singular affine varieties, always defined over the complex field $\mathbb{C}$, and arrive at :

2.8. Theorem If M and $N \in \mathcal{B}(V)$ then $\mathrm{Ext}^{v}_{\mathcal{D}(V)}(M,N)$ are finite-dimensional complex vector spaces for all $0 \leqslant v \leqslant n = \dim(V)$.

Remark We refer to [3, Chapter 3] for the details of the proof.

2.9 Some examples The result in Theorem 2.8 was known before in special cases. In fact, when V is given we can consider the affine p-forms $\Omega^{p}(V)$ for all $p \geqslant 0$. Here we put $\Omega^{o}(V) = A(V)$ and now $\Omega^{1}(V) = \mathrm{Hom}_{A(V)}(\mathcal{J}(V), A(V))$ and so on .

Using the exterior differential mappings from $\Omega^{p}(V) \to \Omega^{p+1}(V)$ we get the complex

$$\Omega^{\star}(V) : 0 \to \Omega^{o}(V) \to \Omega^{1}(V) \to \ldots \to \Omega^{n}(V) \to 0 ,$$

which is called the algebraic de Rham complex on V .

The cohomology groups of this complex are denoted by $H^{p}_{DR}(V)$ and they are called the algebraic de Rham cohomology groups. Using Desingularisation, A.Grothendieck proved that they are finite dimensional complex vector spaces for all $0 \leqslant p \leqslant n$.

In [5] , Monsky found an elementary proof of this finite-dimensionality.

Remark Actually, Grothendieck's result was more precise because he proved that $H^{p}_{DR}(V)$ are isomorphic to the usual Cech cohomology groups $H^{p}(V,C)$ with values in the constant sheaf $\mathbb{C}$ over the manifold V .

Using the ring $\mathcal{D}(V)$ and looking upon $A(V)$ as a left $\mathcal{D}(V)$-module it is easily seen that :

$$H^{p}_{DR}(V) \cong \mathrm{Ext}^{p}_{\mathcal{D}(V)}(A(V), A(V))$$

and besides the left $\mathcal{D}(V)$-module $A(V)$ belong to the Bernstein class $\mathcal{B}(V)$.

So with these traductions we see that Theorem 2.8. gives a far more general finiteness theorem. In particular this is so because the Bernstein class $\mathcal{B}(V)$ contains modules M which need not be finitely generated as modules over the subring $\Sigma_0 = A(V)$ of $\mathcal{D}(V)$. For example, if f is an arbitrary element of $A(V)$ then the ring of fractions : $A(V)[f^{-1}]$ is a left $\mathcal{D}(V)$-module.

Indeed, if $\delta \in \mathcal{D}(V)$ then the $\mathbb{C}$-linear derivation δ extends to a $\mathbb{C}$-linear derivation on this ring of fractions, and so on. It is not a priori obvious that $A(V)[f^{-1}]$ is a finitely generated $\mathcal{D}(V)$-module. However, this can be proved and it also turns out that $A(V)[f^{-1}]$ belongs to the Bernstein class $\mathcal{B}(V)$ and hence Theorem 2.8. applies to such $\mathcal{D}(V)$-modules also.

.10 A problem Is Stafford's Theorem true for the ring $\mathcal{D}(V)$, i.e. is it true that every one-sided ideal in the ring $\mathcal{D}(V)$ can be generated by 2 elements ?

REFERENCES

[1] Bernstein I.N. The analytic continuation of generalized functions with respect to a parameter. Funz. Anal. Akad. Nauk. 6 (1972), 26 - 40

[2] Bernstein I.N. Modules over the ring of differential operators. A study of fundamental solutions with constant coefficients Ibid 5, (1971), 1 - 16.

[3] BJÖRK J.E. Rings of Differential Operators. North-Holland. Math. Library Series Vol. 21 (1979)

[4] ROOS J.E. Determination de la dimension homologique globale des algèbres de Weyl C.R. Acad. Sci. Paris 274 (1972) 23 - 26

[5] MONSKY P. Finiteness of de Rham cohomology. Amer. Math. Journ. 94 (1972), 237-245.

Anneaux d'opérateurs différentiels

par Thierry LEVASSEUR

Nous exposons ici des analogues algébriques de certains théorèmes de J.M. Kantor concernant l'anneau des opérateurs différentiels sur une variété analytique ; principalement le théorème décrivant l'anneau des opérateurs différentiels sur une variété quotient de l'espace affine k^n par un sous groupe fini de $GL(n,k)$, (partie III de l'exposé).
Nous commencerons par rappeler certains résultats classiques en théorie des opérateurs différentiels algébriques, (partie I de l'exposé).

I) Généralités sur les opérateurs différentiels

I.1) Rappels sur les modules de différentielles, d'après [S]

Soit k un anneau commutatif et A une k-algèbre commutative. Si M est un A-module l'isomorphisme $\Phi : \mathrm{Hom}_k(A,M) \longrightarrow \mathrm{Hom}_A(A \otimes_k A, M)$ est défini par $\Phi(u)(a \otimes b) = au(b)$ pour $u \in \mathrm{Hom}_k(A,M)$, a et b dans A ; où $A \otimes_k A$ est un A-module par la multiplication sur la première composante. Notons aussi que $\mathrm{Hom}_k(A,M)$ est muni de la structure de $A \otimes_k A$ module définie par :

$$\{(a \otimes b)u\}(x) = a\,u(bx) \quad \text{pour } a,b,x \in A,\ u \in \mathrm{Hom}_k(A,M).$$

Il existe une suite exacte : $0 \longrightarrow I \xrightarrow{i} A \otimes_k A \xrightarrow{p} A \longrightarrow 0$ où $p(a \otimes b) = ab$ si, $a,b \in A$ et $I = \ker p = \{\sum_i a_i \otimes b_i / \sum a_i b_i = 0\}$. Posons : $j_1 : A \longrightarrow A \otimes_k A$, $j_1(a) = a \otimes 1$ et $j_2 : A \longrightarrow A \otimes_k A$, $j_2(a) = 1 \otimes a$; de sorte que $j_2(a) = a \otimes 1 + (j_2 - j_1)(a)$, nous noterons d l'application $j_2 - j_1$. Il est facile de voir que I est engendré comme idéal par les $\{da,\ a \in A\}$ et que l'on a une décomposition en somme directe de A-modules : $A \otimes_k A = I \oplus (A \otimes_k 1)$.
Si $a \in A$ et $u \in \mathrm{Hom}_k(A,M)$ on notera $[u,a]'$ l'élément de $\mathrm{Hom}_k(A,M)$ défini de la manière suivante : $[u,a]'(x) = u(ax) - au(x) - xu(a)$, pour tout $x \in A$.
Si $M = A$ et $u_1,u_2 \in \mathrm{Hom}_k(A,A)$ on pose $[u_1,u_2] = u_1 \circ u_2 - u_2 \circ u_1$, de sorte que dans ce cas $[u,a]' = u.a - a.u - u(a) = [u,a] - u(a)$.

<u>Définition</u> 1 - <u>Si</u> $q \in \mathbb{Z}$ <u>définissons</u> $\mathrm{Der}_k^{(q)}(A,M)$ <u>en posant</u> :

$\mathrm{Der}_k^{(q)}(A,M) = 0$ <u>si</u> $q \leq 0$; <u>et si</u> $q > 0$, $u \in \mathrm{Der}_k^{(q)}(A,M)$

si $u \in \mathrm{Hom}_k(A,M)$ *et* $[u,a]' \in \mathrm{Der}_k^{(q-1)}(A,M)$ *pour tout* a *dans* A . *Nous dirons que* $\mathrm{Der}_k^{(q)}(A,M)$ *est l'ensemble des dérivations d'ordre inférieur ou égal à* q .

On a : - $0 = \mathrm{Der}_k^{(0)}(A,M) \subset \mathrm{Der}_k^{(1)}(A,M) \subset \ldots \subset \mathrm{Der}_k^{(q)}(A,M) \subset \ldots$

- $\mathrm{Der}_k^{(q)}(A,M)$ est un A-module à gauche et si $u \in \mathrm{Der}_k^{(q)}(A,M)$ alors $u(1) = 0$.

- Les dérivations d'ordre 1 sont évidemment les dérivations usuelles.

Lorsque $M = A$, nous écrirons $\mathrm{Der}_k^{(q)}(A)$ au lieu de $\mathrm{Der}_k^{(q)}(A,A)$.

Proposition 2 - *Soit* $q \geq 0$, *si* $u \in \mathrm{Hom}_k(A,M)$ *les conditions suivantes sont équivalentes* :

(1) $u \in \mathrm{Der}_k^{(q)}(A,M)$

(2) $\Phi(u)(A \otimes 1) = 0$ *et* $\Phi(u)(I^{q+1}) = 0$

(3) $I^{q+1}.u = 0$ *et* $u(1) = 0$

(4) *Il existe un homomorphisme de* A *modules* $f : \frac{I}{I^{q+1}} \longrightarrow M$ *tel que le diagramme suivant soit commutatif* :

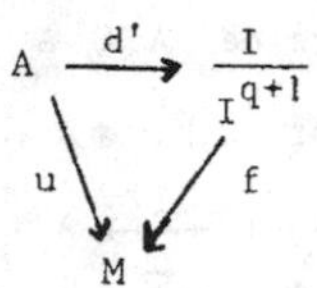

où d' *est la composée de* d *et de l'application* $I \longrightarrow \frac{I}{I^{q+1}}$

(5) *Quelque soit* $(x_o, \ldots, x_q) \in A^{q+1}$, *on a* :

$$u(x_o \ldots x_q) =$$

$$= \sum_{S=1}^{q} (-1)^{S+1} \sum_{i_1 < \ldots < i_S} x_{i_1} \ldots x_{i_S}\, u(x_o \ldots \hat{x}_{i_1} \ldots \hat{x}_{i_S} \ldots x_q)$$

Preuve - cf. [S] prop. A.1.1.

Définition 3 - *On appelle* $\Omega_{A/k}^{(q)} := \frac{I}{I^{q+1}}$ *le A-module des différentielles d'ordre* q, *et l'on pose* $d_{A/k}^{(q)} = d' : A \longrightarrow \Omega_{A/k}^{(q)}$, *où* d' *est définie en prop. 2. (4)*

Ainsi $\Omega^{(q)}_{A/k}$ est un A-module engendré par les $\{d^{(q)}_{A/k}(a),\ a\in A\}$, et pour tout A-module M il existe un isomorphisme de A-modules à gauche :

$$\mathrm{Hom}_A(\Omega^{(q)}_{A/k}, M) \xrightarrow{\sim} \mathrm{Der}^{(q)}_k(A,M),$$

qui à f fait correspondre $f \circ d^{(q)}_{A/k}$.

Signalons les propriétés suivantes :

Proposition 3 -

1) Si S est une partie multiplicative de A on a des isomorphismes de $S^{-1}A$ modules :

$$S^{-1}A \otimes_A \Omega^{(q)}_{A/k} = S^{-1}\Omega^{(q)}_{A/q} \simeq \Omega^{(q)}_{S^{-1}A/k}$$

2) Si $k \longrightarrow A \longrightarrow B$ est une suite d'homomorphismes d'anneaux il existe une suite exacte :

$$\Omega^{(1)}_{A/k} \otimes_A B \longrightarrow \Omega^{(1)}_{B/k} \longrightarrow \Omega^{(1)}_{B/A} \longrightarrow 0$$

3) Si $\mathfrak{a}$ est un idéal de A et $B = \frac{A}{\mathfrak{a}}$, il existe un épimorphisme de B modules : $B \otimes_A \Omega^{(q)}_{A/k} \longrightarrow \Omega^{(q)}_{B/k} \longrightarrow 0$, pour tout $q \in \mathbb{Z}$

4) Si k est un corps et K une extension séparable de type fini de k le degré de transcendance de K sur k est égal à $\dim_K \Omega^{(1)}_{K/k}$, et si K est algébrique sur k, $\Omega^{(q)}_{K/k} = 0$ pour tout q, (en particulier on a $\mathrm{Der}^{(q)}_k(K) = 0$).

Preuve - cf. [S] Appendice A

I.2 - Anneaux d'opérateurs différentiels

Soit A une k-algèbre commutative. Si M est un A-module, nous avons remarqué qu'un élément de $\mathrm{Der}^{(q)}_k(A,M)$ s'annule sur l'élément unité, c'est-à-dire appartient à $\mathrm{Hom}^o_k(A,M) = \{f \in \mathrm{Hom}_k(A,M) / f(1) = 0\}$. Un opérateur différentiel sur A doit être un élément de $\mathrm{Hom}_k(A,A)$; remarquons que $\mathrm{Hom}_k(A,A)$ est un anneau qui contient A et $\mathrm{Hom}_k(A,A) = A \oplus \mathrm{Hom}^o_k(A,A)$, somme directe de A-modules. Le sous A-module à gauche de $\mathrm{Hom}_k(A,A)$, $A \oplus \bigcup_{q=1}^{\infty} \mathrm{Der}^{(q)}_k(A)$, est un sous-anneau pour la composition ; en effet :

Proposition 1 -

(1) Quelques soient a dans A, Δ dans $\mathrm{Der}^{(q)}_k(A)$, on a

$$\Delta . a \in A \oplus \mathrm{Der}^{(q)}_k(A) .$$

(2) Si Δ_1, Δ_2 appartiennent à $\mathrm{Der}^{(q)}_k(A)$, alors $\Delta_1 \Delta_2 \in \mathrm{Der}^{(q+q')}_k(A)$

et $[\Delta_1, \Delta_2] \in \mathrm{Der}_k^{(q+q'-1)}(A)$.

Preuve - (1) Provient de l'égalité : $\Delta.a = [\Delta, a]' + a\Delta + \Delta(a)$

(2) S'obtient par récurrence sur l'ordre à l'aide des formules suivantes :

$$[\Delta_1 \Delta_2, a]' = \Delta_1 [\Delta_2, a]' + [\Delta_1, a]' \Delta_2 + \Delta_1(a)\Delta_2 + \Delta_2(a)\Delta_1 + [\Delta_1, \Delta_2(a)]',$$

et :

$$[\Delta_1 \Delta_2 - \Delta_2 \Delta_1, a]' = \Delta_1 [\Delta_2, a]' - [\Delta_2, a]' \Delta_1 + [\Delta_1, a]'\Delta_2 - \Delta_2 [\Delta_1, a]' + [\Delta_1, \Delta_2(a)]' - [\Delta_2, \Delta_1(a)]' .$$

D'où l'on peut poser :

Définition 2 - L'anneau $A + \bigcup_{q=1}^{\infty} \mathrm{Der}_k^{(q)}(A)$ est appelé anneau des opérateurs différentiels de la k-algèbre A. Nous le noterons $\mathrm{Diff}_k A$ et nous désignerons par $\mathrm{Diff}_k^q(A)$ le A-module à gauche $A \oplus \mathrm{Der}_k^{(q)}(A)$, ensemble des opérateurs différentiels d'ordre inférieur ou égal à q.

Remarques - 1) En tant que A-modules on a : $\mathrm{Diff}_k^q(A) = A \oplus \mathrm{Hom}_A(\Omega_{A/k}^{(q)}, A)$

2) Nous aurions pu définir par récurrence : $P \in \mathrm{Diff}_k^o(A)$ si P est A-linéaire et $P \in \mathrm{Diff}_k^q(A)$ si $[P,a] \in \mathrm{Diff}_k^{q-1}(A)$ pour tout $a \in A$, si $q > 0$, (cf. [K]).

Définition 3 - Nous noterons $P_{A/k}^q$ l'anneau $\dfrac{A \otimes_k A}{I^{q+1}}$, appelé anneau des parties principales d'ordre q.

Propriétés de $\mathrm{Diff}_k A$:

1) Comme A-module $P_{A/k}^q$ est isomorphe à $A \otimes_k 1 \oplus \Omega_{A/k}^{(q)}$, et si l'on note $d_{A/k}^q$ l'homomorphisme d'anneaux : $A \longrightarrow P_{A/k}^q$ défini par $d_{A/k}^q(a)$ = classe de $(1 \otimes a = a \otimes 1 + d_{A/k}^{(q)}(a))$, nous avons : $d_{A/k}^q(a) = a + d_{A/k}^{(q)}(a)$ pour tout a dans A.

De plus les A-modules $\mathrm{Diff}_k^q(A)$ et $\mathrm{Hom}_A(P_{A/k}^q, A)$ sont isomorphes : à tout élément D de $\mathrm{Hom}_A(P_{A/k}^q, A)$ on fait correspondre $D \circ d_{A/k}^q$.

2) Si S est une partie multiplicative de A, on a : $\mathrm{Diff}_k^q(S^{-1}A) \cong S^{-1}A \oplus \mathrm{Hom}_{S^{-1}A}(S^{-1}\Omega_{A/k}^{(q)}, S^{-1}A)$ (cf. I.1 prop. 3). Donc si $\Omega_{A/k}^{(q)}$ est de présentation finie nous aurons un isomorphisme de $S^{-1}A$-modules :

$$\mathrm{Diff}_k^q(S^{-1}A) \cong S^{-1}\,\mathrm{Diff}_k^q(A) \quad .$$

3) L'anneau $P^q_{A/k}$ est filtré par les I^j/I^{q+1}, on a donc un gradué associé, $\mathrm{gr}\,P^q_{A/k} = \coprod_{j=o}^{q} \frac{I^j}{I^{j+1}}$. Notons $\mathrm{gr}\,P_{A/k}$ l'anneau $\coprod_{n\in\mathbb{N}} \frac{I^n}{I^{n+1}}$, et $S^{\cdot}_A(\Omega^{(1)}_{A/k})$ l'algèbre symétrique du A-module $\Omega^{(1)}_{A/k}$. Il existe un homomorphisme surjectif d'algèbres :

(*) $S^{\cdot}_A(\Omega^{(1)}_{A/k}) \longrightarrow \mathrm{gr}\,P_{A/k}$, qui envoie $S^q_A(\Omega^{(1)}_{A/k})$ sur $\mathrm{gr}\,P^q_{A/k}$. En particulier il suffit que $\Omega^{(1)}_{A/k}$ soit de type fini pour que les $\Omega^{(q)}_{A/k}$ le soient pour tout $q\in\mathbb{N}$.

4) L'anneau $\mathrm{Diff}_k\,A$ est gradué par les A-modules $\mathrm{Diff}^q_k(A)$ et d'après la proposition 1 son gradué $\mathrm{gr}\,\mathrm{Diff}_k(A) = \coprod_q \frac{\mathrm{Diff}^q_k(A)}{\mathrm{Diff}^{q-1}_k(A)}$ est commutatif.

5) Reprenons S une partie multiplicative de A, et soit P un élément de $\mathrm{Diff}^q_k(A)$, alors il définit un élément de $\mathrm{Diff}^q_k(S^{-1}A)$; en effet pour $q=0$ c'est clair et pour $q>0$ on pose, si $S^{-1}f\in S^{-1}A$:

$$P(S^{-1}f) = S^{-1}\{P(f) - [P,S]\,(S^{-1}f)\} \qquad \text{(cf. [Sw])}.$$

On vérifie que cette définition ne dépend pas du représentant $S^{-1}f$ choisi et que cette formule donne bien un élément de $\mathrm{Diff}^q_k(S^{-1}A)$.
On a alors de manière évidente un homomorphisme d'anneaux : $\mathrm{Diff}_k(A) \longrightarrow \mathrm{Diff}_k(S^{-1}A)$.
En particulier si S est constituée de non diviseurs de 0, cet homomorphisme est injectif.

I.3) <u>Application</u> : <u>Anneaux de polynômes et algèbres de type fini</u>
Conservons les notations de I.1 ; soit A une k-algèbre et $d : A \longrightarrow I$, $d(a) = 1\otimes a - a\otimes 1$, si $a\in A$. On prouve facilement le lemme :

<u>Lemme</u> 1 : <u>Si</u> $(a_1,\dots,a_n)\in A^n$ <u>alors</u>

$$d(a_1,\dots,a_n) = \sum_{S=1}^{n} \sum_{i_1<\dots<i_n} a_1 \dots \hat{a}_{i_1} \dots \hat{a}_{i_S} \dots a_n\, da_{i_1} \dots da_{i_S} \; .$$

Soit L un ensemble et X_λ ; $\lambda\in L$, des indéterminées, $A = k[X_\lambda,\ \lambda\in L]$.
On peut écrire $A\otimes_k A = k[X_\lambda\otimes 1,\ 1\otimes X_\lambda] = k[X_\lambda,\ dX_\lambda]$, comme l'idéal I est engendré par les $\{d(a),\ a\in A\}$, on a grâce au lemme 1 :

$$d(\textstyle\sum \alpha_I X^I) = \sum \text{(éléments de A) (monômes en } dX_\lambda) \text{ pour } \sum \alpha_I X^I\in A \quad .$$

On a alors facilement dans les hypothèses précédentes :

Proposition 2 - Le A-module $\Omega^{(1)}_{A/k} = \frac{I}{I^2}$ est libre sur A de base $\{d^{(1)} X_\lambda, \lambda \in L\}$, de même des $\Omega^{(q)}_{A/k}$ sont libres de base les classes modulo I^{q+1} des monômes en dX_λ de degré inférieur ou égal à q .

Remarquons que si $F = \sum \alpha_I X^I \in A$, $d^q_{A/k}(F)$ est la classe modulo I^{q+1} de $F(X + dX)$.

Proposition 3 - Si A est une k-algèbre de type fini les $\Omega^{(q)}_{A/k}$ sont des A-modules de type fini, si de plus k est noethérien les A-modules $\mathrm{Diff}^q_k A$ sont aussi de type fini.

Preuve - Si A est quotient de $k[X_1, \dots, X_n]$ la proposition 3 de I.1 montre que les $\Omega^{(q)}_{A/k}$ sont de type fini ; si k est noethérien A l'est aussi donc les $\mathrm{Diff}^q_k(A)$ sont également de type fini.

Remarque 3.1 - Soit A une k-algèbre telle que $\Omega^{(1)}_{A/k}$ soit libre de rang n , ($k[X_1, \dots, X_n]$ par exemple), et soient $Z_1, \dots, Z_n$ tels que $\{d^{(1)} Z_i ; i = 1, \dots, n\}$ soit une base de $\Omega^{(1)}_{A/k}$; notons $\{D_i ; i = 1, \dots, n\}$ la base duale, ainsi $D_i(d^{(1)} Z_j) = \delta_{ij}$. L'isomorphisme $\mathrm{Hom}_A(\Omega^{(1)}_{A/k}, A) \simeq \mathrm{Der}^{(1)}_k(A)$ donne $\frac{\partial}{\partial Z_i} = D_i \circ d^{(1)}_{A/k}$ et toute dérivation s'écrit de manière unique $\sum_{i=1}^{n} a_i \frac{\partial}{\partial Z_i}$ avec a_i dans A , on a alors pour tout $g \in A$: $d^{(1)} g = \sum_{i=1}^{n} \frac{\partial g}{\partial Z_i} d^{(1)} Z_i$.

Examinons les conséquences de cette remarque sur $\mathrm{Diff}_k A$. Pour cela introduisons les notations suivantes ; si A est une k-algèbre et $(Z_\lambda)_{\lambda \in L}$ dans A tels que les $d^{(1)} Z_\lambda$ engendrent $\Omega^{(1)}_{A/k}$, pour $p \in \mathbb{N}^{(L)}$ on notera :

$$|p| = \sum_\lambda p_\lambda \ , \ p! = \prod p_\lambda \ , \ \binom{p}{q} = \frac{p!}{q!(p-q)!} \quad \text{si } q \leq p \text{ et } \binom{p}{q} = 0 \text{ si } q \not\leq p$$

pour $q \in \mathbb{N}^{(L)}$ et enfin $Z^p = \prod Z_\lambda^{p_\lambda}$, $d^{(m)} (Z)^p = (d^m Z - Z)^p = \prod_\lambda d^{(m)}(Z_\lambda)^{p_\lambda}$ dans $P^m_{A/k}$.

Nous poserons $\zeta_\lambda = d^{(m)} Z_\lambda$ et $\zeta^p = d^{(m)}(Z)^p$.

On a le théorème ([G] chap. IV - Th. 16.11.2) :

Théorème 4 - Avec les notations précédentes il y a équivalence entre :

(i) $\Omega^{(1)}_{A/k}$ est libre de base $(d^{(1)} Z_\lambda)_{\lambda \in L}$ et l'homomorphisme (*), de I.2 propriété 3, est bijectif ;

(ii) Il existe une famille $(D_p)_{p\in\mathbb{N}^{(L)}}$ d'éléments de $\mathrm{Diff}_k A$ telle que : $D_p(Z^q) = \binom{p}{q} Z^{q-p}$ pour $p,q\in\mathbb{N}^{(L)}$ (**).

De plus, si (ii) est vérifié, la famille (D_p) est uniquement déterminée par (**) et vérifie $D_p \circ D_q = D_q \circ D_p = \frac{(p+q)!}{p!\,q!} D_{p+q}$, pour $p,q\in\mathbb{N}^{(L)}$.

Si L est fini les D_p, $|p|\leq m$ forment une base du A-module $\mathrm{Diff}_k^m(A)$ pour tout $m\in\mathbb{N}$.

Remarquons que dans le cas (i) les ζ^p pour $|p|\leq m$ forment une base du A-module $P^m_{A/k}$, et si $|p| = 1$ et L fini, les D_p ne sont autres que les $\frac{\partial}{\partial Z_i}$ de la remarque 3.1, mais il n'est pas vrai qu'un opérateur différentiel d'ordre quelconque puisse s'écrire comme une combinaison linéaire de puissances des $\frac{\partial}{\partial Z_i}$. Néanmoins en caractéristique 0 on a le résultat suivant :

Théorème 5 - Soit k un corps de caractéristique 0, A une k-algèbre de type fini telle que $\Omega^{(1)}_{A/k}$ soit un A-module projectif, alors pour tout $P\in\mathrm{Spec}\, A$ il existe $f\notin p$ tel que Af vérifie la condition (ii) du théorème 4.

Preuve - Il existe $f\notin p$ tel que $\Omega^{(1)}_{Af/k}$ soit libre de rang n, si l'on prend $\{d^{(1)} Z_1, \ldots, d^{(1)} Z_n\}$, base de $\Omega^{(1)}_{Af/k}$, les $\frac{\partial}{\partial Z_i}$ de la remarque 3.1 fournissent la famille $\{D_p = \frac{1}{p!} \prod_\lambda (\frac{\partial}{\partial Z_\lambda})^p$ pour $p\in\mathbb{N}^n\}$ qui vérifie le théorème 4 (ii).

Remarquons que dans ce cas on a globalement : $\mathrm{Diff}_k^q(A) = \mathrm{Diff}_k^1(A) \ldots \mathrm{Diff}_k^1(A)$ (q fois) et le gradué de $\mathrm{Diff}_k(A)$ (I.2 propriété 4) est isomorphe à l'algèbre symétrique de $\mathrm{Der}_k^{(1)}(A)$ qui est un A-module projectif (cf. [Sw] théorème 18.2).

Corollaire 6 -

1) Si k est un corps de caractéristique 0 et A une k-algèbre de type fini intègre $\mathrm{Diff}_k(A)$ est un anneau intègre, (cf. [B] page 5).

2) Si $A = k[X_1, \ldots, X_n]$; $\mathrm{Diff}_k(A) = k[X_1, \ldots, X_n, \frac{\partial}{\partial X_1}, \ldots, \frac{\partial}{\partial X_n}]$ est l'algèbre de Weyl à 2n générateurs, $A_n(k)$.

Preuve -

1) Si K est le corps des fractions de A, $\Omega^{(1)}_{K/k}$ est un espace vectoriel de dimension finie, n, donc d'après le théorème 5 il existe $D_1, \ldots, D_n$ tels que $\mathrm{Diff}_k K$ soit engendré sur K par $D_1, \ldots, D_n$; son gradué pour l'ordre étant un

anneau de polynômes à n variables, $\mathrm{Diff}_k K$ est intègre et par la propriété 5) de I.2 on déduit que c'est aussi le cas de $\mathrm{Diff}_k(A)$.

2) Evident.

Donnons une autre description de $\mathrm{Diff}_k(A)$ lorsque A est une k algèbre de type fini sur un corps de caractéristique 0 : $A = \frac{k[X_1,\dots,X_n]}{I}$, I idéal de $R = k[X_1,\dots,X_n]$.

Remarquons que si R est une k-algèbre commutative quelconque et I un idéal de R tout opérateur, Δ, de $\mathrm{Diff}_k^m(R)$ tel que $\Delta(I) \subset I$ définit un opérateur de la k algèbre $\frac{R}{I}$, d'ordre m. Lorsque A est comme ci-dessus $\mathrm{Diff}_k(R) = A_n(k)$ et nous poserons :

$\mathrm{Diff}_k^m(I) = \{P \in \mathrm{Diff}_k^m(R) / P(I) \subset I\}$; $I\,\mathrm{Diff}_k^m(R) = \{P \in \mathrm{Diff}_k^m(R)$ tels que $P = \sum a_\alpha (\frac{\partial}{\partial X})^\alpha$ avec $a_\alpha \in I\}$; $\mathrm{Diff}_k(I) = \bigcup_m \mathrm{Diff}_k^m(I)$ et $I\,\mathrm{Diff}_k(R) = \bigcup_m I\,\mathrm{Diff}_k^m(R)$.

Alors :

<u>Théorème</u> 7 - <u>On a</u> $\mathrm{Diff}_k^m(A) = \frac{\mathrm{Diff}_k^m(I)}{I\,\mathrm{Diff}_k^m(R)}$ <u>pour tout</u> m <u>et</u> $\mathrm{Diff}_k(A) =$

$$\frac{\mathrm{Diff}_k(I)}{I\,\mathrm{Diff}_k(R)} .$$

<u>Preuve</u> - analogue à [K] chap. IV Théorème 2.

Pour terminer rappelons que la condition " $\Omega^{(1)}_{A/k}$ est un A-module projectif" est une condition de lissité sur A :

<u>Théorème</u> 8 - <u>Soit</u> A <u>une</u> k-<u>algèbre de type fini sur un corps</u>, k, <u>de caractéristique</u> 0 <u>algébriquement clos</u>. <u>Alors l'anneau</u> A <u>est régulier de dimension</u> n <u>si et seulement si</u> $\Omega^{(1)}_{A/k}$ <u>est un</u> A-<u>module projectif de rang</u> n .

<u>Preuve</u> - cf. [H] Chap. II - Th. 8.15.

II) <u>Faisceaux d'opérateurs différentiels</u>, <u>et applications</u>

II.1) En ce qui concerne schémas, variétés, faisceaux, $\mathcal{O}_X$-modules on suivra la terminologie de [H] . En particulier si k est un corps algébriquement clos de caractéristique 0 , un schéma $(X, \mathcal{O}_X)$ intègre de type fini est une variété si $(X, \mathcal{O}_X)$ est séparé et l'application diagonale $\Delta : X \longrightarrow X \times_k X$ est une immersion fermée.

Le but de ce paragraphe est de donner une version algébrique des théorèmes du chapitre III de [K] . Nous noterons k un corps algébriquement clos de caractéristique 0 .

Si $(X, \mathcal{O}_X)$ est une variété sur k , l'application diagonale $\Delta : X \longrightarrow X \times_k X$ possède un comorphisme qui sur tout ouvert affine $U = \operatorname{Spec} A$ de X est l'application $A \otimes_k A \xrightarrow{p} A$ de I.1 . On vérifie (cf. [G] Chap. IV, § 16) que pour tout m il existe des faisceaux d'anneaux et de modules $\mathcal{P}^m_{X/k}$ et $\Omega^{(m)}_{X/k}$, qui sont des $\mathcal{O}_X$-modules cohérents tels que sur tout ouvert affine U = Spec A on a :

$$\mathcal{P}^m_{X/k}(U) = \frac{A \otimes_k A}{I^{m+1}} , \quad \Omega^{(m)}_{X/k}(U) = \frac{I}{I^{m+1}} .$$

On note alors $\operatorname{Diff}^m_{X/k} = \operatorname{Hom}_{\mathcal{O}_X}(\mathcal{P}^m_{X/k}, \mathcal{O}_X)$ le faisceau dual, c'est un $\mathcal{O}_X$-module cohérent et $\operatorname{Diff}^m_{X/k}(U) = \operatorname{Diff}^m_k(A)$.

En posant $\operatorname{Diff}_{X/k} = \bigcup_{m=0}^{\infty} \operatorname{Diff}^m_{X/k}$ on obtient un faisceau d'anneaux qui est un $\mathcal{O}_X$-module quasi-cohérent. (En fait $(X, \mathcal{O}_X)$ schéma suffit pour faire ces constructions).

Le théorème 8 de I.3 se traduit alors par :

<u>Proposition</u> 1 - <u>Si</u> $(X, \mathcal{O}_X)$ <u>est une variété sur</u> k , $\Omega^{(1)}_{X/k}$ <u>est un</u> $\mathcal{O}_X$<u>-module localement libre de rang</u> n = dim X <u>si et seulement si la variété</u> X <u>est non singulière</u>.

On a aussi (Th. 5 - I.3).

<u>Proposition</u> 2 - <u>Si</u> $(X, \mathcal{O}_X)$ <u>est une variété non singulière sur</u> k <u>de dimension</u> n , <u>pour tout</u> $x \in X$, <u>il existe un ouvert affine</u> U <u>contenant</u> x <u>tel que</u> $\operatorname{Diff}_{X/k}(U)$ <u>soit engendré sur</u> $\mathcal{O}_X(U)$ <u>par une famille libre</u> $D_1, \ldots, D_n$ <u>d'éléments de</u> $\operatorname{Der}^{(1)}_k(\mathcal{O}_X(U))$.

II.2) <u>Un théorème de prolongement</u>

Soit $\varphi : (X, \mathcal{O}_X) \longrightarrow (Y, \mathcal{O}_Y)$ un isomorphisme entre deux variétés sur k , si V est un ouvert de Y et $P \in \operatorname{Diff}_{Y/k}(V)$ soit $U = \varphi^{-1}(V)$; on définit un élément $\varphi^*(P)$ de $\operatorname{Diff}_{X/k}(U)$ par la formule :

$$(*) \quad \varphi^*(P)\,(g \circ \varphi) = P(g) \circ \varphi \qquad \text{pour tout } g \in \mathcal{O}_Y(V) \quad .$$

Donc tout opérateur différentiel sur Y possède une image inverse sur X définie par (*). Nous allons essayer de construire $\varphi^*(P)$ lorsque φ n'est plus nécessairement un isomorphisme.

Soient $(X, \mathcal{O}_X)$, $(Y, \mathcal{O}_Y)$ deux variétés, $\pi : X \longrightarrow Y$ un morphisme qui est, en dehors d'un sous-schéma fermé propre Z de X , une immersion ; donc $\pi(X \setminus Z)$ est un sous-schéma ouvert d'un sous-schéma fermé, T, de Y .

Si V est un ouvert affine de $\pi(X\setminus Z)$ c'est-à-dire un ouvert affine de T , d'après l'hypothèse il existe un ouvert affine de Y , soit W = Spec B , tel que $V = T \cap W$ est de la forme Spec $\frac{B}{I}$ où I est un idéal de B . Si P est un opérateur différentiel sur Y , on a $P \in \mathrm{Diff}_{Y/k}(W)$ et si P laisse stable l'idéal I on pourra définir P sur V et la condition (*) définira $\pi^*(P)$ sur $U = \pi^{-1}(V)$.

<u>Définition</u> 1 - <u>Si</u> P <u>vérifie l'hypothèse précédente, on dira que</u> P <u>se restreint à</u> $\pi(X\setminus Z)$.

Donc si P se restreint à $\pi(X\setminus Z)$ il possède une image inverse $\pi^*(P)$ définie sur $X\setminus Z$.

Supposons X et Y non singulières et essayons de trouver une écriture locale de $\pi^*(P)$. La question étant locale on peut supposer que X = Spec A(X), Y = Spec A(Y) avec A(X), A(Y) réguliers de dimensions n et m ; si $\pi : X \longrightarrow Y$, on note $\pi^{\#}$ le comorphisme $A(Y) \longrightarrow A(X)$ qui fait de A(X) un A(Y)-module. A l'aide de la proposition 2 de II.1 on peut encore supposer que $\mathrm{Diff}_k\, A(X)$ est engendré par $\partial_1, \dots, \partial_n$ et $\mathrm{Diff}_k\, A(Y)$ par $D_1, \dots, D_m$, en tant qu'algèbres sur A(X) et A(Y) respectivement. Si q est un entier, un élément Δ de $\mathrm{Der}_k^{(q)}(A(Y), A(X)) = \mathrm{Hom}_{A(Y)}(\Omega^{(q)}_{A(Y)/k}, A(X))$ est uniquement déterminé par les images d'une base du A(Y)-module libre $\Omega^{(q)}_{A(Y)/k}$; c'est-à-dire que, avec les notations du Théorème 4 de I.3, si $(d^{(q)}(Z))^p = \zeta^p$ pour $|p| \leq q$ est une base de $P^q_{A(Y)/k}$, Δ est uniquement déterminé par les $x_\mu = \Delta(\zeta^\mu)$ pour $1 \leq |\mu| \leq q$, $\mu \in \mathbb{N}^m$.

Compte tenu de : $\langle \pi^{\#} \circ D^\mu , \zeta^\nu \rangle = \delta_{\nu,u}$ pour tout $\nu, \mu \in \mathbb{N}^m$, on voit que Δ et $\sum\limits_{\substack{|\mu| \leq q \\ \mu \in \mathbb{N}^m}} x_\mu\, \pi^{\#} \circ D^\mu$ sont deux éléments de $\mathrm{Der}_k^{(q)}(A(Y), A(X))$ qui prennent les mêmes valeurs sur les ζ^ν, $\nu \in \mathbb{N}^m$, donc sont égaux.

Si $\beta \in \mathbb{N}^n$, $|\beta| \leq q$, on a $\partial^\beta \circ \pi^{\#} \in \mathrm{Der}_k^{(q)}(A(Y), A(X))$; donc il existe des $c_{\beta,\mu}$ dans $A(X)$ avec $\partial^\beta \circ \pi^{\#} = \sum_\mu c_{\beta,\mu} \pi^{\#} \circ D^\mu$.

Le problème est donc le suivant : si P est un opérateur différentiel sur $A(Y)$ et que l'on veut définir $\pi^*(P)$ sous la forme $\sum b_\beta \partial^\beta$, que sont les b_β ? On sait que ce sont des fonctions régulières sur $X \setminus Z$ et $\pi^*(P)$ est défini par la condition : $(*)$ $\pi^*(P)(\pi^{\#}(f)) = \pi^{\#}(P(f))$ pour $f \in A(Y)$.

Si P s'écrit $\sum a_\alpha D^\alpha$ avec $a_\alpha \in A(Y)$, $(*)$ équivaut à :

$$\sum b_\beta (\partial^\beta \circ \pi^{\#})(f) = \sum_\alpha \pi^{\#} a_\alpha \pi^{\#} \circ D^\alpha (f) ,$$

lorsque l'on cherche $\pi^*(P)$ sous la forme précédente. On obtient ainsi :

$$\sum_\beta b_\beta \sum_\mu c_{\beta,\mu} (\pi^{\#} \circ D^\mu)(f) = \sum_\alpha \pi^{\#} a_\alpha (\pi^{\#} \circ D^\alpha)(f)$$

et par unicité de l'écriture suivant les $\pi^{\#} \circ D^\mu$, on a pour tout $\alpha \in \mathbb{N}^m$:

$$\sum c_{\beta,\alpha} b_\beta = \pi^{\#} a_\alpha .$$

Le problème est ainsi de résoudre un système d'équations dans $A(X)$ qui aura des solutions dans le corps des fractions de $A(X)$, ces solutions étant des fonctions régulières sur $X \setminus Z$.

Pour examiner le cas où X et Y ne sont pas nécessairement non singulières nous aurons besoin de la proposition suivante :

<u>Proposition</u> 2 - <u>Soit</u> X <u>une variété normale</u>, T <u>un sous-schéma fermé tel que</u> : $\mathrm{Codim}_X T \geq 2$; <u>si</u> P <u>est un opérateur différentiel d'ordre</u> q <u>sur</u> $X \setminus T$ <u>il se prolonge de manière unique en un opérateur différentiel d'ordre</u> q <u>sur</u> X.

<u>Preuve</u> - Si l'on prouve la proposition pour $q=0$, il suffira de terminer comme dans [K] page 34. La question étant locale on peut supposer X affine. On a une suite exacte de cohomologie locale à support dans T :

$$0 \longrightarrow H^0_T(X, \mathcal{O}_X) \longrightarrow \Gamma(X, \mathcal{O}_X) \longrightarrow \Gamma(X \setminus T, \mathcal{O}_X) \longrightarrow H^1_T(X, \mathcal{O}_X) \longrightarrow \cdots$$

Puisque X est normale et T de codimension au moins 2 on a :

$$\mathrm{prof}\ \mathcal{O}_{X,x} \geq \inf(2, \dim \mathcal{O}_{X,x}) \geq 2 \text{ pour tout } x \in T .$$

Donc $H^i_T(X, \mathcal{O}_X) = 0$ si $i < 2$, d'où le résultat puisque les opérateurs différentiels d'ordre 0 sur $X \setminus Z$ sont les éléments de $\Gamma(X \setminus Z, \mathcal{O}_X)$.

Revenons au cas où $\pi : X \longrightarrow Y$ est un morphisme de variétés qui est une immersion en dehors d'un sous-schéma fermé propre, Z, de X , et supposons de plus X normale. Notons X^{reg} le sous-schéma ouvert formé des points réguliers de X , en dehors de $Z \cap X^{reg}$, $\pi : X^{reg} \longrightarrow Y$ est encore une immersion. Soit P un opérateur différentiel qui se restreint à $\pi(X \setminus Z)$, on obtient alors $\pi^*(P)$ sur $X \setminus Z$ que l'on veut décrire sur Z . Le problème étant local, on peut supposer Y affine égale à $\operatorname{Spec} \dfrac{k[T_1,\ldots,T_N]}{J}$ et X^{reg} schéma régulier affine. Dans ce cas Y est un sous-schéma fermé de k^N , donc $\pi : X^{reg} \longrightarrow k^N$ est une immersion en dehors de $X^{reg} \cap Z$. L'opérateur P sur Y provient d'un opérateur $\hat{P}$ de k^N (I.3 Théorème 7), et l'image inverse $\pi^*(P)$ coïncide sur X^{reg} avec $\pi^*(\tilde{P})$. D'après ce qui a été fait dans le cas non singulier, $\pi^*(\tilde{P})$ s'écrit localement $\Sigma\, b_\beta \partial^\beta$ où les b_β sont des fonctions régulières sur $X^{reg} \setminus Z$, et rationnelles sur $X^{reg} \cap Z$.

Comme la codimension du lieu singulier, T, de X est supérieure ou égale à 2 , la proposition 2 de II.2 permet d'affirmer que $\pi^*(\tilde{P})$ se prolonge en un opérateur différentiel sur $X \setminus Z$ qui coïncide avec $\pi^*(P)$. Sur Z , localement les dénominateurs des b_β vont aussi admettre un prolongement à Z tout entier.

<u>Définition</u> 3 - <u>Lorsque</u> $\pi^*(P)$ <u>vérifie les propriétés précédentes, on dira que</u> $\pi^*(P)$ <u>"est à coefficients rationnels"</u> <u>sur</u> Z .

D'où :

<u>Théorème</u> 2 - <u>Soient</u> X <u>et</u> Y <u>deux variétés</u>, X <u>normale</u>, <u>et</u> $\pi : X \longrightarrow Y$ <u>un morphisme qui est une immersion en dehors d'un sous-schéma fermé propre</u>, Z , <u>de</u> X . <u>Si</u> P <u>est un opérateur différentiel sur</u> Y <u>qui se restreint à</u> $\pi(X \setminus Z)$, <u>alors</u> P <u>possède une image inverse</u> $\pi^*(P)$ <u>qui est un opérateur différentiel sur</u> $X \setminus Z$ <u>et "est à coefficients rationnels"</u> <u>sur</u> Z .

II.3 - <u>Singularités quotients</u>, <u>d'après</u> [K]

Soit k un corps algébriquement clos de caractéristique 0 . Rappelons quelques définitions et propriétés concernant les morphismes étales (cf. [S.G.A]).
Soient X et Y deux schémas de type fini sur k , f un morphisme $X \longrightarrow Y$, $x \in X$, $y = f(x)$.

<u>Définition</u> 1 - <u>On dit que</u> f <u>est étale en</u> x <u>si</u> $O_{X,x}$ <u>est plat sur</u> $O_{Y,y}$ <u>et</u> $\Omega^{(1)}_{O_{X,x}/O_{Y,y}} = 0$.

On a ([S.G.A] exposé 1) :

<u>Proposition</u> 2 - <u>Soit</u> $f : X \longrightarrow Y$ <u>un morphisme de variétés</u> ; $x \in X$, $y = f(x)$. <u>Si</u> f <u>est étale en</u> x , X <u>est non singulière en</u> x <u>si et seulement si</u> Y <u>l'est en</u> y , <u>et si</u> f <u>est dominant l'ensemble des points où</u> f <u>est étale est un ouvert non vide de</u> X .

On obtient alors :

<u>Proposition</u> 3 - <u>Soit</u> $f : X \longrightarrow Y$ <u>morphisme de variétés</u> ; $x \in X$, $y = f(x)$. <u>Supposons</u> f <u>étale en</u> x <u>et</u> $x \in X^{reg}$, <u>l'ouvert régulier de</u> X . <u>Si</u> $f^{\#}$ <u>est le comorphisme de</u> f, $f^{\#}$ <u>induit des isomorphismes</u> :

$$\Omega^{(1)}_{O_Y,y/k} \otimes_{O_{Y,y}} O_{X,x} \cong \Omega^{(1)}_{O_{X,x/k}}$$

$$\underline{\text{et}} \quad Der^{(1)}_k(O_{X,x}) \cong O_{X,x} \otimes_{O_{Y,y}} Der^{(1)}_k(O_{Y,y}) \quad .$$

<u>Preuve</u> - Si $A = O_{Y,y}$, $B = O_{X,x}$, n = dim A = dim B , $f^{\#}$ $A \longrightarrow B$, $\Omega^{(1)}_{A/k}$ et $\Omega^{(1)}_{B/k}$ sont libres de rang n sur A et B , respectivement. De plus d'après I.1 prop. 3, on a une suite exacte de B modules :

$$\Omega^{(1)}_{A/k} \otimes_A B \xrightarrow{\varphi} \Omega^{(1)}_{B/k} \longrightarrow \Omega^{(1)}_{B/A} = 0 \longrightarrow 0 \quad .$$

Si $N = \ker \varphi$ on voit aisément que N est un B-module de torsion, mais comme B est intègre, N = 0 , donc $\Omega^{(1)}_{B/k} \cong \Omega^{(1)}_{A/k} \otimes_A B$.

En dualisant : $Hom_B(\Omega^{(1)}_{B/k}, B) \cong B \otimes Hom_A(\Omega^{(1)}_{A/k}, A)$, car B est plat sur A , donc $Der^{(1)}_k(B) \cong B \otimes_A Der^{(1)}_k(A)$.

<u>Corollaire</u> 4 - <u>Soit</u> $f : X \longrightarrow Y$ <u>un morphisme de variétés</u>, $x \in X^{reg}$, $y = f(x)$ <u>où</u> f <u>est étale en</u> x . <u>Si</u> $q \in \mathbb{N}$ <u>il existe un ouvert</u> U <u>de</u> X <u>contenant</u> x , <u>un ouvert</u> V <u>de</u> Y <u>contenant</u> y <u>sur lesquels</u> f <u>est étale et</u> :

$$Diff^q_{X/k}(U) \cong O_X(U) \otimes_{O_Y(V)} Diff^q_{Y/k}(V) \quad .$$

<u>Preuve</u> - D'après la proposition 3 : $Der^{(1)}_k(O_{X,x}) \cong O_{X,x} \otimes_{O_{Y,y}} Der^{(1)}(O_{Y,y})$ ce qui implique en particulier que toute dérivation de $O_{Y,y}$ se prolonge par l'intermédiaire de $f^{\#}$ en une dérivation de $\mathcal{O}_{X,x}$. Si $\partial_1, \ldots, \partial_n$ est une base de $Der^{(1)}_k(O_{Y,y})$ sur $O_{Y,y}$, les prolongements ainsi obtenus forment une base de

$Der_k^{(1)}(O_{X,x})$ d'après l'isomorphisme précédent. D'autre part $Der_k^{(1)}(O_{X,x})$ et $Der_k^{(1)}(O_{Y,y})$ engendrent $Diff_k(O_{X,x})$ et $Diff_k(O_{Y,y})$ (cf. I.3 Théorème 4), en particulier on a : $O_{X,x} \otimes_{O_{Y,y}} Diff_k^q(O_{Y,y}) \cong Diff_k^q(O_{X,x})$ pour tout $q \in \mathbb{N}$.

Puisque X^{reg} est un ouvert de X , ainsi que l'ensemble des points où f est étale et que les O_X et O_Y modules $Diff^q_{X/k}$ et $Diff^q_{Y/k}$ sont cohérents on obtient les ouverts U et V de l'énoncé.

Considérons un sous-groupe fini G de $GL(n,k)$ opérant sur l'espace affine k^n , donc sur $R = k[X_1, \dots, X_n]$. Soit $X = Spec\ R = k^n$, l'action de G définit une variété quotient $Y = X/G = Spec\ R^G$ où R^G est l'anneau des invariants. Rappelons que R^G est une k-algèbre de type fini normale. Notons K le corps des fractions de R , alors K^G est le corps des fractions de R^G , l'extension $K \supset K^G$ étant galoisienne de groupe G .

Notons $\pi : X \longrightarrow Y$ l'application quotient ; π est un morphisme fini surjectif, l'ensemble des points où π est étale est un ouvert non vide de X , nous noterons Z le fermé, propre, complémentaire de cet ouvert.

Si $p \in X$, on note $G_i(p) = \{\gamma \in G / \forall a \in R\ \ \gamma(a) - a \in p\}$ le groupe d'inertie de p, si $G_i(p) = \{1\}$ alors π est étale en p .

On a vu que $Diff_k(R) = A_n(k)$, l'algèbre de Weyl, et si $P \in A_n(k)$ on définit, pour tout $\gamma \in G$, un opérateur P^γ en posant : $P^\gamma(f) = \gamma.P(\gamma^{-1}f)$ pour tout f dans R . Ainsi le groupe G opère sur $A_n(k)$, et cette opération respecte l'ordre des opérateurs différentiels.

Avec les notations qui précèdent, le problème est de déterminer les liens qui existent entre $Diff_k(R^G)$ et $A_n(k)^G$. On a le théorème suivant, semblable au th. 4, chap. III de [K] compte tenu du théorème de prolongement algébrique démontré en II.2 :

Théorème 5 -

(a) Pour qu'un élément de $A_n(k)$ induise un opérateur sur R^G il faut et il suffit qu'il appartienne à $A_n(k)^G$

(b) Il existe un homomorphisme injectif : $A_n(k)^G \hookrightarrow Diff_k(R^G)$

(c) L'injection précédente est une surjection si et seulement si le groupe G ne contient pas de pseudo-réflexion différente de l'identité.

Preuve - Commençons par un lemme :

Lemme 6 - Si $P \in Der_k^{(q)}(R)$ et P est R^G-linéaire, alors $P = 0$.

Preuve - Nous savons qu'il y a une injection $Der_k^{(q)}(R) \subset Der_k^{(q)}(K)$. D'autre part on a $Der_{R^G}^{(q)}(R) \subset Der_{R^G}^{(q)}(K)$ et $Der_{K^G}^{(q)}(K) = 0$ si $q \geq 0$ car l'extension $K \supset K^G$ est algébrique de type fini, séparable (cf. I.1, prop. 3). Il suffit donc de vérifier que si $P \in Der_{R^G}^{(q)}(K)$, alors P est K^G linéaire ; mais ceci est évident car si $\beta \in R^G \setminus \{0\}$ et $x \in K$ on a $\beta P(\frac{x}{\beta}) = P(x)$ donc $P(\frac{x}{\beta}) = \frac{1}{\beta} P(x)$.

Démonstration du théorème 5 :

(a) Si $P \in A_n(k)^G$ alors quelques soient $\gamma \in G$ et $f \in R$ on a $P(\gamma.f) = \gamma.P(f)$, donc si $f \in R^G$, $P(f) = \gamma.P(f)$, et $P \in Diff_k(R^G)$.
Réciproquement soit $P \in Diff_k^q(R)$ tel que $P(R^G) \subset R^G$. Posons $P = a + \Delta$ avec $a \in R$, $\Delta \in Der_k^{(q)}(R)$; comme $P(1) = a$, on obtient $a \in R^G$, et $\Delta = P - a$ vérifie $\Delta(R^G) \subset R^G$. Il suffit donc de montrer $\Delta \in A_n(k)^G$. Procédons par récurrence sur q ; si $q = 0$ alors $\Delta = 0$; si $q \geq 1$, pour tout $\gamma \in G$ on a quelques soient $a \in R^G$ et $x \in R$,

$$(\Delta^\gamma - \Delta)(ax) = \gamma\{[\Delta, a]'(\gamma^{-1}x) + a\,\Delta(\gamma^{-1}x) + \gamma^{-1}x\,\Delta(a)\} - [\Delta, a]'(x) - a\,\Delta(x) - x\,\Delta(a) .$$

On vérifie aisément que $[\Delta, a]'(R^G) \subset R^G$ et ainsi par hypothèse de récurrence $[\Delta, a]'^\gamma = [\Delta, a]'$; celà donne $\gamma.[\Delta, a]'(x) = [\Delta, a]'(x)$, d'où $(\Delta^\gamma - \Delta)(ax) = a[\Delta^\gamma - \Delta](x)$. Donc $\Delta^\gamma - \Delta \in Der_{R^G}^{(q)}(R)$, et par le lemme 6, $\Delta \in A_n(k)^G$.

(b) Il s'agit de montrer que si $P \in A_n$ et $P(R^G) = 0$ alors $P = 0$.
Si $P = a + \Delta$ avec $a \in R$, $\Delta \in Der_k^{(q)}(R)$, $P(1) = 0$ implique $a = 0$. Montrons que Δ est R^G linéaire ce qui donnera $\Delta = 0$. On procède par récurrence sur q, on a $\Delta(af) = [\Delta, a]'(f) + a\,\Delta(f) + f\,\Delta(a)$ pour tout $a \in R^G$ et $f \in R$. Par récurrence et puisque $\Delta(a) = 0$, on obtient $\Delta(af) = a\,\Delta(f)$.

(c) Montrons que si $\gamma \in G \setminus \{1\}$ et si γ fixe un hyperplan H, alors $Diff_k(R^G)$ contient strictement $A_n(k)^G$. On peut supposer que $H = \{x = (x_1, \dots, x_n) \in k^n / x_1 = 0\}$, ce qui donne pour γ :
$\gamma(X_1, \dots, X_n) = (\varepsilon X_1, X_2, \dots, X_n)$ où ε est une racine p-ième de l'unité, $p \neq 1$.
Posons $P = \frac{1}{|G|} \sum_{\mu \in G} \frac{1}{\mu(X_1)^{p-1}} (\frac{\partial}{\partial X_1})^\mu$, remarquons que si $f \in R^G$ alors f est fixée par γ, donc il existe $\tilde{f} \in R$ telle que $f = \tilde{f}(X_1^p, X_2, \dots, X_n)$.

Alors $P \notin A_n(k)$, mais :

$$P(f) = \frac{1}{|G|} \sum_{\mu \in G} \frac{1}{\mu(X_1)^{p-1}} \mu.(\frac{\partial}{\partial X_1})(\tilde{f}) =$$

$$= \frac{1}{|G|} \sum_{\mu \in G} \frac{p}{\mu(X_1)^{p-1}} \mu . [X_1^{p-1}(\frac{\partial}{\partial X_1} \hat{f})(X_1^p, \ldots, X_n)]$$

donc $P(f) = \frac{p}{|G|} \sum_{\mu \in G} \mu(h)$ où $h = (\frac{\partial}{\partial X_1} \tilde{f})(X_1^p, X_2, \ldots, X_n)$, donc $P(f) \in R^G$.

Réciproquement si G ne contient pas de pseudo réflexion différente de l'identité montrons que $\mathrm{Diff}_k(R^G) = A_n(k)^G$.

Si $q \in \mathbb{N}$, $x \notin Z$ le corollaire 4 de II.3 implique qu'il existe des ouverts U et V contenant respectivement x et $y = \pi(x)$ tels que :

$$\mathrm{Diff}^q_{X/k}(U) \cong O_X(U) \otimes \mathrm{Diff}^q_{Y/k}(V).$$

Soit $P \in \mathrm{Diff}^q_k(R^G)$, P définit un élément de $\mathrm{Diff}^q_{Y/k}(V)$, donc par l'isomorphisme précédent un élément de $\mathrm{Diff}^q_{X/k}(U)$, ceci étant valable pour tout $x \notin Z$ on a ainsi défini une image inverse $\pi^*(P)$ sur $X \setminus Z$. D'après la proposition 1 de II.2 il suffit de vérifier que $\mathrm{codim}_X Z \geq 2$ pour obtenir $\pi^*(P)$ sur X, donc $\pi^*(P)$ dans $A_n(k)$ qui évidemment coïncidera avec P sur R^G. Il s'agit donc de montrer que Z ne contient pas de points p tels que la hauteur de p soit 1. S'il n'en était pas ainsi $p = fR$; $p \in Z$ impliquerait $G_i(p) \neq \{1\}$, et il existerait $\gamma \in G \setminus \{1\}$ fixant point par point $\frac{R}{p}$; en particulier puisque l'action de G est linéaire il existerait X_i tel que $\gamma(X_i) - X_i \neq 0$, et f diviserait $\gamma(X_i) - X_i \in k X_1 + \ldots + k X_n$, par suite $f \in k X_1 + \ldots + k X_n$ ce qui donnerait que $\gamma \in G \setminus \{1\}$ fixant un hyperplan, d'où la contradiction cherchée.

Références

[B] J. Becker - Higher derivations and the Zariski-Lipman conjecture. Proceedings of Symponia in Pure Mathematics. Vol. 30, (1977).

[G] A. Grothendieck - Eléments de Géométrie algébrique. Chap. IV, quatrième partie. Publ. I.H.E.S. n° 32 (1967).

[H] R. Hartshorne - Algebraic Geometry. Springer Verlag (1977).

[K] J.K. Kantor - Formes et opérateurs différentiels sur les espaces analytiques complexes. Bull. Soc. Math. France. Mémoire 53, (1977).

[S.G.A] Revêtements étales du groupe fondamental - Lec. Notes in Math. n° 224 Springer Verlag (1971).

[S] S. Suzuki - Differentials of Commutative rings. Queen's Paper in pure and applied Math, (1971).

[Sw] M.E. Sweedler - Groups of simple algebras Publ. Math. I.H.E.S. n° 44, (1974).

Université Pierre et Marie Curie
Mathématiques. U.E.R. 47 - Tour 45-46
4, Place Jussieu

75230 PARIS CEDEX 05

FACTORIALITE ET SERIES FORMELLES IRREDUCTIBLES II.

par Marc BAYART

On trouvera ci-après la suite de l'article [1]. Consacrée cette fois aux classes de congruence des séries irréductibles, notre étude résoud par l'affirmative, lorsque l'anneau A contient $\mathbb{Q}$, la conjecture :

"Si A est factoriel, l'ensemble des irréductibles de A [[T]] est ouvert pour la topologie T-adique".

Autrement dit :

si $f = \sum_{n \in \mathbb{N}} a_n T^n$ est irréductible, il existe un entier N tel que toutes les séries g, ayant mêmes coefficients que f jusqu'à l'ordre N inclus, soient également irréductibles, quels que soient leurs autres termes.

Nous renvoyons à la table des matières en ce qui concerne les diverses étapes de la démonstration, notons seulement les expressions particulièrement simples des nombres de classes pour les réductions modulo T^{n+1}, lorsque f(O) est "petit" (§§ IV, V, VI).

Ce résultat montre que, pour les séries formelles, l'irréductibilité est une propriété de "caractère fini", en ce sens qu'elle ne fait intervenir qu'un nombre fini de coefficients. On peut se demander si les anneaux factoriels A tels que A [[T]] soit également factoriel ne sont pas justement ceux pour lesquels la relation de divisibilité vérifie une propriété analogue.

Indiquons, par ailleurs, que les anneaux $\mathbb{B}_n$ (§§ V et VI) semblent jouer un rôle essentiel dans la factorialité de A [[T]]; ils ont en outre l'avantage d'être intègres - contrairement aux A [[T]]/(T^{n+1}) - et de traduire convenablement l'irréductibilité modulo T^{n+1}. Leur étude approfondie devrait, elle aussi, renseigner sur les liens existant entre factorialité de A et de A [[T]], peut-être permettrait-elle également de définir la notion d'"invariant caractéristique" conjecturée par P. SAMUEL dans [7].

TABLE DES MATIERES

§ IV - REDUCTIBILITE MODULO T^{n+1}.

Enoncé du Théorème (T): l'ensemble des irréductibles de A [[T]] est ouvert pour la topologie T-adique.

IV. 1 - Définitions et première réduction.
Séries réductibles, inversibles, irréductibles modulo T^{n+1} ; ordre d'irréductibilité; si A est principal, critère pour que l'ordre d'irréductibilité d'une série soit 1; relations entre ordre d'irréductibilité et ordre réduit lorsque A est un anneau de valuation discrète complet; réductibilité de f modulo T^{n+1} si $f(0) = 0$.

IV. 2 - Conditions suffisantes pour (T).
Définition des énoncés (E_N); la conjonction des (E_N) implique (T); si (E_0) est vrai, tous les (E_N) le sont.

IV. 3 - Localisation et complétion.
(E_0) "passe" du local au global; il suffit de le démontrer pour un anneau de valuation discrète complet ("cas réduit").

§ V - ETUDE DU CAS REDUIT.

V. 1 - Notations et rappels.
Valuation p-adique, théorème de préparation, polynôme quasi-distingué; décomposition en irréductibles quasi-distingués; polygone de Newton, cas d'un polynôme irréductible.

V. 2 - Décomposition du dénominateur.
L'énoncé (E); il implique (E_0), et il suffit de démontrer (E) pour un dénominateur g irréductible et quasi-distingué (énoncé(E')).

V. 3 - Démonstration de (E').
Deux méthodes différentes selon que g est ou non homogène en p(l'uniformisante) et T.

§ VI - CAS PARTICULIERS ET CONSEQUENCES.

VI. 1 - Valeurs explicites de r(F,N).
Valeurs uniformes minimales; elles conviennent lorsque f(0) ne contient pas de facteur bicarré.

VI. 2 - Traductions de (T).

Cardinal de l'ensemble des classes d'irréductibles associées à un irréductible d'ordre N; étude des anneaux $\mathbb{B}_n$; si une infinité d'entre eux sont factoriels, A[[T]] l'est aussi; substitutions de polynômes au lieu de séries.

§ VII - APPENDICE.

VII. 1 - Préliminaires.

Polygone de Newton; propriétés géométriques; définition des ρ-diviseurs; somme, produit et division euclidienne.

VII. 2 - Irréductibilité .

Théorème VII.2.1 : si un polynôme est irréductible, son polygone de Newton a un unique segment non vertical.

BIBLIOGRAPHIE.

§ IV - REDUCTIBILITE MODULO T^{n+1}

Dans ce §, A désigne, sauf mention contraire, un anneau factoriel; on suppose de plus que les éléments de $\mathbb{Z}^*$ sont inversibles dans A : autrement dit, A est une $\mathbb{Q}$-algèbre factorielle.

On se propose de démontrer le résultat suivant :

(T)L'ensemble des éléments irréductibles de A[[T]] est ouvert pour la topologie T-adique.

(T) revient à dire que :

(T')Si la série f est irréductible dans A[[T]], il existe un entier N tel que, pour tout g appartenant à A[[T]], la relation "$g \equiv f\,[T^{N+1}]$" entraîne que g est elle aussi irréductible.

IV. 1 - DEFINITIONS ET PREMIERE REDUCTION.

IV. 1.1 - Définitions.

Si f appartient à A[[T]] et $n \in \mathbb{N}$, nous dirons que f est *réductible* (resp. *inversible*) *modulo* T^{n+1} si, et seulement si, l'ensemble des $f + T^{n+1}.R$ - R appartenant à A[[T]] - contient au moins un élément réductible (resp. inversible); autrement dit si il existe au moins un g, congru à f modulo T^{n+1}, qui soit réductible (resp. inversible).

Comme f(0) est le terme constant commun à tous les $f + T^{n+1}.R$, les énoncés : "f est inversible", "f est inversible modulo T^{n+1}", "tous les $f + T^{n+1}.R$ sont inversibles", sont équivalents.

Par ailleurs, si f est réductible dans A[[T]], elle l'est modulo T^{n+1} pour toute valeur de n. La réciproque est fausse lorsque n est fixé : soit p un élément irréductible de A, et posons $f = p^2 - T$; on voit aisément que f est irréductible, par contre, f est congrue à p^2 modulo T et est donc réductible modulo T.

f est dite *irréductible modulo* T^{n+1} lorsqu'elle n'est ni inversible, ni réductible modulo T^{n+1} ; cela signifie que toutes les séries g telles que $g \equiv f\,[T^{n+1}]$ sont irréductibles. Bien entendu, si f est irréductible modulo T^{n+1}, elle l'est aussi modulo T^{m+1} pour tout $m \geq n$, et f elle même est alors irréductible. Nous appellerons donc *ordre d'irréductibilité* de f le plus petit entier N, si il existe, tel que f soit irréductible modulo T^{N+1} ; cela revient à dire que f est irréductible modulo T^{N+1} mais réductible modulo T^N. Observons que (T) affirme que toute série irréductible a un ordre d'irréductibilité.

Enfin, si f est irréductible (resp. réductible, inversible) modulo T^{n+1}, il en est de même pour u.f, pour tout u inversible dans A[[T]] ; en particulier, si f a un ordre d'irréductibilité, u.f aussi et ils sont égaux.

IV. 1. 2 - Exemples.

Une série a 0 pour ordre d'irréductibilité si, et seulement si, son terme constant est irréductible. En effet, si $f = \sum_{n\in\mathbb{N}} a_n T^n$, toute décomposition de a_0 est une décomposition de f modulo T, et réciproquement; d'où l'équivalence.

Lemme IV. 1.2 - *Soit* A *un anneau principal,* $f = \sum_{n\in\mathbb{N}} a_n.T^n$ *a pour ordre d'irréductibilité* 1 *si, et seulement si :*

α) a_0 est primaire - c'est à dire puissance d'un irréductible - et non irréductible, et

β) a_1 est premier avec a_0.

Démonstration. Condition nécessaire : si f est irréductible modulo T^2, elle est irréductible (cf. IV.1.1), donc a_0 est réduit d'après le lemme I 222 de [1],c'est à dire de la forme : $a_0 = u.p^k$; u inversible dans A, p irréductible dans A, k appartenant à $\mathbb{N}^*$. De plus k est supérieur ou égal à 2, sinon l'ordre d'irréductibilité serait 1. Si a_1 n'est pas premier avec a_0, il est de la forme $a_1 = \alpha.p$, $\alpha \in A$; alors :

$$f \equiv p.(u.p^{k-1} + \alpha.T) \; [T^2]$$

et aucun des termes du second membre n'est inversible : contradiction.

Condition suffisante : avec les notations ci-dessus, supposons que $a_0 = u.p^k$, $k \geqslant 2$, et que a_1 est premier à p. Par l'absurde, si :

$$f \equiv (b_0 + b_1.T)(c_0 + c_1.T) \; [T^2]$$

b_0 et c_0 étant de plus non inversibles, chacun d'eux est donc multiple de p ; par suite $a_1 = b_0.c_1 + c_0.b_1$ l'est également, contradiction : f est irréductible modulo T^2. Par contre, le début de ce n° montre qu'elle est réductible modulo T,ce qui achève la démonstration du lemme.

Cas d'un anneau de valuation discrète.

Dans ce troisième exemple, et jusqu'à la fin de ce n°, nous supposerons,pour simplifier, que A est un anneau de valuation discrète, d'uniformisante p, et qu'il est complet pour la topologie p-adique. Rappelons (cf.[4]) qu'un polynôme distingué est un polynôme unitaire dont tous les autres coefficients appartiennent à l'idéal p.A et que tout f non nul appartenant à A[[T]] s'écrit de manière unique :

$f = p^k u(T).P(T)$, $k \in \mathbb{N}$, u inversible dans A[[T]], P polynôme distingué.

p^k est le contenu de f, c'est à dire le PGCD de ses coefficients; deg(P) est l'ordre réduit de $f.p^{-k}$ - on dira, par abus de langage, ordre réduit de f - c'est à dire la valuation de l'image canonique de $f.p^{-k}$ dans (A/p)[[T]]. Cela étant, on a :

Lemme IV.1.2.2 - *A étant un anneau de valuation discrète complet, soit* f *irréductible dans* A[[T]], *de terme constant non nul et non irréductible. Alors, si l'ordre d'irréductibilité de* f *existe, il est supérieur ou égal à son ordre réduit.*

Démonstration. Remarquons d'abord que le cas où f(0) est irréductible a été étudié en toute généralité au début de ce n°; quant à celui où f(0)=0, il sera traité au IV.13.

Désignant par v_p la valuation p-adique de A, les hypothèses entraînent que $v_p(f(0))$ est supérieur ou égal à 2; par suite, le contenu de f est 1 car, sinon, p diviserait f et le quotient, ayant un terme constant multiple de p, ne serait pas inversible. f est donc de la forme u(T).P(T), u inversible dans A[[T]] et P polynôme distingué, noté $P=p.\beta_0+\ldots+p.\beta_{s-1}.T^{s-1}.+T^s$.

Vu la remarque de la fin du IV.11, il suffit de raisonner sur P. or

$$P \equiv p.(\beta_0+\ldots+\beta_{s-1}.T^{s-1})\ [T^s\ ]$$

et β_0 n'est pas inversible, puisque $v_p(\beta_0)=v_p(a_0)-1$. Ainsi, P est réductible modulo T^s, d'où le lemme.

Montrons que l'inégalité du lemme peut être stricte : soit $f = p^2+2p.T+(1-p).T^2$. Son ordre réduit est 2, puisque 1-p est inversible dans l'anneau local A. Or f est congrue à $(p+T)(p+T-T^2)$ modulo T^3 donc l'ordre d'irréductibilité de f, si il existe, est strictement supérieur à 2.

IV. 1.3 - Cas où f(0)=0.

D'après le IV.11, (T) signifie que toute série irréductible a un ordre d'irréductibilité, soit, par contraposition, que pour tout f appartenant à A[[T]], non nulle et non inversible, on a :

(T'_f) si, pour tout $n \in \mathbb{N}$, f est réductible modulo T^{n+1}, alors f est réductible.

Remarquons, toujours d'après le IV.11, que si u est inversible dans A[[T]], (T'_f) et (T'_{uf}) sont équivalents.

Nous commençons par étudier le cas où f(0)=0:

Lemme IV.1.3 - *Soit* A *un anneau quelconque et soit* f *appartenant à* A[[T]], *avec* f(0) = 0 . f *est irréductible si, et seulement si, elle est de la*

T.u(T), u *inversible dans* A[[T]]. *Dans ce cas, son ordre d'irréductibilité existe et vaut* 1.

Démonstration. D'abord, si f est irréductible et f(0)=0, T divise f donc est associé à f et celle-ci est bien de la forme T.u(T), u inversible. Réciproquement, T est irréductible car l'anneau quotient A[[T]]/T, isomorphe à A, est intègre, et l'idéal engendré par T est premier; par suite, tout élément associé à T est irréductible.

Quant à la seconde assertion, il suffit d'étudier l'ordre d'irréductibilité de T. D'une part, toute série congrue à T modulo T^2 est de la forme $T+b_2.T^2+\dots$ et est donc irréductible d'après la première partie de la démonstration: T est irréductible modulo T^2; d'autre part T^2, qui est réductible, est congru à T modulo T. En définitive, l'ordre d'irréductibilité de T est bien 1, d'où le lemme.

Le lemme IV.13 montre que (T'_f) est vrai pour tous les f de terme constant nul. Il l'est trivialement lorsque le terme constant est inversible, prémise et conclusion étant alors fausses. Seules restent donc à étudier les f telles que f(0) n'est ni nul, ni inversible. Dans ce qui suit, nous désignerons par : A[[T]]' l'ensemble des f ∈ A[[T]] tels que f(0)≠ 0, A[[T]]" celui des f tels que f(0) soit non nul et non inversible.

IV. 2 - CONDITIONS SUFFISANTES POUR (T).

IV. 2. 1 - Soit N ∈ ℕ, notons (E_N) l'énoncé suivant: pour tous F, G, H ∈ A[[T]]', il existe un entier r(F,N)⩾ N et deux parties finies I(F,G,H,N) ⊂ A/G(0), J(F,G,H,N) ⊂ A/H(0) - notés r, I, J quand F, G, H, N sont fixés - tels que, pour g et h ∈ A[[T]], la relation

$$(e)\quad \begin{aligned} F &\equiv g.h\ [T^{r+1}] \\ g &\equiv G\ [T^{N+1}] \\ h &\equiv H\ [T^{N+1}] \end{aligned}$$

entraîne que la classe modulo G(0)(resp. H(0)) du coefficient de T^{N+1} dans g(resp.h) appartient à I (resp. J).

Cela posé, on a :

PROPOSITION IV.2.1. *Si tous les* (E_N) *sont vrais,* (T) *l'est aussi.*

Démonstration. D'après le IV.13, il suffit, pour tout f appartenant à A[[T]]", de montrrr que l'énoncé "pour tout N, (E_N)" implique (T'_f).
Supposons donc que, pour tout entier n, il existe g_n et h_n dans A[[T]]" tels que :

(1) $f \equiv g_n.h_n[T^{n+1}]$; $g_n = \sum_{k \in \mathbb{N}} b_k^n.T^k, h_n = \sum_{k \in \mathbb{N}} c_k^n.T^k$.

Donc $f(0) = b_0^n.c_0^n$ et le lemme II 3.1 de [1] montre qu'il existe une partie infinie I_0 de $\mathbb{N}$ et des u_n inversibles dans A ($n \in I_0$) tels que $u_n^{-1}.b_0^n$ et $u_n.c_0^n$ soient indépendants de n. Remplacer g_n par $u_n.g_n$ et h_n par $u_n^{-1}.h_n$ permet de supposer que :

(2)(1) est vraie et $b_0^n = b_0$, $c_0^n = c_0$ sont indépendants de $n \in I_0$.

Appliquons l'énoncé (E_0) à $F = f, G = b_0$, $H = c_0$ et posons $r_0 = r(f,0)$. Alors, vu que (2) entraîne (e) dès que n est supérieur ou égal à r_0, on a :

pour tout $n \in I_0$, $n \geq r_0$, $\pi_{b_0}(b_1^n) \in I(f,b_0,c_0,0)$ et $\pi_{c_0}(c_1^n) \in J(f,b_0,c_0,0)$.

Comme I_0 est infini, il existe un I_1 infini inclus dans $I_0 \cap \{r_0, r_0+1, \ldots\}$ tel que $\pi_{b_0}(b_1^n)$ et $\pi_{c_0}(c_1^n)$ soient constants pour n appartenant à I_1 ; c'est à direqu'il existe b_1 et c_1 appartenant à A tels que :

(3) pour tout $n \in I_1$, $b_1^n \equiv b_1\ [b_0]$ et $c_1^n \equiv b_1\ [b_0]$ et $c_1^n \equiv c_1 [c_0]$.

Si b_1^n est de la forme $b_1 + \lambda_1^n.b_0$, en multipliant $g_n (n \in I_1)$par $(1 + \lambda_1^n.T)^{-1}$ et h_n par $(1 + \lambda_1^n.T)$, (2) demeure et on a donc, en notant encore g_n et h_n les séries modifiées :

(4) pour tout $n \in I_1$, (1) est vraie et $g_n \equiv b_0 + b_1.T[T^2]$; $h_n \equiv c_0 + c_1.T[T^2]$.

Soit N supérieur ou égal à 1 et supposons trouvés $b_0, \ldots, b_N$ et $c_0, \ldots, c_N$ appartenant à A, une partie infinie I_N de $\mathbb{N}$ et des g_n, $h_n (n \in I_N)$ tels que :

(5) pour tout $n \in I_N, f \equiv g_n.h_n\ [T^{n+1}]$ et

$$g_n \equiv b_0 + \ldots + b_N.T^N[T^{N+1}];\ h_n \equiv c_0 + \ldots + c_N.T^N\ [T^{N+1}].$$

Appliquons alors (E_N) à $F = f, G = b_0 + \ldots + b_N.T^N$, $H = c_0 + \ldots + c_N.T^N$ et posons $r_N = r(t,N)$. On observe ici que (5) entraîne (e) dès que n est supérieur ou égal à r_N, donc :

pour tout $n \in I_N$, $n \geq r_N$, on a : $\pi_{b_0}(b_{N+1}^n) \in I(F,C,H,N)$et $\pi_{c_0}(c_{N+1}^n) \in J(F,G,H,N)$.

Pour la même raison que ci-dessus, quitte à substituer à I_N une partie infinie I_{N+1} de $I_N \cap \{r_N, r_N+1, \ldots\}$ et à remplacer g_n et h_n par des séries associées du type $(1 + \lambda_{N+1}^n.T^{N+1})^{-1}.g_n$, $(1 + \lambda_{N+1}^n.T^{N+1}).h_n$, on aboutit à une relation du type(5), avec N+1 à la place de N. Cette récurrence définit deux séries $g = \sum_{k \in \mathbb{N}} b_k.T^K$ et $h = \sum_{k \in \mathbb{N}} c_k.T^K$; de plus, pour tout entier N, on a, d'après (5):

(6) pour tout $n \in I_N, f \equiv g_n.h_n\ [T^{n+1}]$; $g_n \equiv g\ [T^{N+1}]$, $h_n \equiv h\ [T^{N+1}]$.

Comme I_N est non vide, que ses éléments sont tous supérieurs ou égaux à r_N et que ce dernier est supérieur ou égal à N (cf. (E_N)), (6) entraîne que, pour tout N, f est congru à gh modulo T^{N+1}, autrement dit f = gh. De plus, g(0) et h(0) sont associés aux $g_n(0)$ et $h_n(0)$ respectivement; par suite g et h ne sont pas inversibles, ce qui établit (T_f') et la Proposition IV. 21.

IV.2.2 - Nous montrons maintenant que, dans l'étude des (E_N), il suffit de considérer le cas où N = 0.

PROPOSITION IV.22 . *Si* (E_0) *est vrai, alors tous les* (E_N) *le sont aussi.*

Démonstration. Soient donc $N \geqslant 1$, F, G, H, des éléments de $A[[T]]'$. Soient aussi $r_1 \geqslant 1$ un entier qu'on fixera par la suite et g, h appartenant à $A[[T]]$ tels que :

(1) $F \equiv g.h\ [T^{r_1(N+1)+1}]$; $g \equiv G\ [T^{N+1}]$, $h \equiv H\ [T^{N+1}]$(et donc $F \equiv G.H\ [T^{N+1}]$).

Soient $\omega_1,\ldots,\omega_{N+1}$ les racines (N+1) ièmes de l'unité, toutes distinctes ici puisque, A étant une $\mathbb{Q}$-algèbre, sa caractéristique est nulle, appartenant à Ω, clôture algébrique du corps des fractions de A. D'après (1), g et h sont de la forme :

(2) $g = G + b_{N+1}.T^{N+1} + T^{N+2}.\gamma(T)$,

$h = H + c_{N+1}.T^{N+1} + T^{N+2}.\eta(T)$; γ et $\eta \in A[[T]]$.

De plus, pour tout $i \in \{1,\ldots,N+1\}$, $F(\omega_i T) \equiv g(\omega_i T)h(\omega_i T)[T^{r_1(N+1)+1}]$.

Donc, avec les notations de [1],III 1, et n = N+1

(3) $\hat{F} \equiv \hat{g}.\hat{h}\ [T^{r_1(N+1)+1}]$.

Mais on a vu (cf. Lemme III.11 de [1]) que $\hat{F}$, $\hat{g}$, $\hat{h}$ sont respectivement de la forme $F_1(T^{N+1})$, $g_1(T^{N+1})$, $h_1(T^{N+1})$ avec F_1, g_1 et h_1 appartenant à $A[[T]]$. Par suite, dans la relation (3), le module de la congruence est en fait $(r_1+1)(N+1)$ et, en y substituant T à T^{N+1}, on obtient :

$$(4)\ F_1 \equiv g_1.h_1\ [T^{r_1+1}].$$

D'autre part, $g_1(0) = g(0)^{N+1}$, $h_1(0) = h(0)^{N+1}$, autrement dit, compte tenu de (1) :

$$(5)\ g_1 \equiv G(0)^{N+1}\ [T]\ ;\ h_1 \equiv H(0)^{N+1}\ [T].$$

Quant au coefficient de T dans g_1, c'est celui de T^{N+1} dans $\hat{g}$, c'est à dire, d'après (2), dans le produit des :

$G(\omega_i T) + b_{N+1}T^{N+1} + \omega_i T^{N+2}\gamma(\omega_i T)$. Il ne peut provenir que de :

$$\prod_{i=1}^{N+1} (G(\omega_i T) + b_{N+1}T^{N+1}).$$

On en déduit aisément que ce coefficient vaut :

(6) $\beta + (N+1).b_{N+1}.(G(0))^N$; β étant le coefficient de T^{N+1} dans $\hat{G}$.

De même, si δ est le coefficient de T^{N+1} dans $\hat{H}$, celui de T dans h_1 est :

$$(7)\ \delta + (N+1).\ c_{N+1}.\ (H(0))^N.$$

Appliquons alors (E_0) à F_1, $G(0)^{N+1}$, $H(0)^{N+1}$, g_1, h_1 à la place de F, G, H, g, h respectivement, en observant que la relation (e) découle de (4) et (5) si l'on prend $r_1 = r(F_1,0)$ qui ne dépend effectivement que de F et de N. Posons maintenant $r = r_1(N+1)$. La relation (1) implique, d'après ce qui précède, que

$$(8)\quad \pi_{G(0)^{N+1}}(\beta + (N+1)b_{N+1}.G(0)^N) \in I(F_1, G(0)^{N+1}, H(0)^{N+1}, 0)$$

$$\pi_{H(0)^{N+1}}(\delta + (N+1)c_{N+1}.H(0)^N) \in J(F_1, G(0)^{N+1}, H(0)^{N+1}, 0).$$

Soit ν l'application de A dans lui même qui, à tout $x \in A$, associe $\beta + (N+1).x.G(0)^N$ et remarquons que, vu la définition de β, ν ne dépend que de G, H, N. De plus, les classes modulo $G(0)^{N+1}$ de $\nu(x)$ et $\nu(y)$ sont égales si, et seulement si, G(0) divise (N+1).x, c'est à dire si les classes modulo G(0) de x et y sont égales, puisque N+1 est inversible dans A. On en déduit le diagramme commutatif :

$$\begin{array}{ccc} A & \xrightarrow{\nu} & A \\ \downarrow{\scriptstyle \pi_{G(0)}} & & \downarrow{\scriptstyle \pi_{G(0)^{N+1}}} \\ A/G(0) & \xrightarrow{\overline{\nu}} & A/G(0)^{N+1} \end{array}$$

et $\overline{\nu}$ est injective. Par ailleurs, la première partie de (8) s'écrit - en désignant simplement par I le second membre - : $\overline{\nu}(\pi_{G(0)}(b_{N+1}))$ appartient à I ; autrement dit:

$$\pi_{G(0)}(b_{N+1}) \in \overline{\nu}^{-1}(I).$$

Et ce dernier ensemble est fini, puisque $\overline{\nu}$ est injective. On définit de même μ et $\overline{\mu}$ et, en prenant :

$$r(F,N) = r_1(N+1), \text{ c'est à dire } r(F_1,0)(N+1);$$

$$I(F,G,H,N) = \overline{\nu}^{-1}(I(F_1, G(0)^{N+1}, H(0)^{N+1}, 0)),$$

$$J(F,G,H,N) = \overline{\mu}^{-1}(J(F_1, G(0)^{N+1}, H(0)^{N+1}, 0))$$

on termine la démonstration de (E_N) et de la Proposition IV. 22 .

Etant ainsi ramenés à la plus petite valeur possible pour N, les réductions suivantes vont porter sur l'anneau des coefficients.

IV.3 - LOCALISATION ET COMPLETION.

Lemme IV.3. 1 - *Pour que* (E_0) *soit vrai pour tout anneau - factoriel -, il suffit qu'il soit vrai lorsque* A *est un anneau de valuation discrète.*

<u>Démonstration</u>. Soient A un anneau factoriel et F, G, H appartenant à A[[T]]'. Désignons par $\{p_i \mid i \in I\}$ un système représentatif d'éléments irréductibles de A. Pour tout $i \in I$, soit L_i le localisé de A par l'idéal - premier - Ap_i : L_i est un anneau de valuation discrète, d'uniformisante p_i, c'est une $\mathbb{Q}$-algèbre si A l'est, enfin A[[T]] est un sous-anneau de L_i[[T]].

Soient r_i, $I_i \subset L_i/G(0)L_i$, $J_i \subset L_i/H(0)L_i$ les résultats de l'application de (E_0) à F, G, H dans L_i ; observons que, si p_i ne divise pas F(0), il ne peut diviser, ni G(0), ni H(0), et que, ces derniers étant alors inversibles dans L_i, I_i et J_i sont des singletons {0}. En définitive : les I_i et J_i sont presque tous {0} et les r_i correspondants valent 0 (revenir à la définition).

Cela étant, soit $\delta : A \to \prod_{i\in I} L_i$ l'application diagonale et notons $\pi : \prod_{i\in I} L_i \to \prod_{i\in I} L_i/G(0)L_i$ le "produit" des surjections naturelles. Soit x appartenant à A, $\pi(\delta(x))=0$ signifie que x appartient à tous les $G(0)L_i$, c'est à dire à $G(0)(\bigcap_{i\in I} L_i) = G(0)A$.

On déduit de ceci le diagramme commutatif :

$$\begin{array}{ccc} A & \xrightarrow{\delta} & \prod_{i\in I} L_i \\ \Big\downarrow{\scriptstyle \pi_{G(0)}} & & \Big\downarrow{\scriptstyle \pi} \\ A/G(0) & \xrightarrow{\bar\delta} & \prod_{i\in I} L_i/G(0)L_i \end{array}$$

et $\bar\delta$ est injective.

Posons alors $r = \sup_{i\in I} (r_i)$: r est bien défini puisque les r_i sont presque tous nuls et, d'après la remarque faite au début de la démonstration, il ne dépend que de F. Désignons par b_1 et c_1 les coefficients de T dans g et h respectivement. Si l'on a :

$$F \equiv g.h\ [T^{r+1}];\ g \equiv G[T],\ h \equiv H\ [T],$$

alors, dans chaque L_i la première congruence a lieu modulo T^{r_i+1} et l'énoncé (E_0), appliqué à L_i, montre que la classe de b_1 (resp. c_1) modulo $G(0)L_i$ (resp. $H(0)L_i$) appartient à la partie finie I_i (resp. J_i). Autrement dit, en ce qui concerne b_1 :

$$\pi(\delta(b_1)) \in \prod_{i \in I} I_i ;$$

et, vu le diagramme commutatif précédent :

$$\pi_{G(0)}(b_1) \in \overset{-1}{\bar{\delta}}\left(\prod_{i \in I} I_i\right) ;$$

ce dernier ensemble étant de plus fini, puisque les I_i sont finis, presque tous $\{0\}$ et que $\bar{\delta}$ est injective, et ne dépendant que de F, G, H d'après sa construction. On raisonne de même pour c_1, d'où le lemme :

Lemme IV.3. 2 - *Pour que* (E_0) *soit vrai pour tout anneau de valuation discrète, il suffit qu'il soit vrai sous l'hypothèse supplémentaire que cet anneau est complet.*

Démonstration. Soit A un anneau de valuation discrète, d'uniformisante p ; soient F, G, H appartenant à A[[T]]' et notons $\hat{A}$ le complété de A pour la topologie p-adique. $\hat{A}$ est un anneau de valuation discrète, pour la même uniformisante (cf. [3])et c'est une $\mathbb{Q}$-algèbre si A l'est. Soient $\hat{r}$, $\hat{I} \subset \hat{A}/G(0)\hat{A}$ et $\hat{J} \subset \hat{A}/H(0)\hat{A}$ les résultats de l'application de (E_0) à F, G, H dans $\hat{A}$. Les fonctions naturelles ν et μ de $A/G(0)A$ dans $\hat{A}/G(0)\hat{A}$ et $A/H(0)A$ dans $\hat{A}/H(0)\hat{A}$ respectivement sont injectives car, pour tout a appartenant à A, $a\hat{A} \cap A = aA$ (cf. [6]). Si on pose : $r = r, I = \overset{-1}{\nu}(\hat{I})$, $J = \overset{-1}{\mu}(\hat{J})$, on voit aussitôt qu'ils permettent de vérifier (E_0)pour F, G, H, dans A.

IV.3. 3 - Conclusion.

Les propositions et les lemmes ci-dessus ont ramené, en définitive, la démonstration de (T) à celle de (E_0) dans un anneau de valuation discrète, complet pour sa topologie naturelle : c'est le "cas réduit" dont l'étude va être faite au § suivant.

§ 5 - Etude du cas réduit

V.1 - NOTATIONS ET RAPPELS.

V.1. 1 - Dans ce §, A désignera un anneau de valuation discrète, d'uniformisante p, complet pour la topologie p-adique. On note v la valuation p-adique et $|\;| = e^{-v}$ la valeur absolue associée. Toutes deux se prolongent à K, corps des fractions de A, qui est un corps valué complet. Rappelons que, pour tous x et y appartenant à K, on a : $v(x+y) \geqslant \inf(v(x),v(y))$; $|x+y| \leqslant \sup(|x|,|y|)$ avec égalité lorsque $v(x) \neq v(y)$.
De plus (cf. [1], IV.1. 2) tout élément non nul f de A[[T]] s'écrit de manière unique sous la forme : $f = p^k.u(T).P(T)$; $k \in \mathbb{N}$, u inversible dans A[[T]], P polynôme distingué.
Lorsque k = 0, f est irréductible dans A[[T]] si, et seulement si, P l'est dans A[[T]] et on sait (cf. [4]) que cela revient à dire qu'il l'est dans A[T]. Nous appaellerons *polynôme quasi-distingué* tout élément de A[T] dont le coefficient directeur est inversible, le coefficient constant est une puissance de p et les autres coefficients sont multiples de p.

Lemme V. 11. *Soit* f ∈ A[[T]]' *une série de terme constant non nul. Il existe un entier* k *et des polynômes irréductibles quasi-distingués* $P_1,\dots,P_h$ *tels que* f *soit associée à* $p^k.P_1\dots P_h$.

<u>Démonstration.</u> D'après la "multiplicativité" de la conclusion, on peut supposer que f est irréductible (A étant un anneau de valuation discrète, A[[T]] est factoriel). Deux cas sont alors possibles : ou f est associée à p et il suffit de prendre h=0, ou bien le contenu de f est 1, et f est associée à un polynôme distingué irréductible, lequel à son tour l'est évidemment à un polynôme quasi-distingué, irréductible lui aussi ; d'où le lemme.

V.1. 2 - Polygone de Newton.

Soit $U = \sum_{n \in \mathbb{N}} u_n.T^n$ un élément de K[T]. On note Supp(U) l'ensemble des entiers n tels que u_n soit $\neq 0$: Supp(U) est une partie finie de $\mathbb{N}$; pour tout n appartenant à Supp(U), soit M_n le point du plan $\mathbb{R}^2$ de coordonnées $(n,v(u_n))$. L'intersection de tous les demi-plans supérieurs - c'est à dire de la forme $y \geqslant \lambda.x+\mu$, λ et $\mu \in \mathbb{R}$ - contenant l'ensemble des M_n est, si $U \neq 0$, une partie convexe et fermée dont la frontière est une ligne polygonale appelée *polygone de Newton* de U.

Ceci posé, on a le :

THEOREME V. 12. *Si* U *est irréductible dans* A[T] *et* $U(0) \neq 0$, *son polygone de Newton a un seul côté non vertical.*

Pour ne pas interrompre l'enchaînement des idées, la démonstration est donnée en Appendice. Notons seulement que le théorème de Krasner (cf. [5]) est inapplicable ici, vu que l'hypothèse "A contient $\mathbb{Q}$", qui s'est transmise par localisation et complétion, entraîne que la restriction à $\mathbb{Q}$ de v est triviale. Par ailleurs, les seuls polynômes irréductibles de terme constant nul sont les $\lambda.T$, $\lambda \in K-\{0\}$, leur polygone de Newton est une demi-droite verticale.

V.2. - DECOMPOSITION DU DENOMINATEUR.

V.2. 1 - Préliminaire.

Pour tout $n \in \mathbb{N}$, notons $\mathbb{B}_n$ l'ensemble des f appartenant à K[[T]] dont les coefficients de $T^0,\dots,T^n$ appartiennent à A. On trouvera une étude de $\mathbb{B}_n$ au § VI ; observons pour l'instant que :

Lemme V.2.1 - *Notons* (E) *l'énoncé suivant : "pour tous* $F \in A[[T]]'$ *et* $G_0 \in A-\{0\}$, *il existe un entier* r(F) (*noté* r) *et une partie finie* $I(F,G_0) \subset A/G_0$ (*notée* I) *tels que la relation :* $g \in A[[T]]'$, $g \equiv G_0$ [T] *et* $\frac{F}{g} \in \mathbb{B}_r$, *entraîne que la classe modulo* G_0 *du coefficient de* T *dans* g *appartient à* I". *Alors,* (E) *implique* (E_0).

<u>Démonstration</u>. Soient donc F, G, H, appartenant à A[[T]]'. En appliquant (E) à F et G(0), puis F et H(0), on en déduit un entier r = r(F) et deux parties finies $I \subset A/G(0)$, $J \subset A/H(0)$. Soient alors g et h appartenant à A[[T]] tels que :

$$F \equiv g.h\ [T^{r+1}];\ g \equiv G\ [T],\ h \equiv H\ [T].$$

Par suite, $\frac{F}{g}$ qui est congru à h modulo T^{r+1}, appartient à $\mathbb{B}_r$; de plus g est congru à G(0) modulo T. (E) entraîne donc que la classe modulo G(0) du coefficient de T dans g appartient à I. On raisonne de même pour h, d'ou le Lemme.

V.2. 2 - <u>Multiplicativité</u>.

Notons Ir. l'ensemble des polynômes quasi-distingués irréductibles - dans A[T] ou A[[T]], on a vu au V.11 que cela revient au même -. Pour tout entier $k \geq 1$, soit Ir_k le sous-ensemble de ceux dont le terme constant est p^k ; on conviendra, en outre, de poser Ir_0 = l'ensemble des éléments inversibles de $A[[T]]$.

PROPOSITION V.2.2 - *Soit (E') l'énoncé : "pour tous* $F \in A[[T]]'$ *et* $G_0 \in A-\{0\}$, *il existe un entier* $r' = r'(F)$ *et une partie finie* $I' = I'(F,G_0) \subset A/G_0$ *tels que la relation :*

$g \in Ir.$, $g \equiv G_0[T]$ *et* $\frac{F}{g} \in \mathbb{B}_r$, *entraîne que la classe modulo* G_0 *du coefficient de* T *dans* g *appartient à* I'."
Alors (E') entraîne (E).

Démonstration. Soient donc $F \in A[[T]]'$ et G_0 un élément non nul de A. Notons $\ell = v(G_0)$: G_0 est associé à p^{ℓ} et, par suite, pour tout g appartenant à $A[[T]]$ congru à G_0 modulo T - c'est à dire tel que $g(0) = G_0$ - la décomposition $g = p^k.u(T).P_1...P_h$ du lemme 1 contient au plus ℓ polynômes quasi-distingués irréductibles. En outre chacun d'eux, a priori dans Ir., appartient en fait à $Ir_1 \cup ... \cup Ir_{\ell}$ pour la même raison, savoir la valuation du terme constant.

Pour tout $j \in \{1,...,\ell\}$, soient $I'_j = I'(F,p^j)$ et $r' = r'(F)$ les valeurs données par l'application de (E') à F et p^j.

Posons $r = r'$ et soit un élément g de $A[[T]]'$, décomposé comme ci-dessus. Si

$$(1)\quad g \equiv G_0\ [T]\ ;\ \frac{F}{g} \in \mathbb{B}_r$$

alors, a fortiori, pour tout $i \in \{1,...,h\}$, $\frac{F}{P_i}$ appartient à $\mathbb{B}_r$; si l'on note $j_i = v(P_i(0))$ on a :

$$P_i \equiv p^{j_i}[T]\ ;\ P_i \in Ir_{j_i}$$

donc, d'après (E') appliqué à F et p^{j_i}, la classe modulo p^{j_i} du coefficient de T dans P_i appartient à I'_{j_i}.

Notons λ ; $\mu_1,...,\mu_h$ les coefficients de T dans u ; $P_1,...,P_h$ respectivement. On a donc :

$$(2)\quad \lambda \in A\ ;\ \text{pour tout } i \in \{1,...,h\}, \pi_{p^{j_i}}(\mu_i) \in I'_{j_i}$$

Remarquons maintenant que r est déjà connu - il ne dépend effectivement que de F - ; d'autre part, chacun des j_i appartient à $\{1,...,\ell\}$ et h appartient à $\{0,...,\ell\}$ d'après ce qui précède. Par suite, $(j_1,...,j_h)$ peut prendre au plus $1+\ell+\ell^2+...+\ell^{\ell}$ valeurs possibles - et ce nombre ne dépend que de G_0 - : on peut donc, pour terminer la démonstration, supposer que h et $j_1,...,j_h$ sont fixés ; ayant trouvé $I(F,G_0)$, il ne restera plus qu'à prendre la réunion des parties trouvées.

Cela supposé soit :

$$\nu : A^h \to A \text{ telle que } \nu(x_1,\ldots,x_h) = G_0.\left(\sum_{i=1}^{h} x_i.p^{-j_i}\right).$$

Si chaque x_i appartient à $p^{j_i}A$, alors $\nu(x_1,\ldots,x_h)$ appartient à $G_0.A$. On en déduit un diagramme commutatif :

$$\begin{array}{ccc} A^h & \xrightarrow{\nu} & A \\ {\scriptstyle \text{nat.}}\downarrow & & \downarrow{\scriptstyle \pi_{G_0}} \\ \prod_{i=1}^{h} A/p^{j_i} & \xrightarrow{\overline{\nu}} & A/G_0 \end{array}$$

Or, le coefficient de T dans $g = p^k.u(T).P_1\ldots P_h$ vaut :

$$G_0.(\lambda.(u(0))^{-1} + \sum_{i=1}^{h} \mu_i.(P_i(0))^{-1}),$$

donc sa classe modulo G_0 est exactement $\pi_{G_0}(\nu(\lambda_1,\ldots,\lambda_h))$, puisque u(0) est inversible, donc elle vaut :

$$\overline{\nu}(\pi_{p^{j_1}}(\lambda_1),\ldots,\pi_{p^{j_h}}(\lambda_h))$$

c'est à dire, vu la relation (2), elle appartient à $\overline{\nu}(I'_{j_1} \times \ldots \times I'_{j_h})$.

Comme $\overline{\nu}$ ne dépend que de $j_1,\ldots,j_h$ et G_0, la Proposition est démontrée.

V.3 - DEMONSTRATION DE (E').

THEOREME V.3. 1 - *Soient* F *appartenant à* A[[T]]' *et* G_0 *à* A - { 0 } . *Notons* p^σ *le contenu de* F, s *la valuation de* F(0) *et* ℓ *l'ordre réduit de* $F.p^{-\sigma}$. *Soit* $r = \sup(s, 2\ell + \sigma)$. *Il existe une partie finie* $I' = I'(F,G_0) \subset A/G_0$ *telle que, pour tout* g, *polynôme quasi-distingué irréductible, la relation :*

$$(1)\quad g \equiv G_0\ [T] \text{ et } \frac{F}{g} \in IB_r$$

entraîne que la classe modulo G_0 *du coefficient de* T *dans* g, *appartient à* I'.

La démonstration se fait en deux étapes.

V.3. 2 - *Cas où* g *est homogène, de degré* $v(G_0)$.

D'après les hypothèses, F est de la forme $u.p^s + a_1.T + \ldots$ et $G = p^t + b_1.T + \ldots$, avec u inversible dans A ; on a vu au lemme V.2. 1 que (1) équivaut à l'existence d'un h et d'un R appartenant à A[[T]] tels que :

$$(1')\ g \equiv G_0\ [T] \text{ et } F = g.h + T^{r+1}.R.$$

Supposons, pour l'instant, que g est homogène en p et T, de degré t, c'est à dire que p^t divise tous les coefficients de g(pT), soit encore : contenu $(g(pT))=p^t$. Observons qu'alors b_1 est de la forme $p^{t-1}.\beta$, β appartenant à A.

Substituons, dans (1'), pT à T ; on obtient :

$$F(pT)- p^{r+1}.T^{r+1}.R(pT)= g(pT).h(pT).$$

Soit $p^{s'}$ le contenu du premier membre. On a $s' \leqslant s$, vu le terme constant de F ; par suite, s étant au plus égal à r d'après le choix de r, $p^{s'}$ est exactement de F(pT). Le lemme d'Euclide, valable mutatis mutandis ici, montre que le contenu de h(pT) est $p^{s'-t}$.

Soient F_1, g_1, h_1 les quotients respectifs de $F(pT)-p^{r+1}T^{r+1}R(pT)$, g(pT) et h(pT) par leurs contenus. On a :

$$F_1(T) = g_1(T).h_1(T)$$

et, en notant $\overline{\ldots}$ les séries réduites dans l'anneau (A/p)[[T]], on obtient l'égalité :

$$(2)\ \overline{F}_1 = \overline{g}_1.\overline{h}_1.$$

La division par les contenus montre que $\overline{F}_1$, $\overline{g}_1$ et $\overline{h}_1$ sont toutes trois différentes de 0. D'autre part, elles appartiennent en fait à (A/p)[T], puisque dans chacune les termes en $T^k, k \geqslant s+1$, initialement multiples de p^{s+1}, sont restés multiples de p et ont donné 0 par réduction. Enfin, ce qui précède montre que $\overline{F}_1$ ne dépend que de F, $\overline{g}_1(0)= 1$ et le coefficient de T dans $\overline{g}_1$ est $\overline{\beta}$.

Cela dit, (2) exprime que $\overline{g}_1$ est l'un des diviseurs, à terme constant fixé et égal à 1, du polynôme $\overline{F}_1$ dans l'anneau factoriel (A/p est un corps)(A/p)[T]. On sait qu'il n'existe qu'un nombre fini de tels polynômes ; a fortiori, il existe une partie finie $I_1'' = I_1''(F)$ telle que (1) entraîne que $\overline{\beta}$, c'est à dire $\pi_p(\beta)$, appartient à I_1''.

Or l'application ν de A dans A qui, à x, associe $p^{t-1}.x$, donne par factorisation le diagramme commutatif :

$$\begin{array}{ccc} A & \xrightarrow{\ \nu\ } & A \\ \downarrow{\scriptstyle \pi_p} & & \downarrow{\scriptstyle \pi_{p^t}} \\ A/pA & \xrightarrow{\ \overline{\nu}\ } & A/p^tA \end{array}$$

et on remarque que $\overline{\nu}$ ne dépend que de G_0. De ce qui précède - c'est à dire que (1) entraîne que $\pi_p(\beta) \in I_1''$ - on déduit que, si (1) est vrai, alors $\pi_{p^t}(b_1)=\overline{\nu}(\pi_p(\beta))$ appartient à $\overline{\nu}(I_1'')$.

Alors, $I'_1 = \overline{v}(I''_1)$ ne dépend, comme on l'a vu, que de F et G_0 et convient pour tous les g homogènes appartenant à Ir..

<u>Remarque</u>. On observera que, dans cette première partie, le fait que g est irréductible n'a pas été utilisé.

V.3. 3 - *Cas où* g *n'est pas homogène de degré* $t = v(G_0)$.

Nous revenons à la relation (1) du V.3 1 :

$$(1)\quad g \equiv G_0\ [T]\ \text{et}\ \frac{F}{g} \in \mathbb{B}_r .$$

en supposant que g est quasi-distingué, irréductible et non homogène de degré t. D'après le Théorème V.11, son polygone de Newton comporte un seul segment non vertical. Si :

$$P = p^t + b_1.T + \ldots + b_{\lambda-i}.T^{\lambda-1} + u.T^{\lambda},\ \text{avec u inversible,}$$

$$b_1, \ldots, b_{\lambda-1}\ \text{multiples de p,}$$

ce segment ne peut être que $M_0\ M_\lambda$, avec $M_0 = (0,t)$ et $M_\lambda = (\lambda, 0)$, et on a :

$$(3)\quad 1 \leqslant \lambda < t.$$

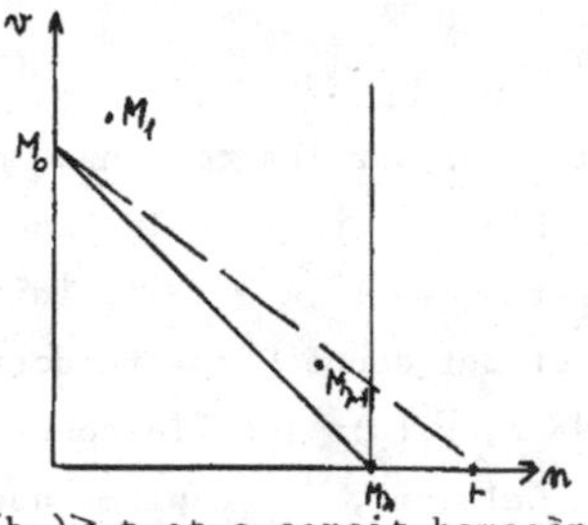

En effet, pour tout n appartenant à $\{0, \ldots, \lambda\}$, M_n est au-dessus de la droite d'équation $\frac{x}{\lambda} + \frac{y}{t} = 1$, donc $v(b_n)$ est supérieur ou égal à $t - \frac{t.n}{\lambda}$, soit $n + v(b_n) \geqslant t + n(1 - \frac{t}{\lambda})$; si λ était supérieur ou égal à t, on aurait donc $n + v(b_n) \geqslant t$ et g serait homogène de degré t, ce qui est exclu.

D'autre part, F s'écrit sous la forme :

$F = p^{\sigma}.u(T).F_1(T)$; u inversible dans $A[[T]]$, F_1 polynôme distingué de degré ℓ.

(ceci résulte de la définition de ℓ, cf. V.1.1 et V.3.1).
Comme $\frac{F}{g}$ appartient à $\mathbb{B}_r$ si, et seulement si, $p^{\sigma}.\frac{F_1}{g}$ appartient à $\mathbb{B}_r$, et comme σ et F_1 ne dépendent que de F, on supposera ci-dessous que $F = p^{\sigma}.F_1$.

Soient alors Ω une clôture algébrique de K, corps des fractions de A et notons E l'anneau des entiers algébriques de Ω.
Désignons par $x_1, \ldots, x_\lambda$ les racines du polynôme g dans Ω. Comme g est quasi-distingué et irréductible, les x_i sont tous distincts et appartiennent à E. Si $y_1, \ldots, y_\ell$ sont les racines de F_1 dans Ω, elles ne sont pas nécessairement distinctes mais

appartiennent à E puisque F_1 est distingué. Soit enfin Ω' l'extension finie :

$$\Omega' = K(x_1,\dots,x_\lambda \; ; \; y_1,\dots,y_\ell).$$

On sait (cf.[2]) que la valuation v de K se prolonge de manière unique en une valuation $\bar{v}$ de Ω' et que l'anneau des entiers algébriques de Ω' est exactement l'anneau de $\bar{v}$, c'est à dire l'ensemble E' des $x \in \Omega'$ tels que $\bar{v}(x) \geqslant 0$. $\bar{v}$ est définie de la façon suivante : si x appartient à Ω' et si P est son polynôme minimal (unitaire) sur K, alors :

$$\bar{v}(x) = \frac{v(P(0))}{\deg(P)} .$$

En particulier, on a :

(4) pour tout $i \in \{1,\dots,\lambda\}$, $\bar{v}(x_i) = \frac{t}{\lambda} > 0$.

Cela étant, le fait que $\frac{F}{g}$ appartient à $\mathbb{B}_r$ signifie que :

$$\frac{F}{u(T-x_1)\dots(T-x_\lambda)} \in \mathbb{B}_r$$

A fortiori, en notant $\mathbb{B}'_r$ l'anneau défini comme $\mathbb{B}_r$ mais avec E' et Ω' à la place de A et K, la quantité ci-dessus appartient à $\mathbb{B}'_r$. Comme les x_i appartiennent à E' et que u, inversible dans A, l'est dans E', on en déduit que pour tout $i \in \{1,\dots,\lambda\}$, $\frac{F}{T-x_i} \in \mathbb{B}'_r$.

C'est à dire par division euclidienne de F par $T-x_i$ que : $\frac{F(x_i)}{T-x_i} \in \mathbb{B}'_r$, soit encore par division suivant les puissances croissantes : pour tout $j \in \{0,\dots,r\}$, $\frac{F(x_i)}{x_i^{j+1}} \in E'$, c'est à dire que $\bar{v}(F(x_i))$ est supérieur ou égal à $(j+1).\bar{v}(x_i)$. Comme $\bar{v}(x_i)$ est > 0 (cf.(4)), on a donc :

(5) pour tout $i \in \{1,\dots,\lambda\}, \bar{v}(F(x_i)) \geqslant (r+1).\bar{v}(x_i)$.

Comme, par définition des y_j, on a : $F = p^\sigma.F_1 = p^\sigma(T-y_1)\dots(T-y_\ell)$, $F(x_i)$ vaut $p^\sigma(x_i-y_1)\dots(x_i-y_\ell)$. Tous les termes de cette égalité sont dans E' et :

(6) $\bar{v}(F(x_i)) = \sigma + \sum_{j=1}^{\ell} \bar{v}(x_i-y_j)$.

Soit y_{j_i} tel que $\bar{v}(x_i-y_{j_i}) = \sup(\{\bar{v}(x_i-y_j) \mid j = 1,\dots,\ell\})$. (6) entraîne que :

(7) $\bar{v}(F(x_i)) \leqslant \sigma + \ell.\bar{v}(x_i - y_{j_i})$.

Donc, d'après (5) et (7), $\ell.\overline{v}(x_i - y_{j_i})$ est supérieur ou égal à $(r+1).\overline{v}(x_i) - \sigma$, donc aussi (cf. (4) et le choix de r) à $(2\ell+\sigma)\frac{t}{\lambda} - \sigma$ et enfin, à l'aide de (3), à $2\ell.\frac{t}{\lambda}$. Autrement dit :

$$(8)\ \overline{v}(x_i - y_{j_i}) \geqslant 2.\overline{v}(x_i).$$

Maintenant, comme $\overline{v}(x_i)$ est strictement positif (cf. (4)) et que $\overline{v}(y_{j_i}) = \inf(\overline{v}(x_i - y_{j_i}), \overline{v}(x_i))$ (les deux termes sont différents à cause de (8)), on a donc démontré que :

$$(9)\ \overline{v}(y_{j_i}) = \overline{v}(x_i).$$

Mais alors $\frac{1}{x_i} - \frac{1}{y_{j_i}} = \frac{y_{j_i} - x_i}{x_i . y_{j_i}}$ a une "$\overline{v}$" positive ou nulle d'après (8) et (9), donc il appartient à E'.

Par ailleurs, la considération de l'équation aux inverses de g montre que : $(x_1)^{-1} + \ldots + (x_\lambda)^{-1} = -b_1 . p^{-t} = -\frac{b_1}{G_0}$, autrement dit :

$$(10)\ b_1 = -G_0\left(\sum_{i=1}^{\lambda}\left(\frac{1}{x_i} - \frac{1}{y_{j_i}}\right)\right) - G_0\left(\sum_{i=1}^{\lambda}\frac{1}{y_{j_i}}\right).$$

Par suite :

$$(11)\ b_1 + G_0\left(\sum_{i=1}^{\lambda}\frac{1}{y_{j_i}}\right) \text{ appartient à } G_0.E'.$$

Or, le nombre de valeurs possibles pour $G_0\left(\sum_{i=1}^{\lambda}\frac{1}{y_{j_i}}\right)$ est inférieur ou égal à $\ell + \ell^2 + \ldots + \ell^{t-1}$ (suivant que $\lambda = 1, \ldots, t-1$, voir (4) et le choix des y_{j_i}), quantité ne dépendant que de F et G_0, ainsi que les valeurs obtenues. Soit $I_2'' \subset E'/G_0.E'$ l'ensemble des classes des $-G_0\left(\sum_{i=1}^{\lambda}\frac{1}{y_{j_i}}\right)$ modulo $G_0.E'$. (11) signifie que :

> "Si (1) est vraie, la classe de b_1 modulo $G_0.E'$ appartient à I_2''".

Or, l'application naturelle ν de $A/G_0.A$ dans $E'/G_0.E'$ est injective car, si x appartient à $G_0.E' \cap A$, on a $\overline{v}(x) \geqslant \overline{v}(G_0)$ donc $v(x) \geqslant v(G_0)$ et x appartient à $G_0.A$.

Posons donc $I_2' = \overline{\nu}^1(I_2'')$. La relation (1) entraîne donc que $\pi_{G_0}(b_1)$ appartient à I_2' ; d'après ce qu'on vient de dire, I_2' ne dépend que de F et G_0, il convient dans le cas étudié ici.

V.3. 4 - Conclusion.

Si l'on pose : $I' = I_1' \cup I_2'$, il est évident d'après les n°s 2 et 3, qui recouvrent tous les cas possibles pour $g \in Ir.$, que I' convient : le Théorème V.3.1 est établi, donc aussi, au moyen des réductions précédentes, le Théorème(T).

§VI - CAS PARTICULIERS ET CONSEQUENCES

VI.1 - VALEURS EXPLICITES DE r(F,N).

VI.1. 1 - On peut, en "remontant" les réductions successives faites aux § IV et V sur l'énoncé (E_N), trouver une majoration de l'entier r(F,N). Elle est, surtout à cause de la méthode utilisée en IV 2.1, très grande. Nous allons plutôt ci-dessous borner inférieurement f(F,N) et trouver dans trois cas particuliers les valeurs optimales. Il s'agira dans tous les cas de valeurs uniformes en ce sens qu'elles seront le minimum possible qui convienne à toutes les F ayant un terme constant donné. Plus précisément :

Lemme V. 1 1. *Soit P un système représentatif d'éléments irréductibles dans l'anneau factoriel A. Soient N et k appartenant à* $\mathbb{N}$. *Si l'entier* r_0*"est un* r(F,N)" *pour tous les* $F \in A[[T]]'$ *tels que : pour tout* $p \in P, v_p(F(0)) \leq k$, *alors* r_0 *est supérieur ou égal à* $(N+1).k$.

<u>Démonstration.</u> Il suffit évidemment de trouver une série F de ce type pour laquelle $(N+1).k - 1$ ne convient pas, c'est à dire telle que la relation (e), puisse être vérifiée pour une infinité de classes du coefficient de T^{N+1} dans g ou h.

Soient, par exemple, p un élément de P et prenons respectivement $F=p^k, g=p$, $H=p^{k-1}$. Pour tout λ appartenant à A, soient :

$$g = p+\lambda.T^{N+1}, h = p^{k-1}+\lambda p^{k-2}T^{N+1}+\ldots+(-1)^{k-1}\lambda^{k-1}T^{(N+1)(k-1)}.$$

On a bien :

$$F \equiv G.H\ [T^{N+1}];\ g \equiv G\ [T^{N+1}],\ h \equiv H\ [T^{N+1}].$$

De plus, g.h - qui vaut $p^k+(-1)^k\lambda^kT^{(N+1)k}$ - est congru à F modulo $T^{(N+1)k}$. Cependant le coefficient de T^{N+1} dans g peut prendre toutes les valeurs possibles modulo p, donc une infinité dès que A/p est infini : l'entier $r = (N+1)k-1$ ne convient pas, d'où le Lemme.

<u>Remarque</u>. Lorsque A est une $\mathbb{Q}$-algèbre, il en est de même pour A/p et ce dernier anneau est infini.

VI.1. 2 - Nous montrons maintenant que, pour les "petites" valeurs de k, on a en fait une égalité.

THEOREME VI.1.2 - *Soit* P *un système représentatif d'éléments irréductibles dans l'anneau factoriel* A. *Soient* F,G, H *appartenant à* A[[T]]' *et* $N \in \mathbb{N}$; *notons* $\alpha_1,\ldots,\alpha_s$ *les exposants non nuls dans la décomposition en facteurs irréductibles de* F(0) *et posons* :
$\alpha = \sup(\alpha_1,\ldots,\alpha_s)$, $\gamma = \alpha_1 \ldots \alpha_s$.
Alors, si α *est inférieur ou égal à* 3 : $r(F,N)=\alpha\,(N+1)$ *convient pour* F *et les cardinaux de* I(F,G,H,N) *et* J(F,G,H,N) *sont inférieurs ou égaux à* γ.

Démonstration. Remarquons d'abord que $\alpha(N+1)$ est ce qu'on a vu être le minimum possible qui ne dépende que de F(0). D'autre part, la démonstration du Lemme IV.3.1 montre que, si les entiers r_i, les parties I_i et J_i correspondant aux localisés de A par les Ap_i, alors $r = \sup(r_1,\ldots,r_s)$ et l'image canonique de ($I_1 x\ldots x I_s$, $J_1 x \ldots x J_s$) conviennent dans A. Or, la définition de γ et celle de I et J est multiplicative, celle de α et r associée à un "sup"; autrement dit, la conclusion du Théorème IV.1.2 "passe du local au global". Comme l'hypothèse que α est inférieur ou égal à 3 se transmet, elle, aux localisés, on conclut de ce qui précède qu'il suffit de démontrer le Théorème IV.1.2 lorsque A est un anneau de valuation discrète, ce qu'on suppose désormais.
Cas où α =0. Cela signifie que F est inversible dans A[[T]]. Si

$$(e_0)\quad F \equiv G.H\ [T^{N+1}],\ g \equiv G\ [T^{N+1}],\ h \equiv H\ [T^{N+1}]$$

$$\text{et } F \equiv g.h\ [T]\ (\text{ici, } r = \alpha(N+1) = 0).$$

alors, g(0)h(0)= F(0) est inversible dans A : G, H, g et h sont des séries inversibles. Dans ce cas, A/G(0)= A/H(0)= { 0 } , a fortiori les cardinaux de I(F,G,H,N) et J(F,G,H,N) sont inférieurs ou égaux à 1 ; d'où le Théorème dans ce cas.
Cas où α = 1. On a, cette fois :

$$(e_1)\quad g \equiv G\ [T^{N+1}],\ h \equiv H\ [T^{N+1}],\ F \equiv g.h\ [T^{N+2}].$$

Si on note a_n, b_n et c_n les termes généraux respectifs de F, g et h respectivement, on a : $a_0 = b_0.c_0$. Comme a_0 est associé à p(l'uniformisante de A) par hypothèse, l'un des deux éléments b_0 ou c_0, par exemple b_0, est inversible. Le raisonnement fait au premier cas montre qu'alors le cardinal de I(F,G,H,N) est inférieur ou égal à 1. Dans ce cas, vu que :

$$b_0.c_{N+1} + (b_1.c_N + \ldots + b_N.c_1) + c_0.b_{N+1} = a_{N+1},$$

et que le terme entre parenthèse est fixe d'après (e_1), ainsi que a_{N+1}, on obtient que $b_0.c_{N+1}$ est constant modulo c_0. Il en est donc de même pour c_{N+1}, puisque $b_0 = H(0)$ est inversible et fixé. Par suite, le cardinal de J(F,G,H,N) est inférieur ou égal à 1, ce qui règle le second cas.

A cause de la longueur des deux démonstrations restantes, nous traitons maintenant uniquement le quatrième cas, celui où $\alpha = 3$, en indiquant ensuite comment retrouver le cas où $\alpha = 2$.

VI.1. 3 - Cas où $\alpha = 3$.

Nous supposons donc que :

$$g \equiv G\ [T^{N+1}],\ h \equiv H\ [T^{N+1}],\ F \equiv g.h\ [T^{3N+4}]$$

et de plus, quitte à modifier G et H à partir de l'ordre N+2, que F est congru à G.H modulo T^{3N+4}. Ecrivons g et h sous les formes respectives :

$$g = G + T^{N+1}.\ \beta\ ;\ h = H + T^{N+1}.\gamma.$$

La relation "$F \equiv g.h\ T^{3N+4}$" s'écrit donc (1) $\beta.H+\gamma.G+T^{N+1}\beta\gamma \equiv 0\ [T^{2N+3}]$.
Notons β_n, γ_n, g_n et h_n les coefficients génériques respectifs de β, γ, G et H. La conclusion revient à dire que le nombre de classes possibles pour β_0 modulo g_0 (resp. γ_0 modulo h_0) est fini. Il suffit de démontrerla première assertion car la relation "$\beta_0.h_0 + \gamma_0.g_0 = 0$" qui provient du terme constant de l'égalité précédente, montre qu'elle entraîne la seconde.

Quitte à multiplier par des éléments inversibles - fixes - de A et à échanger g et h, on peut supposer que :

$$g(0) = p,\ h(0) = p^2.$$

On cherche donc le nombre de valeurs possibles pour β_0 modulo p.

Partons de la relation mentionnée ci-dessus, et qui s'écrit :

$$p.\ \beta_0 + \gamma_0 = 0\ :\ \gamma_0 = -\varepsilon_0.\beta_0 \quad \text{(en posant } \varepsilon_0 = p).$$

L'examen du coefficient de T dans (1) donne :

$$p^2\beta_1 + h_1\beta_0 + p\gamma_1 + g_1\gamma_0 = 0, \text{ soit :}$$

$$\beta_0(h_1 - \varepsilon_0 g_1) + p^2\beta_1 + p\gamma_1 = 0.$$

Si $h_1 - \varepsilon_0 g_1$ n'est pas multiple de p, β_0 doit l'être. Nous exclurons désormaisce cas, puisqu'il ne porte que sur une classe modulo p. On a donc une relation du type :

$$h_1 = \varepsilon_0 g_1 + \varepsilon_1 p, \text{ c'est à dire } h_1 = \varepsilon_0 g_1 + \varepsilon_1 g_0\ .$$

D'où, en reportant :

$$\gamma_1 = -\varepsilon_0\beta_1 - \varepsilon_1\beta_0.$$

Supposons trouvés, en examinant les coefficients de $T,\ldots,T^{j-1}$ (avec $j \leqslant N$), des éléments de A, $\varepsilon_0,\ldots,\varepsilon_{j-1}$ tels que :

$$(2)\ h \equiv (\varepsilon_0 + \varepsilon_1 T + \ldots + \varepsilon_{j-1}T^{j-1}).g\ [T^j]$$

$$(3)\ \gamma \equiv -(\varepsilon_0 + \varepsilon_1 T + \ldots + \varepsilon_{j-1}T^{j-1}).\beta[T^j].$$

Remarquons déjà qu'à cause de (2), $\varepsilon_0,\ldots,\varepsilon_{j-1}$ sont indépendants de β et γ. Le coefficient de T^j dans (1) donne alors :

$$\beta_0 h_j + \sum_{\substack{r+s=j\\ s\leqslant j-1}} \beta_r h_s + g_0\gamma_j + \sum_{\substack{r+s=j\\ s\leqslant j-1}} g_r\gamma_s = 0.$$

Soit, en utilisant (2) et (3) :

$$\beta_0 h_j + g_0\ \ _j + \sum_{\substack{q+k+\ell=j\\ q\geqslant 1}} \beta_q\varepsilon_k g_\ell - \sum_{\substack{q+k+\ell=j\\ \ell\geqslant 1}} g_\ell\varepsilon_k\beta_q = 0.\ \text{D'où :}$$

$$\beta_0(h_j - \varepsilon_0 g_j - \ldots - \varepsilon_{j-1}g_1) + g_0(\gamma_j + \varepsilon_0\beta_j + \ldots + \varepsilon_{j-1}\beta_1) = 0.$$

Comme $g_0 = p$ et que β_0 n'est pas multiple de p, il existe un ε_j tel que :

$$h_j - \varepsilon_0 g_j - \ldots - \varepsilon_{j-1}g_1 = \varepsilon_j p\ ;\ \text{soit :}$$

$$h_j = \varepsilon_0 g_j + \ldots + \varepsilon_j g_0\ ;\ \text{et, en reportant :}$$

$$\gamma_j = -\varepsilon_0\beta_j - \ldots - \varepsilon_j\beta_0.$$

Par conséquent, les relations (2) et (3) sont vraies avec j+1 à la place de j ; en particulier, pour j = N :

$$(4)\ h \equiv (\varepsilon_0 + \ldots + \varepsilon_N T^N).g\ [T^{N+1}]$$

$$(5)\ \gamma \equiv -(\varepsilon_0 + \ldots + \varepsilon_N T^N).\beta\ [T^{N+1}].$$

Considérons le coefficient de T^{N+1} dans (1). Le seul "nouveau" terme est $\beta_0\gamma_0$, provenant de T^{N+1}. On obtient donc :

$$\beta_0(h_{N+1} - \varepsilon_0 g_{N+1} - \ldots - \varepsilon_N g_1) + g_0(\gamma_{N+1} + \varepsilon_0\beta_{N+1} + \ldots$$

$$\ldots + \varepsilon_N\beta_1) - \varepsilon_0\beta_0^2 = 0\ (\text{car } \beta_0\gamma_0 = -\varepsilon_0\beta_0^2).$$

Comme $g_0 = \varepsilon_0 = p$, et que β_0 n'est pas multiple de p, il existe un ε_{N+1} appartenant à A tel que :

$$h_{N+1} = \varepsilon_0 g_{N+1} + \ldots + \varepsilon_{N+1} g_0 \text{ et, par suite :}$$

$$\gamma_{N+1} = -\varepsilon_0 \beta_{N+1} - \ldots - \varepsilon_{N+1}\beta_0 + \beta_0^2.$$

Examinons maintenant le coefficient de T^{N+2} dans (1). On obtient, de même que ci-dessus :

$$\beta_0(h_{N+2} - \varepsilon_0 g_{N+2} - \ldots - \varepsilon_{N+1} g_1) + g_0(\gamma_{N+2} + \varepsilon_0\beta_{N+2} + \ldots$$

$$\ldots + \varepsilon_{N+1}\beta_1) + \beta_0^2 g_1 + (\beta_0\gamma_1 + \beta_1\gamma_0) = 0 \text{ ; c'est à dire :}$$

$$\beta_0^2(g_1 - \varepsilon_1) + \beta_0(h_{N+2} - \varepsilon_0 g_{N+2} - \ldots - \varepsilon_{N+1} g_1 - 2\beta_1\varepsilon_0) +$$

$$+ (\gamma_{N+2} + \varepsilon_0\beta_{N+2} + \ldots + \varepsilon_{N+1}\beta_1).g_0 = 0$$

Il s'agit là d'une équation du second degré dans le corps A/p, dont les coefficients sont constants, puisque $\varepsilon_0 = g_0 = p$ et que les ε_i ne dépendent pas de β et γ. Si les coefficients de cette équation ne sont pas tous multiples de p, elle a au plus deux racines en β_0 modulo p et le résultat cherché est établi. Ecartons donc ce cas et supposons qu'on ait des égalités de la forme :

$$g_1 = \varepsilon_1 - \lambda_1 p \text{ ; } h_{N+2} = \varepsilon_0 g_{N+2} + \ldots + \varepsilon_{N+2} g_0.$$

D'où, en posant $\lambda_0 = 1$ et en reportant :

$$\varepsilon_1 = g_0\lambda_1 + g_1\lambda_0 \text{ ;}$$

$$\gamma_{N+2} = -\varepsilon_0\beta_{N+2} - \ldots - \varepsilon_{N+2}\beta_0 + 2\beta_1\beta_0 + \lambda_1\beta_0^2.$$

Supposons trouvés, en examinant les coefficients de $T^{N+2}, \ldots, T^{N+j}$ $(j \leq N)$ dans (1), des éléments de A : $\varepsilon_{N+2}, \ldots, \varepsilon_{N+j}$; $\lambda_0, \ldots, \lambda_{j-1}$ tels que :

$$(6) \quad h \equiv (\varepsilon_0 + \ldots + \varepsilon_{N+j} T^{N+j}).g \; [T^{N+j+1}];$$

$$(7) \quad \varepsilon_0 + \ldots + \varepsilon_{j-1} T^{j-1} \equiv (\lambda_0 + \ldots + \lambda_{j-1} T^{j-1}).g \; [T^j].$$

Comme (1) entraîne que $\gamma G + \beta H + T^{N+1}\beta\gamma$ est congru à 0 modulo T^{N+j+1}, on déduit de (6) et (7) que :

$$(8) \quad \gamma_0 + \ldots + \gamma_{N+j} T^{N+j} \equiv T^{N+1}(\lambda_0 + \ldots + \lambda_{j-1} T^{j-1})(\beta_0 + \ldots$$

$$\ldots + \beta_{j-1} T^{j-1})^2 - (\beta_0 + \ldots + \beta_{N+j} T^{N+j})(\varepsilon_0 + \ldots + \varepsilon_{N+j} T^{N+j})[T^{N+j+1}].$$

(Pour voir ceci, on remplace H par $(\varepsilon_0 + \ldots + \varepsilon_{N+j}T^{N+j})$ G modulo T^{N+j+1} et $T^{N+1}\beta\gamma$ par $- T^{N+1}$ $\beta\varepsilon\beta$ congru à $- T^{N+1}\beta^2 G(\lambda_0 + \ldots + \lambda_{j-1}T^{j-1})$ modulo T^{N+j+1} ; on utilise les relations (4) et (5), ce qui donne :

$G(\gamma + (\beta_0 + \ldots + \beta_{N+j}T^{N+j})(\varepsilon_0 + \ldots + \varepsilon_{N+j}T^{N+j}) - T^{N+1}(\lambda_0 + \ldots$

$\ldots + \lambda_{j-1}T^{j-1})(\beta_0 + \ldots + \beta_{j-1}T^{j-1})^2)$ congru à 0 modulo T^{N+j+1}, d'où (8), puisque G(0) est non nul).

La nullité du coefficient de T^{N+j+1} dans (1) entraîne successivement que :

$$\beta_0 h_{N+j+1} + g_0\gamma_{N+j+1} + \sum_{\substack{k+\ell=N+j+1\\ k\geq 1}} (\beta_k h_\ell + g_k\gamma_\ell) + \sum_{k+\ell=j} \beta_k\gamma_\ell = 0.$$

$$\beta_0 h_{N+j+1} + g_0\gamma_{N+j+1} + \sum_{\substack{r+s+q=N+j+1\\ r\geq 1}} \beta_r\varepsilon_s g_q - \sum_{\substack{r+s+r'=j\\ r\geq 1}} \beta_r\varepsilon_s\beta_{r'} +$$

$$+ \beta_0\gamma_j + \Big(\sum_{\substack{q+t+r+r'=j\\ q\geq 1}} g_q\lambda_t\beta_r\beta_{r'} - \sum_{\substack{q+r+s=N+j+1\\ q\geq 1}} g_q\beta_r\varepsilon_s\Big) = 0$$

$$\beta_0 h_{N+j+1} + g_0\gamma_{N+j+1} + g_0(\beta_1\varepsilon_{N+j} + \ldots + \beta_{N+j+1}\varepsilon_0) -$$

$$- \beta_0(g_1\varepsilon_{N+j} + \ldots + g_{N+J=1}\varepsilon_0) + \sum_{\substack{q+t+r+r'=j\\ q\geq 1}} g_q\lambda_t\beta_r\beta_{r'} -$$

$$- \sum_{\substack{q+t+r+r'=j\\ r\geq 1}} g_q\lambda_t\beta_r\beta_{r'} + \beta_0\gamma_j = 0$$

$$\beta_0(h_{N+j+1} - \varepsilon_0 g_{N+j+1} - \ldots - \varepsilon_{N+j}g_1) + g_0(\gamma_{N+j+1} + \varepsilon_0\beta_{N+j+1} + \ldots$$

$$\ldots + \varepsilon_{N+j}\beta_1) + \beta_0(\beta_0(\lambda_0 g_j + \ldots + \lambda_{j-1}g_1) + \beta_1(\varepsilon_{j-1} - g_0\lambda_{j-1}) + \ldots$$

$$\ldots + \beta_j(\varepsilon_0 - g_0\lambda_0)) - g_0\Big(\sum_{\substack{t+r+r'=j\\ r\geq 1}} \lambda_t\beta_r\beta_{r'}\Big) - \beta_0\Big(\sum_{k+\ell=j} \beta_k\varepsilon_\ell\Big) = 0$$

$$- \beta_0^2(\varepsilon_j - \lambda_0 g_j - \ldots - \lambda_{j-1}g_1) + \beta_0(h_{N+j+1} - \varepsilon_0 g_{N+j+1} - \ldots$$

$$\ldots - \varepsilon_{N+j}g_1) + (\text{un multiple de } p) = 0$$

Cette dernière équation, modulo p, a ses coefficients indépendants de β et γ, d'où deux valeurs au plus pour β_0 modulo p – et la conclusion cherchée – sauf dans le cas où les coefficients sont tous multiples de p, c'est à dire lorsque l'on a des relations du type suivant :

$$\varepsilon_j = \lambda_0 g_j + \ldots + \lambda_j g_0 \ ; \ h_{N+j+1} = \varepsilon_0 g_{N+j+1} + \ldots + \varepsilon_{N+j+1} g_0$$

d'où l'on déduit (6) et (7) avec N+j+1 à la place de N+j.

Si, par conséquent, l'examen des coefficients de $T^{N+2}, \ldots, T^{2N+1}$ n'a pas permis de conclure, c'est qu'il existe deux polynômes :

$$E = \varepsilon_0 + \ldots + \varepsilon_{2N+1} T^{2N+1} \ ; \ \Lambda = \lambda_0 + \ldots + \lambda_{2N+1} T^{2N+1} \text{ tels que}$$

$$H \equiv E.G\,[T^{2N+2}] \text{ et } E \equiv G.\Lambda[T^{N+1}].$$

Comme ci-dessus, on tire de ces relations :

$$\gamma \equiv T^{N+1}.\Lambda.\beta^2 - \beta.E\,[T^{2N+2}].$$

Considérons alors le coefficient de T^{2N+2} dans (1). Le seul changement par rapport au calcul fait au bas de la page précédente est que le terme $\beta_0\gamma_{N+1}$ devient $\beta_0(-\beta_0\varepsilon_{N+1} - \ldots - \beta_{N+1}\varepsilon_0 + \beta_0^2)$.

On obtient donc :

$$\beta_0^3 - \beta_0^2(\varepsilon_{N+1} - \lambda_0 g_{N+1} - \ldots - \lambda_N g_1) + \beta_0(h_{2N+2} - \varepsilon_0 g_{2N+2} - \ldots$$

$$\ldots - \varepsilon_{2N+1} g_1) + (\text{un multiple de } p) = 0.$$

Il s'agit, modulo p, d'une équation - unitaire - du troisième degré ; elle a au plus trois solutions dans le corps A/p et ses coefficients ne dépendent pas de β et γ, puisque E et Λ ne dépendent que de G et H.

En définitive, suivant les valeurs de G et H, le nombre de solutions possibles pour β_0 modulo p est (0), 1, 2 ou 3 ; ceci achève la démonstration du cas où $\alpha = 3$.

VI.1.4 - Cas où $\alpha = 2$. On prend alors $r = 2N + 3$ et on se ramène au cas où $g_0 = h_0 = p$. Ici, ε_0 vaut 1, le début de la démonstration demeure jusqu'à la première égalité suivant (5) qui, ici, permet de conclure : ε_0 étant égal à 1, on obtient une équation unitaire du second degré modulo p, dont les coefficients ne dépendent que de G et H, et le nombre de solutions possibles pour β_0 est, en définitive, au plus deux.

Le Théorème VI.1.2 est entièrement démontré.

Remarque. On peut se demander si le Théorème VI.1.2. est vrai pour toutes les valeurs de α. On observe par ailleurs, dans la démonstration de VI.1.3 combien le "pire" des cas, c'est à dire celui où seul l'examen du terme en T^{2N+2} permet de conclure, se rapproche d'un des exemples du Lemme VI.1.1, savoir la congruence :

$$p^3 \equiv (p + \lambda.T)(p^2 - \lambda.p.T + \lambda^2.T^2)\ [T^3]$$

VI.2 - TRADUCTION DE (T).

VI.2. 1 - Commençons par une reformulation immédiate. Soit $(a_n)_{n \in \mathbb{N}}$ une suite d'éléments de A, avec a_0 non inversible. Le système infini d'équations aux inconnues $(x_n)_{n \in \mathbb{N}}$ et $(y_n)_{n \in \mathbb{N}}$:

$$(1)\quad \begin{aligned} & x_0 . y_0 = a_0 \\ & x_0 . y_1 + x_1 . y_0 = a_1 \\ & \ldots \\ & x_0 . y_n + x_1 . y_{n-1} + \ldots + x_n . x_0 = a_n \\ & \ldots \end{aligned}$$

n'a pas de solution non triviale (c'est à dire avec x_0 et y_0 non inversibles) si, et seulement si, il existe un entier N tel que le système formé par les N+1 premières équations de (1) n'ait pas de solution non triviale ; ceci traduit le "caractère fini" de l'irréductibilité.

Par ailleurs, (T) montre qu'il y a "beaucoup" d'irréductibles non associés dans A[[T]]. En fait, si f est irréductible et de terme constant non nul, soit N son ordre d'irréductibilité (IV.1.1) et remarquons que, si A est une $\mathbb{Q}$-algèbre, l'anneau A/f(0) est infini (cf. VI.1.1 Remarque). On a :

Lemme VI.2.1 - *Soit* A *une* $\mathbb{Q}$*-algèbre factorielle. Soit* N *l'ordre d'irréductibilité de la série* f , *avec* $f(0) \neq 0$.
L'ensemble des classes d'irréductibles congrus à f *modulo* T^{N+1} *a un cardinal supérieur ou égal à* card$((A/f(0))^{\mathbb{N}})$.

<u>Démonstration</u>. Soit E' un système complet de représentants pour A/f(0), par exemple les éléments d'un supplémentaire dans A du $\mathbb{Q}$-sous-espace f(0).A. Il suffit de montrer que, si F est l'ensemble des idéaux principaux de A[[T]] engendrés par les g congrus à f modulo T^{N+1} et dont les coefficients, à partir du rang N+1, appartiennent à E', alors le cardinal de F est égal à celui de $E'^{\mathbb{N}}$. Soit donc :

$$S = \{g = \sum_{n \in \mathbb{N}} b_n T^n \mid g \equiv f\ [T^{N+1}] \text{ et } b_{N+1}, b_{N+2}, \ldots \in E'\}.$$

et observons que tous les éléments de S sont irréductibles. Soit ν l'application de S dans $E'^{\mathbb{N}}$ qui, à g noté comme ci-dessus, associe la suite $(b_{N+1}, b_{N+2}, \ldots)$. ν est surjective par définition ; quant à l'injectivité, si deux séries ont les mêmes termes à partir de l'ordre N+1 et sont toutes deux congrues à f modulo T^{N+1}, elles sont évidemment égales : ν est une bijection.

Reste à voir maintenant que deux éléments distincts de S ne peuvent être associés. Par l'absurde, supposons que h, de terme général c_n, soit associé à g :

existe un u inversible dans A[[T]] tel que :

$$g = u.h \text{ ; soit k le premier entier tel que } c_k \neq b_k.$$

Nécessairement, k est supérieur ou égal à N+1, et on a :

$$g \equiv h\ [T^k] \text{ et } g \equiv u.h\ [T^k].$$

Comme h(0) = f(0) est différent de 0, on déduit de ceci que u est congru à 1 modulo T^k, autrement dit, est de la forme :

$$u = 1 + u_k T^k + \dots$$

L'égalité des coefficients dans g et u.h pour T^k donne donc :

$$b_k = c_k + u_k . c_0$$

Par suite, b_k et c_k sont congrus modulo c_0, c'est à dire modulo f(0), en contradiction avec le fait qu'ils sont différents et que E' contient un seul représentant de chaque classe.

En définitive, S est lui aussi un système complet de représentants pour les classes d'irréductibles congrus à f modulo T^{N+1} et il a donc même cardinal que F, ce qui achève la démonstration du Lemme.

VI.2. 2 - ETUDE DES ANNEAUX $\mathbb{B}_p$.

Rappelons (cf. V. 2. 1) que, pour tout n ∈ $\mathbb{N}$, si on désigne par K le corps des fractions de A, $\mathbb{B}_n$ est l'ensemble des éléments de K[[T]] dont les coefficients, jusqu'à l'ordre n inclus, appartiennent à A. On voit aisément que $\mathbb{B}_n$ est un sous-anneau de K[[T]] et on a :

$$(1)\quad A[[T]] \underset{\neq}{\subset} \dots \underset{\neq}{\subset} \mathbb{B}_n \underset{\neq}{\subset} \mathbb{B}_{n-1} \underset{\neq}{\subset} \dots \underset{\neq}{\subset} \mathbb{B}_0 \underset{\neq}{\subset} K[[T]].$$

De même, A[[T]], qui est l'intersection des $\mathbb{B}_n$ et qu'on pourra noter $\mathbb{B}_\infty$, s'identifie naturellement à la limite projective $\varprojlim \mathbb{B}_n$ associée aux inclusions de (1).

Remarque. Aucun des anneaux $\mathbb{B}_n$ n'est noethérien : soit en effet a un élément de A non nul et non inversible et, pour tout k ∈ $\mathbb{N}$, soit :

$$I_k = \{ f = \sum_{\ell=n+1}^{\infty} a_\ell T^\ell \mid \text{les } a_\ell \in K \text{ et } a_{n+1}.a^k \in A \}$$

Les I_k sont évidemment des idéaux de $\mathbb{B}_n$, la suite I_k est croissante (parce qu'il en est ainsi des idéaux fractionnaires $a^{-k}.A$) et même strictement croissante, puisque $a^{-k-1}.T^{n+1}$ appartient à I_{k+1} mais pas à I_k. Cela signalé, (T) s'énonce très simplement :

(T") un élément de $A[[T]] = \mathbb{B}_\infty$ est irréductible si, et seulement si, il existe un N tel qu'il soit irréductible dans $\mathbb{B}_N$.

En effet, on voit qu'un élément de $A[[T]]$ est réductible (resp.inversible)modulo T^{N+1} si, et seulement si, il l'est dans $\mathbb{B}_N$.

Corollaire. *Soit* A *une* $\mathbb{Q}$-*algèbre factorielle et supposons que, pour une infinité de* $n \in \mathbb{N}$, $\mathbb{B}_n$ *soit factoriel. Alors* $A[[T]]$ *est factoriel.*

Démonstration. Montrons que tout élément irréductible f de $A[[T]]$ engendre un idéal premier. Soient g et h appartenant à $A[[T]]$ tels que f divise gh. Soit N l'ordre d'irréductibilité de f ; pour chacun des $n \geqslant N$ tels que $\mathbb{B}_n$ soit factoriel, f, qui est irréductible dans $\mathbb{B}_n$, y divise soit g, soit h. L'une de ces deux éventualités, par exemple la première, à lieu une infinité de fois ; autrement dit, le quotient $\frac{g}{f}$ appartient à une infinité de $\mathbb{B}_n$, on en déduit aisément qu'il appartient à $A[[T]]$ et le Corollaire est établi.

VI.2. 3 - Substitutions.

Dans la recherche des substitués $f(g(T),T)$ irréductibles d'un f irréductible dans $A[[X,Y]]$(cf. [1] II.4.2), (T) permet de se restreindre aux polynômes :

Corollaire. *Soit* A *une* $\mathbb{Q}$-*algèbre factorielle. Soit* f *un élément de* $A[[X,Y]]$.*Les conditions suivantes sont équivalentes :*

α) il existe une série $g(g(0)=0)$ *telle que* $f(g(T),T)$ *soit irréductible dans* $A[[T]]$,

β) il existe un polynôme $P(P(0)=0)$ *tel que* $f(P(T),T)$ *soit irréductible dans* $A[[T]]$.

Démonstration. Il suffit de voir que α) ⇒ β). Soit ρ l'application de $A[[T]]$ dans lui même qui, à toute série ϕ, associe $f(T.\phi(T),T)$. La formule de Taylor montre que ρ est continue pour la topologie T-adique. Par hypothèse, l'image réciproque par ρ de l'ensemble I des irréductibles de $A[[T]]$ est non vide ; c'est un ouvert, puisque I est ouvert d'après (T); il rencontre donc $A[T]$ qui est dense dans $A[[T]]$, d'où le résultat.

§ VII - Appendice

VII.1 - PRELIMINAIRE.

VII.1. 1 - Comme en V, A désigne un anneau de valuation discrète, d'uniformisante p, complet pour la topologie p-adique. K est le corps des fractions de A, v la valuation p-adique et $|\ | = e^{-v}$. Rappelons les inégalités, valables pour x et y appartenant à K :

$$(1)\quad v(x+y) \geq \inf(v(x),v(y)) \ ; \ |x+y| \leq |x| + |y| .$$

Avec égalité lorsque $v(x) \neq v(y)$.

A tout polynôme U non nul de K [T] est associé supp(U), ensemble des exposants de T dont le coefficient est non nul dans U, le plus petit (resp. plus grand) élément de supp(U) est la valuation (resp. le degré) de U.

VII.1. 2 - U étant comme ci-dessus, soit ρ un réel strictement positif, on pose :

$$\rho^*(U) = \sup(\{\ |u_n|\rho^n \mid n \in \mathbb{N}\}) .$$

(u_n est le coefficient de T^n dans U). ρ^* (U) existe puisque le support de U est fini, et on a l'interprétation géométrique suivante :

Lemme VII.1.2.1 - *Soient* $\in K[T] - \{0\}$ *et* ρ *un réel* > 0. *Si* $N_{..}$ *est le polygone de Newton de* U , *alors* $-\mathrm{Log}(\rho^*(U))$ *est l'ordonnée à l'origine de l'unique droite d'appui de* N_U *ayant pour pente* $\mathrm{Log}(\rho)$.

Démonstration. Puisque $\rho^*(U) = \sup(\{|u_n|\rho^n | n \in \mathbb{N}\})$, on a :

$$-\mathrm{Log}(\rho^*(U)) = \inf(\{v(u_n) - n.\mathrm{Log}(\rho) \mid n \in \mathbb{N}\}).$$

Autrement dit, $-\mathrm{Log}(\rho^*(U))$ est la plus grande ordonnée à l'origine possible pour la frontière d'un demi-plan supérieur contenant l'ensemble des points $M_n(n,v(u_n))$. Vu la définition de $N_{..}$ (cf. V.1.2), la conclusion en résulte.

Lemme VII.1.2.2 - *Soient* U *et* V *appartenant à* K[T] *et* ρ *un réel strictement positif. On a :*

$$\rho^*(U+V) \leq \sup(\rho^*(U),\rho^*(V)) \ ; \ \rho^*(UV) \leq \rho^*(U).\rho^*(V).$$

Démonstration. La première inégalité est évidente, compte tenu des définitions et de (1). Observons d'ailleurs que, si U = 0, il n'a pas de polygone de Newton, mais que $\rho^*(U)$ est néanmoins défini. Si v_n est le terme général du polynôme V, celui de UV est :

$$w_n = \sum_{i+j=n} u_i.v_j.$$

Si on note encore Ω la clôture algébrique de K, on aura encore, pour tout $x \in \Omega$:

$$(2)\quad w_n.x^n = \sum_{i+j=n} (u_i.x^i)(v_j.x^j).$$

Soit $r = \frac{\alpha}{\beta}$ un rationnel strictement supérieur à $-\mathrm{Log}(\rho)$ et notons z une racine β-ième de p dans Ω. Si $\overline{v}$ et ‖ ‖ désignent les uniques prolongements de v et | | à K' = K(z) (cf. V.3.3), on a, d'après (2) appliqué à $x = z^\alpha$:

$$(3)\quad |w_n|.\|x\|^n \leqslant \sup(\{|u_i|.\|x\|^i\}).\sup(\{|v_j|.\|x\|^j\}).$$

Mais $z^\beta = p$, donc $\overline{v}(x) = \frac{\alpha}{\beta} = r$ et $\|x\| = e^{-r}$. On déduit de (3) que, pour tout rationnel $r > -\mathrm{Log}(\ )$, on a :

$$(4)\quad (e^{-r})^*(UV) \leqslant (e^{-r})^*(U).(e^{-r})^*(V).$$

Lorsque U est fixé, l'application qui à λ associe $\lambda^*(U)$ est évidemment continue. Comme l'ensemble des e^{-r}, pour r rationnel strictement supérieur à $-\mathrm{Log}(\rho)$est dense dans $]0,\rho[$, (4) implique :

$$(5)\quad \rho^*(UV) \leqslant \rho^*(U).\ \rho^*(V).$$

Le Lemme est entièrement établi.

VII.1. 3 - ρ-diviseurs.

<u>Définition</u>. Soient $P = \sum_{n=0} a_n.T^n$ un polynôme de degré d et ρ un réel strictement positif. On dira que P est un *ρ-diviseur* si, et seulement si :

$$|a_0| = |a_d|.\rho^d = \rho^*(P) = 1.$$

En particulier, pour tout $n \in \{0,\dots,d\}$, $|a_n|.\ \rho^n \leqslant 1$. D'autre part, d'après le Lemme VII.1.2.1, la droite d'appui de N_P ayant pour pente Log(ρ) passe par l'origine, c'est à dire par le sommet M_0, de coordonnées 0 et $v(a_0)=0$ (puisque $|a_0| = 1$).Cette droite passe aussi par M_d $(d,v(a_d))$ puisque la définition d'un ρ-diviseur entraîne que $v(a_d) = d.\log(\rho)$. Comme M_0et M_d sont les "points extrêmes" du polygone de Newton N_P, on a démontré le :

Lemme VII.1. 3 - *Soient* $P \in K[T]$, *avec* $P(0) \neq 0$ *et* ρ *un réel* > 0. P *est un ρ-diviseur si, et seulement si, son polygone de Newton a un unique segment non vertical, inclus dans la droite d'équation* $y = x.\mathrm{Log}(\rho)$.

VII. 1. 4 - Division euclidienne

Lemme VII.1:4 - *Soit ρ un réel > 0 . Soient P un ρ-diviseur et U un polynôme quelconque. Si Q et R sont respectivement le quotient et le reste de la division euclidienne de U par P, on a:* $\rho^*(Q) \leqslant \rho^*(U)$ *et* $\rho^*(R) \leqslant \rho^*(U)$.

Démonstration. On a donc : $U = P.Q + R, R = 0$ ou $\deg(R) < \deg(P)$. Si U est nul ou de degré strictement inférieur à celui de P, alors $Q = 0$, $R = U$ et les inégalités sont évidentes. Nous pouvons donc supposer que $\delta = \deg(U) - \deg(P)$ est positif ou nul. On raisonne par récurrence sur δ. Si $u.T^n$ et $v.T^d$ sont les termes directeurs de U et P respectivement, on sait que :

$$Q = \frac{u}{v}.T^{n-d} + Q_1 \text{ (}Q_1\text{, quotient de la division euclidienne de}$$

$$U_1 = U - \frac{u}{v}.T^{n-d}.P \text{ par P).}$$

Or U_1 est nul ou de degré strictement inférieur à celui de U, donc l'hypothèse de récurrence - qu'il était par ailleurs inutile de vérifier pour $\delta = 0$ - montre que :

$$(6)\ \rho^*(Q_1) \leqslant \rho^*(U_1) \text{ et donc } \leqslant \sup(\rho^*(U), \rho^*(\frac{u}{v}.T^{n-d}.P))$$

d'après le Lemme VII.1.2.2

En outre, par définition :

$$(\frac{u}{v}.T^{n-d}) = \frac{|u|.\rho^n}{|v|.\rho^d}$$

vaut aussi $|u|.\rho^n$ puisque P est un ρ-diviseur. Par suite :

$$(7)\ \rho^*(\frac{u}{v}.T^{n-d}) \leqslant \rho^*(U).$$

(6) et (7) montrent que $\rho^*(Q_1) \leqslant \rho^*(U)$. Enfin, $R = U - P.Q_1$, donc :

$$\rho^*(R) \leqslant \sup(\rho^*(U), \rho^*(P).\rho^*(Q_1))$$

d'après le Lemme VII.1.2.2 ; et ceci est inférieur ou égal à $\rho^*(U)$ vu la relation précédente et le fait que P est un ρ-diviseur : la récurrence est établie ; d'où le Lemme .

VII.2 - IRREDUCTIBILITE.

VII.2. 1 - Les notations étant comme ci-dessus, on se propose de démontrer le :

THEOREME VII.2.1 - *Si* $U \in K[T]$ *(avec* $U(0) \neq 0$*) est irréductible, son polygone de Newton a un unique segment non vertical.*

Démonstration. Soit δ le degré de U . Observons d'abord que cet énoncé implique le Théorème V.1.2 vu que, pour un élément de $A[T]$, irréductibilité dans $A[T]$ et dans $K[T]$ sont équivalentes. Par ailleurs, quitte à diviser U par $U(0)$, ce qui ne change pas l'hypothèse et translate son polygone de Newton, on peut supposer - ce qu'on fera désormais - que U est de la forme :

$$U = \sum_{n=0}^{\delta} \lambda_n . T^n \; ; \; \lambda_0 = 1, \lambda_\delta \neq 0, \text{ les } \lambda_n \in K.$$

VII.2.2 - Raisonnant par l'absurde, on suppose que le premier côté non vertical de N_U est le segment $[M_0, M_d]$, avec :

(1) $0 < d < \delta$. On pose :

(2) $P_0 = \lambda_0 + \ldots + \lambda_d . T^d$.

Soit $\rho = \exp(-\text{pente de la droite } M_0 M_d)$; ainsi la pente de cette droite est $-\text{Log}(\rho)$. Le Lemme VII.1.3 montre que P_0 est un ρ-diviseur. De plus, si on pose $\sigma = \rho^*(U - P_0)$, on a :

(3) $0 < \sigma < 1$.

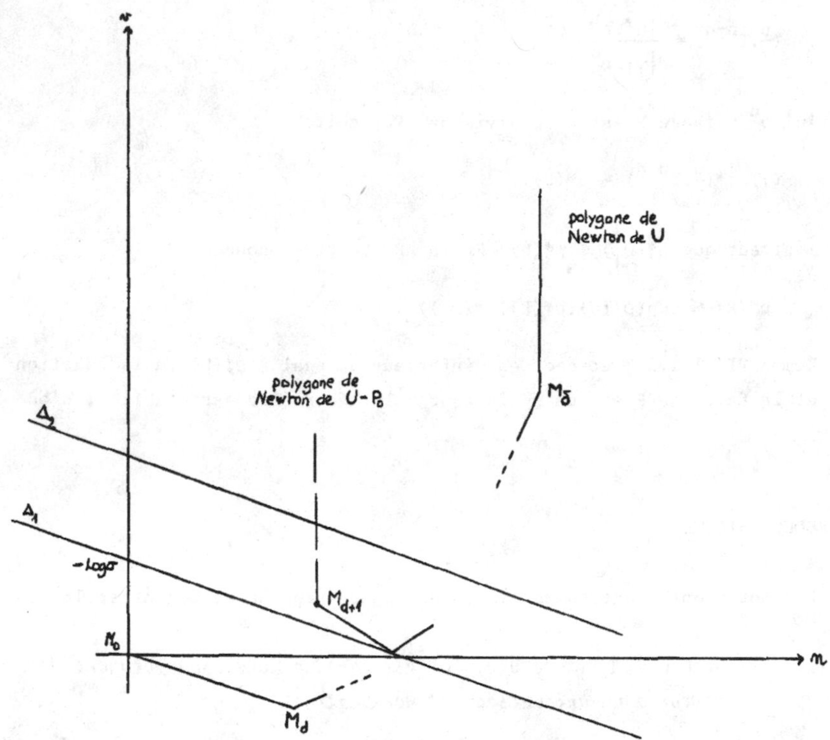

En effet, $U - P_0$ est non nul et, d'autre part, les points M_n, pour n n'appartenant pas à $\{0,\dots,d\}$, étant strictement au-dessus de la droite M_0M_d par définition du polygone de Nexton, le Lemme VII.1.2.1 montre que $-\mathrm{Log}(\sigma)$ est strictement positif.

Pour tout $j \in \mathbb{N}$, soit Δ_j la droite d'équation :

$$y = x.\mathrm{Log}(\rho) - j.\mathrm{Log}(\sigma).$$

Notons M_{n_j} (resp. M_{m_j}) le premier (resp. le dernier) des points représentatifs M_n de U à être strictement au-dessous de Δ_j et posons :

$$(4)\quad I_j = [n_j, m_j]\ ;\ \tilde{U}_j = \sum_{n \in I_j} \lambda_n . T^n .$$

Comme $-j.\mathrm{Log}(\sigma)$ tend vers $+\infty$ avec j, il existe un plus petit entier ℓ tel que :

$$I_\ell = [0,\delta],$$

et on voit aussitôt que :

$$(5)\quad I_0 \subset I_1 \subset \dots \subset I_\ell\ ;\ I_0 = \phi\ ,\ I_1 = [0,d]\ ;$$

$$\tilde{U}_0 = 0,\ \tilde{U}_1 = P_0 \text{ et, pour tout } j \in \{1,\dots,\ell\} :$$

$$\rho^*(\tilde{U}_j) = 1, \rho^*(U - \tilde{U}_j) \leqslant \sigma^j \text{ avec égalité si } j = 1.$$

(Pour la dernière inégalité, on observe que $U - \tilde{U}_j = \sum_{n \notin [n_j, m_j]} \lambda_n . T^n$ et, si $n \notin [n_j, m_j]$, M_n est au-dessus de Δ_j, c'est à dire que $v(\lambda_n)$ est supérieur ou égal à $n.\mathrm{Log}(\rho) - j.\mathrm{Log}(\sigma)$, soit :

$$|\lambda_n| . \rho^n \leqslant \sigma^j).$$

VII.2.3 - Montrons maintenant que, pour tout $j \in \mathbb{N}$, il existe des éléments V_j, W_j et P_j de K[T] tels que :

$$(R_i)\quad \left|\begin{array}{l} U = P_j . V_j + W_j, \\ P_j \text{ a même polygone de Newton que } P_0 \\ \rho^*(W_j) \leqslant \sigma^j, \\ \text{support}(P_j . V_j) \subset I_j \text{ et, si } j \geqslant 1, \\ \rho^*(P_j - P_{j-1}) \leqslant \sigma^{j-1},\ \rho^*(V_j - V_{j-1}) \leqslant \sigma^{j-1}. \end{array}\right.$$

Pour $j = 0$, P_0 est déjà choisi (cf.(2)), posons $V_0 = 0$ et $W_0 = U$.
Comme $\rho^*(U)=1$, les relations (R_0) sont vérifiées.

Pour $j = 1$, prenons $P_1 = P_0$, $V_1 = 1$, $W_1 = U - P_0 = U - \tilde{U}_1$. Comme $\rho^*(W_1)$ est inférieur ou égal à σ d'après (5), (R_1) est vérifié.

Supposons prouvés $(R_0),\dots,(R_j)$, avec $j \geqslant 1$. Soit K_j le tronqué de W_j par I_{j+1}, c'est à dire la somme de ceux des termes de W_j dont l'indice appartient à I_{j+1} ; soit $\overline{W}_j = W_j - K_j$. Ainsi :

$$(6)\quad W_j = \overline{W}_j + K_j \ ;\ \text{support}(K_j) \subset I_{j+1},\ \text{support}(\overline{W}_j) \subset I_\ell - I_{j+1}.$$

D'après (R_j) et (6), on a :

$$U = (P_j.V_j + K_j) + \overline{W}_j \ ;\ \text{support}(P_j.V_j + K_j) \subset I_{j+1}.$$

Par suite :

$$(7)\quad \tilde{U}_{j+1} = P_j.V_j + K_j \ ;\ \rho^*(\overline{W}_j) = \rho^*(U - \tilde{U}_{j+1}) \leqslant \sigma^{j+1}.$$

De plus, comme K_j est un tronqué de W_j, (R_j) montre que :

$$(8)\quad \rho^*(K_j) \leqslant \sigma^j.$$

Cela fait, on a, d'après le Lemme VII.1.4, des relations du type :

$$(9)\quad \left|\begin{array}{l} K_j = P_j.Y_j + R_j \ ;\ Y_j \text{ et } R_j \in K[T], \\ R_j = 0 \text{ ou } \deg(R_j) < \deg(P_j), \text{ donc } < d(\text{cf.}(R_j)), \\ \rho^*(Y_j) \leqslant \rho^*(K_j) \text{ et } \rho^*(R_j) \leqslant \rho^*(K_j), \text{ donc } \leqslant \sigma^j. \end{array}\right.$$

(R_j), (6) et (9) montrent que :

$$U = (P_j + R_j)(V_j + Y_j) + (\overline{W}_j - R_j(V_j + Y_j - 1)).$$

Posons :

$$(10)\quad \left|\begin{array}{l} P_{j+1} = P_j + R_j, \\ V_{j+1} = V_j + Y_j \text{ et} \\ W_{j+1} = \overline{W}_j - R_j(V_j + Y_j - 1). \end{array}\right.$$

VII.2.4 - Vérification de (R_{j+1}).

On a : $P_{j+1} = P_j + R_j$, avec $\deg(R_j) < \deg(P_j)$ et $\rho^*(R_j) \leqslant \sigma^j < 1$ car j est supérieur ou égal à 1 (cf.(3), (9), (10)). Par conséquent, P_j et P_{j+1} ont même degré, même terme directeur ; enfin, tous les points représentatifs de R_j étant au-dessus de Δ_j, P_{j+1} et P_j ont même polygone de Newton. Donc, d'après (R_j), P_{j+1} a même polygone de Newton que P_0.

Quant à $\rho^*(W_{j+1})$, vu que $\rho^*(\overline{W}_j) \leqslant \sigma^{j+1}$ et que $\rho^*(R_j) \leqslant \sigma^j$ (cf. (7) et (9)), on a

$$\rho^*(W_{j+1}) \leqslant \sup(\sigma^{j+1}, \sigma^j.\rho^*(V_j + Y_j - 1)),$$

et il suffit d'établir que $\rho^*(V_j + Y_j - 1)$ est inférieur ou égal à σ, ou encore que $\rho^*(V_j - 1)$ est inférieur ou égal à σ, puisque, d'après (9) :

$$\rho^*(Y_j) \leqslant \sigma^j \leqslant \sigma.$$

Or, on peut écrire :

$$V_j - 1 = V_j - V_1 = \sum_{k=2}^{j} (V_k - V_{k-1}) \text{; donc}$$
$$\rho^*(V_j - 1) \leqslant \sup(\rho^*(V_2 - V_1), \ldots, \rho^*(V_j - V_{j-1})),$$

et, d'après $(R_2), \ldots, (R_j)$ ceci est inférieur ou égal à $\sup(\sigma, \ldots, \sigma^j)$ donc à σ (cf.(3)).

En ce qui concerne le support de $P_{j+1}.V_{j+1}$, comme $P_{j+1}.V_{j+1} = P_j.V_j + R_j.V_j + P_{j+1}.Y_j$ et que, d'après (R_j), le support de $P_j.V_j$ est inclus dans I_j, et a fortiori dans I_{j+1}, il reste à étudier $R_j.V_j$ et $P_{j+1}.Y_j$. Pour le premier, le support de R_j est inclus dans I_1 d'après (9), donc tout intervalle contenant le support d'un polunôme du type $P_j.V$ contient aussi celui de $R_j.V$, en particulier, le support de $R_j.V_j$ est inclus dans I_{j+1}. Enfin, le support de $P_{j+1}.Y_j$ est inclus dans I_{j+1} si, et seulement si, il en est ainsi pour celui de $P_j.Y_j$ (même raison), c'est à dire pour $K_j - R_j$; mais ce dernier point découle de (6) et (9).

Pour terminer, comme $P_{j+1} - P_j = R_j$, on a bien :

$$\rho^*(P_{j+1} - P_j) \leqslant \sigma^{(j+1)-1} \text{ et de même pour } V_{j+1} - V_j = Y_j.$$

Les (R_j) sont ainsi établis par récurrence.

VII.2. 5 - Convergence.

Pour tout $j \in \mathbb{N}$, on a, d'après (R_j) : support $(P_j.V_j) \subset I_\ell$, support $(W_j) \subset I_\ell$ donc aussi support $(V_j) \subset I_\ell$ puisque P_j est de valuation nulle. Ainsi, les P_j, V_j et W_j ont tous un degré inférieur ou égal à δ. Or, si $P = P_0 + \ldots + p_\delta.T^\delta$, on a :

$$\rho^*(P) = \sup(|p_0|, \ldots, |p_\delta|.\rho^\delta) \geqslant M.\sup(|p_0|, \ldots, |p_\delta|),$$
$$\text{avec } M = \inf(1, \rho^\delta).$$

Par suite, la convergence vers 0 pour ρ^* entraîne la convergence uniforme vers 0 pour les coefficients.

De ce fait, les relations : $\rho^*(P_j - P_{j-1}) \leq \sigma^{j-1}$, $\rho^*(V_j - V_{j-1}) \leq \sigma^{j-1}$ montrent que les suites $(P_j)_{j \in \mathbb{N}}$ et $(V_j)_{j \in \mathbb{N}}$ - ou plutôt leurs images canoniques - sont des suites de Cauchy dans l'espace $K^{\delta+1}$ muni de la topologie-produit. K étant complet, ce dernier espace l'est aussi : soient donc P et V les limites respectives de ces suites. De même, le fait que $\rho^*(W_j)$ soit inférieur ou égal à σ^j montre que la suite W_j converge vers 0. Comme la topologie de $K^{\delta+1}$ est séparée et que la multiplication de K est continue, on a, en définitive :

$$U = P.V.$$

Par ailleurs, les conditions sur le polygone de Newton de P_j "passent" à la limite: P a même polygone de Newton que P_0 et est donc un diviseur strict de U, contradiction. Le Théorème VII.2.1 est démontré.

BIBLIOGRAPHIE

[1] M. BAYART.- Factorialité et séries formelles irréductibles I, Lecture Notes in Mathematics n°795. Springer-Verlag 1979.

[2] BOREVITCH et CHAFAREVITCH.- Théorie des Nombres. Gauthier-Villars, Paris,1967.

[3] BOURBAKI.- Algèbre Commutative, Chapitre 6 : Valuations. Hermann, Paris,1964.

[4] BOURBAKI.- Algèbre Commutative, Chapitre 7 : Diviseurs. Hermann, Paris, 1965.

[5] KRASNER.- Essai d'une théorie des fonctions analytiques dans les corps valués complets. C.R. Acad. Sc. Paris 222(1946),p. 37-40, 165-167, 363-365 et 581-583.

[6] NAGATA.- Local Rings. Wiley Interscience, New-York, 1962.

[7] SAMUEL.- On unique factorization domains. *Ill.* J. Math. 5 (1961)p. 1-17.

Marc BAYART
Classes Préparatoires, Lycée Thiers
13232 MARSEILLE Cedex 1

Homologie d'anneaux locaux de

dimension d'immersion 3

par

C. SCHOELLER

Soit $(R,\underline{m},k)$ un anneau local noethérien, pour tout R-module M noethérien, $\mathrm{Tor}^R(M,k)$ est un espace vectoriel sur k, gradué en degrés positifs, de dimension finie en chaque degré. On appelle série de Poincaré du R-module M, la série

$$\beta^R(M) = \sum_{n\geq 0} b_n z^n \quad , \qquad b_n = \dim_k(\mathrm{Tor}_n^R(M,k)).$$

On sait établir un calcul explicite de $\beta^R(k)$ sous la forme d'une fraction rationnelle dans un certain nombre de cas, par exemple lorsque R est un anneau régulier, un anneau d'intersection complète, ou un anneau de Golod, ou lorsque la dimension d'immersion (c.à.d. $\dim_k(\underline{m}/\underline{m}^2)$) de R est 2 .

Divers contre-exemples ont été récemment trouvés, prouvant que $\beta^R(k)$ n'était pas toujours rationnelle [0],[2].

Nous nous proposons ici d'explorer certains cas de rationalité, en particulier lorsque $\dim_k(\underline{m}/\underline{m}^2) = 3$, par l'utilisation de la formule de changement d'anneau

$$\beta^R(k) = \beta^R(R').\beta^{R'}(k).$$

Cette formule a été établie dans [3] lorsque R' <u>est une intersection complète</u> quotient de R par un idéal $\underline{p}$ engendré par une partie d'un système minimal de générateurs de $\underline{m}$. Elle a permis de prouver la rationalité de $\beta^R(M)$ pour tout R'-module M noethérien lorsque $\underline{p}$ était monogène. Nous nous proposons d'étudier le cas où $\underline{p}$ sst engendré par deux éléments d'un système minimal de générateurs de $\underline{m}$.

1 - Généralités et notations.

Dans toute la suite $(R,\underline{m},k)$ sera un anneau local noethérien, $(x_1,\dots,x_n)$ un système minimal de générateurs de son idéal maximal $\underline{m}$ et $\underline{p}$ l'idéal engendré par (x_{n-1},x_n). On désignera par $(R',\underline{m}',k)$ l'anneau local quotient de R par $\underline{p}$. Les images canoniques $x'_1,\dots,x'_{n-2}$, de $x_1,\dots,x_{n-2}$, forment un système minimal de générateurs de $\underline{m}'$.

On supposera que R' est une intersection complète.

Rappelons que, comme R-module, k admet une résolution minimale $\mathfrak{X}$ qui est une R-algèbre au sens de Tate [5] (c.à.d. une algèbre différentielle munie de puissances divisées). La construction de $\mathfrak{X}$ se fait "pas à pas" :

- au premier pas, on pose

$$\mathfrak{X}(1) = R < T_1,\dots,T_n > \quad d^\circ T_i = 1,\ dT_i = x_i\ , \quad i = 1,2,\dots,n\ ;$$

$R < T_1,\dots,T_n >$ est l'algèbre extérieure construite sur $T_1,\dots,T_n$; $\mathfrak{X}(1)$ est le complexe de Koszul de R ;

- au deuxième pas, on pose

$$\mathfrak{X}(2) = \mathfrak{X}(1) < S_1,\dots,S_r > \quad d^\circ S_i = 2\ ,\ dS_i = s_i\ , \quad i = 1,2,\dots,r\ ;$$

les s_i forment un relèvement dans $Z_1(\mathfrak{X}(1))$ d'une base du k-espace vectoriel $H_1(\mathfrak{X}(1))$, et $\mathfrak{X}(1) < S_1,\dots,S_r >$ est l'algèbre des polynômes divisés en $S_1,\dots,S_r$, à coefficients dans $\mathfrak{X}(1)$.

- au m-ième pas, on a construit l'algèbre $\mathfrak{X}(m-1)$ telle que $H_o(\mathfrak{X}(m-1)) \simeq k$ et $H_i(\mathfrak{X}(m-1)) = 0$ pour $i < m-1$; on passe à $\mathfrak{X}(m)$ en rajoutant des variables $U_1,\dots,U_s$, en degré m, de sorte que les dU_j, $j = 1,2,\dots,s$, forment un relèvement dans $Z_{m-1}(\mathfrak{X}(m-1))$ d'une base de $H_{m-1}(\mathfrak{X}(m-1))$ et l'on pose

$$\mathfrak{X}(m) = \mathfrak{X}(m-1) < U_1,\dots,U_s >,$$

cette notation représentant

- l'algèbre extérieure à coefficients dans $\mathfrak{X}(m-1)$ construite sur les U_i, si m est impair,
- l'algèbre de polynômes divisés en $U_1,\dots,U_s$, à coefficients dans $\mathfrak{X}(m-1)$, si m est pair.

Nous n'aurons besoin ici que de la construction explicite des deux premiers pas.

D'après [3], et selon nos hypothèses, il est possible de construire par la méthode de Tate, une R-algèbre $\mathfrak{Y}$ qui est une résolution minimale de R' et s'identifie à une sous-algèbre différentielle de $\mathfrak{X}$ (au premier pas $\mathfrak{Y}(1)$ s'identifie clairement à $R < T_{n-1}, T_n >$. On posera $F = \mathfrak{Y}(1)$.

Par ailleurs, k, considéré comme R'-module, admet une résolution minimale $\mathfrak{X}'$ qui est une R'-algèbre ; en fait, R' étant une intersection complète, on a $\mathfrak{X}' = \mathfrak{X}'(2)$.

Enfin, toujours selon [3], il existe un morphisme surjectif $\varphi : \mathfrak{X} \to \mathfrak{X}'$ d'algèbres différentielles sur R, induisant l'identité sur k. Alors on peut construire une sous-algèbre $\tilde{\mathfrak{X}}$ de $\mathfrak{X}$ ($\tilde{\mathfrak{X}}$ n'est pas une sous algèbre différentielle) et des isomorphismes

$$\tilde{\mathfrak{X}} \underset{R}{\otimes} R' \xrightarrow{\sim} \mathfrak{X}'$$

$$\mathfrak{X} \xrightarrow{\sim} \mathfrak{Y} \underset{R}{\otimes} \tilde{\mathfrak{X}}.$$

Dans ces conditions, pour tout R'-module M, on a

$$H(\mathfrak{X} \otimes M) \xrightarrow{\sim} (\mathfrak{Y} \otimes k) \underset{k}{\otimes} H(\mathfrak{X}' \otimes M) \quad ,$$

c'est à dire

$$\mathrm{Tor}^R(k,M) \xrightarrow{\sim} \mathrm{Tor}^R(R',k) \underset{k}{\otimes} \mathrm{Tor}^{R'}(k,M).$$

Lorsque M est noethérien il en résulte la relation :

$$\beta^R(M) = \beta^R(R').\beta^{R'}(M).$$

2 - Etude du cas où $\dim_k \underline{m}/\underline{m}^2 = 3$.

2.1. Dans ce cas R' est un anneau de valuation discrète. On peut supposer R' non régulier (s'il était régulier, R/x_3R serait soit régulier, soit intersection complète, et l'on pourrait conclure, d'après [3] que $B^R(M)$ est rationnelle pour tout R'-module M). Donc il existe $m \in \mathbb{N}$ tel que $x_1'^m \neq 0$ et $x_1'^{m+1} = 0$.

Nous préciserons d'abord la construction de $\mathfrak{Y}(2)$, $\mathfrak{X}(2)$ et $\mathfrak{X}'(2)$. On pose

$$E' = \mathfrak{X}'(1) = R' < T_1' > \quad , \quad d^oT_1' = 1 \quad , \quad dT_1' = x_1' \quad ,$$

$$E = \mathfrak{X}(1) = R < T_1, T_2, T_3 > \, , \quad d^oT_i = 1 \quad , \quad dT_i = x_i \quad , \; i = 1,2,3 \; ;$$

alors $F = \mathfrak{Y}(1) = R < T_2, T_3 >$, est la sous algèbre de E engendrée par T_2 et T_3. Il est clair que l'on peut prendre,

$$\mathfrak{X}' = \mathfrak{X}'(2) = R' < T_1', S' > \ , \ \partial^\circ S' = 2 \ , \ dS' = s' = x_1'^m \, T_1'$$

Par ailleurs, puisque $x_1^{m+1} \in \underline{p}$, il existe dans R une relation du type

$$x_1^{m+1} + \lambda x_2 + \mu x_3 = 0 \quad \text{avec} \quad \lambda \in \underline{m}, \ \mu \in \underline{m} \ ,$$

de sorte que $s = x_1^m T_1 + \lambda T_2 + \mu T_3$ est un cycle (et non un bord) de E qui se projette en s'. D'après [3] on peut de plus, modulo un bord de E, supposer λ et μ dans $\underline{p}$.

Soit $(s_1, \dots, s_r)$ un relèvement dans $Z_1(F)$ d'une base de $H_1(F) \otimes k$, on a, pour tout $i = 1, 2, \dots, r$, $s_i = a_i T_2 + b_i T_3$ avec $a_i x_2 + b_i x_3 = 0$. On sait (cf.[3]) que l'on peut prendre

$$\mathfrak{Y}(2) = F < S_1, \dots, S_r > \ , \ \partial^\circ S_i = 2 \ , \ dS_i = s_i \ ,$$

$$\text{et} \quad \mathfrak{X}(2) = E < S_1, \dots, S_r, S > \ , \ dS_i = s_i \ , \ dS = s \ ,$$

de sorte que l'algèbre $\widetilde{\mathfrak{X}}(2)$ n'est autre que $R < T_1, S >$, et $\varphi : \mathfrak{X}(2) \to \mathfrak{X}'(2)$ le morphisme d'algèbres différentielles envoyant T_1 sur T_1', S sur S' et les autres générateurs sur 0.

2.2. Remarques sur l'homologie des R'-modules.

Soit N un R'-module, on calcule $\mathrm{Tor}^{R'}(N,k)$ au moyen de $\mathfrak{X}' \otimes N$. Pour tout $p \geq 0$ on a, en appelant $\mathrm{Ker}_N x_1^m$ le noyau de la multiplication par x_1^m dans N :

$$Z_{2p+2}(\mathfrak{X}' \otimes N) = S'^{(p+1)} \otimes \mathrm{Ker}_N x_1^m \quad , \quad B_{2p+2}(\mathfrak{X}' \otimes N) = S'^{(p+1)} \otimes x_1 N \ ,$$

$$Z_{2p+1}(\mathfrak{X}' \otimes N) = S'^{(p)} T_1' \otimes \mathrm{Soc}\, N \quad , \quad B_{2p+1}(\mathfrak{X}' \otimes N) = S'^{(p)} T_1' \otimes x_1^m N \ .$$

Il en résulte que les isomorphismes naturels $S'^{(p+1)} \otimes y \mapsto S' \otimes y$ de $\mathfrak{X}'_{2p+2} \otimes N$ sur $\mathfrak{X}'_2 \otimes N$, et $S'^{(p)} T_1' \otimes y \mapsto T_1' \otimes y$ de $\mathfrak{X}'_{2p+1} \otimes N$ sur $\mathfrak{X}'_1 \otimes N$ induisent des isomorphismes :

$$\theta_{2p+2} : \mathrm{Tor}^{R'}_{2p+2}(N,k) \xrightarrow{\sim} \mathrm{Tor}^{R'}_2(N,k)$$

$$\text{et} \quad \theta_{2p+1} : \mathrm{Tor}^{R'}_{2p+1}(N,k) \xrightarrow{\sim} \mathrm{Tor}^{R'}_1(N,k).$$

Enfin, lorsqu'on a fait choix d'une décomposition de N en somme directe

$$N \simeq N^L \oplus N^T$$

où N^L est un R'-module libre et N^T un R'-module annulé par x_1^m, on a

$$\mathrm{Tor}^{R'}_{2p+2}(N,k) \simeq \mathrm{Tor}^{R'}_{2}(N,k) \simeq \mathrm{Tor}^{R'}_{2}(N^T,k) \simeq N^T \otimes k$$

$$\text{et} \quad \mathrm{Tor}^{R'}_{2p+1}(N,k) \simeq \mathrm{Tor}^{R'}_{1}(N,k) \simeq \mathrm{Tor}^{R'}_{1}(N^T,k) \simeq \mathrm{soc}\, N^T$$

où soc N^T désigne le socle de N^T.

2.3. Pour prouver la rationalité de $\mathcal{P}^R(k)$, sous certaines hypothèses, nous utiliserons les deux R'-modules

$$A = \mathrm{ann}\, x_2 \cap \mathrm{ann}\, x_3 \simeq H_2(F)$$

$$\text{et} \quad M = H_1(F) \quad ,$$

et des relations entre leurs homologies.

Nous désignerons par Z et B respectivement, les modules $Z_1(F)$ et $B_1(F)$ de sorte que $M \simeq Z/B$. Comme $F = R < T_2, T_3 >$, Z et B sont les sous-modules de R^2 définis par

$$Z = \{(a,b) \in R^2 \ , \ ax_2 + bx_3 = 0\} \ ,$$

$$B = \{\lambda(-x_3, x_2) \ , \ \lambda \in R\}.$$

Enfin, comme R' est de valuation discrète, on utilisera une décomposition

$$M \simeq \bigoplus_{i=1}^{r} M_i$$

de M en somme directe de modules monogènes $M_i \simeq R'\bar{\sigma}_i \simeq R'/(x_1'^{\alpha_i+1})$, avec $\alpha_i \leq m$.

De l'exactitude des suites

2.3.1 $\quad 0 \to \underline{p} \to R \to R' \to 0$,

2.3.2 $\quad 0 \to Z \to R^2 \to \underline{p} \to 0$,

2.3.3 $\quad 0 \to A \to R \to B \to 0$,

2.3.4 $\quad 0 \to B \to Z \to M \to 0$,

il résulte, d'une part les isomorphismes

$$\mathrm{Tor}^R_{n+2}(R',k) \xrightarrow{\sim} \mathrm{Tor}^R_n(Z,k) \qquad , \qquad \forall n \geq 0 \ ,$$

$$\tilde{\delta}_{n+1} : \quad \mathrm{Tor}^R_{n+1}(B,k) \xrightarrow{\sim} \mathrm{Tor}^R_n(A,k) \qquad , \qquad \forall n \geq 0 \ ,$$

d'autre part l'exactitude de la suite longue

$$2.3.5. \quad \ldots \to \mathrm{Tor}^R_{n+1}(M,k) \xrightarrow{\partial_{n+1}} \mathrm{Tor}^R_n(B,k) \to \mathrm{Tor}^R_n(Z,k) \to \mathrm{Tor}^R_n(M,k) \to \ldots$$

La partie de degré 0 de 2.3.5 donne en fait la suite exacte

$$0 \to \mathrm{Tor}^R_o(B,k) \to \mathrm{Tor}^R_o(Z,k) \to \mathrm{Tor}^R_o(M,k) \to 0 \ ;$$

en effet, d'une part $(-x_3,x_2)$ engendre B, et d'autre part on a $Z \subset \underline{m}\, R^2$ donc $(-x_3,x_2) \notin \underline{m}\, Z \subset \underline{m}^2 R^2$, d'où l'injectivité de $B \otimes k \to Z \otimes k$.

Nous nous proposons d'étudier le morphisme

$$\delta : \mathrm{Tor}^R(M,k) \to \mathrm{Tor}^R(A,k)$$

de degré -2, défini par $\delta_{n+2} = \tilde{\delta}_{n+1} \circ \partial_{n+2}$ pour tout $n \geq 0$. En particulier, il est clair que, si δ est nul, la suite 2.5. se scinde et que, des relations

$$\mathrm{Tor}^R_n(Z,k) \xrightarrow{\sim} \mathrm{Tor}^R_{n-1}(A,k) \oplus \mathrm{Tor}^R_n(M,k) \qquad , \qquad \forall n > 0 \ ,$$

$$\mathrm{Tor}^R_{n+2}(R',k) \xrightarrow{\sim} \mathrm{Tor}^R_n(Z,k) \qquad , \qquad \forall n \geq 0 \ ,$$

$$\mathrm{Tor}^R(A,k) \xrightarrow{\sim} \mathrm{Tor}^R(R',k) \otimes \mathrm{Tor}^{R'}(A,k) \ ,$$

$$\mathrm{Tor}^R(M,k) \xrightarrow{\sim} \mathrm{Tor}^R(R',k) \otimes \mathrm{Tor}^{R'}(M,k) \ ,$$

il résulte une relation rationnelle entre $P^R(R')$, $P^{R'}(A)$ et $P^{R'}(M)$ et, comme ces deux dernières séries sont rationnelles il en est de même de la première donc de $P^R(k)$.

Rappelons que $\mathrm{Tor}^R(M,k) \xrightarrow{\sim} H(\mathfrak{X} \otimes M) \xrightarrow{\sim} (\mathfrak{Y} \otimes k) \otimes H(\mathfrak{X}' \otimes M)$.

2.4. <u>LEMME</u>.

a. <u>La restriction de</u> δ <u>à</u> $\mathfrak{Y}_o \otimes H(\mathfrak{X}' \otimes M)$ <u>se factorise à travers</u> $\mathfrak{Y}_o \otimes H(\mathfrak{X}' \otimes A)$.

b. <u>Si on appelle</u> $\tilde{\delta} : \mathrm{Tor}^{R'}(M,k) \to \mathrm{Tor}^{R'}(A,k)$ <u>le morphisme de degré</u> (-2) <u>ainsi défini, les séries associées à</u> Ker $\tilde{\delta}$ et Im $\tilde{\delta}$ <u>sont</u>

des fractions rationnelles. Plus précisément, on a, pour tout $p \geq 2$:

$$\mathrm{Ker}\,\tilde{\delta}_{2p} \simeq \mathrm{Ker}\,\tilde{\delta}_2 \ \ \underline{\mathrm{et}} \ \ \mathrm{Im}\,\tilde{\delta}_{2p} \xrightarrow{\sim} \mathrm{Im}\,\tilde{\delta}_4 \ ,$$

$$\mathrm{Ker}\,\tilde{\delta}_{2p+1} \simeq \mathrm{Ker}\,\tilde{\delta}_3 \ \ \underline{\mathrm{et}} \ \ \mathrm{Im}\,\tilde{\delta}_{2p+1} \xrightarrow{\sim} \mathrm{Im}\,\tilde{\delta}_3 \ .$$

On calcule δ au moyen de la suite exacte de complexes

2.4.1 $$0 \to \mathfrak{X} \otimes B \to \mathfrak{X} \otimes Z \to \mathfrak{X} \otimes M \to 0.$$

A cause de la décomposition $M \xrightarrow{\sim} \otimes M_i$ en somme directe de modules monogènes, on peut se contenter de prouver le lemme lorsque M est monogène et non isomorphe à R' (le cas $M \simeq R'$ étant trivial).
Supposons donc M monogène, engendré par $\bar{\sigma}$ image canonique de $\sigma = (a,b) \in Z$, avec $a\,x_2 + b\,x_3 = 0$, et tel que $x_1^{\alpha}\,\bar{\sigma} \neq 0$, $x_1^{\alpha+1}\,\bar{\sigma} = 0$ pour un $\alpha < m$. Un calcul simple utilisant la description de $\mathfrak{X}'$ ci-dessus, montre que

$$Z_{2p}(\mathfrak{X}' \otimes M) = R'S'^{(p)} \otimes M \xrightarrow{\sim} M \ , \quad B_{2p}(\mathfrak{X}' \otimes M) = R'x_1 S'^{(p)} \otimes M \xrightarrow{\sim} x_1 M \ ,$$

$$Z_{2p+1}(\mathfrak{X}' \otimes M) \xrightarrow{\sim} R'x_1^{\alpha}\, S'^{(p)} T_1' \otimes M \xrightarrow{\sim} \mathrm{soc}(M) \ , \qquad \text{et} \quad B_{2p+1} = 0.$$

Le calcul de la restriction de δ à $\mathfrak{U}_0 \otimes H(\mathfrak{X}' \otimes M)$ se réduit donc

- en degré $2p$ au calcul de $\delta(\dot{\zeta})$, où $\dot{\zeta}$ est la classe d'homologie de $\zeta = S^{(p)} \otimes \bar{\sigma} \in \mathfrak{X} \otimes M)$
- en degré $2p+1$ au calcul de $\delta(\dot{\zeta}')$, où $\dot{\zeta}'$ est la classe d'homologie de $\zeta' = S^{(p)} T_1 \otimes x_1^{\alpha} \bar{\sigma}$.

Calcul de $\delta_{2p}(\dot{\zeta}) = \tilde{\partial}_{2p-1} \circ \partial_{2p}(\dot{\zeta})$.

Par la méthode classique, dans la suite 2.4.1 on relève $\dot{\zeta}$ en un cycle z de $\mathfrak{X}_{2p} \otimes Z$; on peut prendre $z = S^{(p)} \otimes \sigma$, et l'on a

$$dz = (x_1^m T_1 + \lambda\, T_2 + \mu\, T_3) \otimes \sigma \ ;$$

Comme $x_1^{\alpha+1}\bar{\sigma} = 0$, on sait que $x_1^{\alpha+1}\sigma \in B$, autrement dit, il existe $\rho' \in R$ tel que

$$x_1^{\alpha+1}(a,b) = \rho'(-x_3, x_2) \ ;$$

et l'on a, puisque $\alpha < m$, $x_1^m(a,b) = \rho(-x_3,x_2)$ avec $\rho = \rho' x_1^{m-\alpha-1}$. De plus, comme λ est dans $\underline{p}$ on l'écrit $\lambda = \lambda' x_2 + \lambda'' x_3$, et, compte tenu de $ax_2 + bx_3 = 0$, il vient :

$$\lambda(a,b) = \lambda' x_2(a,b) + \lambda'' x_3(a,b) = (\lambda' b - \lambda'' a)(-x_3, x_2)$$

On opère de même pour μ et l'on obtient

$$dz = [\rho T_1 + (\lambda' b - \lambda'' a)T_2 + (\mu' b - \mu'' a)T_3]S^{(p-1)} \otimes (-x_3, x_2)$$

Alors dz est un cycle de $\mathfrak{X} \otimes B$ et $\partial(\dot{\zeta})$ est la classe $\dot{Z}$ de dz dans $H_{2p-1}(\mathfrak{X} \otimes B)$. Il reste à calculer $\tilde{\delta}_{2p-1}(\dot{Z})$ au moyen de la suite exacte de complexes

$$0 \to \mathfrak{X} \otimes A \to \mathfrak{X} \otimes R \to \mathfrak{X} \otimes B \to 0.$$

On relève dz dans $\mathfrak{X} \otimes R$ en $Z = wS^{(p-1)}$, où

$w = \rho T_1 + (\lambda'' b - \lambda'' a)T_2 + (\mu' b - \mu'' a)T_3$; alors $dZ = dw.S^{(p-1)} + w\, s\, S^{(p-2)}$.

On vérifie aisément que $ws = 0$, et il reste

$$dZ = dw.S^{(p-1)} = [\rho x_1 + (\lambda' b - \lambda'' a)x_2 + (\mu' b - \mu'' a)x_3]S^{(p-1)} ,$$

soit encore

$$dZ = (\rho x_1 + \lambda b - \mu a)S^{(p-1)} = vS^{(p-1)}$$

Alors $v = \rho x_1 + \lambda b - \mu a$ est dans A et $dZ \in \mathfrak{Y}_o \otimes (\tilde{\mathfrak{X}} \otimes A)$ est un cycle de degré $2p-2$ de $\mathfrak{X} \otimes A$,

2.4.2. $\qquad \delta_{2p}(\dot{\zeta}) = \mathrm{cls}(S^{(p-1)} \otimes v)$ dans $H(\mathfrak{X} \otimes A)$.

On déduit de ce calcul les résultats proposés dans le lemme pour les degrés pairs :

a) pour $p \geq 1$, on a $\delta_{2p}(\mathfrak{Y}_o \otimes \mathrm{Tor}^{R'}_{2p}(M,k)) \subset \mathfrak{Y}_o \otimes \mathrm{Tor}^{R'}_{2p-2}(A,k)$, donc δ_{2p} induit $\tilde{\delta}_{2p} : \mathrm{Tor}^{R'}_{2p}(M,k) \to \mathrm{Tor}^{R'}_{2p-2}(A,k)$.

b) l'élément v de A étant indépendant de p, le diagramme ci-dessous est commutatif pour $p \geq 3$:

$$\begin{array}{ccc} \mathrm{Tor}^{R'}_{2p}(M,k) & \xrightarrow{\tilde{\delta}_{2p}} & \mathrm{Tor}^{R'}_{2p-2}(A,k) \\ \wr\downarrow \theta_{2p} & & \wr\downarrow \theta_{2p-2} \\ \mathrm{Tor}^{R'}_{4}(M,k) & \xrightarrow{\tilde{\delta}_{4}} & \mathrm{Tor}^{R'}_{2}(A,k). \end{array}$$

<u>Calcul de</u> $\delta_{2p+1}(\dot{\zeta}') = \tilde{\delta}_{2p} \circ \partial_{2p+1}(\dot{\zeta}')$.

De façon analogue et avec les mêmes notations, on relève $\dot{\zeta}'$ dans $\mathfrak{X} \otimes Z$ en $z' = S^{(p)}T_1 \otimes x_1^{\alpha}\sigma$ et l'on obtient l'expression de dz' comme cycle de $\mathfrak{X} \otimes B$ sous la forme :

$$dz' = \{\rho' S^{(p)} - x_1^{\alpha}[(\lambda' b - \lambda'' a)T_1 T_2 + (\mu' b - \mu'' a)T_1 T_3]S^{(p-1)}\} \otimes (-x_3, x_2).$$

Alors $\partial_{2p+1}(\dot{\zeta}')$ est la classe $\dot{Z}'$ de dz' dans $H(\mathfrak{X} \otimes B)$; on calcule $\tilde{\partial}(\dot{Z}')$ en relevant $\dot{Z}'$ dans $\mathfrak{X} \otimes R$ en

$$Z' = \rho' S^{(p)} - x_1^{\alpha}[(\lambda' b - \lambda'' a)T_1 T_2 + (\mu' b - \mu'' a)T_1 T_3]S^{(p-1)}$$

Le calcul de dZ', compte tenu des relations entre les divers coefficients, donne

$$dZ' = vx_1^{\alpha} S^{(p-1)} T_1 \quad , \quad \text{avec} \quad v = \rho x_1 + \lambda b - \mu a \quad ,$$

d'où

2.4.3. $$\delta_{2p+1}(\dot{\zeta}') = \mathrm{cls}\,(S^{(p-1)} T_1 \otimes x_1^{\alpha} v) \quad \text{dans} \quad H(\mathfrak{X} \otimes A).$$

L'élément $x_1^{\alpha} v$ de A étant indépendant de p, on a des conclusions analogues aux précédentes

Ceci achève la démonstration du lemme, les séries associées à $\mathrm{Ker}\,\tilde{\delta}$ et à $\mathrm{Im}\,\tilde{\delta}$ étant alors du type

$$f(z) = P(z) + \frac{K}{1-z^2} + \frac{K'z}{1-z^2}$$

où $P(z)$ est un polynôme de degré 2 à coefficients dans $\mathbb{Z}$, et (K,K') un couple d'entiers.

2.4.4. <u>Remarque sur le calcul de</u> $\tilde{\delta}$

Désignons encore par $\sigma = a\,T_2 + b\,T_3$ le cycle de E correspondant à l'élément (a,b) de Z.

Lorsqu'on effectue le produit $s.\sigma$ dans E on obtient, en tenant compte de l'égalité $x_1^m(a,b) = \rho(-x_3, x_2)$,

$$s.x_1^{\alpha}\sigma = x_1^{\alpha} v T_2 T_3 - x_1^{\alpha} d(\rho T_1 T_2 T_3) \quad , \quad \forall \alpha = 0,1,\ldots,m.$$

Par ailleurs, on vérifie aisément que, A étant identifié à $Z_2(F) \subset Z_2(E)$ par l'application $x \mapsto x\,T_2 T_3$, on a, pour tout $\alpha = 0,1,\ldots,m$,

$$x_1^{\alpha}\, B_2(E) \cap A = x_1^{\alpha+1} A.$$

De même, un calcul simple prouve que, si la classe $\bar{\sigma}$ de (a,b) dans M est annulée par x_1^{α} , alors on a $x_1^{\alpha}\, v = 0$.

De ces diverses remarques, il résulte que la multiplication par s induit deux morphismes

$$s_o^* : \quad \mathrm{Ker}_M x_1^m / x_1 M \to \mathrm{Ker}_A x_1^m / x_1 A$$

$$s_1^* : \quad \mathrm{Soc}\ M/x_1^m M \to \mathrm{Soc}\ A/x_1^m A.$$

qui ne dépendent que de la classe d'homologie $\bar{s}$ de s.

La description donnée ci-dessus pour $\tilde{\delta}$ prouve alors que, pour $p \geq 1$, $\tilde{\delta}_{2p+2}$ s'identifie à s_o^* et $\tilde{\delta}_{2p+1}$ à s_1^*, compte tenu des isomorphismes

$$H_{2p}(\mathfrak{X}' \otimes N) \simeq \mathrm{Ker}_N x_1^m / x_1 N \quad \text{et} \quad H_{2p+1}(\mathfrak{X}' \otimes N) \simeq \mathrm{Soc}\ N/x_1^m N \quad \text{pour} \quad N = M$$

et $N = A$.

Nous exprimerons ce résultat en disant que, en degré $n \geq 2$, <u>$\tilde{\delta}_n$ est la multiplication par</u> $\bar{s}$.

2.5. Le lemme suivant donne le calcul de δ à partir de $\tilde{\delta}$. Rappelons que $H(\mathfrak{X} \otimes M) \simeq (\mathfrak{U} \otimes k) \otimes H(\mathfrak{X}' \otimes M)$; il suffit donc de calculer δ pour tout élément de la forme $u \otimes \dot{\zeta}$, où $u \in \mathfrak{U} \otimes k$ et $\dot{\zeta} \in H(\mathfrak{X}' \otimes M)$.

La multiplication dans F permet de définir le sous-module A' de A par

$$Z_1(F).Z_1(F) = A'T_2T_3 \quad ;$$

autrement dit A' est lènsemble des éléments de A de la forme $ab'-ba'$ où (a,b) et (a',b') vérifient $ax_2 + bx_3 = 0$ et $a'x_2 + b'x_3 = 0$.

<u>LEMME</u>

<u>Pour tout</u> $\dot{U} \in \mathfrak{U}_q \otimes k$ <u>et tout</u> $\dot{\zeta} \in H_n(\mathfrak{X}' \otimes M)$ <u>on a</u>

$$\delta(\dot{U} \otimes \dot{\zeta}) = \dot{U} \otimes \tilde{\delta}\ \dot{\zeta} + \Delta(\dot{U} \otimes \dot{\zeta})$$

<u>où</u> $\Delta : (\mathfrak{U} \otimes k) \otimes H(\mathfrak{X}' \otimes M) \to H(\mathfrak{X} \otimes A)$ <u>est un morphisme de degré</u> (-2) <u>possédant les propriétés suivantes</u> :

a) <u>Si</u> $\dot{U} \in \mathfrak{U}(q) \otimes k$, <u>alors</u> $\Delta(\dot{U} \otimes \dot{\zeta}) \in \mathfrak{U}(q-2) \otimes H(\mathfrak{X}' \otimes A)$

b) <u>Pour tout</u> $\dot{U} \in \mathfrak{U}_q \otimes k$ <u>et</u> $\dot{U}' \in \mathfrak{U}_{q'} \otimes k$,

$$\Delta(\dot{U}\dot{U}' \otimes \dot{\zeta}) = \dot{U}\ \Delta(\dot{U}' \otimes \dot{\zeta}) + (-1)^{qq'} \dot{U}' \Delta(\dot{U} \otimes \dot{\zeta})$$

c) <u>il existe un représentant de</u> $\Delta(\dot{U} \otimes \dot{\zeta})$ <u>dans</u> $Z_{n+q-2}(\mathfrak{X} \otimes A') \subset Z_{n+q-2}(\mathfrak{X} \otimes A)$.

Effectuons le calcul de $\delta(\dot{U} \otimes \dot{\zeta})$ lorsque $n = 2p$ et $\dot{\zeta} = \mathrm{cls}(S^{(p)} \otimes \bar{\sigma})$ où, comme précédemment, $\bar{\sigma}$ est l'image dans M de l'élément (a,b) de Z.

Soit U un relèvement de $\dot{U}$ dans $\mathfrak{U}$. On sait, d'après [3] qu'il existe dans $\mathfrak{X} \otimes_R R' \simeq \mathfrak{U} \otimes_R \mathfrak{X}'$ un cycle $\bar{z}_U$ de la forme $\bar{z}_U = U \otimes 1 + \sum_{i=1}^{K} U_i$, où, pour tout i, $U_i \in \mathfrak{U}_{q_i} \otimes \mathfrak{X}'_{q'_i}$ avec $q_i + q'_i = q$ et $q_i < q$. Alors, on peut choisir un relèvement z_U de $U \otimes 1$ dans $\mathfrak{X}_q = \bigoplus_{m+m'=q} (\mathfrak{U}_m \otimes \widetilde{\mathfrak{X}}_{m'})$ tel que

$$z_U \equiv U \ \text{modulo}\{(\underline{p}\, \mathfrak{U}_q \otimes R) \oplus (\sum_{m<q} \mathfrak{U}_m \otimes \widetilde{\mathfrak{X}}_{m'})\}$$

$$dz_U \in \underline{p}\, \mathfrak{X}_{q-1}$$

Soit $(V_i)_{i\in I}$ une base du R-module libre $\mathfrak{X}_{q-1}$, on pose

$$dz_U = \sum_i (\lambda_i x_2 + \mu_i x_3) V_i \quad , \quad \lambda_i \in R,\ \mu_i \in R.$$

Alors $\dot{U} \otimes \dot{\zeta}$ se relève dans $\mathfrak{X} \otimes Z$ en $z_U . z$, où $z = S^{(p)} \otimes (a,b)$, et $\partial(\dot{U} \otimes \dot{\zeta})$ est la classe dans $H(\mathfrak{X} \otimes B)$ de $d(z_U.z) = dz_U.z + (-1)^q z_U.dz$. Comme $dz_U.z = \sum_i V_i\, S^{(p)}(\lambda_i x_2 + \mu_i x_3)(a,b) = \sum_i V_i\, S^{(p)}(\lambda_i b - \mu_i a) \otimes (-x_3, x_2)$ et $z_U dz = z_U . w S^{(p-1)} \otimes (-x_3, x_2)$ (où w est l'élément défini en 2.4 dans le calcul de $\delta_{2p}(\dot{\zeta})$), on écrit $d(z_U.z) = u \otimes (-x_3, x_2) \in Z(\mathfrak{X} \otimes B)$

avec $u = \sum_i V_i\, S^{(p)}(\lambda_i b - \mu_i a) + (-1)^q z_U\, w S^{(p-1)}$

et $\partial(\dot{U} \otimes \dot{\zeta}) = \mathrm{cls}(du)$ dans $H(\mathfrak{X} \otimes A)$.

Le calcul de du donne, après simplification,

$$du = \sum_i dV_i\, S^{(p)}(\lambda_i b - \mu_i a) + z_U d(w S^{(p-1)}) \quad ,$$

et, dans $H(\mathfrak{X} \otimes A) \simeq (\mathfrak{U} \otimes k) \otimes H(\mathfrak{X}' \otimes A)$

$$\delta(\dot{U} \otimes \dot{\zeta}) = \dot{U} \otimes \widetilde{\delta}(\dot{\zeta}) + \Delta(\dot{U} \otimes \dot{\zeta})$$

où $\Delta(\dot{U} \otimes \dot{\zeta})$ est la classe d'homologie de $\sum_i dV_i\, S^{(p)}(\lambda_i b - \mu_i a)$, qui est bien un élément de $\mathfrak{X} \otimes A$ comme le prouve le calcul ci-dessous :

on décompose dV_i sur une base $(W_j \ ; \ j \in J)$ de $\mathfrak{X}_{q-2}$, soit

$$dV_i = \sum_j \rho_{ij} W_j \ ;$$

comme $d^2 z_U = 0$, on a $\sum_i \rho_{ij} \lambda_i x_2 + \sum_i \rho_{ij} \mu_i x_3 = 0 \ , \ \forall j \in J$;

si l'on pose $c_j = \sum_i \rho_{ij} \lambda_i$ et $d_j = \sum_i \rho_{ij} \mu_j$, on voit que $(c_j,d_j) \in Z$, et par suite

$$\sum_i dV_i\, S^{(p)}(\lambda_i b - \mu_i a) = \sum_j W_j\, S^{(p)} \otimes (c_j b - d_j a) \in \mathfrak{X}\otimes A' \subset \mathfrak{X}\otimes A .$$

En résumé, on a

$$\delta(\dot{U} \otimes \dot{\zeta}) = \dot{U} \otimes \tilde{\delta}(\dot{\zeta}) + \Delta(\dot{U} \otimes \dot{\zeta})$$

avec $\Delta(\dot{U} \otimes \dot{\zeta}) = \mathrm{cls}\ [\sum_j W_j\, S^{(p)} \otimes (c_j b - d_j a)]$

2.5.1.

où $\Psi_j\ ,\ W_j\, S^{(p)} \in \mathcal{Y}(q-2) \otimes \tilde{\mathfrak{X}}$

et $c_j b - d_j a \in A'$.

Ces formules restent valables lorsque $p = 0$, $\dot{\zeta}$ est alors la classe dans $M \otimes k$ de l'élément $\bar{\sigma}$ de M , lui-même projection du cycle (a,b) de Z.

Par un calcul analogue, si $z' = S^{(p)}T_1(a',b')$ est un relèvement dans $\mathfrak{X} \otimes Z$ d'un élément $\dot{\zeta}'$ de $H_{2p+1}(\mathfrak{X}' \otimes M)$, pour $p \geqslant 0$, alors, pour $\dot{U} \in \mathcal{Y}_m \otimes k$ et en utilisant les mêmes notations que ci-dessus on trouve :

2.5.2.

$$\delta(\dot{U} \otimes \dot{\zeta}') = \dot{U} \otimes \tilde{\delta}(\dot{\zeta}') + \Delta(\dot{U} \otimes \dot{\zeta}')$$

avec $\Delta(\dot{U} \otimes \dot{\zeta}') = \mathrm{cls}\ [\sum_j W_j\, S^{(p)}T_1 \otimes (c_j b'-a'd_j)]$,

$$\Psi_j\ \sum W_j\, S^{(p)}T_1 \in \mathcal{Y}(q-2) \otimes \tilde{\mathfrak{X}} ,$$

$$c_j b' - a'd_j \in A' \cap \mathrm{soc}\ A .$$

Les parties a) et c) du lemme résultent clairement des formules 2.5.1. et 2.5.2.

La démonstration de l'assertion b) est laissée au lecteur : il suffit de calculer $\delta(\dot{U}\dot{U}' \otimes \dot{\zeta})$ par la méthode précédente en remarquant que l'on peut utiliser le produit des cycles z_U et $z_{U'}$ pour relever $\dot{U}\,\dot{U}' \otimes 1$ dans $\mathfrak{X}_{q+q'}$.

2.6. Le lemme et les calculs ci-dessus conduisent à une meilleure description de δ . On remarque d'abord que si, pour $m \geqslant 2$,

$$\Psi_m : \mathcal{Y}_m \otimes k \simeq \mathrm{Tor}^R_m(R',k) \Rightarrow \mathrm{Tor}^R_{m-2}(Z,k)$$

est l'isomorphisme canonique, alors, avec les notations ci-dessus

$$\Psi_m(U) = \sum_j W_j(c_j,d_j) .$$

On posera $\Psi = (\Psi_o\ ,\ \Psi_1,\ldots,\ \Psi_m\ ,\ldots)$, avec $\Psi_o = \Psi_1 = 0$, et l'on appellera

$\psi : \mathfrak{U} \otimes k \to \mathrm{Tor}^R(M,k)$ le composé de Ψ et de la projection canonique $\mathrm{Tor}^R(Z,k) \to \mathrm{Tor}^R(M,k)$; ainsi ψ est de degré -2 et $\psi_o = \psi_1 = 0$.

D'un autre côté, si $(a,b) \in Z$ et $(c,d) \in B$, i.e $(c,d) = \rho(-x_3, x_2)$, alors $ad-bc = 0$, de sorte que la multiplication de E induit une application de $M \otimes M$ dans A : si $\bar{\sigma}$ est l'image canonique de (a,b) et $\bar{\sigma}'$ celle de (a',b') on posera $\bar{\sigma} \wedge \bar{\sigma}' = ab' - ba'$. Il en résulte, en particulier, une application de degré 0 que l'on notera

$$\mu : H(\mathfrak{X} \otimes M) \otimes H(\mathfrak{X}' \otimes M) \to H(\mathfrak{X} \otimes A) ;$$

si $z = \sum_j W_j \bar{\sigma}_j$ est un cycle de $\mathfrak{X}_m \otimes M$ de classe $\dot{z}$ et $z' = G_{m'} \bar{\sigma}'$ un cycle de $\mathfrak{X}'_{m'} \otimes M$ de classe $\dot{z}'$ ($G_{m'} = S^{(p)}$ ou $S^{(p)} T_1$ suivant que $m' = 2p$ ou $2p+1$), alors $\mu(\dot{z} \otimes \dot{z}') = \mathrm{cls}(\sum_j W_j G_{m'} \otimes \bar{\sigma}_j \wedge \bar{\sigma}')$.

Avec ces notations on a, pour tout $\dot{U} \in \mathfrak{U} \otimes k$ et $\dot{\zeta} \in H(\mathfrak{X}' \otimes M)$

$$\Delta(\dot{U} \otimes \dot{\zeta}) = \mu[\psi(\dot{U}) \otimes \dot{\zeta}]$$

<u>Proposition</u>

<u>Si l'on note</u>

$$\Delta : (\mathfrak{U} \otimes k) \otimes H(\mathfrak{X}' \otimes M) \longrightarrow H(\mathfrak{X} \otimes A) \simeq \mathrm{Tor}^R(A,k)$$

<u>l'homomorphisme de degré</u> -2 <u>composé de</u>
$\psi \otimes \mathrm{Id} : \mathfrak{U} \otimes k \otimes H(\mathfrak{X}' \otimes M) \to H(\mathfrak{X} \otimes M) \otimes H(\mathfrak{X}' \otimes M)$ <u>et de</u>
$\mu : H(\mathfrak{X} \otimes M) \otimes H(\mathfrak{X}' \otimes M) \to H(\mathfrak{X} \otimes A)$, <u>on a, pour tout</u> $\dot{U} \in \mathfrak{U} \otimes k$ et $\dot{\zeta} \in H(\mathfrak{X}' \otimes M)$

$$\delta(\dot{U} \otimes \dot{\zeta}) = U\,\delta(\dot{\zeta}) + \Delta(\dot{U} \otimes \dot{\zeta}).$$

2.7. <u>Etude de divers cas de rationalité de</u> $\mathfrak{P}^R(k)$

2.7.1. <u>Proposition</u>

<u>La série</u> $\mathfrak{P}^R(k)$ <u>est une fraction rationnelle dans les deux cas suivants</u> :

1. <u>Lorsque la restriction</u> $\tilde{\delta}$ <u>de</u> δ <u>à</u> $\mathrm{Tor}^{R'}(M,k)$ <u>est une surjection sur</u> $\mathrm{Tor}^{R'}(A,k)$;
2. <u>Lorsque, pour tout couple</u> $(\bar{z}_1, \bar{z}_2)$ <u>d'éléments de</u> $H_1(F) \otimes k$, <u>il existe des représentants</u> z_1 <u>et</u> z_2 <u>dans</u> $Z = Z_1(F)$ <u>tels que</u> $z_1 z_2 = 0$

Corollaire

Dans les deux cas ci-dessus $\beta^R(N)$ est une fraction rationnelle pour tout R'-module N noethérien.

Démonstration.

La première hypothèse implique la surjectivité de δ. En effet, on prouve, par récurrence sur m, que l'image de δ contient $\mathcal{U}_m \otimes \mathrm{Tor}^{R'}(A,k)$ pour tout m. Par hypothèse, cette propriété est vraie pour $m = 0$; on la suppose vraie pour tout $m' < m$. Soient alors $\dot{U} \in \mathcal{U}_m \otimes k$, $\dot{\xi} \in \mathrm{Tor}_r^{R'}(A,k)$ et $\dot{\zeta}$ un élément de $\mathrm{Tor}^{R'}(M,k)$ tel que $\tilde{\delta}(\dot{\zeta}) = \dot{\xi}$. Comme $\delta(\dot{U} \otimes \dot{\zeta}) = \dot{U} \otimes \dot{\xi} + \Delta(\dot{U} \otimes \zeta)$ avec $\Delta(\dot{U} \otimes \dot{\zeta}) \in \sum_{m' \leq m-2} \mathcal{U}_{m'} \otimes \mathrm{Tor}^{R'}(A,k)$, utilisant l'hypothèse de récurrence on en déduit que $\dot{U} \otimes \dot{\xi} \in \mathrm{Im}\,\delta$.

Il en résulte que, pour tout $m \geq 2$, la suite

$$0 \to \mathrm{Tor}_m^R(Z,k) \to \mathrm{Tor}_m^R(M,k) \xrightarrow{\delta_m} \mathrm{Tor}_{m-2}^R(A,k) \to 0$$

déduite de 2.3.5. est exacte. Compte tenu de l'isomorphisme $\mathrm{Tor}_m^R(Z,k) \simeq \mathrm{Tor}_{m-2}^R(R',k)$; on obtient l'égalité

$$z^2[\beta^R(R') + \beta^R(A)] = \beta^R(M) - c_0 - c_1 z$$

où $c_0 = \dim_k M \otimes k$, $c_1 = \dim_k \mathrm{Tor}_1^R(M,k)$.

Comme $\beta^R(M) = \beta^R(R').\beta^{R'}(M)$ et $\beta^R(A) = \beta^R(R').\beta^{R'}(A)$, il vient

$$\beta^R(R') = (c_0 + c_1 z)/[\beta^{R'}(M) - z^2(1 + \beta^{R'}(A))].$$

Ainsi, $\beta^{R'}(M)$ et $\beta^{R'}(A)$ étant rationnelles, il en est de même de $\beta^R(R')$ donc de $\beta^R(k)$ et de $\beta^R(N)$ pour tout R'-module N noethérien.

Dans la deuxième hypothèse proposée, on a évidemment $\Delta = 0$. Alors $\delta(\dot{U} \otimes \dot{\zeta}) = \dot{U} \otimes \tilde{\delta}(\dot{\zeta})$ pour tous $\dot{U} \in \mathcal{U}_m \otimes k$ et $\dot{\zeta} \in \mathrm{Tor}_{m'}^{R'}(M,k)$. Il en résulte

$$\mathrm{Im}(\delta_m) \simeq [\mathcal{U} \otimes \mathrm{Im}\,\tilde{\delta}]_{m-2}, \quad \forall m \geq 2$$

$$\text{et } \mathrm{Ker}(\delta_m) \simeq [\mathcal{U} \otimes \mathrm{Ker}\,\tilde{\delta}]_m, \quad \forall m \geq 0$$

Comme, par ailleurs, la suite exacte 2.3.5. nous donne en chaque degré $m \geq 0$ la suite exacte courte

$$0 \to \mathrm{Tor}_{m+1}^R(A,k)/\mathrm{Im}\,\delta_{m+3} \to \mathrm{Tor}_m^R(R',k) \to \mathrm{Ker}\,\delta_{m+2} \to 0 ,$$

on a :

$$z^2\,\beta^R(R') = \beta^R(R')[f(z) + z\,\beta^{R'}(A)/g(z)] - c_0 - c_1 z - d_1 z$$

où $f(z)$ et $g(z)$ sont les séries associées aux espaces vectoriels $\operatorname{Ker} \tilde{\delta}$ et $\operatorname{Im}\tilde{\delta}$, $c_o = \dim_k M \otimes k$, $c_1 = \dim_k \operatorname{Tor}_1^R(M,k)$ et $d_1 = \dim_k(A \otimes k/\operatorname{Im} \delta_2)$. Les séries $f(z)$ et $g(z)$ étant rationnelles (cf 2.4.) la rationalité de $\mathcal{B}^R(R')$ s'en déduit, ce qui achève la démonstration de la proposition et de son corollaire.

2.7.2 <u>Exemples d'anneaux de dimension d'immersion</u> 3 <u>dont la série de Poincaré est rationnelle</u>.

Nous conservons les notations des paragraphes précédents. $(R,\underline{m},k)$ est le quotient d'un anneau $(\mathcal{R},\mathfrak{m},k)$, régulier, par un idéal $\mathcal{I} < \mathfrak{m}$; on appelle (X_1,X_2,X_3) une famille de générateurs de $\mathfrak{m}$ dont la projection canonique est (x_1,x_2,x_3), et $\mathcal{P}$ l'idéal (X_2,X_3). Alors, par hypothèse, $\mathcal{I} = G\mathcal{R} + \mathcal{I}'$, où $\mathcal{I}' = \mathcal{I} \cap \mathcal{P}$ et $G = X_1^{m+1} + \lambda' X_2 + \mu' X_3$, $\lambda' \in \mathfrak{m}$, $\mu' \in \mathfrak{m}$, $m \geq 1$.

<u>Exemple 1</u> : Pour tout $r \geq 1$, si on a à la fois $\mathcal{I} \supset \mathfrak{m}^{2n}$ et $\mathcal{I}' \subset \mathfrak{m}^n\mathcal{P}$ alors la série $\mathcal{B}^R(k)$ est rationnelle. En effet, la partie 2 de la proposition 2.7.1 s'applique puisque l'on peut choisir comme relèvements des générateurs $\bar{\sigma}_i$, $i = 1,\ldots,m$, de M des cycles de la forme $s_i = a_i T_2 + b_i T_3$, avec a_i et b_i dans $\underline{m}^n$ de sorte que $s_i s_j = 0$, $\forall (i,j)$.

<u>Exemple 2</u> : Si $\mathcal{I} \supset \mathfrak{m}^3$, $\mathcal{B}^R(k)$ est rationnelle. Ce résultat est prouvé dans [1]. Une démonstration peut en être faite en utilisant à la fois les résultats ci-dessus et ceux de [3] suivant les différents cas possibles (il y en a 15)

<u>Exemple 3</u> : Dans cet exemple, l'application Δ est nulle bien que la condition sur les cycles de 2.7.1. ne soit pas remplie.

On prend $R = \mathcal{R}/\mathcal{I}$ avec $\mathcal{I} = (X_1^3 + X_3(X_2+X_3), X_2^2, X_1X_2, X_1X_3)$. Il est clair que $\mathcal{I} \supset \mathfrak{m}^4$. On choisit comme générateurs de $Z_1(E)$

$$s = x_1^2 T_1 + (x_2+x_3)T_3$$
$$s_1 = x_2T_2$$
$$s_2 = x_1T_2$$
$$s_3 = x_1T_3$$

Les s_i sont dans $Z_1(F)$ il leur correspond un système minimal de générateurs pour M : $\bar{\sigma}_1 = \overline{(x_2,0)}$, $\bar{\sigma}_2 = \overline{(x_1,0)}$, $\bar{\sigma}_3 = \overline{(0,x_1)}$ et l'on a $x_1\bar{\sigma}_1 = 0$, $x_1^2\,\bar{\sigma}_2 \neq 0$ et $x_1^2\bar{\sigma}_3 \neq 0$. On vérifie aisément que l'on a :

$$M \xrightarrow{\sim} k\,\bar\sigma_1 \oplus R'\bar\sigma_2 \oplus R'\bar\sigma_3 ,$$

$$A \xrightarrow{\sim} R'e_1 \oplus k\,e_2 \quad , \text{ avec } e_1 = x_1 \text{ et } e_2 = x_2x_3 .$$

et par suite $\mathrm{Tor}^R(M,k) \xrightarrow{\sim} \mathrm{Tor}^R(k,k) \oplus \mathrm{Tor}^R(R',k) \oplus \mathrm{Tor}^R(R',k)$ et $\mathrm{Tor}^R(A,k) \xrightarrow{\sim} \mathrm{Tor}^R(R',k) \oplus \mathrm{Tor}^R(k,k)$. Enfin, pour les calculs de $\tilde\delta$ et Δ on utilisera

$$ss_1 = -\,x_2x_3\,T_2T_3 ,$$
$$\bar\sigma_1 \wedge \bar\sigma_2 = 0 = \bar\sigma_1 \wedge \bar\sigma_3 ,$$
$$\bar\sigma_2 \wedge \bar\sigma_3 = x_1^2 .$$

<u>Calcul de</u> $\tilde\delta$. On appelle $\dot\sigma_i$ l'image canonique de $\bar\sigma_i$ dans $M \otimes k$ et l'on a

$$H_o(\mathfrak{X}' \otimes M) \xrightarrow{\sim} k\,\dot\sigma_1 \oplus k\,\dot\sigma_2 \oplus k\,\dot\sigma_3 ,$$
$$H_{2p}(\mathfrak{X}' \otimes M) \xrightarrow{\sim} k\,S^{(p)}\dot\sigma_1 \quad , \forall p > 0 \quad ,$$
$$H_{2p+1}(\mathfrak{X}' \otimes M) \xrightarrow{\sim} k\,S^{(p)}T_1\dot\sigma_1 \quad , \forall p \geq 0 .$$

Alors, pour $p > 0$, on a $\delta(S^{(p)}\dot\sigma_1) = -\,S^{(p-1)}e_2$ et $\delta(S^{(p)}T_1\dot\sigma_1) = -S^{(p-1)}T_1e_2$. Ainsi, $\tilde\delta_o = \tilde\delta_1 = 0$ et $\tilde\delta_p$ est injective pour tout $p \geq 2$.

<u>Calcul de</u> Δ. Comme $\bar\sigma_i \wedge \bar\sigma_1 = 0$ pour $i = 1,2,3$, on voit que $\Delta(\dot U \otimes \dot\zeta) = \mu(\psi(\dot U) \otimes \dot\zeta)$ est nul lorsque $\dot\zeta = S^{(p)}\dot\sigma_1$ ou $S^{(p)}T_1\dot\sigma_1$, $p \geq 0$. Donc Δ est nulle sur $\mathfrak{U} \otimes \sum_{p>o} H_p(\mathfrak{X}' \otimes M)$. Il reste à calculer $\Delta(\dot U \otimes \dot\sigma_i)$ pour $\dot U \in \mathfrak{U} \otimes k$, $i = 2,3$. Comme $\bar\sigma_1 \wedge \bar\sigma_2 = \bar\sigma_1 \wedge \bar\sigma_3 = 0$ et $\bar\sigma_2 \wedge \bar\sigma_3 = x_1^2$, $\Delta(\dot U \otimes \dot\sigma_i)$ est la classe d'un cycle z de $\mathfrak{X} \otimes A$ dont les coefficients sont en fait dans x_1^2R'. La classe d'un tel cycle est toujours nulle (utiliser les isomorphismes $x_1R \xrightarrow{\sim} R'$ et $\mathrm{Tor}^R(x_1R,k) \xrightarrow{\sim} \mathfrak{U} \otimes \mathrm{Tor}^{R'}(x_1R,k) = \mathfrak{U} \otimes (x_1R/x_1^2R))$.
En résumé, Δ est nulle et, pour tout $n \geq 0$, on a

$$\mathrm{Ker}\ \delta_{n+2} \xrightarrow{\sim} (\mathrm{Tor}^R_{n+2}(R',k) \otimes M) \oplus (\mathrm{Tor}^R_{n+1}(R',k) \otimes k\,\dot\sigma_1) ,$$
$$\mathrm{Im}\ \delta_{n+2} \xrightarrow{\sim} \mathrm{Tor}^R_n(k,k).$$

<u>Calcul de la série</u> $\mathfrak{P}^R(k)$. On pose

$$\mathfrak{P}^R(k) = \sum_{n\geq o} b_n z^n$$
$$\mathfrak{P}^R(R') = \sum_{n\geq o} a_n z^n$$

$$\mathbb{B}^{R'}(k) = \sum_{n\geq o} z^n$$

et l'on a $\Sigma\, b_n z^n = (\Sigma\, a_n z^n)(\Sigma\, z^n)$

Par ailleurs, $\dim_k \operatorname{Tor}_n^R(Z,k) = a_{n+2}$. Alors, compte tenu de ces diverses relations et des relations de dimensions données par la suite 2.3.5., on trouve

$$2a_{n+1} = a_{n+4} - 3a_{n+2} \quad , \quad \forall n \geq 0 \, ,$$

d'où $2\,z^3[\mathbb{B}^R(R')-a_o] = \mathbb{B}^R(R') - a_o - a_1 z - a_2 z^2 - a_3 z^3 - 3z^2[\mathbb{B}^R(R') - a_o - a_1 z]$.

On trouve aisément : $a_o = 1$, $a_1 = 2$, $a_2 = \dim_k(Z \otimes k) = 4$, et $a_3 = 8$ (cette dernière valeur est obtenue à partir de $\dim_k(\operatorname{Im}\delta_2) = 1$, $\dim_k \operatorname{Tor}_1^R(M,k) = 7$ et de la suite en acte

$$\operatorname{Tor}_2^R(M,k) \xrightarrow{\delta_2} A \otimes k \to \operatorname{Tor}_1^R(Z,k) \to \operatorname{Tor}_1^R(M,k) \to 0).$$

On en déduit :

$$\mathbb{B}^R(R') = 1/1\text{-}2z$$

$$\text{et} \quad \mathbb{B}^R(k) = 1/(1-2z)(1-z).$$

<u>Exemple 4</u> : dans ce cas $R = \mathcal{R}/\mathcal{J}$ avec

$$\mathcal{J} = (X_1^3+(X_2+X_3)X_3, X_1X_2^2, X_1X_2(X_1+X_3), X_3(X_1+X_3), X_3(X_1^2-X_2-X_3), X_1^4X_3, X_2^5).$$

On constate que $\underline{m}^5 = 0$. Avec le notations précédentes on peut prendre ;

$s = x_1^2 T_1 + (x_2+x_3)T_3$,

$s_1 = x_1 x_2 T_2$,

$s_2 = x_1(x_1+x_3)T_2$,

$s_3 = (x_1+x_3)T_3$,

$s_4 = (x_1^2-x_2-x_3)T_3$,

$s_5 = x_1^4 T_3$,

$s_6 = x_2^4 T_2$,

$M = k\,\bar\sigma_1 \oplus M'$, avec $M' = k\,\bar\sigma_2 \oplus k\,\bar\sigma_5 \oplus k\,\bar\sigma_6 \oplus (R'/x_1^2R')\bar\sigma_3 \oplus (R'/x_1^2R')\bar\sigma_4$,

$A \xrightarrow{\sim} k\,e_1 \oplus \underline{m}^4$, avec $e_1 = x_1x_2x_3$.

On a alors

$$s\,s_1 = -x_1x_2x_3\,T_2T_3 \quad \text{et} \quad ss_i = 0 \quad \forall i \geq 2 \quad ,$$

$$\bar\sigma_1 \wedge \bar\sigma_4 = -\,x_1x_2x_3 \quad \text{et} \quad \bar\sigma_i \wedge \bar\sigma_j = 0 \text{ pour } (i,j)\neq(1,4) \text{ et } (4,1).$$

Il résulte de là, en utilisant les décompositions

$$\mathrm{Tor}^R(M,k) \simeq \mathrm{Tor}^R(k,k)\bar{\sigma}_1 \oplus \mathrm{Tor}^R(M',k)$$

$$\text{et} \quad \mathrm{Tor}^R(A,k) \simeq \mathrm{Tor}^R(k,k)e_1 \oplus \mathrm{Tor}^R(\underline{m}^4,k),$$

que $\tilde{\delta}$ est nulle sur $\mathrm{Tor}^{R'}(M',k)$ et induit un isomorphisme de degré -2 de $\sum_{m\geq 2} \mathrm{Tor}^{R'}(k,k)\bar{\sigma}_1$ sur $\mathrm{Tor}^{R'}(k,k)e_1$.

Par ailleurs, la restriction de δ à $\mathrm{Tor}^R(k,k)\bar{\sigma}_1$ a pour image $\mathrm{Tor}^R(k,k)\bar{e}_1$: en effet, on sait d'une part que $\tilde{\delta}(\mathrm{Tor}^{R'}(k,k)\bar{\sigma}_1) = \mathrm{Tor}^{R'}(k,k)e_1$, d'autre part que l'image de la multiplication $M \otimes M \to A$ est le sous module ke_1, il suffit alors d'adapter le raisonnement fait pour démontrer la première partie de la proposition 2.7.1. pour obtenir le résultat annoncé. On voit que δ induit un isomorphisme de degré -2 de $V \otimes \sum_{m\geq 2} \mathrm{Tor}^{R'}(k,k)\bar{\sigma}_1$ sur $\mathrm{Tor}^R(k,k)\bar{e}_1$.

Enfin, comme l'image de Δ est contenue dans $\mathrm{Tor}^R(k,k)\bar{e}_1$, on aboutit aux conclusions suivantes valables pour tout $n \geq 0$

$$\mathrm{Im}(\delta_{n+2}) = \mathrm{Tor}^R_n(k,k)e_1$$

$$\mathrm{Ker}(\delta_{n+2}) \simeq \mathrm{Tor}^R_{m+2}(M,k)/\mathrm{Tor}^R_{n+2}(k,k)\bar{\sigma}_1 .$$

Il en résulte, en utilisant la suite exacte longue 2.3.5., que $B^R(k)$ est une fraction rationnelle.

3. Etude du cas où $\dim_k \underline{m}/\underline{m}^2 = n > 3$

Les résultats du paragraphe 2 se généralisent au moyen de calculs analogues. La seule difficulté provient de ce que $R' = R/\underline{p}$ n'est plus de valuation discrète ; on a donc des expressions moins simples pour $\mathrm{Tor}^{R'}(N,k)$ lorsque N est un R'-module.

Avec des notations analogues, on a deux R'-modules $M = H_1(F)$ et $A = \mathrm{ann}\, x_{n-1} \cap \mathrm{ann}\, x_n$, et l'on retrouve la suite exacte 2.3.5.

On prouve comme au paragraphe 2 que l'homomorphisme

$$\delta : \mathrm{Tor}^R(M,k) \simeq \mathrm{Tor}^R(R',k) \otimes \mathrm{Tor}^{R'}(M,k) \to \mathrm{Tor}^R(R',k) \otimes \mathrm{Tor}^{R'}(A,k) \simeq \mathrm{Tor}^R(A,k)$$

de degré -2 possède les propriétés suivantes :

3.1. La restriction de δ à $\mathrm{Tor}^{R'}(M,k)$ définit, en fait, un homomorphisme

$$\tilde{\delta} : \mathrm{Tor}^{R'}(M,k) \to \mathrm{Tor}^{R'}(A,k)$$

3.2. Pour tout $\dot{U} \in \mathfrak{U} \otimes k$ et $\dot{\zeta} \in H(\mathfrak{X}' \otimes M)$ on a

$$\delta(\dot{U} \otimes \dot{\zeta}) = \dot{U} \otimes \tilde{\delta}\, \dot{\zeta} + \Delta(\dot{U} \otimes \dot{\zeta})$$

où Δ est encore le composé de

$$\psi \otimes \mathrm{Id} : (\mathfrak{U} \otimes k) \otimes H(\mathfrak{X}' \otimes M) \to H(\mathfrak{X} \otimes M) \otimes H(\mathfrak{X}' \otimes M)$$

et de l'application

$$\mu : H(\mathfrak{X} \otimes M) \otimes H(\mathfrak{X}' \otimes M) \to H(\mathfrak{X} \otimes A)$$

induite par la multiplication dans Ξ.

3.3. Si, pour tout couple $(\bar{z}_1, \bar{z}_2)$ d'éléments de $H_1(F)$ il existe un couple (z_1, z_2) de représentants dans $Z_1(F)$ tels que $z_1 z_2 = 0$, alors $\Delta = 0$.

Enfin, il est possible de prouver la propriété suivante :

3.4. Les séries associées aux espaces vectoriels gradués $\mathrm{Ker}\,\tilde{\delta}$ et $\mathrm{Im}\,\tilde{\delta}$ sont rationnelles. C'est uniquement la démonstration de cette propriété qui ne peut être calquée sur celle du §2., elle peut être faite par les méthodes de [3] mais ne sera pas donnée ici.

La proposition ci-dessous résulte de ces diverses propriétés.

3.5 Proposition.

Avec les hypothèses du paragraphe 1 la série $\mathrm{B}^R(k)$ est une fraction rationnelle lorsque, pour tout couple $(\bar{z}_1, \bar{z}_2)$ d'éléments de $H_1(F)$ il existe un couple (z_1, z_2) de représentants dans $Z_1(F)$ tels que $z_1 z_2 = 0$.

Dans ces conditions, $\mathrm{B}^R(N)$ est aussi une fraction rationnelle pour tout R'-module N noethérien.

Une généralisation de certains de ces résultats, par des méthodes différentes, sera donnée ultérieurement [4].

Bibliographie

[0] D. ANICK - Construction d'espaces de lacets et d'anneaux locaux à séries de Poincaré-Betti non rationnelles. C.R. Acad. Sc. Paris 162 (1980).

[1] J. BACKELIN et R. FROBERG - Studies on some k-Algebras giving the Poincaré series of graded k-algebras of length ≤ 7 and local rings of embedding dimension 3 with $\underline{m}^3 = 0$. Preprint series Stockholm Univ. n°9. 1978.

[2] C. LOFWALL et J.E. ROOS - Cohomologie des algèbres de Lie graduées et séries de Poincaré-Betti non rationnelles. C.R. Acad. Sc. Paris 162 (1980).

[3] C. SCHOELLER - Rationalité de certaines séries de Poincaré - Séminaire d'Algèbre Paul Dubreil Proc. Paris 1977-78. Lecture Notes in Math 740 p 323-384.

[4] C. SCHOELLER - Anneauxsemi-golodiens (à paraître)

[5] J. TATE - Homology of noetherian rings ans locals rings. Ill. J. of Math. 1 (1957) p 14-27.

C. SCHOELLER
Université des Sciences et
Techniques du Languedoc
Institut de Mathématiques
Place Eugène Bataillon
34.060 Montpellier Cedex

DIMENSIONS COHOMOLOGIQUES RELIÉES AUX FONCTEURS $\varprojlim^{(i)}$.

L. Gruson et C.U. Jensen

Si quid tamen olim scripseris
- - , nonumque prematur in annum,
membranis intus positis.

0. Introduction. Soient R un anneau unitaire et I un ensemble ordonné filtrant à droite (en abrégé f.à d.). Un I-système projectif de R-modulus (à gauche) est une famille de R-modules $\{M_\alpha\}$, $\alpha \in I$, est de R-homomorphismes $\{f_{\alpha\beta}:M_\beta \to M_\alpha\}$, $\alpha \leq \beta$, tels que $f_{\alpha\alpha} = 1_{M_\alpha}$ pour tout $\alpha \in I$ et $f_{\alpha\gamma} = f_{\alpha\beta}f_{\beta\gamma}$ si $\alpha \leq \beta \leq \gamma$.

Si $\{M_\alpha, f_{\alpha\beta}\}$ et $\{N_\alpha, g_{\alpha\beta}\}$ sont deux I-systèmes projectifs, on entend par une application (morphisme) de $\{M_\alpha, f_{\alpha\beta}\}$ dans $\{N_\alpha, g_{\alpha\beta}\}$ une famille de R-homomorphismes $\{u_\alpha\}$ $u_\alpha: M_\alpha \to N_\alpha$ tels que $u_\alpha f_{\alpha\beta} = g_{\alpha\beta}u_\beta$ si $\alpha \leq \beta$.

Si R et I sont fixes, les I-systèmes projectifs et les morphismes introduits ci-dessus forment une catégorie abélienne ayant assez d'injectifs.

Le foncteur limite projective $\varprojlim$ est un foncteur exact à gauche de cette catégorie dans la catégorie des R-modules. On désigne par $\varprojlim^{(n)}$ le $n^{\text{ième}}$ foncteur dérivé à droite de $\varprojlim$. Un aperçu de la théorie de ces foncteurs se trouve dans [14]. Une question fondamentale de ce domaine est d'obtenir des conditions pour l'annulation de ces foncteurs. Si aucune condition n'est imposée aux modules M_α on a le résultat suivant: Soient I un ensemble ordonné f.à d. et n un entier ≥ 0 ; alors $\varprojlim^{(i)} M_\alpha = 0$ pour tout $i > n$ et pour tout I-système projectif de R-modules M_α si et seulement si I contient un sous-ensemble cofinal de puissance $\leq \aleph_{n-1}$. (Ici $\aleph_{-1}$ signifie le cardinal des ensembles finis.) [10, 14, 16].

Cette situation se change totalement si l'on impose la restriction que les modules M_α intervenant dans les systèmes projectifs soient de type fini (ou de présentation finie). Par exemple, si $R = \underline{Z}$, l'anneau des entiers rationnels, alors $\varprojlim^{(i)} M_\alpha = 0$ pour tout $i > 1$ et tout système projectif $\{M_\alpha\}$ de $\underline{Z}$-modules de type fini [18]. Plus généralement, Roos [16] a démontré pour tout anneau commutatif et noethérien R de dimension globale finie que $\varprojlim^{(i)} M_\alpha = 0$ pour tout $i >$ Krull-dim R et pour change système projectif de R-modules de type fini. La question a été posée de savoir si ce théorème reste vrai sans la condition: "R est de dimension globale finie". Les résultats plus généraux de ce papier montrent que la condition de régularité sur R est inutile pour la validité du théorème de Roos. On va donner de plus une généralisation non-commutative.

Dans cet ordre d'idées nous allons obtenir d'autres résultats qui disent, grosso modo, qu'un système projectif M_α ne peut avoir "trop de" foncteurs $\varprojlim^{(i)} M_\alpha$ non-nuls si les modules M_α (ou les groupes abéliens sous-adjacents) ne sont pas "trop larges".

En tant qu'applications explicites nous donnerons des résultats en théorie des modules que l'on ne peut pas (probablement) prouver sans la théorie de L-dimension. Par exemple, au paragraphe 7 nous allons démontrer pour un anneau R noethérien à droite que de dimension projective de tout R-module à gauche plat est au plus égale à la dimension de Krull-Gabriel (à droite) de R.

§10 contient des résultats caractérisant les anneaux de représentation finie parmi les anneaux artiniens.

La plupart des résultats de ce papier date des années soixante-dix: pour des raisons différentes nous avons suivi le conseil sage mais quelque peu hasardeux d'Horace cité plus haut.

Table des matières.

§1. La catégorie D(R).

§2. Sous-groupes de définition finie.

§3. La L-dimension et l'annulation de $\varprojlim^{(i)}$.

§4. Caractérisations dans le cas cohérent.

§5. Catégories localement cohérentes.

§5A. Une application concernant les modules de Mittag-Leffler et la dimension finitiste d'un anneau noethérien.

§6. Dimension injective de foncteurs exacts.

§7. Bornes explicites de la L-dimension.

§8. Résultats supplémentaires concernant l'annulation de $\varprojlim^{(i)}$ et les cardinaux de groupes.

§9. La L-dimension des anneaux complets.

§10. La L-dimension globale d'un anneau.

1. La catégorie D(R).

Soient R un anneau unitaire et Pf(R) la catégorie des R-modules à droite de présentation finie, considérée comme sous-catégorie pleine de la catégorie de tous les R-modules à droite. Les fonteurs additifs de Pf(R) dans la catégorie Ab des groupes abéliens forment une catégorie de Grothendieck D(R) avec un générateur. On dit qu'un objet A de D(R) est de type fini si la réunion de toute famille croissante de sous-objets propres de A est elle-même un sous-objet propre de A . Il revient au même de dire que tout ensemble filtrant croissant de sous-objets de A , de borne supérieure A , contient A . Les objets représentables de D(R), c.à d. les foncteurs de la forme $\bar{M} = \mathrm{Hom}_R(M,-)$, $(M \in Pf(R))$, sont des objets projectifs de type fini de D(R) et ils engendrent D(R). Réciproquement, tout objet projectif de type fini de D(R) est le quotient d'un objet représentable, par conséquent facteur direct d'un objet représentable. C'est facile de verifier que chaque facteur direct d'un objet représentable est lui-même représentable. Donc, les objets représentable sont exactement les objets projectifs de type fini de $D(R)$.

Un objet A de $D(R)$ est dit cohérent (dans la terminologie de Roos [19]) s'il est de type fini et si, pour tout morphisme u d'un objet de type fini B de $D(R)$ dans A, le noyau $\mathrm{Ker}(u)$ est de type fini. En vertu des remarques précédentes il s'ensuit qu'un objet A de $D(R)$ est cohérent si et seulement s'il existe une suite exacte de $Pf(R)$:

$$0 \to X \to Y \to Z \to 0$$

telle que la suite correspondante de $D(R)$

$$0 \to \overline{Z} \to \overline{Y} \to \overline{X} \to A \to 0$$

est exacte (pour un morphisme convenable: $\overline{X} \to A$). Donc les objets cohérents de $D(R)$ sont exactement les foncteurs cohérents introduits dans [1]. En particulier, $D(R)$ possède une famille de générateurs cohérents et est donc une catégorie "localement cohérente" dans terminologie de [19]. (Nous revendrons plus tard à une déscription plus détaillée des catégories localement cohérentes.)

A plusieurs occasions nous aurons besoin d'une suite spectrale, explicitée pour les catégories des modules en [14]. Soit (A_α), $a \in I$, un petit système inductif filtrant de $D(R)$, de limite A; alors pour tout objet F de $D(R)$ il existe une suite spectrale

$$E_2^{p,q} = \varprojlim{}^{(p)} \mathrm{Ext}^q_{D(R)}(A_\alpha, F) \underset{p}{\Rightarrow} \mathrm{Ext}^n_{D(R)}(A, F).$$

En particulier, si (N_α), $a \in I$, est une I-système projectif f.à d. de R-modules de $Pf(R)$, les foncteurs $\overline{N}_\alpha = \mathrm{Hom}_R(N_\alpha, -)$ forment un système inductif de $D(R)$. Puisque tout objet $\overline{N}_\alpha$ est projectif, la suite spectrale ci-dessus dégénère en des isomorphismes

$$\varprojlim^{(n)} F(N_\alpha) \simeq \mathrm{Ext}^n_{D(R)}(\varinjlim \overline{N}_\alpha, F) \quad (n \geq 0),$$

où l'on a utilisé l'isomorphisme canonique de Yoneda

$$\mathrm{Hom}_{D(R)}(\overline{N}_\alpha, F) \simeq F(N_\alpha) .$$

Si M est un R-module à gauche, le foncteur $\overline{\overline{M}} = - \otimes_R M$ de $Pf(R)$ dans Ab est un objet de $D(R)$. Comme $D(R)$ a suffisament d'objets injectifs on peut introduire la définition suivante:

<u>Définition</u>. Pour un R-module (à gauche) M on appelle L-dimension (notation $L\text{-}\dim_R(M)$) la dimension injective de $\overline{\overline{M}}$ dans $D(R)$: c'est un entier ou le symbole $+\infty$.

(Remarque. La dénomination L-dimension provient du rapport à l'annulation des foncteurs $\varprojlim^{(i)}$ que nous allons établir plus loin.)

Donnons une autre interpretation de la L-dimension. Rappelons qu'une suite exacte de R-modules à gauche

$$0 \to N_1 \underset{u}{\to} N_2 \underset{v}{\to} N_3 \to 0 \qquad (*)$$

est dite pur-exacte ou universellement exacte si la suite

$$0 \to X \otimes N_1 \xrightarrow[1_X \otimes u]{} X \otimes_R N_2 \xrightarrow[1_X \otimes v]{} X \otimes_R N_3 \to 0$$

est exacte pour tout R-module à droite X. On dit qu'un R-module à gauche M est pur-injectif (ou relativement injectif) si toute suite pur-exacte (*) ayant $N_1 = M$ est scindée. De même un R-module à gauche P est dit pur-projectif (ou relativement projectif) si toute suite pur-exacte (*) ayant $N_3 = P$ est scindée. Il se trouve queles modules pur-projectifs sont exactement les facteurs directs des sommes directes de modules de présentation finie et les modules pur-injectifs sont les modules algébriquement compacts [20], dont nous donnerons plus tard une déscription détaillée.

Soit T le foncteur contravariant $\mathrm{Hom}_Z(-, \mathbb{Q}/\mathbb{Z})$ de la catégorie des R-modules à gauche dans la catégorie des R-modules à droite. Alors pour tout R-module à gauche M l'application canonique $M \to T(T(M))$ est un monomorphisme pur et $T(T(M))$ est un module pur-injectif. Donc tout module est sous-module d'un module pur-injectif convenable, et l'on peut introduire la dimension pur-injective d'un R-module M comme le plus petit n (un entier ou ∞) pour lequel il existe une suite exacte

$$0 \to M \to I_o \to I_1 \to \cdots \to I_n \to 0 \qquad (**)$$

où $I_o, I_1, \cdots, I_n$ sont pur-injectifs et les suites exactes courtes dont (**) se compose, sont pur-exactes au sens défini plus haut. Dans ce cas on dit que (**) est une résolution pur-injective de M. Bien entendu, la dimension pur-projective d'un module est définie de la manière duale.

La relation entre L-dimension et dimension pur-injective est donnée dans

Proposition 1.1. Pour tout R-module (à gauche) M la L-dimension est égale à la dimension pur-injective.

Démonstration. En vertu d'un argument standard ("dimension shifting") l'assertion de la proposition est une conséquence immédiate de la caractérisation suivante des objets injectifs de $D(R)$.

Proposition 1.2. Un objet F de $D(R)$ est injectif si et seulement si F est de la forme $F = \overline{\overline{M}}$ pour un R-module à gauche pur-injectif M.

Démonstration. Supposons d'abord que F est un objet injectif de $D(R)$. D'après [8] F est alors un foncteur exact à droite de $Pf(R)$ dans Ab et donc de la forme $\overline{\overline{M}}$ pour un R-module à gauche M. On peut prendre $M = F(R)$ avec la structure de R-module définie par

$$r\cdot y = F(\mu_r)[y] \ , \ y \in F(R) = M, r \in R \ ,$$

où μ_r est l'homothétie: $\mu_r(r') = r\,r'$, $r' \in R$. Si $u: M \to N$ est un monomorphisme pur, alors $\bar{\bar{u}}: \bar{\bar{M}} \to \bar{\bar{N}}$ est un monomorphisme de D(R); $\bar{\bar{M}}$ etant injectif $\bar{\bar{u}}$ est inversible à gauche. L'inverse à gauche correspondant a la forme $\bar{\bar{v}}$ pour un homomorphisme $v: N \to M$. Ici l'on a $v \circ u = 1_M$, autrement dit u est un monomorphisme scindé, et il s'ensuit que M est pur-injectif.

Réciproquement, soit F un objet de D(R) de la forme $F = \bar{\bar{M}}$, où M est un R-module pur-injectif. Puisque D(R) a suffisament d'objets injectifs, $F = \bar{\bar{M}}$ peut être plongé dans un objet injectif. En utilisant le résultat plus haut nous concluons qu'il existe un monomorphisme pur $u: M \to N$ tel que $\bar{\bar{N}}$ soit un objet injectif de D(R). Comme M est pur-injectif, u a un inverse à gauche v; mais la relation $v \circ u = 1_M$ implique que $\bar{\bar{v}} \circ \bar{\bar{u}} = 1_{\bar{\bar{M}}}$. Par conséquent, $\bar{\bar{M}}$ étant facteur direct d'un objet injectif est lui-même un objet injectif.

Bien que nous n'en ayons pas besoin danc ce papier nous faisons mention de la généralisation suivante, dont nous omettons la démonstration.

Proposition 1.3. Pour tout couple de R-modules à gauche M et N on a un isomorphisme $\mathrm{Ext}^n_{D(R)}(\bar{\bar{M}}, \bar{\bar{N}}) \cong \mathrm{Pext}^n_R(M,N)$. Ici $\mathrm{Pext}^n_R(M,N)$ est défini comme le $n^{\text{ième}}$ groupe du complexe $\mathrm{Hom}_R(M,\underline{Q})$, ou $\underline{Q}$ est une résolution pur-injective de N.

Nous terminons cette section par le résultat suivant

Proposition 1.4. Pour tout R-module à gauche M et tout objet cohérent F de D(R) on a $\mathrm{Ext}^n_{D(R)}(F,\bar{\bar{M}}) = 0$ pour chaque $n > 0$.

Démonstration. F a une résolution projective dans la catégorie D(R) de la forme

$$0 \to \overline{X}_2 \to \overline{X}_1 \to \overline{X}_o \to F \to 0 \qquad (***)$$

qui provient d'une suite exacte

$$X_o \to X_1 \to X_2 \to 0$$

de Pf(R).

En appliquant le foncteur $\mathrm{Hom}_{D(R)}(-,\overline{\overline{M}})$ à (***) on obtient l'assertion de la proposition car $\overline{\overline{M}}$ est exact à droite.

2. Sous-groupes de définition finie.

Avant de formuler nos premiers résultats concernant la L-dimension et les foncteurs dérivés de $\varprojlim$ nous aurons besoin de la notion de "sous-groupe de définition finie" qui s'avère tres utile sous plusieurs rapports.

Pour un quotient cohérent C de $\overline{R} = \mathrm{Hom}_R(R,-)$ dans la catégorie D(R) et un R-module à gauche M on a un monomorphisme $\mathrm{Hom}_{D(R)}(C,\overline{\overline{M}}) \to \mathrm{Hom}_{D(R)}(\overline{R},\overline{\overline{M}})$ de groupes abéliens. En identifiant $\mathrm{Hom}_{D(R)}(\overline{R},\overline{\overline{M}})$ et le groupe abélien sous-adjacent de M on peut regarder $\mathrm{Hom}_{D(R)}(C,\overline{\overline{M}})$ comme un sous-groupe additif de M. <u>On dit qu'un sous-groupe additif de M est de R-définition finie s'il intervient de cette manière lorsque C parcourt les quotients cohérents de $\overline{R}$.</u>

Nous allons donner une déscription plus explicite de cette notion.

Proposition 2.1. Soit V un sous-groupe additif du R-module à gauche M. Pour que V soit de R-definition finie dans M, il faut et il suffit qu'il exist deux familles finies d'éléments b_i, $1 \leq i \leq \mu$, b_{ij}, $1 \leq j \leq \nu$, dans R vérifiant la condition suivante: V est l'ensemble des $m \in M$ tels que le système linéaire

$$b_i m = \sum_{j=1}^{\nu} b_{ij} x_j \quad (1 \leqq i \leqq \mu)$$

admette au moins une solution $(x_j) \in P^{\nu}$, $(1 \leq j \leq \nu)$

Démonstration. Soit C un quotient cohérent de $\overline{R}$. Il existe une suite exacte de $D(R)$

$$\overline{A} \underset{\overline{\varphi}}{\to} \overline{R} \to C \to 0 \qquad (*)$$

où A est un R-module à droite de présentation finie et $\overline{\varphi}$ provient d'un R-homomorphisme $\varphi: R \to A$. Il existe une suite exacte courte de $Pf(R)$

$$0 \to G \underset{\alpha}{\to} L \underset{\beta}{\to} A \to 0$$

où L est un R-module à droite libre de type fini et G est un R-module à droite de type fini. Soient (e_i), $1 \leq i \leq \mu$, une base de L et les éléments $y_j = \sum_{i=1}^{\mu} e_i b_{ij}$, $1 \leqq j \leqq \nu$, $b_{ij} \in R$, un système de générateurs de G.

A partir de (*) nous obtenons une suite exacte

$$0 = \mathrm{Hom}_{D(R)}(C,\overline{\overline{M}}) \to \mathrm{Hom}_{D(R)}(\overline{R},\overline{\overline{M}}) \to \mathrm{Hom}_{D(R)}(\overline{A},\overline{\overline{M}})$$

$$R \otimes_R M \underset{\varphi \otimes 1_M}{\to} A \otimes_R M .$$

Donc il reste à calculer le groupe additif $V = \mathrm{Ker}(\varphi \otimes 1_M) = \{m \in M | (\varphi \otimes 1_M)(1 \otimes m) = 0\}$. Soit $\widetilde{\varphi}: R \to L$ un R-homomorphisme tel que $\varphi = \beta\widetilde{\varphi}$. Si $\widetilde{\varphi}(1) = \sum_{i=1}^{\mu} e_i b_i$, $b_i \in R$, alors
$V = \{m \in M | (\varphi \otimes 1_M)(1 \otimes m) \in \mathrm{Im}(\widetilde{\varphi} \otimes 1_M) = \{m \in M | \sum_{i=1}^{\mu} e_i b_i) \otimes m =$
$\sum_{j=1}^{\nu} \left(\sum_{i=1}^{\mu} e_i b_{ij}\right) \otimes \overline{m}_j$ pour des $\overline{m}_j \in M$, $1 \leq j \leq \nu\}$
$\{m \in M | b_i m_i = \sum_{j=1}^{\nu} b_{ij} m_j$, $1 \leq i \leq \mu$, pour des $\overline{m}_j \in M$, $1 \leqq j \leqq \nu\}$.

Ceci prouve la nécessité de la proposition 2.1. La suffisance peut être démontrée de façon analogue. Nous en laissons les détails au lecteur.

Remarque 2.2. La déscription des sous-groupe de définition finie donnée en [11] n'est qu'une transformation immédiate de la caractérisation de la proposition 2.1.

Nous allons donner maintenant une déscription plus explicite des sous-groupes de définition finie en certains cas. Mais d'abord, pour illustrer les notions nous mentionnous des exemples triviaux.

Exemple 1. Soient R un anneau intègre et K son corps des fractions. Si M est un K-module quelconque, (0) et M sont les seuls sous-groupes de définition finie. Ici M peut être indifféremment consideré comme un R-module ou un K-module.

Exemple 2. Soient R un anneau commutatif et I un idéal de R. Si I est de type fini, I est un sous-groupe de définition finie. La réciproque n'est pas vraie. (L'annulateur d'un élément arbitraire est un sous-groupe de définition finie.)

Exemple 3. Soient R un anneau quelconque et M un R-module à gauche. Pour tout idéal à droite de type fini I le sous-groupe additif IM de M est de définition finie.

Dans cet ordre d'idées nous allons prouver le résultat suivant:

Proposition 2.3. Soient R un anneau cohérent à droite et P un R-module à gauche plat; alors les sous-groupes de R-définition finie de P sont précisément ceux de la forme IP, I parcourant les idéaux à droite de type fini de R.

Démonstration. D'après l'exemple 3 il suffit de vérifier que tout sous-groupe V de R-définition finie de P a la forme IP pour un idéal à droite de type fini I de R.

En vertu de la proposition 2.1 il existent des éléments b_i, $1 \leqq i \leqq \mu$, b_{ij}, $1 \leqq j \leqq \nu$, de R tels que V soit l'ensemble des éléments $p \in P$ pour lesquels les équations

$$b_i p = \sum_{j=1}^{\nu} b_{ij} x_j \quad (1 \leqq i \leqq \mu)$$

admettent une solution (x_j), $1 \leqq j \leqq \nu$, $\in P^{\nu}$.

Soient L un R-module à droite libre de type fini de base (e_i), $1 \leqq i \leqq \mu$, et $\varphi: R \to L$ le R-homomorphisme défini par $\varphi(r) = \sum_{i=1}^{\mu} e_i b_i r$. De plus considérons le sous-module de type fini G de L engendré par $\left(\sum_{i=1}^{\mu} e_i b_{ij}\right)$, $1 \leqq j \leqq \nu$, et soit κ l'homomorphisme canonique $L \to L/G$. Puisque L/G est de présentation finie et que R est cohérent à droite, le noyau $\mathrm{Ker}(\kappa\varphi)$ est un idéal à droite de type fini I de R.

Soient α, (resp. β), le monomorphisme naturel de I dans R, (resp. de G dans L), et soit ψ l'isomorphisme canonique $P \to R \otimes_R P$.

De l'exactitude de la suite

$$G \otimes_R P \xrightarrow[\beta \otimes 1_P]{} L \otimes_R P \xrightarrow[\kappa \otimes 1_P]{} (L/G) \otimes_R P$$

on déduit:

$\mathrm{Ker}(\kappa\varphi \otimes 1_P)\psi = \{p \in P \mid \varphi \otimes 1_P)(1 \otimes p) \in \mathrm{Im}(\beta \otimes 1_P)\} =$
$\{p \in P \mid \left(\sum_{i=1}^{\mu} e_i b_i\right) \otimes p = \sum_{j=1}^{\nu} \left(\sum_{i=1}^{\mu} e_i b_{ij}\right) \otimes p_j$ pour des $p_j \in P$,
$1 \leq j \leq \nu\} = \{p \in P \mid b_i p = \sum_j b_{ij} p_j$, $1 \leq i \leq \mu$, pour des $p_j \in P$,
$1 \leq j \leq \nu\} = V$.

Comme P est plat, la suite

$$I \otimes_R P \xrightarrow[(\alpha \otimes 1_P)]{} R \otimes_R P \xrightarrow[(\kappa\varphi \otimes 1_P)]{} (L/G) \otimes_R P$$

est exacte, et l'on en conclut $V = \mathrm{Ker}(\kappa\varphi \otimes 1_P) \circ \psi) = \mathrm{Im}(\psi \circ^{-1} (\alpha \otimes 1_P)) = IP$. Ceci achève la démonstration de la proposition 2.3.

Nous terminons cette section par une remarque dont nous aurons besoin au paragraphe suivant.

Remarque 2.4. Soient $\varphi_1 : R \to A_1$ et $\varphi_2 : R \to A_2$ des R-homomorphismes de R dans les R-modules à droite de présentation finie A_1 et A_2, et soient C_1 et C_2 les quotients cohérents correspondants de $\overline{R}$ dans la catégorie D(R). Si $\varphi^* : R \to A_1 \oplus A_2$ est l'application diagonale $\varphi^*(r) = (\varphi(r_1), \varphi(r_2))$ et C^* est le quotient correspondant de $\overline{R}$ on a (dans le groupe additif du R-module à gauche M) la relation:

$$\mathrm{Hom}_{D(R)}(C, \overline{\overline{M}}) = \mathrm{Hom}_{D(R)}(C_1, \overline{\overline{M}}) \cap \mathrm{Hom}_{D(R)}(C_2, \overline{\overline{M}}).$$

De plus, nous ordonnons les quotients cohérents de $\overline{R}$ en posant C' C" (C' et C" étant des quotients cohérents de $\overline{R}$) s'il existe un morphisme γ de D(R) tel que le diagramme

$$\begin{array}{ccc} \overline{R} & \to & C' \\ & \searrow_{\gamma} & \uparrow \\ & & C'' \end{array}$$

soit commutatif.

Si $\mathcal{V}$ est un ensemble filtrant décroissant de sous-groupes de R-définition finie du R-module à gauche M, les observations précédentes montrent que les quotients cohérents C de $\overline{R}$ pour lesquels $\mathrm{Hom}_{D(R)}(C, \overline{\overline{M}}) \in \mathcal{V}$ forment un système inductif filtrant.

Remarque 2.5. En [12] on a donné une caractérisation des modules qui satisfont à la condition des chaînes descendantes pour les sous-groupes de définition finie. On arrive par là aux modules M pour lesquels $\overline{\overline{M}}$ est un objet Σ-injectif de D(R).

3. La L-dimension et l'annulation de $\varprojlim^{(i)}$.

Dans cette section nous allons établir des résultats importants, qui relient la L-dimension aux foncteurs dérivés de $\varprojlim$ pour certains systèmes projectifs de modules sur un anneau donné.

<u>Théorème 3.1</u>. Soient R un anneau quelconque, M un R-module à gauche et n un entier. Alors les conditions suivantes, (i) - (iv), sont équivalentes:

(i) $L\text{-dim}_R(M) \leq n$.

(ii) Pour tout système inductif filtrant $(C_\alpha), \alpha \in I$, d'objets cohérents de la catégorie $D(R)$ et tout entier $p > n$, on a $\varprojlim^{(p)} \mathrm{Hom}_{D(R)}(C_\alpha, \overline{\overline{M}}) = 0$.

(iii) Pour tout système inductif filtrant $(F_\alpha), \alpha \in I$, de R-modules à gauche de présentation finie et tout entier $p > n$, on a $\varprojlim^{(p)} \mathrm{Hom}_R(F_\alpha, M) = 0$.

(iv) Pour tout ensemble filtrant décroissant $(V_\alpha), \alpha \in I$, de sous-groupes additifs de R-définition finie de M et tout entier $p > n$, on a $\varprojlim^{(p)} V_\alpha = 0$.

De plus, les conditions (i) - (iv) impliquent la suivante:

(v) Pour tout système projectif filtrant $(F_\alpha), \alpha \in I$, de R-modules à droite de présentation finie et tout entier $p > n$, on a $\varprojlim^{(p)}(F_\alpha \otimes_R M) = 0$.

<u>Démonstration</u>. Soit $(C_\alpha), \alpha \in I$, un système inductif d'objets cohérents de $D(R)$. En vertu de la proposition 1.4 la suite spectrale du §1 dégénère en des isomorphismes:

$$\varprojlim^{(p)} \mathrm{Hom}_{D(R)}(C_\alpha, \overline{\overline{M}}) \simeq \mathrm{Ext}^p_{D(R)}(\varinjlim C_\alpha, \overline{\overline{M}}) \qquad (*)$$

L'implication (i) $\Rightarrow$ (ii) en résulte immédiatement.

L'implication (ii) $\Rightarrow$ (iii) est une conséquence des deux lemmes suivants.

<u>Lemme 3.2</u>. Si F et M sont des R-modules à gauche on a un isomorphisme naturel

$$\mathrm{Hom}_R(F, M) \simeq \mathrm{Hom}_{D(R)}(\overline{\overline{F}}, \overline{\overline{M}}) .$$

Démonstration. Il est facile de voir directement que l'homomorphisme canonique $\mathrm{Hom}_R(F,M) \simeq \mathrm{Hom}_{D(R)}(\overline{\overline{F}},\overline{\overline{M}})$ et bijectif lorsque F est libre. On obtient le résultat général en considérant une présentation libre de F et appliquant un "diagram chasing" standard.

Lemme 3.3. Si F est un R-module à gauche de présentation finie, l'objet $\overline{\overline{F}}$ de D(R) est cohérent.

Démonstration. Si L est un R-module à gauche libre de type fini, le module dual $L^* = \mathrm{Hom}_R(L,R)$ est un R-module à droite et l'on a un isomorphisme canonique $\mathrm{Hom}_R(L^*,Y) \simeq Y \otimes_R L$ pour tout R-module à droite Y ; en particulier $\overline{L^*} = \overline{\overline{L}}$ dans D(R). Puisque F admet une présentation finie il s'ensuit que $\overline{\overline{F}}$ est un objet cohérent de D(R).

Revenons à la démonstration du théorème 3.1. De façon pareille l'implication (ii) ⇒ (v) est une conséquence de l'isomorphisme naturel $\mathrm{Hom}_{D(R)}(\overline{F},\overline{\overline{M}}) \simeq F \otimes M$, où F parcourt les R-modules à droite de présentation finie. De même on obtient l'implication (ii) ⇒ (iv) en considérant le système inductif des quotients cohérents C de $\overline{R}$ tels que $\mathrm{Hom}_{D(R)}(C,\overline{\overline{M}}) \in (V_\alpha)$, (cf. la remarque 2.4).

Pour terminer la démonstration du théorème 3.1 il suffit de prouver les implications (iii) ⇒ (i) et (iv) ⇒ (i). Pour cela on choisit une résolution injective de $\overline{\overline{M}}$ dans D(R):

$$0 \to \overline{\overline{M}} \to I_o \to I_1 \to \cdots \to I_p \to \qquad (**)$$

Pour tout $p \geqq 0$ posons $K_p = \mathrm{Ker}(I_{p+1} \to I_{p+2})$. Comme tous les objets qui apparaissent ici, sont des foncteurs exacts à droite de Pf(R) dans Ab, il s'ensuit que $I_p = \overline{\overline{I_p(R)}}$ et $K_p = \overline{\overline{K_p(R)}}$, où $I_p(R)$ et $K_p(R)$ sont munis par la structure naturelle de R-modules à gauche. $K_n(R)$ peut être écrit comme limite inductive de R-modules de présentation finie (F_β), et l'on en obtient

$K_n = \varinjlim \overline{\overline{F}}_\beta$, puisque le foncteur tensoriel commute aux limites inductives. La proposition 1.4, le lemme 3.3 et l'isomorphisme (*) impliquent que la condition (iii) entraîne $\mathrm{Ext}^{n+1}_{D(R)}(K_n,\overline{\overline{M}}) = 0$. Par décalage on en déduit $\mathrm{Ext}^1_{D(R)}(K_n,K_{n-1}) = 0$. En particulier la suite exacte de $D(R)$

$$0 \to K_{n-1} \to I_n \to K_n \to 0$$

est scindée, donc K_{n-1} est injectif et $\text{L-dim}_R\, M \leqq n$. Nous avons ainsi établi l'implication (iii) ⇒ (i).

Il reste à verifier (iv) ⇒ (i). Soit (**) une résolution injective minimale de $\overline{\overline{M}}$ et conservons les notations plus haut. Il suffit alors de vérifier que $K_n = 0$. Supposons que $K_n \neq 0$. Puisque $K_n = \overline{\overline{K_n(R)}}$, ceci entraîne $K_n(R) \neq 0$. Soit $x \neq 0$ un élément de $K_n(R)$. Soit $\mu \in \mathrm{Hom}_{D(R)}(\overline{R},K_n)$ le morphisme défini par $\mu(f) = K_n(f)x$, f étant un homomorphisme $\in \mathrm{Hom}_R(R,A)$, $A \in Pf(R)$. Ici l'on a évidemment $\mathrm{Im}\,\mu = \overline{R}/\mathrm{Ker}\,\mu \neq 0$. Tout objet de $D(R)$, en particulier $\mathrm{Ker}\,\mu$ est la réunion filtrante d'une famille (Y_α) de sous-objets de type fini. En tant que sous-objet de l'objet cohérent $\overline{R}$ Y_α est lui-même cohérent. Par suite, l'on a $\mathrm{Im}\,\mu = \varinjlim(\overline{R}/Y_\alpha)$ où les quotients $\overline{R}/Y_\alpha$ sont des objets cohérents. Les groupes $\mathrm{Hom}_{D(R)}(\overline{R}/Y_\alpha,\overline{\overline{M}})$ forment un ensemble filtrant décroissant de sous-groupes additifs de R-définition finie de M.

La condition (iv) et l'isomorphisme (*) impliquent

$$\mathrm{Ext}^{n+1}_{D(R)}(\mathrm{Im}\,\mu,\overline{\overline{M}}) = \mathrm{Ext}^{n+1}_{D(R)}(\varinjlim \overline{R}/Y_\alpha,\overline{\overline{M}}) = 0 .$$

Par décalage on en obtient $\mathrm{Ext}^1_{D(R)}(\mathrm{Im}\,\mu,K_{n-1}) = 0$. Parce que $\mathrm{Im}\,\mu \subseteqq K_n$ et que I_n est une extension essentielle de K_{n-1} (par la minimalite de la resolution injective (**)), on conclut $\mathrm{Im}\,\mu = 0$. Donc l'hypothèse $K_n \neq 0$ est contradictoire, et l'on a établi l'implication (iv) ⇒ (i).

Remarque 3.4. En général, la condition (v) n'entraîne pas les conditions (i) - (iv). Il existe des contreexemples même dans le cas où R est l'anneau des entiers $\underline{\underline{Z}}$.

Nous donnerons maintenant une déscription plus détaillée des modules M pour lesquels $L\text{-dim}\, M = 0$.

Théorème 3.5. Soient R un anneau et M un R-module à gauche. Les conditions suivantes sont équivalentes:

(i) $L\text{-dim}_R(M) = 0$, i.e. M est pur-injectif.

(ii) Pour tout système inductif filtrant $(N_\alpha)_{\alpha\in I}$ de R-modules à gauche et tout entier $p > 0$ on a $\varprojlim^{(p)} \mathrm{Hom}_R(N_\alpha, M) = 0$.

(iii) Tout système d'équations linéaires

$$\sum_{f\in J} a_{ij} x_j = m_i \quad (i \in I)$$

où les a_{ij} sont des scalaires presque tous nuls pour i fixé et les m_j sont des éléments de M) admet une solution $(x_j)_{j\in J} \in M^J$ dès que, pour chaque partie finie I' de I, le système formé des équations d'indice appartenant à I' admet une solution.

(iv) Tout ensemble filtrant décroissant de variétés $\underline{\underline{Z}}$-linéaires de M, dont les directions sont de R-définition finie dans M, est d'intersection non vide.

Démonstration. Supposons (i) vérifié et prouvons (ii). Si l'on pose $N = \varinjlim N_\alpha$ on a $\bar{\bar{N}} = \varinjlim \bar{\bar{N}}_\alpha$. Le lemme 3.2 et la suite spectrale

$$E_2^{p,q} = \varprojlim^{(p)} \mathrm{Ext}^q_{D(R)}(\bar{\bar{N}}_\alpha, \bar{\bar{M}}) \underset{p}{\Rightarrow} \mathrm{Ext}^n_{D(R)}(\bar{\bar{N}}, \bar{\bar{M}})$$

entraînent $\varprojlim^{(p)} \mathrm{Hom}_R(N_\alpha, M) = 0$ pour tout $p > 0$, c.à d. (ii) est vérifié.

Supposons (ii) vérifié et prouvons (iii). Pour une partie K de I on note N_K le conoyau de l'application R-linéaire $d_K\colon R^{(K)} \to R^{(J)}$

définie par la matrice (a_{ij}). En appliquant le foncteur $\mathrm{Hom}_R(-,M)$ on obtient une suite exacte

$$0 \to \mathrm{Hom}_R(N_K,M) \to M^J \xrightarrow[\mathrm{Hom}(d_K,1_M)]{} M^{(K)}$$

où l'image de $\mathrm{Hom}_R(d_K,1_M)$ est le sous-groupe de M^K formé des familles $(m_i)_{i\in K}$ tel que le système d'équations

$$\sum_{j\in J} a_{ij}x_j = m_i \quad (i \in K)$$

ait une solution $(x_j) \in M^J$.

Lorsque K parcourt les parties finies de I les groupes M^K, $\mathrm{Hom}_R(N_K,M)$ et les images $\mathrm{Im}(\mathrm{Hom}_R(d_K,1_M))$ forment des systèmes projectifs (avec les applications évidentes). Sous l'hypothèse (ii) on a $\varprojlim^{(1)} \mathrm{Hom}_R(N_K,M) = 0$; par conséquent l'application canonique

$$M^J \to \varprojlim \mathrm{Im}(\mathrm{Hom}_R(d_K,1_M))$$

est surjective. L'interpretation plus haut de $\mathrm{Im}(\mathrm{Hom}_R(d_K,1_M))$ implique que la condition (iii) est vérifiée.

Supposons (iii) vérifié et prouvons (iv). La condition (iv) signifie que pour tout ensemble filtrant décroissant (V_α), $\alpha \in I$, de sous-groupes additifs de R-définition finie de M, l'application canonique $M \to \varprojlim M/V_\alpha$ est surjective. Pour la vérifier, choisissons un élément $y = (m_\alpha + V\ ) \in \varprojlim M/V_\alpha$; il s'agit de trouver une solution x du système de congruences $x - m_\alpha \in V_\alpha$, $\alpha \in I$. En vertu de la proposition 2.1 chacune de ces congruences équivaut à un système fini d'équations linéaires du type considéré en (iii). D'après (iii), pour vérifier que ces systèmes ont une solution commune, il suffit de vérifier que toute conjonction finie de ces systèmes a une solution, ce qui est garanti par l'hypothèse que (V_α) est filtrant.

Finalement l'implication (iv) ⇒ (i) n'est qu'un cas spécial de l'implication (iv) ⇒ (i) du théorème 3.1.

Remarque 3.6. L'équivalence (i) $\Longleftrightarrow$ (iii) est bien connue (nous ne l'avons redémontrée ici que pour donner un exemple d'utilisation des foncteurs dérivés de $\varprojlim$). À cause de (iii) on appelle aussi algébriquement compact un module verifiant les conditions du théorème 3.5. L'equivalence (i) $\Longleftrightarrow$ (iv) est (probablement) nouvelle; elle traduit la condition de compacité algébrique pour un R-module M en termes d'une condition de complétion de M relativement à certaines topologies $\underline{Z}$-linéaires.

Nous terminons cette section par quelques conséquences immédiates du théorème 3.1.

Proposition 3.7. Soient t un entier, R un anneau quelconque et M un R-module à gauche. Si $\text{L-dim}_R(M) \leq t$, alors $\text{Ext}^n_R(Q,M) = 0$ pour tout R-module à gauche plat Q et tout entier $n > t$. En particulier, si M est algébriquement compact et Q est plat on a $\text{Exr}^n_R(Q,M) = 0$ pour tout entier $n > 0$.

Démonstration. Q peut être écrit sous la forme $Q = \varinjlim L_\alpha$ pour un système inductif (L_α) de R-modules à gauche libres de type fini (15). Il y a une suite spectrale (cf. 14)

$$E_2^{p,q} = \varprojlim{}^{(p)} \text{Ext}^q_R(L_\alpha,M) \underset{p}{\Rightarrow} \text{Ext}^n_R(Q,M),$$

qui, vu la liberté des modules L_α, dégénère en des isomorphismes

$$\varprojlim{}^{(p)} \text{Hom}_R(L_\alpha,M) \simeq \text{Ext}^p_R(Q,M).$$

L'assertion de la proposition 3.7 résulte maintenant du théorème 3.1.

Proposition 3.8. Soit R un anneau quelconque, et soit t un entier tel que la L-dimension de tout R-module à gauche plat est $\leq t$. Alors la dimension projective de tout R-module à gauche plat et $\leq t$.

Démonstration. Soit P un R-module à gauche plat et choisissons une suite exacte:

$$0 \to K \to L_{t-1} \to \cdots \to L_1 \to L_o \to P \to 0 \qquad (***)$$

où les modules $L_o, L_1, \cdots, L_{t-1}$ sont libres. K est alors automatiquement plat. Évidemment, on a pour tout R-module X

$$\mathrm{Ext}^{t+1}_R(P,X) \simeq \mathrm{Ext}^1_R(K,X) .$$

D'après la proposition 3.7 et l'hypothèse on en conclut que $\mathrm{Ext}^1_R(K,X) = 0$ pour tout R-module à gauche plat X . K peut être écrit comme quotient d'un module libre:

$$0 \to N \to L \to K \to 0$$

où L est un R-module libre et N un R-module plat. Puisque $\mathrm{Ext}^1_R(K,N) = 0$ cette suite exacte est scindée, et N est facteur direct de L , en particulier un R-module projectif. Donc (***) est une résolution de P de longueur t , et la dimension projective de P est $\leqq t$. C.Q.F.D.

4. Caractérisations explicites dans le cas cohérent.

Dans cette section nous allons préciser des résultats du paragraphe précédent dans le cas, où l'anneau R est cohérent à droite c.à d. tout idéal à droite de type fini est de présentation finie. Lorsque R est cohérent à droite il est facile de voir (et bien connu) que la catégorie Pf(R) des R-modules à droite de présentation finie est une catégorie abélienne. Avant de formuler le théorème principal de cette section nous donnerons un résultat préliminaire.

Proposition 4.1. Soient R un anneau cohérent à droite et P un R-module à gauche plat. Alors il existe un monomorphisme pur $P \to \check{P}$, ou $\check{P}$ est un R-module à gauche plat et pur-injectif (algébriquement compact).

Démonstration. Comme nous avons observé plus haut la catégorie Pf(R) est abélienne, et par suite (8) la catégorie $\mathcal{D}$ = Sex(Pf(R),Ab) est une catégorie de Grothendieck. Le foncteur T_P: Pf(R) → Ab défini par $T_P(X) = X \otimes_R P$, $(X \in Pf(R))$, appartient à $\mathcal{D}$ et possède une enveloppe injective $\check{T}$ dans $\mathcal{D}$. $\mathcal{D}$ est une sous-catégorie réflexive de D(R) puisque le foncteur canonique $\mathcal{D} \to D(R)$ admet un adjoint à gauche. Par suite $\check{T}$ est injectif en tant qu'objet de D(R) et donc, en vertu de la proposition 1.2, de la forme $\check{T}(X) = X \otimes_R \check{P}$, $(X \in Pf(R))$, $\check{P}$ étant un R-module à gauche pur-injectif. Parce que $\check{T}$ est un objet de $\mathcal{D}$, il s'ensuit que $\check{P}$ est plat. Le foncteur canonique de $\mathcal{D}$ dans D(R) préserve les monomorphismes; on en conclut qu'il existe un monomorphisme pur $P \to \check{P}$.

C.Q.F.D.

Remarque 4.2. On pourrait donner une autre démonstration de la proposition 4.1 en prouvant que le bidual $Hom_Z(Hom_Z(P,\underline{\underline{Q}}/\underline{\underline{Z}}), \underline{\underline{Q}}/\underline{\underline{Z}})$ d'un R-module à gauche plat P est plat dès que R est cohérent à droite.

Nous sommes maintenant à même de démontrer un résultat qui précise le théorème 3.1 au cas cohérent.

Théorème 4.3. Soient R un anneau cohérent à droite, P un R-module à gauche plat et n un entier. Alors les conditions suivantes sont équivalentes:

(i) $L\text{-}dim_R(P) \leqq n$.

(ii) Pour tout système projectif filtrant (F_α), $\alpha \in I$, de R-modules à droite de présentation finie et tout entier $p > n$, on a $\varprojlim^{(p)}(F_\alpha \otimes_R P) = 0$.

(iii) Pour tout système projectif filtrant (L_α), $\alpha \in I$, de R-modules à droite libres de type fini et tout entier $p > n$, on a $\varprojlim^{(p)}(L_\alpha \otimes_R P) = 0$.

(iv) Pour tout R-module à gauche plat Q et tout entier $p > n$, on a $\mathrm{Ext}_R^p(Q,P) = 0$.

Démonstration. En vertu du théorème 3.1 il suffit de prouver les implications (iii) ⇒ (iv) et (iv) ⇒ (i).

Considérons d'abord l'implication (iii) ⇒ (iv). Le R-module plat Q peut être écrit $Q = \varinjlim L_\alpha$ pour un système inductif (L_α) de R-modules à gauche libres de type fini. En utilisant une suite spectrale standard on obtient des isomorphismes

$$\varprojlim^{(p)} \mathrm{Hom}_R(L_\alpha,P) \simeq \mathrm{Ext}_R^p(Q,P) \quad , \; p > 0 \; .$$

Nous employons maintenant l'isomorphisme canonique $\mathrm{Hom}_R(L_\alpha,P) \simeq \mathrm{Hom}_R(L_\alpha,R) \otimes_R P$ et observons que pour tout $\alpha \in I$ le R-module à droite $\mathrm{Hom}_R(L_\alpha,R)$ est libre de type fini. Grâce à ces isomorphismes l'implication (iii) ⇒ (iv) est claire. Finalement supposons que (iv) est vérifiée et prouvons (i). Par usage itératif de la proposition 4.1 on conclut qu'il existe une suite universellement exacte (pur-exacte):

$$0 \to P \to I_o \to I_1 \to \cdots \to I_{n-1} \to C_n \to 0 \qquad (*)$$

où les modules I_t, $0 \leqq t \leqq n-1$, sont pur-injectifs et plats, et le module C_n est plat. Vu la proposition 3.7 on a $\mathrm{Ext}_R^m(Q,I_t) = 0$, $0 \leqq t \leqq n-1$, pour tout $m > 0$ et tout R-module à gauche plat Q. Par conséquent, sous l'hypothèse (iv) on obtient par décalage pour tout R-module plat Q:

$$\mathrm{Ext}_R^1(Q,C_n) = \mathrm{Ext}_R^{n+1}(Q,P) = 0 \qquad (**)$$

Puisque C_n est plat, il résulte de la proposition 4.1 qu'il existe une suite pur-exacte

$$0 \to C_n \to \check{C}_n \to C' \to 0 \qquad (***)$$

où $\check{C}_n$ est pur-injectif et plat et C' est plat.

D'après (**) $\mathrm{Ext}^1_R(C',C_n) = 0$, et donc (***) et scindée. Par suite, C_n en tant que facteur direct de $\check{C}_n$ est pur-injectif. Ceci entraîne que (*) est une résolution pur-injective de P de longueur n, i.e. $L\text{-}\dim_R(P) \leqq n$. C.Q.F.D.

Pour terminer ce paragraphe, nous faisons mention du cas spécial, où R est commutatif et cohérent et P = R. Alors on voit que $L\text{-}\dim_R(R) \leqq n$ si et seulement si $\varprojlim^{(p)} M_\alpha = 0$ pour tout $p > n$ et tout système projectif filtrant (M_α) de R-modules de présentation finie. En particulier, le théorème ci-dessus montre que $L\text{-}\dim_R(R) \leqq$ la dimension injective de R en tant que module sur lui-même. Donc, si R est un anneau de Gorenstein de dimension n, on a $L\text{-}\dim_R(R) \leqq n$. Pour un anneau régulier næthérien on retrouve ainsi les résultats de Roos [17, 18].

5. Catégories localement cohérentes.

Dans la section 1 nous avons introduit pour la catégorie D(R) les notions d'objet de type fini et d'objet cohérent. Ces définitions se traduisent mot à mot dans une catégorie générale de Grothendieck $\mathcal{D}$. Aussi, comme dans le cas classique, on dit qu'un objet X de $\mathcal{D}$ est noethérien (resp. artinien) si toute suite croissante (resp. décroissante) de sous-objets de X est stationnaire. Alors un objet X de $\mathcal{D}$ est noethérien si et seulement si tout sous-objet de X est de type fini.

On dit qu'une catégorie de Grothendieck est <u>localement cohérente</u> s'il existe une famille de générateurs cohérents.

Nous mentionnous le résultat suivant (Roos [19], Lazard [15])), qui nous sera utile:

Proposition 5.1. Soit $\mathcal{D}$ une catégorie localement cohérente.

(i) Pour qu'un objet X de $\mathcal{D}$ soit cohérent, il faut et il suffit que le foncteur $\mathrm{Hom}(X,-): \mathcal{D} \to (Ab)$ commute aux limites inductives filtrantes.

(ii) Tout objet de $\mathcal{D}$ est limite inductive filtrante d'objets cohérents.

(iii) Pour tout objet cohérent X de $\mathcal{D}$ et tout sous-objet Y de X, l'ensemble des sous-objets cohérents de Y est filtrant croissant de borne supérieure égale à Y. (En particulier, tout sous-objet de type fini d'un objet cohérent est lui-même cohérent.)

Pour décrire les catégories localement cohérentes nous considérons une petite catégorie abélienne $\mathcal{C}$ et la catégorie $\mathcal{D}$ = $\mathrm{Sex}(\mathcal{C},Ab)$ des foncteurs exacts à gauche de $\mathcal{C}$ dans (Ab). Comme démontré en [7], $\mathcal{D}$ est une catégorie de Grothendieck. De plus, $\mathcal{D}$ est localement cohérente ; en effet, les objets cohérents de $\mathcal{D}$ sont exactement les objets représentables $\overline{C} = \mathrm{Hom}\ (C,-)$, $(C \in \mathcal{C})$, qui forment une famille de générateurs de $\mathcal{D}$, et la correspondance $C \to \overline{C}$ fournit une équivalence entre la catégorie opposée $\mathcal{C}^{op}$ et la sous-catégorie pleine de $\mathcal{D}$ formée des objets cohérents de $\mathcal{D}$.

Réciproquement, si $\mathcal{D}$ est une catégorie localement cohérente, la sous-catégorie pleine $\tilde{\mathcal{C}}$ formée des objets cohérents de $\mathcal{D}$ est une petite catégorie abélienne telle que $\mathcal{D}$ est équivalente à $\mathrm{Sex}(\tilde{\mathcal{C}}^{op},Ab)$. (L'équivalence $\mathcal{D} \simeq \mathrm{Sex}(\tilde{\mathcal{C}}^{op},Ab)$ est obtenue par la correspondance $F \to \mathrm{Hom}\ (-,F)$, $F \in \mathcal{D}$.)

Maintenant, soit $\mathcal{C}$ une catégorie abélienne noethérienne (c.à d. tout objet de $\mathcal{C}$ est noethérien). $\mathcal{D} = \mathrm{Sex}(\mathcal{C}^{op},Ab)$ est une catégorie localement cohérente pour laquelle la sous-catégorie pleine des objets cohérents est équivalente à $\mathcal{C}$, donc noethérienne. Puisque les objets cohérents de $\mathcal{D}$ forment une famille de générateurs de $\mathcal{D}$, la catégorie $\mathcal{D} = \mathrm{Sex}(\mathcal{C}^{op},Ab)$ est localement noethérienne, i.e. une catégorie de Grothendieck ayant une famille de générateurs noethériens. D'autre part, si $\mathcal{D}$ est une catégorie localement noethérienne, $\mathcal{D}$ (en tant que catégorie localement cohérente) est équivalente à Sex($\mathcal{C}^{op}$,Ab), où $\mathcal{C}$ est la catégorie des objets cohérents de $\mathcal{D}$. La catégorie $\mathcal{D}$ étant localement noethérienne les objets cohérents sont noethériens, et par suite $\mathcal{C}$ est noethérienne et $\mathcal{C}^{op}$ est artinienne. Nous obtenons ainsi la

<u>Proposition 5.2</u>. [8] Une petite catégorie abélienne $\mathcal{C}$ est artinienne si et seulement si Sex($\mathcal{C}$,Ab) est localement noethérienne.

La catégorie D(R) introduite au §1 est localement cohérente; car les objets représentables (et cohérents) $\overline{A}$,A ∈ Pf(R), forment un système de générateurs de D(R). Si l'on note C(R) la sous-catégorie pleine des objets cohérents de D(R), les remarques ci-dessus montrent qu'il y a une équivalence D(R) ≃ Sex[(C(R))^{op},Ab]. Cette équivalence est obtenue en faisant correspondre à tout objet T de D(R) le foncteur contravariant $Hom_{D(R)}(-,T)$ de C(R) dans Ab. Nous allons étudier cette équivalence plus précisément.

<u>Lemme 5.3</u>. Pour qu'un objet T de D(R) soit exact à droite, il faut et el suffit que $Hom_{D(R)}(-,T)$ soit un foncteur (contravariant) exact de C(R) dans Ab.

<u>Démonstration</u>. Condition suffisante. À partir d'une suite exacte de Pf(R):

$$A_1 \to A_2 \to A_3 \to 0$$

nous arrivons à une suite exacte de C(R):

$$\overline{A}_3 \to \overline{A}_2 \to \overline{A}_1 \to 0 .$$

En vertu du théorème général de Yoneda il y a un isomorphisme canonique $Hom_{D(R)}(\overline{A},T) \simeq T(A)$, A ∈ Pf(R); donc, l'exactitude de

$\mathrm{Hom}_{D(R)}(-,T)$ entraîne l'exactitude de

$$TA_1 \to TA_2 \to TA_3 \to 0 .$$

Condition nécessaire. Il suffit de montrer que le foncteur dérivé à droite $R^1\mathrm{Hom}_{D(R)}(U,T)$ s'annulle pour tout U de $C(R)$ et tout objet exact à droite T de $D(R)$. Puisque U est cohérent il existe une suite de $Pf(R)$:

$$B_o \to B_1 \to B_2 \to 0$$

telle que la suite

$$0 \to \overline{B}_2 \to \overline{B}_1 \to \overline{B}_o \to U \to 0 \qquad (*)$$

soit exacte pour un morphisme convenable: $\overline{B}_o \to U$. Ici $(*)$ est une résolution projective de U dans $D(R)$, donc $R^1\mathrm{Hom}_{D(R)}(U,T)$ est le premier groupe de cohomologie du complexe:

$$0 \to \mathrm{Hom}_{D(R)}(\overline{B}_o,T) \to \mathrm{Hom}_{D(R)}(\overline{B}_1,T) \to \mathrm{Hom}_{D(R)}(\overline{B}_2,T) \to 0 \to \cdots$$

Le foncteur T étant exact à droite, l'isomorphisme de Yoneda $\mathrm{Hom}_{D(R)}(\overline{B}_i,T) \simeq T(B_i), i = 0,1,2,$ implique immédiatement $R^1\mathrm{Hom}_{D(R)}(U,T) = 0.$ C.Q.F.D.

Les objets représentables $\overline{A}$, $A \in Pf(R)$, sont exactement les objets projectifs de $C(R)$. Par conséquent on a

Lemme 5.4. La catégorie $C(R)$ a suffisamment d'objets projectifs, et la correspondance $A \to \overline{A}$ fournit une équivalence entre la sous-catégorie pleine des objets projectifs de $C(R)$ et la catégorie opposée $Pf(R)^{op}$.

Pour décrire les objets injectifs de $C(R)$ nous considérons un R-module à gauche de présentation finie M. Il y a une suite exacte

$$P_1 \to P_o \to M \to 0$$

où P_o et P_1 sont des R-modules à gauche projectifs de type fini. En général, pour un R-module à gauche P désignons par P^* le R-module à droite $\mathrm{Hom}_R(P,R)$. Alors P_o^* et P_1^* sont des R-modules à droite projectifs de type fini, et l'on a un isomorphisme canonique $\mathrm{Hom}_R(P_i^*,X) \simeq X \otimes_R P_i$, $i = 1,2$, $X \in \mathrm{Pf}(R)$. En vertu de l'exactitude à droite du foncteur produit tensoriel nous en déduisons la suite exacte:

$$\mathrm{Hom}_R(P_1^*,X) \to \mathrm{Hom}_R(P_o,X) \to X \otimes_R M \to 0 .$$

Ceci montre que $\overline{\overline{M}} = - \otimes_R M$ est un objet cohérent de $D(R)$. De plus, le lemme 5.3 implique que $\mathrm{Hom}_{D(R)}(-,\overline{\overline{M}})$ est un foncteur exact de $C(R)$ dans Ab, i.e. $\overline{\overline{M}}$ est un objet injectif de $C(R)$.

Inversement, nous allons démontrer que tout objet injectif de $C(R)$ est de la forme $\overline{\overline{M}}$, $M \in \mathrm{Pf}(R)$. D'après ce qui précède on voit facilement que tout objet cohérent U de $D(R)$ admettant une présentation

$$\overline{P}_1 \to \overline{P}_o \to U \to 0$$

où P_o et P_1 sont des R-modules à droite projectifs de type fini, est de la forme $\overline{\overline{M}}$, $M \in \mathrm{Pf}(R)$. (On prend pour M le conoyau de l'homomorphisme $P_1^* \to P_o^*$.)

En général, soit U un objet injectif de $C(R)$. En tant qu'objet de $C(R)$, U a une présentation

$$\overline{A}_1 \overset{\overline{\alpha}}{\to} \overline{A}_o \to U \to 0$$

où A_o et A_1 appartiennent à $\mathrm{Pf}(R)$. Pour l'homomorphisme correspondant $\alpha\colon A_o \to A_1$ on fabrique aisément un carré cocartésien de $\mathrm{Pf}(R)$

$$\begin{array}{ccc} P_o & \overset{\beta}{\to} & P_1 \\ \varepsilon \downarrow & & \downarrow \sigma \\ A_o & \overset{\alpha}{\to} & A_1 \end{array}$$

où les R-modules à droite P_o et P_1 sont projectifs de type fini.

Le foncteur contravariant: $A \to \bar{A}$ de Pf(R) dans C(R) transforme le carré ci-dessus dans un carré cartésien:

$$\begin{array}{ccc} \bar{P}_1 & \overset{\bar{\beta}}{\to} & \bar{P}_o \\ \bar{\varepsilon}\downarrow & & \downarrow\bar{\sigma} \\ \bar{A}_1 & \overset{\bar{\alpha}}{\to} & \bar{A}_o \end{array}$$

qui induit un monomorphisme du conoyau $\mathrm{Coker}(\bar{\alpha}) = U$ dans $\mathrm{Coker}(\bar{\beta})$. D'après les remarques précédentes on a $\mathrm{Coker}(\bar{\beta}) \simeq \bar{\bar{M}}$ pour un module convenable M de Pf(R). Puisque U est un objet injectif, U est un facteur direct de $\bar{\bar{M}}$ et par suite lui-même de la forme $\bar{\bar{N}}$, $N \in \mathrm{Pf}(R)$.

Dans la proposition suivante nous formulons les résultats obtenus.

Proposition 5.5. La catégorie C(R) a suffisamment d'objets injectifs et la correspondance $M \to \bar{\bar{M}}$ définit une équivalence entre Pf(R) et la sous-catégorie pleine des objets injectifs de C(R).

Il est maintenant facile de déduire le théorème de dualité suivant:

Théorème 5.6. ("Théorème de dualité"). Pour tout anneau R il y a une équivalence des catégories $C(R)^{op} \simeq C(R^{op})$, où R^{op} est l'anneau opposé de R.

Remarque 5.7. Dans la terminologie de Roos [19] le théorème dit que les catégories D(R) et $D(R^{op})$ sont "conjuguées".

Démonstration du théorème 5.6. Les catégories $C(R)^{op}$ et $C(R^{op})$ sont abéliennes et admettent suffisamment d'objets injectifs. En vertu des résultats précédents les sous-catégories pleines correspondantes formées des objets injectifs sont toutes deux équivalentes à Pf(R). On en conclut par [8] que $C(R)^{op}$ et $C(R^{op})$ sont équivalentes. C.Q.F.D.

5A. Une application concernant les modules de Mittag-Leffler et la dimension finitiste d'un anneau noethérien.

Rappelons qu'un R-module (à gauche) M est un module de Mittag-Leffler si tout homomorphisme de tout R-module de présentation finie dans M admet un stabilisateur (cf. [12]). D'abord nous donnerons une caractérisation exprimée dans le langage de la catégorie D(R).

Proposition 5.8. Soit R un anneau quelconque. Pour un R-module à gauche M les conditions suivantes sont équivalentes:

1) M est un module de Mittag-Leffler.

2) Pour tout R-module à gauche N de présentation finie et tout R-homomorphisme $\alpha: N \to M$ il existe un R-module à gauche G de présentation finie et un R-homomorphisme $\beta: N \to G$ tel que les morphismes $\bar{\bar{\alpha}}: \bar{\bar{N}} \to \bar{\bar{M}}$ et $\bar{\bar{\beta}}: \bar{\bar{N}} \to \bar{\bar{G}}$ de D(R) ont même noyau.

3) $\bar{\bar{M}}$ est un objet pseudo-cohérent de D(R); c.à d. tout sous-objet de $\bar{\bar{M}}$ de type fini est cohérent.

Démonstration. L'équivalence 1) $\Longleftrightarrow$ 2) est évidente. Supposons 2) vérifié et prouvons 3). Soit A un R-module à droite de Pf(R) et considérons un morphisme $\mu: \bar{A} \to \bar{\bar{M}}$. Il faut démontrer que Ker μ est un objet de type fini de D(R). Le morphisme μ se factorise à travers un R-module à gauche libre de type fini L, c.à d. il existe un morphisme $\nu: \bar{A} \to \bar{\bar{L}}$ et un R-homomorphisme $\alpha: L \to M$ tel que $\mu = \bar{\bar{\alpha}}\nu$. En vertu de l'hypothèse 2) il y a un R-module G de Pf(R) et un R-homomorphisme $\beta: L \to G$ pour lequel $\mathrm{Ker}(\bar{\bar{\alpha}}) = \mathrm{Ker}(\bar{\bar{\beta}})$. Donc, Ker μ = Ker $\bar{\bar{\beta}}\nu$ et Ker μ est de type fini puisque $\bar{\bar{G}}$ et un objet cohérent de D(R) (cf. Prop. 5.5).

Supposons 3) vérifié et considérons un R-homomorphisme $\alpha: F \to M$, $F \in Pf(R)$. L'image Im $\bar{\bar{\alpha}}$ est un sous-objet de $\bar{\bar{M}}$ de type fini, et comme $\bar{\bar{M}}$ est pseudo-cohérent, Im $\bar{\bar{\alpha}}$ est un objet cohérent. D'après la proposition 5.5 Im $\bar{\bar{\alpha}}$ peut être plongé dans un objet injectif de C(R), qui est de la forme $\bar{\bar{G}}$, $G \in Pf(R)$. Donc, il existe un R-homomorphisme $\beta: F \to G$ tel que $\mathrm{Ker}\,\overline{\alpha} = \mathrm{Ker}\,\overline{\overline{\beta}}$, et nous avons prouvé 2).

En [13] on a démontré

Théorème 5.9. Pour tout noethérien commutatif R on a $FPD(R) = K\text{-}dim(R)$, où $FPD(R)$ est la dimension finitiste de R, c.à d. le plus petit entier n tel que tout R-module de dimension projective finie soit de dimension projective $\leq n$ et $K\text{-}dim(R)$ désigne la dimension de Krull de R.

L'inégalité $FPD(R) \geq K\text{-}dim(R)$ est établie par Bass [3]. L'inégalité inverse est prouvée en [13]; nous en donnerons ici une démonstration qui nous semble plus facile que celle de [13] et qui utilise les résultats de cette section.

Comme il est montré dans [13] il suffit de voir que si M est un R-module de type dénombrable de dimension projective finie, et si

$$0 \to A \to P_{d-1} \to \cdots \to P_o \to M \to 0 \qquad (*)$$

est une résolution de M telle que P_i soit projectif de type dénombrable pour $i < d = K\text{-}dim(R)$, alors A est projectif.

Dans cette situation la dimension faible ("Tor-dimension") de M est finie et donc d'après [2] au plus égale à $\sup_{\mathfrak{p} \in \mathrm{Spec}(R)} \mathrm{prof}(R_{\mathfrak{p}}) \leq d$. On en conclut que A est plat; comme le module est de type dénombrable, il suffit de montrer que c'est un module de Mittag_Leffler [13, (2.2.2)]. D'après la proposition 5.8 il faut vérifier que $\overline{A}$ est un objet pseudo-cohérent de $D(R)$. Puisque $\mathrm{Tor}^R_{d+1}(M,-) = 0$ on obtient de (*) par décalage une suite exacte

$$0 \to \mathrm{Tor}^R_d(M,-) \to A \otimes_R (-) \to P_{d-1} \otimes_R (-)$$

qui montre qu'il suffit de prouver que le foncteur exact à gauche $\mathrm{Tor}^R_d(M,-)$ est un objet pseudo-cohérent de $D(R)$.

Pour un R-module X de type fini désignons par X' le plus grand sous-module de X de longueur finie et par $X" = X/X'$ le quotient correspondant. Si $\mathfrak{P}_i$, $1 \leq i \leq t$, sont les idéaux premiers associés a $X"$, la hauteur de chaque $\mathfrak{P}_i$ est au plus égale à d-1. Pour la dimension faible on en déduit que

$\text{Tor-dim}_{R_{\mathfrak{P}_i}}(M_{\mathfrak{P}_i}) \leq d-1$ pour $1 \leq i \leq t$, en particulier $(\text{Tor}_d^R(M,X"))_{\mathfrak{P}_i} = 0$ pour $1 \leq i \leq t$.

Ceci entraîne que $\text{Tor}_d^R(M,X") = 0$; en effet, supposons qu'il existait un élément $\xi \neq 0$, $\xi \in \text{Tor}_d^R(M,X")$. Pour tout i, $1 \leq i \leq t$, il existe un élément $s_i \in R$, $s_i \notin \mathfrak{P}_i$ tel que $s_i\xi = 0$. Par conséquent, il y a un élément $s \in \sum_{=1}^{t} Rs_i$ tel que $s \notin \mathfrak{P}_i$, $1 \leq i \leq t$. [(4), prop. 2, p. 70]. L'homothétie de $X"$, de rapport s, est alors injective, et puisque $\text{Tor}_d^R(M,-)$ est exact à gauche, l'homothétie de $\text{Tor}_d^R(M,X")$ de rapport s, est injective, ce qui donne la contradiction désirée.

Nous sommes maintenant à même de démontrer que $\text{Tor}_d^R(M,-)$ est pseudo-cohérent. Pour cela il faut prouver que le noyau de tout morphisme $\mu: \overline{B} \to \text{Tor}_d^R(M,-)$, $B \in Pf(R)$, est de type fini. Puisque $\text{Tor}_d^R(M,-)$ est exact à gauche, d'après la remarque précédente on peut supposer que B est de longueur finie. De plus on note que $\text{Ker}(\mu)$ est exact à gauche. Désignons par $\mathcal{C}$ la sous-catégorie pleine de $Pf(R)$ formée des modules de longueur finie.

Dans la catégorie $\text{Add}(\mathcal{C},\text{Ab})$ tout sous-foncteur exact à gauche de $\text{Hom}_R(B,-)$ est représentable par un module C de $\mathcal{C}$ (en vertu de la proposition 5.2). On en conclut aisément que

$$0 \to \text{Hom}_R(C,-) \to \text{Hom}_R(B,-) \underset{\mu}{\to} \text{Tor}_d^R(M,-)$$

est une suite exacte de $D(R)$.

Ceci achève la démonstration que $\text{Tor}_d^R(M,-)$, et donc $\overline{\overline{A}}$ est pseudo-cohérent.

6. Dimension injective de foncteurs exacts.

Soient $\mathcal{C}$ une petite catégorie abélienne et $\mathcal{D}$ = Sex($\mathcal{C}$,Ab) la catégorie des foncteurs exacts à gauche de $\mathcal{C}$ dans Ab. Comme déjà utilisé au paragraphe précédent $\mathcal{D}$ est une catégorie de Grothendieck. Il est facile de voir que tout objet injectif de $\mathcal{D}$ est un foncteur exact de $\mathcal{C}$ dans Ab. Si $\mathcal{C}$ est artinien il est prouvé dans [8] que tout foncteur exact de $\mathcal{C}$ dans Ab est un objet injectif de $\mathcal{D}$. Inversement, si tout foncteur exact de $\mathcal{C}$ dans Ab est un objet injectif de $\mathcal{D}$ il résulte aisément de [8] que $\mathcal{D}$ est localement noethérien et donc d'après la proposition 5.2, $\mathcal{C}$ est une catégorie artinienne. Nous donnerons une formulation différente de ce fait.

Pour la catégorie $\mathcal{C}$ on pose

$$d(\mathcal{C}) = \sup(\dim.\mathrm{inj}_{\mathcal{D}}(T))$$

la borne supérieure étant prise sur l'ensemble des foncteurs <u>exacts</u> $T: \mathcal{C} \to \mathrm{Ab}$.

Les résultats ci-dessus peuvent être reformulés comme suit:

<u>Théorème 6.1</u>. Une petite catégorie abélienne $\mathcal{C}$ est artinienne si et seulement si $d(\mathcal{C}) = 0$.

Nous allons généraliser la partie "seulement si" du théorème 6.1 en considérant des catégories dont les objets satisfont à une condition plus faible pour les suites décroissantes de leurs sous-objets.

<u>Définition 6.2</u>. Soient n un entier $\geqq - 1$ et X un objet de la petite catégorie abélienne $\mathcal{C}$. On dit que X est $\aleph_n$-artinien si tout ensemble filtrant décroissant de sous-objets de X admet une partie cofinale de cardinal $\aleph_n$.

(Pour $n = -1$ on retrouve la notion d'objet artinien en vertu de la convention générale que les ensembles finis sont de cardinal $\aleph_{-1}$.)

Avant de formuler notre résultat principal de cette section concernant les objets $\mathcal{K}_n$-artiniens nous aurons besoin des considérations préliminaires.

Soit $T: \mathcal{C} \to Ab$ un foncteur exact. Comme $\mathcal{D} = Sex(\mathcal{C}, Ab)$ est une catégorie de Grothendieck, tout objet de $\mathcal{D}$ admet une enveloppe injective. On peut donc construire une résolution injective minimale de T dans $\mathcal{D}$, i.e. une longue suite exacte

$$0 \to T \to I_o \to I \to \cdots \to I_n \to \cdots \qquad (*)$$

telle que I_o soit une enveloppe injective de T et que I_{n+1} soit une enveloppe injective de $T_n = Ker(I_{n+1} \to I_{n+2})$ pour tout entier $n \geq 0$. Puisque T est un foncteur exact, la suite (*) de foncteurs de $\mathcal{C}$ dans Ab est "exacte par points", i.e. le complexe de groupes abéliens

$$0 \to T(X) \to I_o(X) \to I_1(X) \to I_n(X) \to \qquad (**)$$

est acyclique pour tout objet de $\mathcal{C}$. De plus, I_n et T_n sont des foncteurs exacts pour tout entier n. Compte tenu de l'isomorphisme canonique de Yoneda $Hom_{\mathcal{D}}(\overline{X}, T) \simeq T(X)$ pour tout objet X de $\mathcal{C}$, l'exactitude de (**) implique que $Ext^n_{\mathcal{D}}(\overline{X}, T) = 0$ pour tout entier $n > 0$.

Maintenant nous sommes à même de formuler

<u>Théorème 6.3</u>. Avec les notations ci-dessus on a $T_{n+1}(A) = 0$ pour tout objet $\mathcal{K}_n$-artinien A de $\mathcal{C}$ et tout entier $n \geq -1$.

Avant de prouver ce théorème nous insérons un lemme.

<u>Lemme 6.4</u>. Soient A un objet de $\mathcal{C}$ et T un foncteur exact de $\mathcal{C}$ dans Ab. Si ξ est un élément de $T(A)$, on désigne par μ le morphisme de Yoneda dans $\mathcal{D} = Sex(\mathcal{C}, Ab)$ de $\overline{A}$ vers T défini par $\mu_X(f) = Tf(\xi)$ où $f \in Hom_{\mathcal{C}}(A, X)$ et X parcourt les objets de . Alors $Im\ \mu = \varinjlim \overline{B}_\alpha$ où B_α parcourt l'ensemble filtrant décroissant de sous-objets de A pour lesquels $\xi \in Im(Ti_\alpha)$, i_α étant le

monomorphisme canonique: $B_\alpha \to A$.

Démonstration du lemme 6.4. On voit aisément que les objets B_α décrits dans le lemme forment un ensemble filtrant décroissant. De plus, évidemment il suffit de montrer que $\operatorname{Ker} \mu = \varinjlim(\overline{A/B_\alpha})$.

Si l'on note par κ_α le morphisme canonique $A \to A/B_\alpha$ on définit un homomorphisme

$$\varphi_X\colon \varinjlim \operatorname{Hom}_{\mathcal{C}}(A/B_\alpha,X) \to \operatorname{Ker} \mu_X, \quad X \in \mathcal{C},$$

en posant $\varphi_X(f_\alpha) = f_\alpha\kappa_\alpha$ pour $f_\alpha \in \operatorname{Hom}_{\mathcal{C}}(A/B_\alpha,X)$. En vertu de la définition des sous-objets B_α il s'ensuit que φ_X est un isomorphisme pour tout objet X de $\mathcal{C}$. Ceci termine la démonstration du lemme vu le fait que les limites inductives de foncteurs de $\operatorname{Sex}(\mathcal{C},\mathrm{Ab})$ peuvent être formées "par points" .

Démonstration du théorème 6.3. Soient A un objet $\aleph_n$-artinien de $\mathcal{C}$ et ξ un élément de $T_{n+1}A$. On désigne par μ le morphisme de $\overline{A}$ vers T_{n+1} défini au lemme 6.4. Nous allons déduire une contradiction de l'hypothèse $\operatorname{Im} \mu \neq 0$.

Comme I_{n+1} est une extension essentielle de T_n (on convient ici que $T_{-1} = T$), on a une suite exacte non-scindée

$$0 \to T_n \to G \to \operatorname{Im} \mu \to 0$$

où G est l'image réciproque de $\operatorname{Im} \mu$ dans I_{n+1}. Par conséquent $\operatorname{Ext}^1_{\mathcal{D}}(\operatorname{Im} \mu,T_n) \neq 0$, et par décalage on en conclut $\operatorname{Ext}^{n+2}_{\mathcal{D}}(\operatorname{Im} \mu,T) \neq 0$. D'après le lemme 6.4 on peut écrire $\operatorname{Im} \mu = \varinjlim \overline{B}_\alpha$ pour un ensemble filtrant décroissant de sous-objets $\{B_\alpha\}$ de A. En vertu de la remarque précédent le théorème 6.3 la suite spectrale

$$\varprojlim{}^{(p)} \operatorname{Ext}^q_{\mathcal{D}}(\overline{B}_\alpha,T) \underset{p}{\Rightarrow} \operatorname{Ext}^m_{\mathcal{D}}(\varinjlim \overline{B}_\alpha,T)$$

dégénère en des isomorphismes $\operatorname{Ext}^m_{\mathcal{D}}(\operatorname{Im} \mu,T) \simeq \varprojlim{}^{(m)} T(B_\alpha)$ pour tout entier m ; en particulier on a $\operatorname{Ext}^{n+2}_{\mathcal{D}}(\operatorname{Im} \mu,T) \simeq \varprojlim{}^{(n+2)} T(B_\alpha)$.

L'objet A étant $\aleph_n$-artinien l'ensemble filtrant décroissant (B_α) a une partie cofinale de cardinal $\leq \aleph_n$. D'après un théorème de Goblot [10] ceci implique que $0 = \varprojlim^{(n+2)} T(B_\alpha) \simeq \mathrm{Ext}^{n+2}_{\mathcal{D}}(\mathrm{Im}\,\mu, T)$, d'où l'on obtient la contradiction cherchée.

Corollaire 6.5. Soit n un entier ≥ -1. Si tout objet de $\mathcal{C}$ est $\aleph_n$-artinien, on a $d(\mathcal{C}) \leq n+1$.

Démonstration. Le théorème 6.3 montre que pour tout foncteur exact $T: \mathcal{C} \to Ab$, la résolution injective minimale de T est de longueur $\leq n+1$.

C.Q.F.D.

Avant d'énoncer une autre conséquence du théorème 6.3 introduisons une définition. On considère comme toujours une petite catégorie abélienne $\mathcal{C}$. Si u est un ordinal, on définit par récurrence transfinie sur u une sous-catégorie épaisse $\mathcal{C}_u$ de la manière suivante: $\mathcal{C}_o$ est formée des objets de longueur finie de $\mathcal{C}$; si u a un prédécesseur v, $\mathcal{C}_v$ est formée des objets de $\mathcal{C}$ qui sont de longueur finie dans la catégorie quotient $\mathcal{C}/\mathcal{C}_v$; si u est un ordinal limite, $\mathcal{C}_u$ est la réunion des $\mathcal{C}_v$ pour v parcourant u.

Définition 6.6. On dit que la dimension de la petite catégorie abélienne $\mathcal{C}$ est définie si $\mathcal{C} = \mathcal{C}_u$ pour tout ordinal u assez grand. Si cette condition est vérifiée, on appelle dimension de $\mathcal{C}$ et l'on note par $\dim(\mathcal{C})$ le plus petit des ordinaux u tels que $\mathcal{C} = \mathcal{C}_u$. Dans le cas contraire on écrit $\dim(\mathcal{C}) = \infty$.

Remarque 6.7. Si R est un anneau noethérien à gauche et $\mathcal{C}$ est la catégorie des R-modules à gauche de type fini, $\dim(\mathcal{C})$ coïncide avec la dimension de Krull à gauche (au sens de Gabriel) de R.

On montre aisément que, si $\mathcal{C}$ est artinienne, $\dim(\mathcal{C})$ coïncide avec la dimension de Grull-Gabriel de la catégorie localement noethérienne $\mathcal{D} = \mathrm{Sex}(\mathcal{C}, Ab)$, cf. §5). Il est clair que $\dim(\mathcal{C}) = \dim(\mathcal{C}^{op})$.

Théorème 6.8. Soient $\mathcal{C}$ une petite catégorie abélienne et T un foncteur exact de $\mathcal{C}$ dans Ab. Soit

$$0 \to T \to I_o \to I_1 \to \cdots \to I_n \to \cdots$$

une résolution injective minimale de T dans $\mathrm{Sex}(\mathcal{C},\mathrm{Ab})$, et posons $T_p = \mathrm{Ker}(T_{p+1} \to T_{p+2})$, $p \geqq 0$. Alors les foncteurs T_p et I_{p+1} s'annulent sur les sous-catégories $\mathcal{C}_t$ introduites plus haut pour tout $p \geqq t$, t étant un entier $\geqq 0$.

Démonstration. En vertu du théorème 6.1 les foncteurs T_p et I_{p+1} s'annulent sur $\mathcal{C}_o$ pour tout $p \geqq 0$. Puisque tous les foncteurs T_p et I_{p+1}, $p \geqq 0$, sont exacts ils induisent des foncteurs T'_p et I'_{p+1} sur la catégorie quotient $\mathcal{C}/\mathcal{C}_o$ et

$$0 \to T'_o \to I'_1 \to \cdots \to I'_n \to \cdots$$

est une résolution injective minimale du foncteur exact T'_o dans la catégorie $\mathrm{Sex}(\mathcal{C}/\mathcal{C}_o,\mathrm{Ab})$. En considérant la catégorie $\mathcal{C}/\mathcal{C}_o$ on obtient l'assertion du théorème par récurrence sur t.

Corollaire 6.9. On a $d(\mathcal{C}) \leqq \dim(\mathcal{C})$ pour toute petite catégorie abélienne .

7. Bornes explicites de la L-dimension.

Nous allons donner des applications des résultats du paragraphe précédent pour obtenir des bornes supérieures pour la L-dimension en certains cas explicites.

Théorème 7.1. Soit R un anneau noethérien à droite, dont la dimension de Krull-Gabriel (à droite) est un entier n. Alors $\text{L-dim}(P) \leqq n$ pour tout R-module à gauche plat P.

Démonstration. Puisque P est plat, $T(X) = X \otimes_R P$, $X \in Pf(R)$, est un foncteur exact de $Pf(R)$ dans Ab. Comme R est noethérien à droite, la catégorie $Pf(R)$ est abélienne, et T est un objet de $Sex(Pf(R),Ab)$. Soit

$$0 \to T \to I_o \to I_1 \to \cdots \qquad (*)$$

une résolution injective minimale de T dans $Sex(Pf(R),Ab)$. Comme nous avons remarqué au §6 la suite (*) est exacte "par points". Donc, elle est exacte en tant que suite de foncteurs additifs de $Pf(R)$ dans Ab, c.à d. (*) est une suite exacte de la catégorie $D(R)$ (voir §1). Le foncteur canonique $Sex(Pf(R),Ab) \to D(R)$ admet un adjoint à gauche [8], par conséquent tout objet injectif de $Sex(Pf(R),Ab)$ est injectif en tant qu'objet de $D(R)$. Ceci implique que (*) est une résolution injective de T dans la catégorie $D(R)$. Puisque $\dim(Pf(R)) = n$ le corollaire 6.9 entraîne que $\dim.\text{inj.}_{D(R)} T \leqq n$. i.e. $L\text{-}\dim(P) \leqq n$. C.Q.F.D.

À l'aide de la proposition 3.8 on déduit du théorème 7.1:

Corollaire 7.2. Soit R un anneau noethérien à droite, dont la dimension de Krull-Gabriel (à droite) est un entier n. Alors la dimension projective de tout R-module à gauche plat est $\leqq n$.

Remarque. Dans l'assertion de ce corollaire la notion de L-dimension ne figure pas. Nous ne connaissons aucune démonstration directe du corollaire. Dans le cas commutatif le résultat était montré dans [13], mais la preuve de [13] ne s'étend pas au cas non-commutatif, parce qu'elle utilise le fait que la dimension de Krull d'un anneau commutatif noethérien R coïncide avec la dimension finitiste FPD(R).

Le corollaire suivant n'est qu'un cas particulier du théorème 7.1 (cf. le théorème 3.5):

Corollaire 7.3. Si R est un anneau artinien à droite, tout R-module à gauche plat (projectif) est algébriquement compact.

Donnons encore une application des résultats de §6.

<u>Théorème 7.4</u>. Soient R un anneau noethérien à droite et (M_α) un système projectif filtrant de R-modules à droite de type fini. Si la dimension de Krull-Gabriel de tout R-module M_α est $\leqq n$, pour un entier n fixe, alors $\varprojlim^{(i)} M_\alpha = 0$ pour $i > n$. En particulier, si la dimension de Krull-Gabriel à droite de R est un entier d, alors $\varprojlim^{(i)} M_\alpha = 0$ pour tout système projectif filtrant (M_α) de R-modules à droite de type fini.

<u>Démonstration</u>. Soit T le foncteur "oublieux" de la catégorie $\mathcal{C} = \mathrm{Pf}(R)$ dans Ab. Nous considérons une résolution injective minimale de T dans $\mathrm{Sex}(\mathcal{C},\mathrm{Ab})$

$$0 \to T \to I_0 \to I_1 \to \cdots$$

et posons $T_p = \mathrm{Ker}(I_{p+1} \to I_{p+2})$ pour $p \geqq 0$. Chacun des groupes abéliens $T_p(R), I_p(R)$ et $T(R)$ peut être muni de facon canonique d'une structure de R-module à gauche. (De cette manière $T(R) = R$ obtient sa structure donnée en tant que R-module à gauche.) Les foncteurs ci-dessus étant exacts, il existe des isomorphismes naturels $T_p(X) \simeq X \otimes_R T_p(R)$, resp. $I_p(X) \simeq X \otimes_R I_p(R)$, $X \in \mathcal{C}$. En vertu de l'hypothèse sur le système (M_α) on a $M_\alpha \in \mathcal{C}_n$ pour chaque α, (on utilise toujours les notations du §6). Donc, grâce au théorème 6.8 on a $T_p(X) = I_{p+1}(X) = 0$ pour tout $p > n$, et l'on obtient une suite exacte de systèmes projectifs:

$$0 \to (M_\alpha) \to (M_\alpha \otimes_R I_0(R)) \to (M_\alpha \otimes_R I_1(R)) \to \cdots \to (M_\alpha \otimes_R I_n(R)) \to 0 \qquad (**)$$

Puisque tout objet injectif de $\mathrm{Sex}(\mathcal{C},\mathrm{Ab})$ est injectif en tant qu'objet de la catégorie $D(R)$, la proposition 1.2 montre que $I_p(R)$ est pur-injectif (algébriquement compact) pour $p \geqq 0$. Par suite, la proposition 3.1 implique que $\varprojlim^{(i)}(M_\alpha \otimes_R I_p(R)) = 0$ pour $i > 0$ et $p \geqq 0$. De la suite exacte (**) on donc obtient par décalage que

$\varprojlim^{(i)} M_\alpha = 0$ pour tout $i > n$. C.Q.F.D.

Le corollaire suivant n'est qu'une traduction dans le langage des anneaux commutatifs:

Corollaire 7.5. Soient R un anneau commutatif noethérien et (M_α) un système projectif filtrant de R-modules de type fini. Alors $\varprojlim^{(i)} M_\alpha = 0$ pour tout $i > \sup(\dim(\operatorname{supp} M))$. En particulier $\varprojlim^{(i)} M_\alpha = 0$ si $i > \text{K-dim}(R)$.

De même on obtient pour les anneaux commutatifs:

Proposition 7.6. Soient R un anneau commutatif noethérien et P un R-module plat. Si

$$0 \to P \to A_o \to A_1 \to \cdots \to A_n \to \cdots$$

est une résolution pur-injective minimale de P, alors ${}_{\mathfrak{a}}A_p = A_p$ pour tout ideal $\mathfrak{a}$ de R pour lequel $\dim \mathfrak{a} \leqq p-1$, $(p \geqq 1)$.

Les applications suivantes des résultats du paragraphe 6 concernent des rapports entre L-dimension et certains cardinaux associés a l'anneau. D'abord nous allons introduire des notions emsemblistes pour une catégorie. Soit $\aleph$ un cardinal infini. Un ensemble ordonné I est dit $\aleph$ -filtrant si toute partie de I de cardinal $\aleph$ est majorée, i.e. pour toute partie I' avec $|I'| \leqq \aleph$ il existe un element β de I tel que $\alpha \leqq \beta$ pour chaque $\alpha \in I'$. Nous considérons une catégorie de Grothendieck $\mathcal{D}$ localement de type fini, i.e. $\mathcal{D}$ possède une famille de générateurs de type fini (par exemple la catégorie $D(R)$). Tout objet de $\mathcal{D}$ est réunion filtrante de ses sous-objets de type fini. Un objet M de $\mathcal{D}$ est dit de type $\aleph$ si tout ensemble $\aleph$ -filtrant de sous-objets de M, de borne supérieure M, contient M. Il revient au même que dire que M est la réunion d'un ensemble filtrant, ayant au plus le cardinal $\aleph$, de sous-objets de type fini.

La démonstration du lemme suivant sera laissée au lecteur.

Lemme 7.7. Soit M un objet d'une catégorie de Grothendieck localement de type fini. Les conditions suivantes sont équivalentes:

(i) Tout sous-objet de M est de type $\aleph$.

(ii) Tout ensemble filtrant croissant de sous-objets de M a une partie cofinale de cardinal $\aleph$.

Définition 7.8. Un objet d'une catégorie de Grothendieck localement de type fini est dit $\aleph$-noethérien s'il satisfait aux conditions du lemme 7.7.

Pour la catégorie $\mathcal{D}$ on voit facilement que la sous-catégorie pleine formée des objets $\aleph$-noethériens est épaisse et stable par sommes directes indexées par . *On dit que $\mathcal{D}$ est localement $\aleph$-noethérienne si elle est engendrée par ses objets $\aleph$-noethériens.* Dans une catégorie localement $\aleph$-noethérienne tout objet de type $\aleph$ est $\aleph$-noethérien.

Théorème 7.9. Pour un anneau quelconque R les conditions suivantes sont équivalentes:

(i) D(R) est localement $\aleph$-noethérienne.

(ii) Le foncteur $\bar{R} = \mathrm{Hom}_R(R,-)$ de Pf(R) dans Ab est un objet $\aleph$-noethérien de D(R).

(iii) Le foncteur $\bar{M} = \mathrm{Hom}_R(M,-)$ de Pf(R) dans Ab est un objet $\aleph$-noethérien de D(R) pour tout R-module à droite de présentation finie M.

(iv) Pour tout objet cohérent T de D(R) tout ensemble filtrant croissant de sous-objets cohérents de T dans D(R) a une partie cofinale de cardinal $\aleph$.

(v) Tout R-module à gauche est réunion filtrante de sous-modules purs admettant une $\aleph$-présentation.

Les conditions (i)-(v) sont vérifiées si card(R) $\leq \aleph$.

<u>Démonstration</u>. L'équivalence des conditions (i)-(iv) est une conséquence de la proposition 5.1 et des remarques précédentes. Nous laissons la démonstration au lecteur.

Supposons maintenant (i)-(iv) vérifiés et prouvons (v). Soit A un R-module à gauche; nous allons voir que toute partie X de A telle que $\mathrm{card}(X) \leq \aleph$ est contenue dans un sous-module pur de A admettant une $\aleph$ -présentation. A peut être écrit comme la limite d'un système inductif (F_α), $\alpha \in I$, de R-modules à gauche de présentation finie. Soit S l'ensemble des parties de I, filtrant pour l'ordre induit, de cardinal $\leq \aleph$. C'est facile de voir que S est $\aleph$ -filtrant. Pour tout élément $J \in S$ on pose $A_J = \varinjlim_{\alpha \in J} F_\alpha$. Alors $A = \varinjlim_{J \in S} A_J$ et $\overline{\overline{A}} = \varinjlim_{J\in S} \overline{\overline{A}}_J$. De plus, soit $Q_J = \mathrm{Ker}(\overline{A}_J \to \overline{\overline{A}}_J)$. Si J et K sont deux éléments de S tels que $J \subseteq K$, on pose $Q_{JK} = \mathrm{Ker}(\overline{\overline{A}}_J \to \overline{\overline{A}}_K)$.

Sous l'hypothèse (i)-(iv) $\overline{\overline{A}}_J$ est $\aleph$ -noethérien pour tout $J \in S$; donc Q_J est de type $\aleph$. Comme $Q_J = \bigcup_{K \supseteq J} Q_{JK}$ et S est $\aleph$ -filtrant, on a $Q_J = Q_{JK}$ pour un K convenable. Pour tout $J \in S$, choisissons un majorant $s(J)$ de J tel que $Q_J = Q_{J,s(J)}$. Soit J_o un élément de S assez grand pour que X soit contenu dans la réunion des images des F_α $(\alpha \in J_o)$. On pose par récurrence $J_{n+1} = s(J_n)$ et $K = \bigcup_{n \in \mathbb{N}} J_n$; évidemment K est un élément de S. D'après la construction de K il est immédiat que $Q_K = 0$. Donc $A_K \to A$ est un monomorphisme pur; d'autre part A_K admet manifestement une $\aleph$ -présentation, et son image dans A contient X, cqfd.

Inversement, supposons (v) vérifié; nous allons voir que tout sous-objet Y de $\overline{R}$ est de type $\aleph$. L'enveloppe injective de $\overline{R}/Y$ est de la forme $\overline{\overline{M}}$, où M est un R-module à gauche pur-injectif. Il existe un diagramme commutatif

$$\begin{array}{ccc} \overline{R} & \overset{\mu}{\rightarrow} & \overline{\overline{M}} \\ \downarrow & \nearrow \nu & \\ \overline{R}/Y & & \end{array}$$

où μ est le monomorphisme canonique de $\overline{R}/Y$ dans $\overline{\overline{M}}$ et μ est le morphisme de Yoneda $\mathrm{Hom}_R(R,X) \underset{\mu_x}{\rightarrow} X \otimes_R M$, $\mu_x(u) = u(1) \otimes m$ $u \in \mathrm{Hom}_R(R,X)$ et m est un élément de M. Par hypothèse il existe un sous-module pur N de M contenant m et admettant une $\aleph$ -présentation; on peut donc trouver un système inductif (F_α), $\alpha \in I$, de R-modules à gauche de présentation finie de limite N tel que $\mathrm{card}(I) \leqq \aleph$. Il existe un $\alpha \in I$ et un élément $m_\alpha \in F_\alpha$ tel que $m = f_\alpha(m_\alpha)$, où f_α est l'application canonique de F_α dans $N = \varinjlim F_\alpha$. Si $f_{\beta\alpha}$, $\beta > \alpha$, désigne l'application canonique $F_\alpha \rightarrow F_\beta$, on pose $m_\beta = f_{\beta\alpha}(m_\alpha)$. Pour $\beta > \alpha$, soit μ_β le morphisme $\overline{R} \rightarrow \overline{\overline{F}}_\beta$ défini par $\mu_\beta(v) = v(1) \otimes m_\beta$, $v \in \mathrm{Hom}_R(R,X)$. Alors $\mathrm{Ker}(\mu_\beta)$ est on objet de type fini de $D(R)$ et le fait que $Y = \mathrm{Ker}(\mu) = \bigcup_{\beta \geqq \alpha} \mathrm{Ker}\, \mu_\beta$ montre que Y est de type $\aleph$, cqfd.

Finalement, si $\mathrm{card}(R) \leqq \aleph$ il est clair que l'ensemble des objets cohérents de $D(R)$ est de cardinal $\leqq \aleph$, donc (iv) est vérifié.

<u>Théorème 7.10</u>. Soir R un anneau satisfaisant aux conditions équivalentes du théorème 7.9 avec $\aleph = \aleph_n$, n étant un entier $\geqq 0$. (Par exemple R peut être un anneau de cardinal $\leqq \aleph_n$.) Alors L-dim $M \leqq n+1$ pour tout R-module à gauche M. En particulier, la dimension projective de tout R-module à gauche plat est $\leqq n+1$.

<u>Démonstration</u>. Pour tout R-module à gauche M le foncteur $\overline{\overline{M}} = - \otimes_R M$ est un objet de $D(R)$, et en vertu de l'équivalence $D(R) \simeq \mathrm{Sex}(C(R)^{op},\mathrm{Ab})$ (cf. §5) la dimension injective de $\overline{\overline{M}}$ dans $D(R)$ est égale à la dimension injective de $\mathrm{Hom}_{D(R)}(-,\overline{\overline{M}})$ dans $\mathrm{Sex}(C(R)^{op},\mathrm{Ab})$. Puisque $\overline{\overline{M}}$ est exacte à droite, le foncteur

$\mathrm{Hom}_{D(R)}(-,\overline{\overline{M}})$ est exact (voir le lemme 5.3). La condition (iv) implique que la catégorie $C(R)^{op}$ est $\aleph_n$-artinienne. Le fait que L-dim $M \leqq n+1$ est maintenant une conséquence du corollaire 6.5.

Parfois la propriété suivante des anneaux décrits au théorème 7.9 est utile. Pour un cardinal infini $\aleph$ un R-module à gauche M est dit $\aleph$ -algébriquement compact si tout système d'équations linéaires

$$\sum_j a_{ij}x_j = m_i \ (i \in I), \ (j \in J), \ \mathrm{card}(J) \leqq \aleph,$$

où les a_{ij} sont des scalaires presque tous nuls pour i fixé et les m_i sont des éléments de M, admet une solution $(x_j)_{j\in J} \in M^J$ dès que, pour chaque partie finie I' de I, le système formé des équations d'indice appartenant a I', admet une solution. En utilisant les arguments de la démonstration du théorème 3.5 on voit que M est $\aleph$ -algébriquement compact si et seulement si l'application canonique $M \to \varprojlim_{\alpha\in I} M/V_\alpha$ est surjective pour tout ensemble filtrante décroissant (V_α), $\alpha \in I$, $\mathrm{card}(I) \leqq \aleph$, de sous-groupes additifs de R-définition finie de M. Évidemment cette condition signifie que $\varprojlim^{(1)} V_\alpha = 0$ pour tout ensemble filtrant décroissant de sous-groupes de R-définition de M, où $\mathrm{card}(I) \leqq \aleph$.

Maintenant, soit R un anneau satisfaisant aux conditions du théorème 7.9, et considérons un ensemble filtrant décroissant (V_α) de sous-groupes additifs de R-définition finie du R-module M à gauche $\aleph$ -algébriquement compact. Les sous-objets cohérents (Y) de $\overline{R}$ dans D(R) pour lesquels $\mathrm{Hom}_{D(R)}(\overline{R}/Y,\overline{\overline{M}}) \in (V_\alpha)$, forment un ensemble filtrant décroissant. Donc, grâce à la condition (iv) du théorème 7.9 l'ensemble (Y) a une partie cofinale du cardinal $\leqq \aleph$, et par conséquent (V_α) a une partie cofinale de cardinal $\leqq \aleph$. En utilisant le théorème 3.1 et le fait que le foncteur $\varprojlim^{(1)}$ s'annulle lorsqu'il s'annulle pour un sous-système cofinale du système donné, on obtient le résultat suivant:

Théorème 7.11. Soit R un anneau satisfaisant aux conditions équivalentes du théorème 7.9. Alors tont R-module à gauche $\aleph$ -algébriquement compact est algébriquement compact, (pur-injectif).

Remarque 7.12. Le résultat ci-dessus est d'un interêt particulier dans le cas $\aleph = \aleph_0$. En effet, le théorème 7.11 implique que tout ultraproduit, par rapport à un ultrafiltre non-principal, d'une famille quelconque de R-modules à gauche est algébriquement compact sous l'hypothèse que R satisfait aux conditions du théorème 7.9 avec $\aleph = \aleph_0$. (Pour être rigoureux il faut supposer que tout ultrafiltre non-principal est "ω-incomplet", ce qui est un axiome compatible avec les axiomes usuels de la théorie des ensembles.)

Il est une question ouverte de savoir si la réciproque du théorème 7.11 est vrai. En particulier, il pourrait être interessant de savoir si un anneau R satisfait aux conditions du théorème 7.9 dans le cas où tout ultraproduit, par rapport à un ultrafiltre non-principal, de R-modules à gauche est un R-module algébriquement compact.

8. Résultats supplémentaires concernant l'annulation de $\varprojlim^{(i)}$ et les cardinaux des groupes.

La proposition suivante est une conséquence immédiate des résultats du paragraphe précédent et du théorème 3.1:

Proposition 8.1. Si R est un anneau quelconque de cardinal $\leq \aleph_n$, alors $\varprojlim^{(i)} M_\alpha = 0$ pour tout système projectif filtrant (M_α) de R-modules (à droite) de présentation finie et tout $i \geqq n+2$.

Dans cette section nous allons considérer des systèmes projectifs filtrants de groupes abéliens arbitraires (A_α), tels que card$(A_\alpha) \leq \aleph_n$ pour tout α, et nous allons chercher des bornes pour le nombre de foncteurs $\varprojlim^{(i)} A_\alpha$ non-nuls.

Théorème 8.2. Soit (A_α) un système projectif filtrant de groupes abéliens de cardinal $\leq \aleph_t$, t étant un entier fixé ≥ 0. De plus, soit T un foncteur exact de Ab vers Ab. En supposant l'hypothèse généralisée du continu $2^{\aleph_t} = \aleph_{t+1}$, on a $\varprojlim^{(i)} TA_\alpha = 0$ pour tout $i \geq t+3$. En particulier, sous l'hypothèse du continu $2^{\aleph_0} = \aleph_1$, pour tout système projectif filtrant (A_α) de groupes abéliens (au plus) dénombrables on a $\varprojlim^{(i)} A_\alpha = 0$ pour chaque $i \geq 3$.

Démonstration. Soit $\mathcal{C}$ la catégorie des groupes abéliens de cardinal $\leq \aleph_t$. Si $2^{\aleph_t} = \aleph_{t+1}$ toute groupe abélien de $\mathcal{C}$ contient au plus $\aleph_{t+1}$ sous-groupes, et donc tout objet de $\mathcal{C}$ et $\aleph_{t+1}$-artinien. Soit $\hat{\mathcal{C}} = \text{Sex}(\mathcal{C}, \text{Ab})$ la catégorie des foncteurs exacts à gauche de $\mathcal{C}$ dans Ab. En vertu du corollaire 6.5 on a dim.inj.$_{\hat{\mathcal{C}}}(T) \leq t+2$. Par conséquent $\text{Ext}^i_{\hat{\mathcal{C}}}(X,T) = 0$ pour tout $X \in \mathcal{C}$ et tout $i \geq t+3$.

Si (A_α) est un système projectif filtrant de groupes abéliens de $\mathcal{C}$, les foncteurs exacts à gauche $\bar{A}_\alpha = \text{Hom}(A_\alpha, -)$ forment un système inductif d'objets de $\hat{\mathcal{C}}$. Or, il y a une suite spectrale (cf. §1):

$$E_2^{p,q} = \varprojlim^{(p)} \text{Ext}^q_{\hat{\mathcal{C}}}(\bar{A}_\alpha, T) \underset{p}{\Rightarrow} \text{Ext}^n_{\hat{\mathcal{C}}}(\varinjlim \bar{A}_\alpha, T).$$

Comme $\bar{A}_\alpha$ est un objet projectif de $\hat{\mathcal{C}}$ pour tout α, la suite spectrale dégénère en des isomorphismes:

$$\varprojlim^{(n)} \text{Hom}_{\hat{\mathcal{C}}}(\bar{A}_\alpha, T) \simeq \text{Ext}^n_{\hat{\mathcal{C}}}(\varinjlim \bar{A}_\alpha, T).$$

Donc pour tout $n \geq t+3$ on obtient:

$$\varprojlim^{(n)} \text{Hom}_{\hat{\mathcal{C}}}(\bar{A}_\alpha, T) = 0 ,$$

et puisque $\mathrm{Hom}_{\mathfrak{C}}(\bar{A}_\alpha, T) \simeq TA_\alpha$ (par des isomorphismes naturels), il s'ensuit que $\varprojlim^{(n)} TA_\alpha = 0$ pour $n \geqq t+3$. C.Q.F.D.

C'est une question ouverte de savoir si l'hypothèse (généralisée) du continu est en effet nécessaire pour la vérité de l'assertion du théorème 8.2. D'ailleurs, nous ignorons si les bornes des nombres de foncteurs $\varprojlim^{(i)}$ non-nuls sont les meilleures possibles. Nous allons seulement prouver dans le cas dénombrable:

<u>Théorème 8.3</u>. Pour tout corps K il existe un système projectif filtrant (A_α) d'espaces vectoriels sur K de dimension dénombrable tel que $\varprojlim^{(2)} A_\alpha \neq 0$.

<u>Démonstration</u>. Nous commençons par des préparations ensemblistes. Soit $\underline{\underline{N}}$ l'ensemble des entiers positifs. On dit que deux parties X et Y de $\underline{\underline{N}}$ sont "presque disjointes", si l'intersection $X \cap Y$ est finie ou vide. Nous aurons besoin du lemme suivant, dont la démonstration sera laissée au lecteur.

<u>Lemme 8.4</u>. Soient $X_1, \cdots, X_n$ et $Y_1, \cdots, Y_m$ des parties presque disjointes de $\underline{\underline{N}}$. Si $\bigcup_{i=1}^{n} X_i \subseteq \bigcup_{j=1}^{m} Y_j$, alors $n \leqq m$ et toute partie X_i, $(1 \leqq i \leqq n)$ est égale a une partie Y_j, $(1 \leqq j \leqq m)$. En particulier, si $\bigcup_{i=1}^{n} X_i = \bigcup_{j=1}^{m} Y_j$, les parties X_i, $(1 \leqq i \leqq n)$ ne sont qu'une permutation des parties Y_j, $(1 \leqq j \leqq m)$, et $n = m$. Soit maintenant $\mathfrak{F} = (X_\alpha)$ une famille de parties presque disjointes de $\underline{\underline{N}}$, et soit $\mathfrak{U}$ la famille des réunions finies des parties dans $\mathfrak{F}$. Les éléments de $\mathfrak{U}$ forment - avec l'ordre évident - un ensemble filtrant à droite. Si I_1 et I_2 appartiennent à $\mathfrak{F}$, $I_1 \subseteq I_2$, il y a une application linéaire canonique $K^{(\underline{\underline{N}} - I_2)} \to K^{(\underline{\underline{N}} - I_1)}$, où $K^{(J)}$ est considéré comme un sous-espace d'un K-espace fixe $K^{(\underline{\underline{N}})}$. Par ce moyen on obtient un système projectif filtrant $\{K^{(\underline{\underline{N}} - I)}\}$, $I \in \mathfrak{F}$. Les espaces complémentaires

$K^{(I)}$, $I \in \mathfrak{F}$, forment avec les applications canoniques (restrictions) encore un système projectif. Ceci donne lieu à une suite exacte de systèmes projectifs

$$0 \to \{K^{(\underline{\underline{N}} - I)}\} \to \{K^{(\underline{\underline{N}})}\} \underset{\{\varphi_I\}}{\to} \{K^{(I)}\} \to 0$$

où $\{K^{(\underline{\underline{N}})}\}$ est le système projectif constant formé par $\{K^{(\underline{\underline{N}})}\}$, dont toutes les applications sont l'identité $1_{K^{(\underline{\underline{N}})}}$. $\varphi_{(I)}$ désigne l'application canonique (restriction) de $K^{(\underline{\underline{N}})}$ sur $K^{(I)}$. Par décalage on en obtient un isomorphisme $\varprojlim^{(2)} K^{(\underline{\underline{N}} - I)} \simeq \varprojlim^{(1)} K^{(I)}$.

Pour une choix convenable de $\mathfrak{F}$ nous allons démontrer que $\varprojlim^{(1)} K^{(I)} \neq 0$.

Dans ce but nous remarquons que le système $\{K^{(I)}\}$, $I \in \mathfrak{F}$, est un sous-système de $\{K^I\}$, $I \in \mathfrak{F}$, dont les applications sont les projections naturelles.

Dans ce qui suit nous appelons le système $\{K^{(I)}\}$, $I \in \mathfrak{F}$, le système de sommes et le système $\{K^I\}$, $I \in \mathfrak{F}$, le système de produits.

De plus, nous rappelons que le foncteur dérivé $\varprojlim^{(n)} A_\alpha$ d'un système projectif $\{A_\alpha, f_{\alpha\beta}\}$ peut être calculé comme le $n^{\text{ième}}$ groupe de cohomologie du complexe

$$\to A_n = \prod_{\alpha_o \leq \cdots \leq \alpha_n} A_{\alpha_o \cdots \alpha_n} \overset{\partial^n}{\to}$$

où

$$\partial^n(\underline{a})(\alpha_o, \alpha_1, \cdots, \alpha_{n+1}) = \sum_{n=1}^{n+1} (-1)^n a_{\alpha_o \hat{\alpha}_\mu \alpha_{n+1}} + f_{\alpha_o \alpha_1}(a_{\alpha_1 \cdots \alpha_{n+1}}).$$

Désignons les 1-cocycles du système de sommes, resp. du système de produits, par $c^s_{(I,T)}$, $(I \subseteq J)$, resp. par $c^p_{(I,T)}$ $(I \subseteq J)$.

De même, les 0-cochaînes sont désignées par $c^s(I)$, resp. $c^p(I)$. Dans les deux cas on a $\partial^1(c)(I,J) = c(I) - \text{res}_I c(J)$, $(c = c^s$ ou $c^p)$.

Si c^p est une 0-cochaîne du système de produits telle que $\partial^1(c^p)$ soit une 1-cochaîne du système de sommes, alors $\partial^1(c^p)$ est un 1-cocycle du système de sommes. Si c est une 0-cochaîne du système de produits, alors pour tout $I \in \mathcal{F}$ $c(I)$ est un élément de $K^{\underline{\underline{N}}}$ et les supports, (c.à d. les parties de $\underline{\underline{N}}$ sur lesquelles $c(I)$ a des coordonnées non-nulles) sont des parties de I. Donc, pour une 0-cochaîne c^p du système de produits $\partial^1(c)$ est un 1-cocycle de sommes si et seulement si $c(I) - \mathrm{res}_I c(J)$ est de support fini pour tout couple $I \subseteq J$. Ainsi, pour prouver que $\varprojlim^{(1)} K^{(I)} \neq 0$ il suffit de construire une 0-cochaîne c du système de produits telle que $\partial^1(c)$ ait des supports finis et $\partial^1(c)$ n'est pas le cobord d'aucune cochaîne du système de sommes.

Pour tout $X \in \mathcal{F}$ soit - pour un instant - $c(X)$ un élément quelconque de $K^{\underline{\underline{N}}}$ dont le support est une partie de X. Tout $I \in \mathcal{U}$ peut être écrit $I = \bigcup_{\mu=1}^{n} X_\mu$, $X_\mu \in \mathcal{F}$. En vertu du lemme 8.4 $c(I) = \sum_{\mu=1}^{n} c(X_\mu)$ est une 0-cochaîne du système de produits. De plus, le lemme 8.4 montre que $\partial^1(c)$ a des supports finis. Deux 0-cochaînes du système de produits ont le même cobord si et seulement si leur différence appartient à $\varprojlim K^I$, donc nous allons choisir $\mathcal{F}$ et les $c(X)$ d'une telle façon qu'il existe pour tout élément $\underline{\underline{b}}$ de $\varprojlim K^I$ une partie $X \in \mathcal{F}$ pour laquelle $c(X) - b(X)$ soit de support infini. Pour cela nous aurons besoin du lemme suivant.

<u>Lemme 8.5</u>. Il existe deux familles distinctes $\mathcal{A}$ et $\mathcal{B}$ de parties presques disjointes de $\underline{\underline{N}}$ telles que pour toute partie T de $\underline{\underline{N}}$ l'intersection $A \cap T$ soit infinie pour un A convenable de $\mathcal{A}$ ou l'intersection $B \cap (\underline{\underline{N}} \setminus T)$ soit infinie pour un B convenable de $\mathcal{B}$.

Supposons le lemme démontré et finissons la preuve du théorème 8.3. Dans les notations du lemme 8.5 nous posons $\mathcal{F} = \mathcal{A} \cup \mathcal{B}$. Ceci étant, soit $C(X)$ la fonction caractéristique de X si $X \in \mathcal{A}$ et soit $C(X)$ la fonction identique $\underline{\underline{0}}$ si $X \in \mathcal{B}$. Il est facile de vérifier que l'on en obtient un cocycle avec les propriétés désirées.

Démonstration du lemme 8.5. Soit $\beta\underline{\underline{N}}$ le compactifié de Stone-Čech de l'ensemble $\underline{\underline{N}}$ muni de la topologie discrète. Pour toute partie A de $\underline{\underline{N}}$ soit $A' = \bar{A} \setminus \underline{\underline{N}}$ où $\bar{A}$ désigne l'adhérence de A dans $\beta\underline{\underline{N}}$. Pour deux parties A et B de $\underline{\underline{N}}$ on a $A' = B'$ si et seulement si la différence symétrique de A et B est un ensemble fini. Les ensembles A' forment une base des parties ouvertes de $\beta\underline{\underline{N}} \setminus \underline{\underline{N}}$. De plus le sous-espace fermé $\beta\underline{\underline{N}} \setminus \underline{\underline{N}}$ de l'espace extrêmement discontinu $\beta\underline{\underline{N}}$ n'est pas extrêmement discontinu (voir [9] 6R); par conséquent il existe deux parties ouvertes disjointes $\mathcal{O}\!\iota$ et $\mathcal{L}$ de $\beta\underline{\underline{N}} \setminus \underline{\underline{N}}$ dont les adhérences ont une intersection non vide. Soit $\mathcal{A}$ une famille de parties de $\underline{\underline{N}}$ telle que $\mathcal{O}\!\iota = \bigcup_{A \in \mathcal{A}} A'$, où l'on peut supposer que les parties A, $A \in \mathcal{A}$, sont presques disjointes. De même, soit $\mathcal{B}$ une famille de parties presque disjointes de $\underline{\underline{N}}$ telle que $\mathcal{L} = \bigcup_{B \in \mathcal{B}} B'$. Évidemment, $\mathcal{A}$ et $\mathcal{B}$ sont des familles distinctes de parties de $\underline{\underline{N}}$. Soit $\underline{\underline{N}} = S \cup T$, $S \cap T = \emptyset$ et supposons que $A \cap T$ soit fini pour toute partie $A \in \mathcal{A}$ et $B \cap S$ soit fini pour toute partie $B \in \mathcal{B}$. Ceci entraînerait que $(A \cap S)' = A'$ pour tout $A \in \mathcal{A}$ et $(B \cap T)' = B'$ pour tout $B \in \mathcal{B}$. Donc $S' \supseteq \mathcal{O}\!\iota$ et $T' \supseteq \mathcal{L}$. Ainsi S' et T' seraient deux parties fermées disjointes contenant $\mathcal{O}\!\iota$ et $\mathcal{L}$, ce qui contredit le fait que les adhérences de $\mathcal{O}\!\iota$ et $\mathcal{L}$ ont une intersection non vide.

Remarque 8.6. Une modification facile de la démonstration du théorème 8.3 (on considère les valuations de $\underline{\underline{Q}}$) montre qu'il existe un système projectif filtrant (A_α) de sous-groupes du groupe additif $\underline{\underline{Q}}$ des nombres rationnels tel que $\varprojlim^{(2)} A_\alpha \neq 0$.

De même on obtient le résultat plus général, dont nous omettons la démonstration:

Proposition 8.7. R soit un anneau intègre commutatif noethérien. Alors R est semi-local et de dimension de Krull $\leqq 1$ si et seulement si $\varprojlim^{(i)} A_\alpha = 0$ pour tout système projectif filtrant (A_α) de R-modules sans torsion de rang fini et tout entier $i > 1$.

9. La L-dimension des anneaux complets.

Dans cette section nous allons considérer des rapports entre la L-dimension d'un anneau et son complété pour une topologie adique. Nous commençons en rappelant le résultat suivant de [14].

Théorème 9.1. Pour un anneau noethérien commutatif R les conditions suivantes sont équivalentes:

1) L-dim $R = 0$.

2) $\varprojlim^{(i)} M_\alpha = 0$ pour tout système projectif filtrant (M_α) de R-modules de type fini et tout $i > 0$.

3) R est un produit direct d'un nombre fini d'anneaux locaux complets.

4) L'application canonique $\mathrm{Hom}_R(R^{\underline{\underline{N}}}, R) \to \mathrm{Hom}_R(R^{(\underline{\underline{N}})}, R)$ est surjective.

Remarque. La formulation suivante curieuse de la condition 4) est due à I. Beck. On dit qu'un anneau R a des sommes infinies bien définies, si l'on peut attacher à toute suite d'éléments de R, (r_i), $i \in \underline{\underline{N}}$, un élément $\sum_{i=1}^{\infty} r_i \in R$ tel que les règles ordinaires des sommes infinies restent valables et tel que $\sum_{i=1}^{\infty} r_i$ coïncide avec la somme algébrique $\sum_{i=1}^{\infty} r_i$, lorsque $\{i \mid r_i \neq 0\}$ est un ensemble fini. Ceci étant, R a des sommes infinies bien définies si et seulement si la condition 4) est remplie.

Dans le cas général aucune caractérisation des anneaux de L-dimension zéro n'est connue. Nous ne faisons mention que de deux résultats spéciaux:

Proposition 9.2. Soit R un anneau régulier au sens de von Neumann (absolument plat). Alors L-dim(R) = 0 si et seulement si R est auto-injectif à gauche.

Proposition 9.3. Soit R un anneau de valuation. Alors L-dim(R) = 0 si et seulement si R est maximal, c.à d., aucun anneau de valuation S dominant R, $S \neq R$, a même groupe des valeurs et même corps résiduel que R. [5].

Soient R un anneau de valuation et v la valuation correspondante. Rappelons que R est "ω-pseudocomplet" si pour toute suite $(a_n)_{n \in \underline{\underline{N}}}$ d'éléments de R telle que, pour $n_1 < n_2 < n_3$ on ait $v(a_{n_1} - a_{n_2}) < v(a_{n_2} - a_{n_3})$, il existe un élément $a \in R$ tel que $v(a - a_n) = v(a_n - a_m)$ pour chaque couple d'entiers m,n tels que $m > n$. La démonstration du résultat suivant n'est qu'une application facile du théorème 3.1. Nous laissons les détails au soin du lecteur.

Proposition 9.4. Si R est un anneau de valuation ω-pseudocomplet, on a $L\text{-dim}(R) \leqq 1$.

Exemple 9.5. Soit R l'anneau des fonctions entières à valeurs complexes d'une variable. Tout idéal de R de type fini est principal; par suite, pour chaque idéal maximal m de R la localisation R_m est un anneau de valuation. Il y a deux espèces d'idéaux maximaux de R; pour tout $\alpha \in \underline{\underline{C}}$ l'ensemble des fonctions qui s'annullent en α est un idéal maximal principal de R, et l'on obtient ainsi tous les idéaux maximaux principaux de R. La deuxième espèce se présente de la manière suivante: Soit U un ultrafiltre non-principal sur $\underline{\underline{C}}$ auquel appartient une partie fermée discrète de $\underline{\underline{C}}$; alors les fonctions entières dont l'ensembles des zéros appartient à U forment un idéal maximal non-principal de R, et l'on obtient ainsi tous les idéaux maximaux non-principaux de R. En utilisant les théorèmes de Weierstrass et Mittag-Leffler on démontre que la localisation R_m est ω-pseudocomplet si m est un idéal maximal non-principal. De plus, c'est évident que la localisation R_m est un anneau de valuation discrete (de rang 1) non-complet si m est principal. Ceci étant, on voit que $\text{L-dim}(R_m) = 1$ pour tout idéal maximal m de R. Nous ignorons la valeur exacte de $\text{L-dim}(R)$; nous ne savons que $1 \leqq \text{L-dim}(R) \leqq 2$ si l'on suppose l'hypothèse du continu. Remarquons que les résultats plus haut restent vrais si l'on remplace R par l'anneau des fonctions analytiques dans un ouvert connexe de $\mathbb{C}$.

Nous formulons maintenant le résultat principal de cette section:

<u>Théorème 9.6</u>. Soient R un anneau cohérent à droite et I un idéal de type fini, engendré par un ensemble fini d'éléments centraux. Si R est complet pour la topologie I-adique (c. à d. l'application canonique $R \to \varprojlim R/I^\nu$ est surjective), alors $\text{L-dim}(R) = \text{L-dim}(R/I)$.

<u>Démonstration</u>. Il est facile de voir que R/I est cohérent à droite. Soit $\text{L-dim}(R/I) = n$ et supposons d'abord que $n < \infty$. En vertu du théorème 4.3 il existe un système projectif filtrant (F_α) de (R/I)-modules à droite de présentation finie tel que $\varprojlim^{(n)} F_\alpha \neq 0$. Puisque (F_α) peut être considéré comme un système projectif de R-modules à droite de présentation finie nous en concluons $\text{L-dim}(R) \geqq n$. Si $n = \infty$ il existe un ensemble infini d'entiers i pour lesquels $\varprojlim^{(i)} F_\alpha \neq 0$ pour un système projectif (F_α) convenable de (R/I)-modules à droite de présentation finie. En considérant (F_α) comme un système de R-modules on obtient $\text{L-dim}(R) = \infty$.

Il nous reste de montrer que $\text{L-dim}(R) \leqq \text{L-dim}(R/I)$. Évidemment on peut supposer que $\text{L-dim}(R/I) = n$ est un nombre fini. Pour tout système projectif filtrant $(A_\alpha, f_{\alpha\beta})$ de R-modules à droite libres de type fini nous allons prouver $\varprojlim^{(i)} A_\alpha = 0$ pour chaque $i > n$.

Par [14] $\underleftarrow{\lim}^{(i)} A_\alpha$ peut être calculé comme le $i^{\text{ième}}$ groupe de cohomologie du complexe

$$\rightarrow \Pi^k = \prod_{(\alpha_o \leq \cdots \leq \alpha_k)} A_{\alpha_o \cdots \alpha_k} \overset{\delta^k}{\rightarrow} \qquad (*)$$

où

$$\delta^k(a_{\alpha_o \quad \alpha_{k+1}}) = f_{\alpha_o \alpha_1}(a_{\alpha_1 \quad \alpha_{k+1}}) + \sum_{\nu=1}^{k+1} (-1)^\nu a_{\alpha_o \cdots \hat{\alpha}_\nu \quad \alpha_{k+1}} .$$

Il faut démontrer que $\operatorname{Ker} \delta^i = \operatorname{Im} \delta^{i-1}$ pour tout $i > n$. Pour plus de brièveté nous allons employer la notation $\delta = \delta^i, \delta' = \delta^{i-1}$ et par a nous désignons un élément de Π^i et par b un élément de Π^{i-1}.

Soit a un élément dans $\operatorname{Ker} \delta$. Les (R/I)-modules à droite de présentation finie (A_α/IA_α) forment un système projectif, dont le foncteur dérivé $\underleftarrow{\lim}^{(i)}(A_\alpha/IA_\alpha)$ est le $i^{\text{ième}}$ groupe de cohomologie du complexe qui provient de (*) en remplaçant les modules $A_{\alpha_o \cdots \alpha_k}$ par $A_{\alpha_o \cdots \alpha_k}/IA_{\alpha_o \cdots \alpha_k}$. L'élément a modulo I est alors un $i^{\text{ième}}$ cocycle et, puisque $\text{L-dom}(R/I) = n < i$, a modulo I sera un cobord. Donc $a - \delta' b_o \in I\Pi^i$ pour un b_o convenable. Ensuite, en considérant le système projectif $(IA_\alpha/I^2 A_\alpha)$ de (R/I)-modules à droite de présentation finie, on voit comme plus haut, qu'il y a un élément $b_1 \in I\Pi^{i-1}$ tel que $a - \delta' b_o - \delta' b_1 \in I^2 \Pi^i$. En continuant de cette manière nous obtenons $a = \delta' b$, où $b = b_o + b_1 + \ldots$ est un élément bien défini de Π^{i-1} vu le fait que R est complet pour la topologie I-adique.

<u>Corollaire 9.7</u>. Si l'anneau $R[[X_1, \cdots, X_n]]$ de séries formelles est cohérent à droite, alors $\text{L-dim}(R[[X_1, \cdots, X_n]]) = \text{L-dim}(R)$. En particulier, $\text{L-dim}(R[[X_1, \cdots, X_n]]) = \text{L-dim}(R)$ pour tout anneau R noethérien à droite.

<u>Remarque 9.8</u>. Dans le cas $\text{L-dim}(R) = 0$ on voit directement - en utilisant le théorème 3.5 (iii) - que l'assertion du théorème 9.6

reste vraie sans aucune condition de cohérence. Plus généralement, pour toute famille de variables (X_β), $\beta \in J$, soit $R[[(X_\beta)]]$ l'anneau des séries formelles non restreintes, c.à d. le groupe additif sous-jacent de $R[[(X_\beta)]]$ est (par abus de langage) $(R^I)^{\underline{N}}$. Alors l'anneau $R[[(X_\beta)]]$ est algébriquement compact en tant que module (à gauche) sur lui-même si et seulement si R est algébriquement compact comme R-module (à gauche).

10. La L-dimension globale d'un anneau.

Jusqu'ici nous avons principalement considéré la L-dimension d'un module particulier sur un anneau R. Nous définissons la L-dimension globale de R, notée L-gl-dim(R), comme sup(L-dim(M)) la borne supérieure étant prise sur les R-modules à gauche M. (Cette dimension est parfois connue comme la dimension globale pure à gauche; la dimension globale pure à droite de l'anneau R sera alors dans notre notation L-gl.dim(R^{op}).) Avec les notions de §1 il s'ensuit que pour tout entier $t \geqq 0$ on a L-gl.dim(R) $\leqq t$ si et seulement si $\mathrm{Pext}^{t+1}_R(M,N) = 0$ pour tous les R-modules à gauche M et N. Bien entendu, L-gl-dim(R) peut être introduit - alternativement - à l'aide de la dimension pur-projective (voir §1). De facon précise la dimension pur-projective d'un R-module à gauche N est définie comme suit: Pour un entier $s \geqq 0$ la dimension pur-projective de N est $\leqq s$ s'il existe une suite exacte pure

$$0 \to P_s \to \cdots \to P_1 \to P_o \to N \to 0$$

où les modules P_i, $0 \leqq i \leqq s$, sont pur-projectifs. La L-dimension globale de R sera alors la borne supérieure des dimensions pur-projectives des R-modules à gauche.

La L-dimension globale L-gl.dim(R) a des rapports à la dimension globale de la catégorie D(R) = Add(Pf(R),Ab) introduite en §1.

Proposition 10.1. Pour tout anneau R on a les inégalités:

$$\text{L-gl.dim}(R) \leqq \text{gl.dim } D(R) \leqq 2 + \text{L-gl.dim}(R).$$

Démonstration. Évidemment il suffit de prouver la seconde inégalité. Tout objet $T \in D(R)$ a une enveloppe injective de la forme $\overline{\overline{M}}$, M étant un R-module à gauche algébriquement compact. Par conséquent il existe une suite exacte dans D(R):

$$0 \to T \to \overline{\overline{M}}_o \to \overline{\overline{M}}_1 \to T' \to 0,$$

où T' est exact à droite et donc de la forme $\overline{\overline{A}}$ pour un R-module à gauche A convenable. Puisque $\text{dim.inj.}_{D(R)}\overline{\overline{A}} = \text{L-dim}(A) \leqq \text{L.gl.dim}(R)$, il s'ensuit par décalage que $\text{dim.inj.}_{D(R)}T \leqq 2 + \text{L-gl.dim}(R)$. On en obtient l'inégalité désirée.

En général, les inégalités ci-dessus n'admettent pas d'amélioration; on a des égalités en des cas particuliers seulement. Par exemple, pour les anneaux artiniens on a le résultat suivant:

Théorème 10.2. R soit un anneau artinien à droite. Si R est semisimple on a $\text{L-gl.dim}(R) = \text{gl.dim } D(R) = 0$. Si R n'est pas semisimple, alors $\text{gl.dim } D(R) = 2 + \text{L-gl.dim}(R)$.

Démonstration. Si R est semisimple, l'assertion est évidente. Donc, soit R un anneau non-semisimple et supposons d'abord que $\text{L-gl.dim}(R) = 0$. En vertu de la proposition 10.1 on a $\text{gl.dim } D(R) \leqq 2$. Pour obtenir l'inégalité réciproque nous considérons un R-module à gauche M, qui n'est pas projectif. Soit

$$L_1 \underset{\beta}{\to} L_o \underset{\alpha}{\to} M \to 0 \qquad (*)$$

une suite exacte, où L_o et L_1 sont des R-modules à gauche libres. La suite (*) donne lieu à une suite exacte de D(R):

$$\bar{\bar{L}}_1 \underset{\bar{\bar{\beta}}}{\rightarrow} \bar{\bar{L}}_o \underset{\bar{\bar{\alpha}}}{\rightarrow} \bar{\bar{M}} \rightarrow 0 .$$

Puisque M n'est pas projectif, les applications α et $\bar{\bar{\alpha}}$ ne sont pas scindées, et par conséquent $T = \mathrm{Ker}(\bar{\bar{\alpha}})$ n'est pas un objet injectif de $D(R)$. Puisque $L\text{-gl.dim}(R) = 0$ le R-module L_1 est algébriquement compact et $\bar{\bar{L}}_1$ est on objet injectif de $D(R)$. Si $T' = \mathrm{Ker}(\bar{\bar{\beta}})$, il y a une suite exacte

$$0 \rightarrow T' \rightarrow \bar{\bar{L}}_1 \rightarrow T \rightarrow 0$$

qui n'est pas scindée. Le fait que T n'est pas injectif implique que $\mathrm{dim.inj.}_{D(R)} T' \geqq 2$. Par suite, $\mathrm{gl.dim}\, D(R) = 2$ si $L\text{-gl.dim}(R) = 0$ et R n'est pas semisimple.

Considérons maintenant le cas $L\text{-gl.dim}(R) > 0$. Bien entendu, nous pouvons supposer que $L\text{-gl.dim}(R)$ est un nombre fini t. Soit M un R-module à gauche tel que $L\text{-dim}(M) = \mathrm{dim.inj.}_{D(R)} \bar{\bar{M}} = t$. Soit

$$L_1 \underset{\beta}{\rightarrow} L_o \underset{\alpha}{\rightarrow} M \rightarrow 0$$

une présentation libre de M et soit

$$\bar{\bar{L}}_1 \underset{\bar{\bar{\beta}}}{\rightarrow} \bar{\bar{L}}_o \underset{\bar{\bar{\alpha}}}{\rightarrow} \bar{\bar{M}} \rightarrow 0 \qquad (**)$$

la suite exacte correspondante de $D(R)$.

Comme R est artinien, il s'ensuit de la proposition 2.3 et du théorème 3.1 que tout R-module à gauche plat est algébriquement compact. En particulier, les modules L_o et L_1 sont des modules algébriquement compacts, et $\bar{\bar{L}}_o$ et $\bar{\bar{L}}_1$ sont des objets injectifs de $D(R)$. Puisque $\mathrm{dim.inj.}_{D(R)} \bar{\bar{M}} = t > 0$, on déduit de (**) par décalage que $\mathrm{dim.inj.}_{D(R)} \mathrm{Ker}(\bar{\bar{\beta}}) \geqq 2 + t$. Donc, $\mathrm{gl.dim}\, D(R) \geqq 2 + t = 2 + L\text{-gl.dim}(R)$. La proposition 10.1 implique l'inégalité réciproque. Ceci achève la démonstration du théorème 10.2.

Il y a encore un cas, où la L-dimension globale d'un anneau R est déterminée par $\mathrm{gl.dim}\, D(R)$. En effet, si R est régulier au sens de von Neumann tout objet de D(R) = Add(Pf(R),Ab) est exact, et l'on en déduit:

Proposition 10.3. Si R est un anneau régulier au sens de von Neumann, alors gl.dim D(R) = L.gl.dim(R) = l.gl.dim(R).

Dans cet ordre d'idées nous faisons mention des résultats suivants, qui sont essentiellement équivalents au théorème de Kulikof [7]:

Proposition 10.4. Soit R un anneau de Dedekind (qui n'est pas un corps) et soit D(R), resp. D'(R), la catégorie des foncteurs additifs covariants, resp. contravariants de Pf(R) dans Ab. Alors gl.dim.D(R) = 3 et gl.dim.D'(R) = 2.

En dehors des cas ci-dessus on connait peu d'estimations générales de la L-dimension globale. Sous ce rapport nous signalons une conséquence immédiate du théorème 7.10:

Proposition 10.5. Pour tout entier non négatif n et tout anneau R de puissance $\aleph_n$, on a L-gl.dim(R) $\leqq$ n + 1.

Exemple 10.6. Si R désigne l'anneau des fonctions entières (à valeurs complexes) d'une variable on voit facilement à l'aide du théorème 4.3 que L-gl.dim(R) $\geqq$ 2 ; si l'on suppose l'hypothèse du continu, la proposition 10.5 implique L-gl.dim(R) = 2 .

Nous donnerons maintenant une caractérisation des anneaux R pour lesquels L-gl.dim(R) = 0 .

Proposition 10.7. Pour tout anneau R les conditions suivantes sont équivalentes:

(i) L-gl.dim(R) = 0 .

(ii) La catégorie D(R) est localement noethérienne.

(ii') $\overline{R}$ est un objet noethérien de D(R).

(iii) Tout R-module à gauche est somme directe de R-modules de présentation finie.

(iv) Il existe un cardinal $\aleph$ tel que tout R-module à gauche est somme directe de R-modules engendrés par au plus $\aleph$ éléments.

(v) Tout R-module à gauche est somme directe de R-modules indécomposables.

(vi) Tout R-module à gauche est facteur direct d'une somme directe de R-modules indécomposables.

<u>Démonstration</u>. L'équivalence (i) $\Leftrightarrow$ (ii) est une conséquence de la proposition 5.2, du lemme 5.3 et du théorème 6.1. L'implication (ii) $\Rightarrow$ (ii') est triviale, et sa réciproque vient du fait que la sous-catégorie épaisse de D(R) engendrée par $\overline{R}$ contient les objet cohérents. Si (i) est verifié, tout R-module M est pur-projectif, i.e. M est facteur direct d'une somme directe $\Sigma \oplus M_\alpha$ de R-modules M_α de présentation finie. Alors l'objet injectif $\overline{\overline{M}}$ de D(R) est facteur direct de $\Sigma \oplus \overline{\overline{M}}_\alpha$; d'autre part (ii) est verifié, donc en vertu du théorème d'échange pour les objets injectifs d'une catégorie localement noethérienne il s'ensuit que $\overline{\overline{M}}$ est une somme directe $\Sigma \oplus \overline{\overline{F}}_\alpha$ de facteurs directs de $\overline{\overline{M}}_\alpha$. Chaque F_α est un facteur direct de M_α et donc de présentation finie; M est alors la somme directe des modules F_α et (iii) est établi. Si (iii) est vérifié, tout R-module à gauche est pur-projectif, donc (i) est vérifié. Les implications (iii) $\Rightarrow$ (iv) et (iii) $\Rightarrow$ (v) et (iii) $\Rightarrow$ (vi) sont evidentes, et leurs réciproques sont des conséquences de la caractérisation de [19] des catégories localement noethériennes. Ceci achève la démonstration de la proposition 10.7.

<u>Remarque 10.8</u>. Si les conditions de la proposition 10.7 sont vérifiées, l'anneau R est artinien à gauche. En effet, R est noethérien à gauche, car tout R-module à gauche de type fini est de présentation finie; de plus, R est parfait à gauche, puisque tout R-module

à gauche plat est projectif. Il est bien connu que ces conditions impliquent que R est artinien à gauche.

Remarque 10.9. Par la méthode de la démonstration de la proposition 10.7 on voit qu'un anneau R cohérent à droite est parfait à gauche si et seulement s'il existe un cardinal $\aleph$ tel que tout R-module à gauche plat est somme directe de R-modules engendrés par au plus $\aleph$ éléments. De même, un anneau R cohérent à droite est parfait à gauche si et seulement si tout R-module à gauche plat est somme directe de R-modules indécomposables. En effet, au lieu de D(R) on considère la catégorie des foncteurs exacts à gauche et utilise les caractérisations de [19].

Théorème 10.10. Pour un anneau R les conditions suivantes sont équivalentes:

(i) $\text{L-gl.dim}(R) = 0 = \text{L-gl.dim}(R^{op})$.

(ii) La catégorie D(R) est localement finie, (c.à d., D(R) a une famille de générateurs de longueur finie.)

(ii') $\bar{R}$ est un objet de longueur finie de D(R).

(iii) R est artinien à gauche et l'ensemble des classes d'isomorphisme de R-modules à gauche indécomposables de type fini est fini.

Démonstration. En vertu de la proposition 5.2, du lemme 5.3 et du théorème 6.1 la condition (i) exprime que les catégories abéliennes C(R) et $C(R^{op})$ sont artiniennes. La condition (ii) exprime que C(R) consiste d'objets de longueur finie, i.e. C(R) est noethérienne et artinienne. Comme les catégories $C(R)^{op}$ et $C(R^{op})$ d'après le théorème de dualité (théorème 5.6) sont équivalentes, les conditions (i) et (ii) sont équivalentes.

Comme en 10.7 l'équivalence (ii) $\Longleftrightarrow$ (ii') vient du fait que la sous-catégorie épaisse de D(R) engendrée par $\bar{R}$ contient les objets cohérents.

Supposons (ii) vérifié. On choisit une suite de Jordan-Hölder de $\overline{R}$; tout objet simple de D(R) est isomorphe à l'un des quotients de cette suite; l'ensemble des types d'objets simples de D(R) est donc fini. Le passage à l'enveloppe injective définit une bijection de l'ensemble des types d'objets simples de D(R) sur l'emsemble des types d'injectifs indécomposables de D(R), et le foncteur: Pf(R) → D(R), défini par $M \to \overline{\overline{M}}$, induit une bijection de ce dernier ensemble sur l'ensemble des classes d'isomorphisme de R-modules indécomposables (nécessairement de type fini). Ceci prouve (iii) compte tenu de la remarque 10.8.

L'implication (iii) ⇒ (i) est une conséquence d'un résultat de Dlab et Ringel [6].

Remarque 10.11. L'hypothèse dans la condition (iii) que R soit artinien est importante pour la validité du théorème 10.9. En effet, comme nous a communiqué S. Jøndrup il existe un anneau R qui n'est artinien ni à gauche, ni à droite tel que R ne possède que deux modules (à gauche) indécomposables de type fini. Cependant, dans le cas commutatif il est facile de voir qu'un anneau R est forcément artinien si R ne possède qu'un nombre fini de modules indécomposables de type fini.

Remarque 10.12. C'est toujours une question ouverte de savoir si le théorème 10.10 reste vrai lorsque la condition (i) est remplacée par la condition "unilatérale" L-gl.dim(R) = 0 .

B I B L I O G R A P H I E

1. Auslander, M: Coherent functors, Proc. Conf. on Categorical algebra, La Jolla 1965, Springer-Verlag, 189-231.

2. Auslander, M. and Buchsbaum, D.: Homological dimension in Noetherian rings. II Trans. Amer. Math. Soc. 88 (1958), 194-206.

3. Bass, H.: Injective dimension in Noetherian rings. Trans. Amer. Math. Soc. 102 (1962), 18-29.

4. Bourbaki, N.: Algèbre commutative, Chap. I-II, Hermann, Paris 196 .

5. Bourbaki, N.: Algèbre commutative, Chap. V-VI, Hermann, Paris 1964.

6. Dlab, V. and Ringel, C.M.: Decomposition of modules over right uniserial rings, Math. Z. 129 (1972), 207-230.

7. Fuchs, L.: Infinite abelian groups, Vol. 1. Academic Press, New York, 1970.

8. Gabriel, P.: Catégories abéliennes, Bull. Soc. Math. France 90 (1962), 323-448.

9. Gillman, L. and Jerison, M.: Rings of continuous functions, Van Nostrand, Princeton, 1960.

10. Goblot, R.: Sur les dérivés de certaines limites projectives. Applications aux modules, Bull. Sci. Math. 94 (1970), 251-255.

11. Gruson, L. et Jensen, C.U.: Modules algébriquement compacts et foncteurs $\varprojlim^{(i)}$, C.R. Acad. Sci. Paris 276 (1973), 1651-1653.

12. Gruson, L. et Jensen, C.U.: Deux applications de la notion de L-dimension, C.R. Acad. Sci. Paris 282 (1976), 23-24.

13. Gruson, L. et Raynaud, M.: Critères de platitude et de projectivité, Invent. Math. 13 (1971), 1-89.

14. Jensen, C.U.: Les foncteurs dérivés de $\varprojlim$ et leurs applications en théorie des modules, Lect. Notes in Math. 254 (1972).

15. Lazard, D.: Autour de la platitude, Bull. Soc. Math. France 97 (1969), 81-128.

16. Mitchell, B.: Rings with several objects, Advances in Math. 8 (1972), 1-161.

17. Roos, J.E.: Bidualité et structure des foncteurs dérivés de $\varprojlim$, C.R. Acad. Sci. Paris 254 (1962), 1556-1558.

18. Roos, J.E.: Ibid. C.R. Acad. Sci. Paris 254 (1962), 1720-1722.

19. Roos, J.E.: Locally Noetherian categories, Lect. Notes in Math. 92 (1969), 197-277.

20. Warfield, R.B.: Purity and algebraic compactness for modules, Pac. J. Math. 28 (1969), 699-719.

21. Zimmermann, W.: Rein injektive direkte Summen von Moduln. Comm. Algebra 5 (1977), 1083-1117.

L. Gruson
3 Avenue des Chalets
F-75016 Paris
France.

C.U. Jensen
Matematisk Institut
Universitetsparken 5
DK-2100 København Ø
Danemark.

GENUS AND A RIEMANN-ROCH THEOREM FOR NON-COMMUTATIVE FUNCTION FIELDS IN ONE VARIABLE

-:-:-:-

J.P. VAN DEUREN
Université Catholique de Louvain, Belgium

J. VAN GEEL
University of Antwerp, UIA, Belgium

F. VAN OYSTAEYEN
University of Antwerp, UIA, Belgium

0. Introduction

After F.K. Schmidt proved the Riemann-Roch theorem over arbitrary groundfields, the theorem became one of the pillars of modern algebraic geometry. Recently, essential parts of the machinery producing the interplay between ring theoretical methods and geometric properties in the commutative case, have been generalized to certain non-commutative rings. Let us mention M. Artin's work on the geometry of P.I. rings, cf. (2), which has been complemented by results of F. Van Oystaeyen and A. Verschoren cf. (13), (14), on the structural sheaves and "non-commutative varieties". On the other hand, J.P. Van Deuren observed that certain skewfields contain maximal commutative subfields of arbitrary genus over the groundfiels, cf. (10). The combination of all these results calls for a theory of divisors, genus and a Riemann-Roch-type theorem linked to prime ideal spectra and non-commutative valuations.

In this paper we consider skewfields which are finite dimensional over a function field in one variable. The geometric application we have in mind, only hinted at here, requires a more general set-up, i.e. the study of central simple algebras over an algebraic function field. But this study hinges upon the theory of primes in algebras, cf. (11), (13), and we aim to come back to this in another paper.

In a relatively unknown paper, E. Witt produced a Riemann-Roch theorem, with applications to the zeta-function, for central simple algebras over an algebraic function field in one variable but with perfect groundfield. His results are based upon the definition of the canonical class by means of the different D of the algebra under consideration. Our set-up, with an eye to geometry instead of arithmetic, is based upon "valuations". Whereas in E. Witt's paper, cf (16), the role played by the ring of constants and non-commutative valuations is not explicited, our paper is not restricted to the perfect case but it contains not much information on the canonical class. We do show that in the perfect case both theories coïncide.

Important examples are obtained by considering skewfields of twisted polynomials $D(X,\varphi)$, where D is a skewfield, finite dimensional over its center k' and φ is an automorphism of D such that φ^e is inner for some $e \in \mathbb{N}$. If k is the fixed field of φ in k' then the genus of $D(X,\varphi)$ is $1-n$ where $n = [D : k]$. In this case too our "geometrical" genus coïncides with E. Witt's genus. The classical example $\mathbb{C}(X,-)$, where $-$ stands for complex conjugation, has genus equal to -1 (this example is a nice easy example of a skewfield with many valuations on it). The fact that negative genera occur is explained by the existence of non-central commutative subfields.

The case $K = D(X,\varphi)$ leads the way to finding a relation between the genus g_K of K and the genus $g_{Z(K)}$ of the center $Z(K)$ of K. We obtain :

$$g_K \leqslant Ng_{Z(K)}-n+1 \ .$$

where n is the k-dimension of a maximal k-algebraic subring in K and $N = [K:Z(K)]$.

1. Algebraic Function Fields of One Variable

Let K be an arbitrary field. A skewfield K containing k in its center is said to be an algebraic function (skew-)field of one variable over k if there is an $x \in K$ which is not-algebraic over k and such that : $[K:k(x)] < \infty$.

Proposition 1.1

Let K be an algebraic function skewfield of one variable over k, then :

1 - K is finite dimensional over its center $Z(K)$.

2 - $Z(K)$ is an algebraic function field of one variable.

3 - The transcendence degree of any commutative subfield of K is at most equal to 1.

Proof

1) Pick a non-algebraic $x \in K$. Then $k(x)$ is a commutative field such that $[K:k(x)] < \infty$. Since K is a free $k(x)$-module of finite rank it follows that K satisfies the polynomial identities of some matrix ring over $k(x)$, then by Posner's theorem, cf (3), it follows that K is finite dimensional over its center. Note that for every $y \in K$ which is not k-algebraic we have that $[K:k(y)] < \infty$.

2) By 1, $Z(K)$ contains an element t which is not k-algebraic, hence $[Z(K):k(t)] < \infty$, i.e. $Z(K)$ is an algebraic function field of one variable.

3) Obvious. Note also that every maximal commutative subfield of K is an algebraic function field of one variable.

If K is an algebraic function skewfield over k then also over k_1, the algebraic closure of k in $Z(K)$. Henceforth we assume that k is algebraically closed in $Z(K)$ (this is not a real restriction). However, the set of k-algebraic elements of K does not form a ring and we will have to find a suitable substitute for it. Recall that a subring Λ of k is said to be a <u>total</u> subring of K if for each $x \in K$ either x or x^{-1} is in Λ. A <u>valuation ring of K</u> is a total subring which is invariant under all inner automorphisme of K. For details on valuation rings and maximal orders the reader is referred to (8) and (11).

Lemma 1.2.

Let Λ be a maximal order in K over a discrete valuation ring O_v in $Z(K)$. The following statements are equivalent to one another :

1) $\Lambda/\text{rad}\ \Lambda$ is a skewfield.

2) If $x \in K$ then either x or x^{-1} is in Λ.

3) Λ is a valuation ring in K.

4) Λ is the unique maximal O_v-order in K.

5) Every left ideal of Λ is two-sided.

6) Let $\hat{Z}$ be the completion of $Z(K)$ with respect to the v-topology, then $\hat{K} = \hat{Z} \underset{Z(K)}{\otimes} K$ is a skewfield.

Proof

Known. Recall that, if we have that $\hat{K} \cong M_r(E)$, where E is a $\hat{Z}$-central skewfield, then : $\Lambda/\text{rad}\,\Lambda \cong M_r(\Omega/\text{rad}\,\Omega)$ where Ω is the unique maximal $\hat{O}_v$-order in $\hat{K}$.

Remark 1.3.

a) Although in this paper the valuations on the center will indeed be discrete, it may be worthwhile to point out some results for arbitrary valuation rings O_v. In his Ph.D. thesis, F. Van Oystaeyen mentioned that total subrings of a skewfield (finite dimensional over its center) are valuation rings if they are domains of unramified pseudo-places ; J. Van Geel conjectured that all total subrings of such skewfields are valuation rings and this has recently been proved by P.M. Cohn, (5).

b) A valuation ring R_v in any skewfield K yields a valuation function $v : K \rightarrow \Gamma$, where Γ is an ordered but possibly non-commutative group, such that : $v(xy) = v(x) \star v(y)$ ($\star$ the grouplaw of Γ), $v(x + y) \geqslant \inf \{v(x), v(y)\}$.

Conversely, to such a valuation function v on K there corresponds a valuation ring $R_v = \{x \in K, v(x) \geqslant o\}$ with maximal ideal (the Jacobson radical of R_v) $M_v = \{x \in K, v(x) > o\}$, where o denotes the zero element of the group Γ.

Let $\overline{k}$ stand for the set of k-algebraic elements in K. Amongst the subrings of K which are contained in $\overline{k}$ we may consider maximal subrings, say $\{l_i, i \in I\}$. We shall fix a certain l_i from hereon and we will denote it by l. If $x \in l$ then from the minimal equation for x over k it follows that $x^{-1} \in k[x]$ hence $x^{-1} \in l$, i.e. l is a skewfield. Moreover, as a subring of $\overline{k}$, l satisfies a polynomial identity too, hence l is finite dimensional over its center $Z(l) = k'$, say. If $t \in Z(K) - k$ then $[K:k(t)] < \infty$ and hence $[l(t):k(t)] = [l:k] < \infty$. Let us now fix notations as follows :

$$N = [K:Z(K)]$$
$$n = [l:k] .$$

Some light is shed upon the relation between n and N by the following :

Proposition 1.4.

Choose $t \in Z(K)$ such that $[K:k(t)]$ is minimal, say equal to m. Put $r = [Z(K)k':k'(t)]$, note that $r \leq m$. Then we have that:

$$N = n.\frac{r^2}{m^2}.s^2.\ [k':k] ,$$

where s^2 is the dimension of the centralizer of l in k over its center.

Proof

Consider $R = l \underset{k}{\otimes} Z(K)$. Clearly, R is an Azumaya algebra with center $C = K' \underset{k}{\otimes} Z(K)$. Since k'/k is algebraic whereas k is algebraically closed in Z(K) it follows that all zero-divisors of C are nilpotent and the prime radical m of C is the unique prime ideal of C. Consequently Rm is the unique prime ideal of R, i.e. R has constant rank equal to the p.i. degree of Rm and this also equals $[l:k']$ the rank of R over the local ring C, thus :

$$[R/Rm : C/m] = [l.Z(K):k'.Z(K)] = [l:k'] = \frac{n}{[k':k]} .$$

By definition of r : $[k'.Z(K):k(t)] = r.[k':k]$, and thus :

$$[k'.Z(K):Z(K)] = \frac{r}{m}[k':k] .$$

On the other hand, since $1.Z(K)$ is a simple subalgebra of K we find that : $N = [1.Z(K) : Z(K)]\ [Z_K(1.Z(K)) : Z(K)]$, where $Z_K(1.Z(K)) = Z_K(1)$ is the centralizer of 1 in K. Now $[1.Z(K) : Z(K)] = [1.Z(K) : k'Z(K)]\ [k'.Z(K) : Z(K)] = \frac{n}{[k':k]} \cdot \frac{r}{m} \cdot [k':k] = \frac{nr}{m}$. Combination of these results yields the desired equality :

$$N = n.s^2.\frac{r^2}{m^2}.[k':k]. \quad \square$$

Remark 1.5.

In case either K or k' is separable over k it follows that $C = k' \underset{k}{\otimes} Z(K)$ is a finite integral extension of $Z(K)$, therefore C is a field and the free composite $k'.Z(K) \cong C$, $1.Z(K) \cong 1 \underset{k}{\otimes} Z(K)$. In this case $r=m$ and $N=n.s^2.[k':k]$. In case $K = D(X,\varphi)$ cf. Section 2, k is the fixed field of φ within k' hence k' is separable over k and the above remark applies. Moreover, with $1=D$, one easily verifies that $s^2 = 1$, and $N=n.[k':k]$. Note that in general, n,r,s depend on the choice of 1 ; after we have proved the Riemann-Roch theorem it will show that in reality the choice of 1 does not interfer much.

An $x \in K$ is integral over some valuation ring O_v of $Z(K)$ if and only if the minimal polynomial of x over K has coefficients in O_v (cf. (8) for example). Hence if we choose a maximal O_v-order Λ_v in K for each valuation ring O_v of $Z(K)$ then $\underset{v}{\cap} \Lambda_v$ consists of k-algebraic elements. With notations as before

Proposition 1.6

We have : $1 = \cap \{\Lambda, \Lambda$ a maximal order over a k-valuation ring of K such that $1 \subset \Lambda\}$.

Proof

Suppose there exists $t \in \cap \Lambda$ which is not k-algebraic. Pick a k-valuation ring O in $Z(K)(t)$ such that $t \notin O$ and put $O_v = O \cap Z(K)$. Since $A = Z(K)(t).1$ is finite dimensional over $Z(K)$ it is a skewfield in K.

Pick a k-basis $\{l_1,\ldots,l_n\}$ for l. The order $O[l_1,\ldots,l_n]$ is contained in a maximal O_v-order Λ_1 of A.

Now pick a (left) basis for K over A, $\{v_1,\ldots,v_m\}$ say. Up to multiplying the $v_j, j=1,\ldots,m$, by a suitable central element in m_v, we may assume that the structural constants over A are in Λ_1, i.e. $v_iv_j = \Sigma\, \lambda^k_{ij} v_k$ with all $\lambda^k_{ij} \in \Lambda_1$.

Then $\Lambda_1[v_1,\ldots,v_m]$ is an O_v-order in K and as such, it is contained in a maximal O_v-order Λ_o of K. Since $\Lambda_o \cap A = \Lambda_1$ and $t^{-1} \in \text{rad}\,\Lambda_1$, it follows that $t \notin \Lambda_o$ whilst $\Lambda_o \supset l$. □

<u>Remark 1.7</u>

By the proposition and the remark preceding it, it follows that amongst the maximal O_v-orders appearing in $l = \bigcap_v \Lambda_v$ we may select one O_v-order for each k-valuation ring O_v of $Z(K)$, without enlarging the intersection (l is a maximal k-algebraic subring in $k!$). Depending upon the choice of l we will now fix our choice for the Λ_v.

Pick $t \in Z(K)-k$ and denote by O_v^{α}, $\alpha \in \mathcal{A}$, the k-valuation rings of $Z(K)$ containing $k[t]$. Let O_v^{β}, $\beta \in \mathcal{B}$, be the k-valuation rings of $Z(K)$ containing $k[t^{-1}]$ but not t. Note that B is a finite set since its cardinality is bounded by $[k(t)_s : k(t)]$ where $k(t)_s$ is the separable closure of $k(t)$ in $Z(K)$. Put $R = \bigcap_{\alpha\in\mathcal{A}} O_v^{\alpha}$; then R is a Dedekind ring and the quotient field of R is equal to $Z(K)$ (note that this uses the classical Riemann-Roch theorem on $Z(K)$. One easily constructs a maximal R-order in K, Λ say, such that $\Lambda \supset l$. Further, for each $\beta \in \mathcal{B}$ one, equally easily, constructs a maximal O_v^{β}-order Λ_{β} containing l. Put Λ_{α} equal to the central localization of Λ at the prime ideal p_{α} of R corresponding to O_v^{α}, $\alpha \in \mathcal{A}$. We have that

$$l = \bigcap_{\alpha\in\mathcal{A}} \Lambda_{\alpha} \cap \bigcap_{\beta\in\mathcal{B}} \Lambda_{\beta} = \Lambda \cap \bigcap_{\beta\in\mathcal{B}} \Lambda_{\beta}$$

If O_v is a k-valuation ring of $Z(k)$ such that there exists a valuation ring R_v of K with $R_v \cap Z(K) = O_v$ then R_v is amongst the Λ_α and Λ_β selected since, by Lemma 1.2., it has to be unique maximal O_v-order in K and $1 \subset \bar{k} \subset R_v$. This shows that the decomposition of 1 we have chosen contains all valuation rings of K.

For further use we will fix a $Z(K)$-basis $\{u_1,\ldots,u_n\}$ for K which is contained in Λ.

2. Valuations of an Algebraic Function Field of one Variable

Let R_v a k-valuation ring in K. The residue field $R_v/\text{rad}\,R_v$ will de denoted by k_v. It is clear that k_v is a finite extension of k and $f_v = [k_v:k]$ is called the residue class degree.

The following lemma may be considered "well-known" :

Lemma 2.1

If Δ is a skewfield which is algebraic over a commutative field F then any non-trivial valuation on Δ induces a non-trivial valuation on F.

So in the situation we are facing, every k-valuation of K induces a k-valuation v^c on $Z(K)$ and by Proposition 1.1., v^c is a discrete valuation. By Lemma 1.2., the valuation ring R_v of v in K is the unique maximal O_{v^c}-order in K, where O_{v^c} is the valuation ring of v^c in $Z(K)$. Let us denote by M_v the maximal ideal of R_v and by M_v^c the maximal ideal of O_{v^c}.

From [8] we recall that there esists an $e_v \in N$, such that $M_v^{e_v} = R_v.M_v^c$ and it is clear that e_v is exactly the index of the value group of v^c in the value group of v. If $e_v = 1$ then v^c is said to be unramified, otherwise e_v is called the ramification index of v^c.

The relative residue class degree is $\psi_{v^c} = [k_v:k_{v^c}]$;

from [8] we recall:

1° $e_v \psi_{v^c} = N$

2° $\psi_{v^c} f_{v^c} = f_v$, where $f_{v^c} = [k_{v^c} : k]$.

We also retain that e_v , ψ_{v^c} , f_v and f_{v^c} are finite numbers.

It is a well-known fact that a totally ordered group Γ containing an ordered subgroup of finite index which is isomorphic to Z it itself isomorphic to Z . Hence it follows that <u>every k-valuation of K is a discrete rank one valuation.</u>

Now some explicit calculations in the case $K = D(X,\varphi)$ where D is a skewfield with center $Z(D) = k'$, φ an automorphism of D such that φ^e is inner for some e (minimal as such) in N, and k is the field fixed by φ within k'. $D(X,\varphi)$ is the field of fractions of the skew polynomial ring $D[X,\varphi]$; details about some of the results frequently mentioned or used in the sequel may be found in the papers by G. Cauchon, cf. (3), or E. Nauwelaerts, F. Van Oystaeyen, cf. (7).

The ring $D[X,\varphi]$ is a Dedekind prime P.I. ring and also a left and right principal ideal domain. From [7] we retain that it is also a Zariski central ring, consequently all localizations at prime ideals will be central localizations at the corresponding prime ideals of the centre. In particular, $D(X,\varphi)$ is obtained from $D[X,\varphi]$ by central localization of $D[X,\varphi]$ at 0, i.e. $K = \{xc^{-1}, x \in D[X,\varphi], c \in Z(R)\}$, (denoting R for $D[X,\varphi]$). The centre of R, as determined by G. Cauchon, is equal to $k[T]$ where $T = \lambda X^e$ for some unit λ in D inducing the inner automorphism φ^{-e} of D . It is obvious that $D(X,\varphi)$ is an algebraic function skewfield of one variable with $1 = D$ and $Z(K) = K(T)$, a purely transcendental field extension of k, i.e. of genus o !

Every ideal of R is of the form RcX^m, $m \in N$ and $c \in Z(R)-(T)$. Now if R_v is a k-valuation ring of K then R_v contains either $R = D[X,\varphi]$ or $R^{-1} = D[X^{-1}, \varphi^{-1}]$. The theory of Dedekind P.I. rings yields that O_v has to be a localization of R or R^{-1} at a prime ideal of R or R^{-1} resp. In R, prime ideals are either (X) or an ideal generated by a central irreductible polynomial $c(T)$; in R^{-1}, prime ideals are either (X^{-1}) or an ideal generated by a central irreductible polynomial $c(T^{-1}) \in k[T^{-1}]$. It is clear that the localizations of $R,(R^{-1})$, at central prime ideals will contain X^{-1}, (X) ; hence we obtain all localizations at prime ideals of R or R^{-1} in the list : $R_{(X)}$; $R^{-1}_{(X-1)}$; $R_{(p)}$, p central irreducible in $k[T]$. If now p is such that Rp is a completely prime ideal of R , i.e. $R-Rp$ is multiplicatively closed, then $R_{(p)}/\text{rad}\, R_{(p)}$ is a skewfield and by Lemma 1.2. it follows then that $R_{(p)}$ is a k-valuation ring of K. Conversely, since any k-valuation ring of K corresponds to a ring listed above it follows that all k-valuation rings of $D(X,\varphi)$ are given by : $R_{(X)}$, $R^{-1}_{(X^{-1})}$ and $R_{(p)}$ where p is a central irreducible polynomial in $k[T]$ such that Rp is a completely prime ideal.

1° Let v_X be the valuation associated to $R_{(X)}$,
Then $k_{v_X} = R_{(X)}/(X)R_{(X)} \cong R/(X) \cong D$ and therefore we obtain $\psi_{v_X^c} = n$, $f_{v_X^c} = 1$, $f_{v_X} = n$.

The ramification index $e_{v_X} = e$ and we have $en = N$ where $n = [D:k]$.

2° Let $v_{X^{-1}}$ be the valuation associated to $R^{-1}_{(X^{-1})}$.

This case may be treated as v_X in 1°. However a more explicit description is possible here.
Write $z \in K$ as $z = fg^{-1}$ with $f,g \in R$, i.e.

$$z = \left(\sum_{i=1}^{\alpha} a_i X^i\right)\left(\sum_{j=1}^{\beta} b_j X^j\right)^{-1} \quad \text{with } a_i, b_j \in D \text{ and}$$

$a_\alpha \neq 0$, $b_\beta \neq 0$. Up to changing b_j, $j=1 \ldots \beta$, (due to some commutations with certain powers of X) we may rewrite this as :

$$z = \left(\Sigma_{i=1}^{\alpha} a_i X^{-\alpha+i} \right) . \left(\Sigma_{j=1}^{\beta} b'_j X^{-\beta+j} \quad X^{\alpha-\beta} . \right)$$

with $a_\alpha \neq 0$ and $b'_\beta \neq 0$ since φ is an automorphism. From this one easily derives that :

$$v_{X^{-1}}(z) = \beta - \alpha = \deg_X g - \deg_X f .$$

Let us write v_∞ instead of $v_{X^{-1}}$; then :

$$R_\infty = \{fg^{-1} \in K, \deg_X g \geqslant \deg_X f\}$$

$$M_\infty = \{fg^{-1} \in K, \deg_X g > \deg_X f\}.$$

3° The case v_p .

Write $P = Rp$, $R_{v_p} = R_p = R_{(p)}$,

$M_{v_p} = PR_p = pR_p$ i.e. the valuation v_p is unramified.

Furthermore : $k_{v_p} \cong R/P$. The residue class degree of v_p^c is equal to $[k[T]/(p) : k] = \deg_T p$ and thus :

$$\psi_{v_p^c} = N , e_{v_p} = 1 , f_{v_p^c} = \deg_T p , f_{v_p} = N.\deg_T p .$$

<u>Example 2.2.</u>

Take $K = \mathbb{C}(X,-)$. A valuation of the center $\mathbb{R}(X^2)$ extends to a valuation of K if and only if it corresponds to a prime ideal $(X^2 + c)$ in $\mathbb{R}(X^2)$ with $c > 0$.

3. <u>Divisors of Algebraic Functions Skewfields of one Variable</u>

With notations as in Section 1 , we have fixed a maximal k-algebraic subring Λ in K and we have that $\Lambda = \cap\{\Lambda_j, j \in \mathcal{A} \cup \mathcal{B}\}$ the Λ_j being maximal O_{v_j}-orders in K such that the O_{v_j} run through all the non-trivial k-valuations rings of Z(K) . Let $\mathcal{E}$

be the set $\{\Lambda_j, j \in A \cup B\}$ and let $V(K)$ be the set of k-valuation rings of K ; note that $V(K) \subset E$. In this section we assume that $V(K) \neq \phi$. We were unable to prove that $V(K) \neq \phi$ but we conjecture that this is always the case in tne situation we have here ; all examples we have considered do satisfy this property. Since in the special case of a perfect groundfield k, E.Witt's theory also applies, we have defined geometric divisors (and not just merely divisors) as follows : a geometric divisor d of K is an element of the free abelian group generated by $\mathcal{E}$ such that $d = \Sigma n_j \Lambda_j$ with $n_j = 0$ if $\Lambda_j \in \mathcal{E} - V(K)$. If v is a valuation of K then $v(d)$ will denote the coefficient of the valuation ring Λ_v of v in d. The integer $\deg d = \Sigma n_v f_v$ is called the degree of the divisor d . Let E be the intersection of those Λ_j in $E-V(K)$; then E is an R_1-order of K where R_1 is a Dedekind ring with field of fractions equal to $Z(K)$, R_1 is the intersection of those k-valuation rings of $Z(K)$ which do not extend to a valuation of K . If all k-valuations of $Z(K)$ extend to K then we put $R_1 = Z(K)$ and $E = K$.

To any $y \in E^{*}$ we may associate a (principal) divisor by putting $d(y) = \Sigma_v v(y)\Lambda_v$. The divisor of zeroes of $y \in E$ is defined by putting $z(y) = \sum_{v(y)>o} v(y),\Lambda_v$, and the divisor of poles of y may be defined as $p(y) = \sum_{v(y)<o} v(y),\Lambda_v$.

A divisor d_1 divides a divisor d_2 (written : $d_1 | d_2$) if and only if $v(d_1) \leqslant v(d_2)$ for all $v \in V(K)$.

If S is a subset of $V(K)$, put :

$\Gamma(d \mid S) = \{z \in K , v(z) \geqslant v(d)$ for all $v \in S\}$.

It is obvious that $\Gamma(d \mid S)$ is a k-vectorspace and if $S \subset S'$ then $\Gamma(d \mid S') \subset \Gamma(d \mid S)$.

Lemma 3.1

Let $S \subset V(K)$ be non-empty and finite, let $d_1 | d_2$, then :

$$\dim \frac{\Gamma(d_1|S)}{\Gamma(d_2|S)} = \deg d_2^S - \deg d_1^S$$

where $d_i^S = \sum_{v\in S} v(d_i)\Lambda_v$, $i = 1,2$.

Proof

Since the analogue of the approximation theorem does hold for valuations on K we may repeat the classical proof given in the commutative case as it may be found for example in (6), p.21.

To a divisor d of K we associate a k-vectorspace

$L(d) = \{ z \in E\ ,\ v(z) \geq v(d) \text{ for all } v \in V(K) \}$.

Lemma 3.2.

For each divisor d of K we have that $l(d) = \dim_k L(d)$ is finite and if d_1 divides d_2 then :

$l(d_1) + \deg d_1 \leq l(d_2) + \deg d_2$.

Proof

Let S be the subset of elements of V(K) appearing in d_1 or d_2 ; clearly $L(d_i) \subset \Gamma(d_i|S)$ and also $L(d_2) = L(d_1) \cap \Gamma(d_2|S)$.

From $\frac{L(d_1)}{L(d_2)} \to \frac{\Gamma(d_1|S)}{\Gamma(d_2|S)}$ it follows that $l(d_1) - l(d_2)$ $\leq \deg {}^S d_2 - \deg {}^S d_1 = \deg d_2 - \deg d_1$. On the other hand if we choose d_2 to be a multiple of the zero-divisor z and different from it, then $L(d_2) = (0)$ and thus $l(d_1) < \infty$. □

Lemma 3.3.

Take $t \in E^\star - 1$ and put $m = [K:k(t)]$,

Then $\deg z(t) = \deg p(t) = m$; in particular $\deg d(t) = 0$.

Proof

First note that if $x \in K$ is arbitrary then x is in almost all Λ_j selected as in Section 1. Indeed, write $x = \sum_{i=1} c_i u_i$, where u_i is the Z(K)-basis for K selected in Section 1.

1°) Let S be the subset of V(K) appearing in z(t).

Take any m+1 elements $z_1,\ldots,z_{m+1}$ of $\Gamma(z|S)$. Since $[K:k(t)] = m$ we have that $\sum_{j=1}^{m+1} q_j(t)z_j = 0$ with $q_j(t) \in k[t]$

and at least one $q_j(t)$ having non-zero constant term (otherwise multiply the equation by t^{-1} on the left). Write

$q_j(t) = q_j(0) + tq_j'(t)$, then :

$$(\star) \quad \Sigma_{j=1}^{m+1} q_j(0) z_j = - t \; \Sigma_{j=1}^{m+1} q_j'(t) z_j \; ; \text{ somme } \; q_{j_o}(0) = 0 \; .$$

Calculating $v_\nu(\star)$ for each $v_\nu \in S$ yields that

$$v_\nu (\Sigma_{j=1}^{m+1} q_j(0) z_j) = v_\nu \; (t) + v_\nu \; (\Sigma_{j=1}^{m+1} q_j'(t) z_j) \geqslant v_\nu (t) \; .$$

Therefore $\Sigma_{j=1}^{m+1} q_j(0) z_j \in \Gamma(z(t)|S)$. Now using Lemma 3.1. on

$d_1 = z$, $d_2 = z(t)$ we obtain :

$$\deg z(t) = \dim_k \frac{\Gamma(z|S)}{\Gamma(z(t)|S)} \leqslant m \; .$$

The same inequality holds for $p(t)$ (change from t to t^{-1} !)

2°) Put $\Lambda_t = \cap\{\Lambda_j, \Lambda_j \supset t\}$, $R_t = \Lambda_t \cap Z(K)$ and note that almost all Λ_j are being used in the intersection yielding Λ_t . We know that R_t is a Dedekind ring with field of fractions equal to $Z(K)$ since t is non-algebraic (this is the Riemann-Roch theorem in $Z(K)$!) and that Λ_t is a maximal R_t-order in K . An arbitrary left $k(t)$-basis for K, $\{y_1', \ldots, y_m'\}$ say, may be multiplied by an element λ of R_t such that $\lambda y_i' \in \Lambda_t$, $i = 1,\ldots,m$. Since $\lambda \in Z(K)^\star$ it is easily seen that, putting $y_i = \lambda y_i'$, we find a left $k(t)$-basis $\{y_1,\ldots,y_m\}$ for K which is contained in Λ_t .

Obviously $t^i y_j \in L(p(t)^{-k})$ if $i = 0,\ldots,k$ and $j = 1,\ldots,m$. Since the $t^i y_j$ are $m(k+1)$ in number and k-linearly independent we find :

(★) $m(k+1) \leqslant l(p(t)^{-k}) \leqslant l(p(t)) + \deg p(t) - \deg p(t)^{-k}$ (using Lemma 3.2. on $p(t)^{-k} \mid p(t))$.

So, as k grows arbitrarily large we see that (★) yields $\deg p(t) \geqslant m$. Together with 1 we have now established that $\deg p(t) = m$ and $\deg d(t) = 0$.

Corollary 3.4

We have an exact sequence of groupe :

$1 \to 1^\star \to E^\star \to div_o \to K_o \to 0$ where Div_o is the (additively written) group of divisors of K of degree 0 and K_o is called the group of divisor classes of degree 0 .

Proof

Since for every $z \in E^\star$, $\lambda \in 1^\star$, $z \lambda z^{-1} \in E$ is k-algebraic i.e. in $1^\star$ it follows that $1^\star$ is a normal subgroup of $E^\star$. Moreover it is clear that a commutator $uvu^{-1}v^{-1}$ with $u,v \in E^\star$ is in all k-valuation rings of K and in E hence in $1^\star$, therefore $1^\star$ contains the commutator subgroup of $E^\star$ (thus $E^\star/1^\star$ fits in Div_o !). □

If K is commutative $E^\star = K^\star$. If $K = \mathbb{C}(X,-)$ for example, as a matter of fact this holds for any $D(X,\varphi)$ as considered in Section 2, then $E^\star$ contains at least X since X and X^{-1} are contained in all localizations at prime ideals of $D[X,\varphi]$ or $D[X^{-1}, \varphi^{-1}]$ with exception only at $D[X,\varphi]_{(X)}$ and $D[X^{-1}, \varphi^{-1}]_{(X^{-1})}$ but the latter rings are D-valuation rings of $D(X,\varphi)$.

Note that $X \in K$ is in E if its minimal polynomial over Z(K) is of the form $T^n + a_1 T^{n-1} + \ldots + a_n = 0$ with $a_i \in R$; and if a_n is a unit of R then X is in $E^\star$, the converse follows from $-a_n^{-1}(x^{n-1} + \ldots + a_1) = X^{-1} \in E^\star$. (Because if a_n^{-1} is in one of the maximal ideals of the k-valuation rings over R then X^{-1} is in one of the maximal ideals of the selected orders over E , contradiction).

Thus elements of E with norm in R^* are in E^*.

The invariant subgroup R^*/k^* of $E^*/1^*$ is the subroup of the central divisors of $Z(K)$ of degree 0 which "extend" to divisors of degree 0 of K. In Riemann's theorem we only need the existence of transcendental elements in $E^*/1^*$ and this is of course garantueed by the above inclusion of R^*/k^* in $E^*/1^*$.

Theorem 3.5 (non-commutative version of Riemann's theorem).

There exists an integer g such that for each divisor d of K we have $1(d)+\deg d \geq 1-g$ and if $\deg d$ is "large enough" then we obtain $1(-d)+\deg(-d) = 1-g$.

Proof

From (*) in the proof of Lemma 3.3. one easily derives that for each $t \in E^* - 1^*$ there exists an integer, Q say, depending only on t such that $1(p(t)^{-m}) + \deg p(t)^{-m} \geq -Q$ for every $m \in Z$. Denote by $1-g$ the lower bound of $1(p(t)^{-m}) + \deg p(t)^{-m}$ for $m \in Z$. From here on the proof runs along the same lines as the proof of the Riemann theorem in the commutative case (of course for principal divisors we use the $d(t)$ with $t \in E^*$).

The integer g defined above depends on the choice of 1, at least so it seems. Let us denote by g_1 the genus as definded above with respect to the chosen 1, let us put $g = \text{ing}\{g_1$, 1 maximal k-algebraic in $K\}$, we call g the geometric genus of K. Note that we can always suppose 1 is chosen such that $g_1 = g$.

Theorem 3.6.

If the field $Z(K)$ has class number 1 then every maximal k-algebraic subring of K is conjugated to 1. Moreover g_1 does not depend on 1, i.e. $g = g_1$ and g does not depend upon the choice of the maximal orders Λ_j for which $1 = \bigcap_j \Lambda_j$.

Proof

If $Z(K)$ has class number 1 and R_1 is a Dedekind ring of $Z(K)$ with field of fractions $Z(K)$ then all maximal R_1-orders of K are isomorphic (the number of isomorphy-classes divides the class number !) and therefore conjugated.

First suppose that 1 is fixed and let $\{\Lambda'_j, j \in A \cup B\}$ be a second choice of maximal O_j-orders such that $\bigcap_j \Lambda'_j = 1$. The substitution $\Lambda_j \to \Lambda'_j$ comes down to substituting a maximal R_1-order E' for E, since the k-valuation rings of K, Λ_v, are the unique maximal O_v-orders of K and hence these remain fixed under the change $\Lambda_j \to \Lambda'_j$. By our opening statement: $E' = \lambda E \lambda^{-1}$ for some $\lambda \in K^*$ and divisors with respect to E are transformed into divisors with respect to E'. Obviously all k-dimensions we study remain unaltered and thus $1-g$ remains the same. Secondly let 1_1 be another choice for 1 and write $1_1 = \bigcap_j \Lambda_j^{(1)}$ for well-chosen maximal O_j-orders $\Lambda_j^{(1)}$. If E_1 is the intersection of those $\Lambda_j^{(1)}$ which are not k-valuation rings of K then, as before, $E_1 = \lambda E \lambda^{-1}$ for some $\lambda \in K^*$. However $\lambda 1 \lambda^{-1}$ is then a k-algebraic subring of E_1, but k-algebraic elements are also in all k-valuation rings, thus $\lambda 1 \lambda^{-1} \subset 1_1$ follows.

But $1 \subset \lambda^{-1} 1_1 \lambda$, where $\lambda^{-1} 1_1 \lambda$ is a k-algebraic ring, implies that $1 = \lambda^{-1} 1_1 \lambda$. As before it is easily verified that "conjugation by λ" will transform divisors with respect to E into divisors with respect to E_1, i.e. for all divisors d, $1(d)$, $\deg d$ remain unaltered and so does g. □

We are now able to formulate the non-commutative version of Riemann-Roch's theorem using repartitions. A <u>repartition</u> P of K is a mapping of the set $\mathcal{P}$ into K such that $P(\Lambda_j) \in \Lambda_j$ for almost all $j \in A \cup B$. The set $\mathcal{R}$ of repartitions of K is a k-algebra and K may be identified with a subfield of R by identification of $z \in K$ with $P_z \in R$ which is defined by $P_z(\Lambda_j) = z$. The k-valuations of K extend to "valuations" of R in a natural way : $v(P) = v(P(\Lambda_v))$. We say that a <u>divisor d divides a repartion</u> P if and only if $v(P) \geq v(d)$ for all $v \in V(K)$ and $P(\Lambda)$ $P(\Lambda) \in \Lambda$ for every $\Lambda \in \mathcal{P} - V(K)$. Put $\Lambda(d) = \{P \in R, d | P\}$. Then $\Lambda(d)$ is a k-subspace of R. If $d_1 | d_2$ then $\Lambda(d_1) \supset \Lambda(d_2)$

and it is easily verified that $\frac{\Lambda(d_1)}{\Lambda(d_2)}$ is k-isomorphic $\frac{\Gamma(d_1|S)}{\Gamma(d_2|S)}$

where S is the set of k-valuations appearing in d_1 or d_2.
The following lemma follows easily from this and the fact that k-valuations of K can be calculated with (approximation theorem) as in the commutative case.

Lemma 3.7

Let $d_1 | d_2$ be divisors of K, then :

1°) $\dim_k \frac{\Lambda(d_1)}{\Lambda(d_2)} = \deg d_2 - \deg d_1$

2°) $\dim_k \frac{\Lambda(d_1)+K}{\Lambda(d_2)+K} = l(d_2) + \deg d_2 - l(d_1) - \deg d_1$.

(2° follows in a straightforward way from $(\Lambda(d_2) + K) \cap \Lambda(d_1) = \Lambda(d_2) + L(d_1)$).

Theorem 3.9

Let d be any divisor of L then :

$$\dim_k \frac{R}{\Lambda(d)+K} = l(d) + \deg d + g-1$$

In particular for the zero-divisor z we obtain

$$\dim_k \frac{R}{\Lambda(z)+K} = g + n - 1$$

Proof

Now similar to the proof on p. 34 of (6). Note that we have denoted g_l by $g, n = [l:k]$, but the above relations hold for every choice of l. □

Finally let us consider the perfect case.

Theorem 3.10

If k is perfect then $g_l = g$, for each choice of l and the maximal O_j-orders Λ_j such that $l = \bigcap_j \Lambda_j$.

Proof

From E.Witt's Riemann-Roch theorem applied to a positive divisor we find that $g_1 = g_w$ for each choice of 1 . In other words, since in (16) the definition of g_w does not depend on the choice of the maximal orders used in the construction we may choose these orders in the way we did in this paper and thus we find $g_1 = g_w$.

4 The case $K = D(X,\varphi)$

We use the notations of Section 2 . It has been established that $D(X,\varphi)$ allows non-trivial k-valuations e.g. $D[X,\varphi]_{(X)}$ and $D[X^{-1},\varphi^{-1}]_{(X^{-1})} = \Lambda_\infty$.

Since k is the fixed field of $k' = Z(D)$ under the automorphism φ it follows that k'/k is a separable extension. Therefore the formula of Proposition 1.4. simplifies to : $N = n[k':k]$. Put $d_m = -m\Lambda_\infty$.

Due to the fact that K is obtained from $D[X,\varphi]$ by inverting central elements it is easy to adapt C . Chevalley's original proof of the fact that the k-vectorspaces $L(d_m)$ are exactly given by $\{f \in K, f \in D[X,\varphi], \deg_X f \leqslant m\}$ to this case.

Hence $l(d_m) = n(m+1)$. Since $\deg d_m = nm$ we find, for large m : $1-g = n(m+1)-nm = n$, and thus $g = 1-n$. For example, if $D = F$ is a commutative field, φ an automorphism of F such that $\varphi^e = 1$ for $e \in \mathbb{N}$ (chosen minimal as such), then we have $n = \sqrt{N}$ and $g = 1 - \sqrt{N}$.

5 Genus Versus Genus of the Centre

If $v \in V(K)$ then e_v is the ramification index of v over v^c (cf. Section 1).

Lemma 5.1

Let $d = \Sigma n_v v$ be a divisor of K with $n_v \in e_v Z$. Let d^c be the divisor of $Z(K)$ given as $\Sigma n_v^c v^c$ with $n_v^c = \frac{n_v}{e_v}$.

Then $\deg d = N.\deg d^c$.

Proof

$\deg d = \Sigma e_v n_v^c f_v = \Sigma e_v n_v^c \Psi_{v^c} f_{v^c} = N.\ \Sigma n_{v^c}^c f_{v^c}$;

for the latter equality one uses $N = e_v \Psi_{v^c}$.

Clearly $\deg d = N.\deg d^c$ follows. □

Let $\mathcal{R}_{Z(K)}$ be the k-space of all repartitions of $Z(K)$, while the k-space of repartitions of K is now denoted by $\mathcal{R}_K$. If d is a divisor of K which has a divisor d^c associated to it in the above way, then we let $\Lambda_K(d)$ and $\Lambda_{Z(K)}(d^c)$ be the corresponding k-subspaces of $\mathcal{R}_K$ and $\mathcal{R}_{Z(K)}$ resp.

Note that for the zero-divisor z of K, z^c is the zero-divisor of $Z(K)$. Fix a $Z(K)$-basis $\{u_1,\ldots,u_N\}$ for K .

It is possible to define a k-linear map $\Pi : \mathcal{R}^N_{Z(K)} \to \mathcal{R}_K$, by putting $P = \Pi(P_1^c,\ldots,P_N^c)$ where : $P(\Lambda_j) = \Sigma_{i=1}^N u_i P_i^c(O_j)$.

It is clear that P is a repartition of K . Put $R^\Pi = \operatorname{Im}\Pi$. Because of our definition of "repartitions" it is easily checked that Π is injective (note that this would have failed if repartitions were defined using valuations of K alone since not every k-valuation of $Z(K)$ extends to a k-valuation of K !).

Lemma 5.2

Let d be any divisor of K then $R_K = R^\Pi + \Lambda_K(d)$.

Proof

Write $d = \Sigma n_v v$ and let $\{\Lambda_1,\ldots,\Lambda_s\}$ be the finite number of maximal orders in $\{\Lambda_j , j \in \mathcal{A} \cup \mathcal{B}\}$ where $P(\Lambda_j) \notin \Lambda_j$ for some given $P \in R_K$. Now define an assignement $\Lambda_j \to y_j \in K$ as follows : if $\Lambda_j = \Lambda_v$, i.e. a k-valuation ring of K choose $y_j = y_v \in K$ such that $v(y_v - P(\Lambda_v)) \geq n_v$, if Λ_j is not a valuation ring but in the set $\Lambda_1,\ldots,\Lambda_s$ then choose y_j such that $y_j - P(\Lambda_j) \in \Lambda_j$, if Λ_j is not of the form Λ_v nor in the set $\{\Lambda_1,\ldots\Lambda_s\}$ choose $y_j = 0$. It is clear that this assignement defines a repartition, M say, of K such that $M-P \in \Lambda_K(d)$. It remains to be verified that $M \in R^{\Pi}$. If Λ_i is a k-valuation ring or in the set $\{\Lambda_1,\dots\Lambda_s\}$ we put $y_i = \sum_{k=1}^{N} \lambda_k^i u_k$ with $\lambda_k^i \in Z(K)$.

We may define repartitions M_k^c of $Z(K)$ as follows :

$M_k(O_i) = \lambda_k^i$, and $M_k(O_1) = 0$ if Λ_1 is not in $\{\Lambda_1,\dots,\Lambda_s\} \cup$ k-valuation rings of K .

Obviously $\Pi(M_1^c,\dots,M_N^c) = M$ and $M \in R^{\Pi}$. □

Writting down Riemann-Roch's theorem for $Z(K)$, and for K as we established it, yields :

$g_k + n-1 = \dim_k R_k / \Lambda_K(z) + K$

$g_{Z(K)} = \dim_k R_{Z(K)} / \Lambda_{Z(K)}(z) + Z(K)$.

Taking the direct sum of N-times the exact sequence :

$0 \to \Lambda_{Z(K)}(z) + Z(K) \to R_{Z(K)} \to R_{Z(K)} / \Lambda_{Z(K)}(z) + Z(K) \to 0$, and counting k-dimensions afterwards, yields :

$Ng_{Z(K)} = \dim_k R^N_{Z(K)} / \Lambda^N_{Z(K)}(z) + Z(K)^N$.

Since $\Pi(\Lambda^N_{Z(K)}(z) + Z(K)^N) = \Lambda^{\Pi}_{Z(K)}(z) + K$ we obtain :

$$\begin{array}{ccccccccc}
0 \to & \Lambda^N_{Z(K)}(z) + Z(K)^N & \to & R^N_{Z(K)} & \to & R^N_{Z(K)} / \Lambda^N_{Z(K)}(z) + Z(K)^N & \to 0 \\
 & \downarrow \text{res} & & \downarrow \Pi & & \downarrow & \\
0 \to & \Lambda^{\Pi}_{Z(K)}(z) + K & \to & R^{\Pi} & \to & R^{\Pi} / \Lambda^{\Pi}_{Z(K)}(z) + K & \to 0
\end{array}$$

with exact rows, where $\Lambda^{\Pi}_{Z(K)}(z)$ is $\Pi\,\Lambda_{Z(K)}(z)$.

Calculating k-dimensions yields :

(*) $\quad Ng_{Z(K)} = \dim_k R^{\Pi} / \Lambda^{\Pi}_{Z(K)}(z) + K$

and on the other hand

$$g_K + n-1 = \dim_k R_K/\Lambda_K(z) + K = \dim_k\left(R^{\Pi} + \Lambda_K(z) + K \Big/ \Lambda_K(z) + K\right)$$

$$= \dim_k\left(R^{\Pi}\Big/\Lambda_K^{\Pi}(z) + K\right) - \dim_k\left(R^{\Pi} \cap (\Lambda_K(z) + K)/\Lambda_K^{\Pi}(z) + K\right).$$

Combination of (*) and (**) yields :

$$Ng_{Z(K)} - g_k - n + 1 = \dim_k\left(R^{\Pi} \cap (\Lambda_K(z) + K)\Big/\Lambda_K^{\Pi}(z) + K\right)$$

So we have proved :

Theorem 5.3

The genus g_K of K is smaller than or equal to $Ng_{Z(K)}-n+1$.

Corollary 5.4

$Ng_{Z(K)}-n+1 \geqslant g_K \geqslant 1-n$.

Hence if $g_{Z(K)} = 0$ (as in the case $D(X,\varphi)$) then we have that $g_K = 1-n$.

Remark 5.5

We do not know of any example where $Ng_{Z(K)}-n+1 \neq g_K$ and we feel tempted to conjecture that equality holds in general but we will not do that here.

Remark 5.6

A "completely" irreducible curve in the sense of M. Artin (2) or A. Verschoren (14), may be viewed as the set of maximal ideals of some affine P.I. ring R which is obtained as $k\{\xi_1,\xi_2\}/P$, where ξ_1 and ξ_2 are generic mxm matrices for some m , and P is a completely prime ideal of the generic matrix ring $k\{\xi_1,\xi_2\}$. If we put K equal to the field of functions of R then the above Riemann-Roch theorem has a geometric meaning. In a forthcoming paper the Riemann-Roch theorem and its consequences will be derived for arbitrary central simple algebras over algebraic function fields of one variable (extending E. Witt's results to the non-perfect case) using primes of algebras (11), (13). Then it will be possible to treat irreducible curves (i.e. P is just a prime ideal).

Acknowledgement

J. VAN GEEL thanks the Department of Mathematics of Bedford College (London) for the hospitality enjoyed there and, in particular, Prof. Dr. P.M. COHN for his encouragement and stimulating conversations.

References

(1) A. Albert, Structure of Algebras, A.M.S. Colloquim Publ., New York, 1939.

(2) M. Artin, On Azumaya Algebras and Finite Dimensional Representation of Rings, J. of Algebra 11, 1969, pp.532-563.

(3) G. Cauchon, Les T-anneaux et les anneaux à identités polymiales Noetheriens, Ph.D. Thesis, Université de Paris-Sud XI, Centre d'Orsay, 1977.

(4) C. Chevalley, Introduction to the theory of Algebraic Functions of One Variable, A.M.S. Math. Surveys nr. VI, 1951.

(5) P. Cohn, Total subring in division algebras, Prepint, Bedford College, London 1979.

(6) M. Deuring, Lectures on the Theory of Algebraic Functions of One Variable. LNM 314, Springer-Verlag Berlin, 1973.

(7) E. Nauwelaerts, F. Van Oystaeyen, Birational Hereditary Noetherian Prime Rings, to appear soon in Communications in Algebra.

(8) I. Reiner, Maximal Orders, LMS monographs, Academic Press, London, 1975.

(9) F.K. Schmidt, Zur Aritmetischen Theorie der Algebraïschen Funktionen I, Math. Zeitschrift 41, 1936.

(10) J.P. Van Deuren, Paramétrisation non-commutative, rapport no, 71, Séminaire de Math. Pure, Louvain La Neuve.

(11) J. Van Geel, Primes in Algebras and the Arithemetic in Central Simple Algebras, to appear in Communication in Algebra.

(12) F. Van Oystaeyen, Zariski Central Rings, Communications in Algebra, 6(8), pp. 799-821, 1978.

(13) F. Van Oystaeyen, Prime Spectra in Non-commutative Algebra, LNM 444, Springer Verlag, Berlin, 1975.

(14) A. Verschoren, Some ideas in Non-commutative Algebraic Geometry, Ph.D. Thesis, University of Antwerp, UIA, 1979.

(15) A. Weil, Basic Number Theory, Springer-Verlag Berlin, 1973.

(16) E. Witt, Riemann-Rochser Satz un Z-Funktion im Hyperkomplexen, Math. Ann., Bd, 110, pp.12-28, 1934.

July 1979.

POUR UNE GEOMETRIE ALGEBRIQUE NON-COMMUTATIVE

A. VERSCHOREN (*)

Université d'Anvers.

UIA.

Cette note reprend quelques résultats récents en géométrie algébrique dite "non-commutative", qui sont dus en grande partie à PROCESI, ARTIN, SCHELTER, VAN OYSTAEYEN et l'auteur. Nous ne donnons aucune démonstration, mais nous renvoyons chaque fois aux sources du résultat et à [43] pour un exposé plus complet.

1 - Les représentations d'un anneau non-commutatif, d'après Artin.

(1.1) Rappelons d'abord quelques réflexions exprimées par ARTIN dans [3,5]. L'étude des variétés algébriques dans la théorie classique se réduit essentiellement à celle des algèbres affines commutatives sur un corps. Soit k un corps (commutatif !) algébriquement clos. Une algèbre affine sur k est une algèbre R qui est associative et de type fini sur k, c'est-à-dire qui a une présentation de la forme :

$$R = k\{X_1,\ldots,X_n\}/I$$

où I est un idéal bilatère de l'algèbre $k\{X_1,\ldots,X_n\}$ des polynômes non commutatifs en n variables sur k. Afin d'étudier ces algèbres, nous aurons à considérer leurs représentations.

(*) L'auteur bénéficie d'une bourse de recherche à l'N.F.W.O-CNRB

(1.2) - Au premier abord, on se limitera aux représentations :

$$\pi : R \longrightarrow M_n(k)$$

de l'algèbre affine R dans l'algèbre des matrices de dimension n sur k. Une telle représentation est dite <u>irréductible</u> si l'application π est surjective. D'autre part, deux représentations sont <u>équivalentes</u> si elles ne diffèrent que par un automorphisme de $M_n(k)$. Le point essentiel est de remarquer que les classes d'équivalence de représentations irréductibles de R correspondent biunivoquement à certains de ses idéaux maximaux. En effet, il est facile de voir que la classe d'équivalence d'une représentation irréductible π est complètement déterminée par son noyau $\mathrm{Ker}\,\pi$, qui est évidemment un idéal maximal de R. De cette façon l'on obtient une partition :

$$\Omega(R) = \Omega_1(R) \cup \ldots \cup \Omega_\infty(R)$$

où $\Omega(R)$ est l'ensemble des idéaux maximaux de R et où pour chaque entier positif n, l'ensemble $\Omega_n(R)$ consiste des idéaux maximaux M de R ayant la propriété que R/M est isomorphe à $M_n(k)$ et où $\Omega_\infty(R)$ contient les idéaux maximaux de R n'ayant cette propriété pour aucun n.

(1.3) - On peut mettre sur $\Omega(R)$ la topologie de Zariski induite par celle de Spec(R), l'ensemble des idéaux premiers de R. En particulier, les ensembles ouverts de $\Omega(R)$ sont de la forme :

$$X_I = \{M \in \Omega(R) \ ; \ I \not\subset M\}$$

où I est un idéal (bilatère !) de R. Si nous dénotons par $\Lambda_p(R)$ l'union $\bigcup_{n \leq p} \Omega_n(R)$, on peut démontrer que $\Lambda_p(R)$ est un sous-ensemble fermé de $\Omega(R)$ pour tout entier positif p. Ceci implique que chaque $\Omega_n(R)$ est localement fermé dans $\Omega(R)$. Mike ARTIN a démontré dans [4] que chaque $\Omega_n(R)$ possède la structure d'une variété algébrique (commutative, bien entendu).

(1.4) - Nous voulons étudier $\Omega(R)$ de façon topologique, afin d'en déduire des propriétés structurales de R, comme on le fait dans le cas commutatif. Néanmoins, ce point de vue est loin de donner des résultats convaincants dans tous les cas. Citons d'abord quelques problèmes qui peuvent se présenter à ce sujet.

(1.4.1) Dans certains cas, tous les ensembles $\Omega_n(R)$ sont vides, à l'exception de $\Omega_\infty(R)$ près. Exemple : l'algèbre de Weyl $k\{X,Y\}/(XY-YX-1)$. L'étude de ces exemples du point de vue des représentations est inepte.

(1.4.2) Il est tout à fait évident que nous ne pouvons étudier qu'un nombre fini de $\mathcal{R}_n(R)$ à la fois, et on sait qu'en général $\mathcal{R}(R)$ ne peut être décrit comme un espace de dimension finie. Exemple : si $R = k\{X,Y\}$, on trouve que $\dim \mathcal{R}_n(R) = n^2+1$. En contraste, si $R = k\{\xi_1^{(p)},\dots,\xi_n^{(p)}\}$, l'algèbre des matrices génériques de dimension p, alors $\dim \mathcal{R}(R) = np^2-(p^2-1)$, cf.[29,31]. Nous reviendrons plus tard sur cet exemple.

(1.4.3) Finalement, il est difficile, sinon <u>très</u> difficile, de déterminer la structure exacte des $\mathcal{R}_n(R)$, même dans les cas les plus élémentaires. ARTIN et SCHELTER ont surtout étudié ces espaces à l'aide de la notion de courbe non-commutative, comme nous le verrons plus loin.

(1.5) - C'est à ce point-là qu'interviennent de façon naturelle les algèbres à identité polynomiale. En effet, il est bien connu qu'il existe des identités polynomiales qui sont satisfaites par toute algèbre de matrices de dimension n. Ce sont des polynômes à coefficients dans $\mathbb{Z}$, tels que toute substitution de matrices, à la place des variables, du polynôme considéré se réduit à zéro. Un exemple type de tels polynômes est le <u>polynôme standard</u> :

$$S_{2n}(X_1,\dots,X_{2n})$$

où

$$S_m(X_1,\dots,X_m) = \sum_{\sigma\in\mathfrak{S}_m} (-1)^{\sigma} X_{\sigma(1)}\cdot\ldots\cdot X_{\sigma(m)} .$$

Il a la propriété d'être une identité polynomiale pour les matrices de dimension inférieure ou égale à n et ne l'est pas pour les algèbres de matrices de dimension supérieure à n. Ainsi, si nous dénotons par J_n l'idéal de R engendré par toutes les substitutions d'éléments de R dans S_{2n}, on voit que :

$$V(\mathfrak{J}_n) = \mathcal{R}(R) - X_{\mathfrak{J}_n} = \mathcal{R}(R/J_n) = \mathcal{R}_1(R)\cup\ldots\cup\mathcal{R}_n(R) .$$

Donc si nous voulons étudier $\mathcal{R}_1(R)\cup\ldots\cup\mathcal{R}_n(R) = \Lambda_n(R)$, c'est-à-dire les n premières composantes de $\mathcal{R}(R)$, nous pouvons remplacer R par $\overline{R} = R/J_n$, qui est une algèbre à identité polynomiale. D'après cette remarque évidente, il est clair que les algèbres à identité polynomiale joueront un rôle de premier plan dans l'étude des algèbres affines du point de vue géométrique.

(1.6) - Bien que nous connaissons (théoriquement, bien sûr !) la structure des différentes composantes $\mathcal{R}_n(R)$ du spectre maximal $\mathcal{R}(R)$ -ce sont des variétés algébriques ordinaires- le point essentiel est de décrire comment ces pièces sont collées ensemble. Le plan d'attaque moderne pour ce genre de problème, consiste à recoller des faisceaux structurals canoniques sur ces composantes ou à mettre sur $\mathfrak{M}(R)$ lui-même (ou sur une partie $\Lambda_n(R)$, dans notre point de vue) un faisceau structural globalisant l'information connue sur chaque $\mathfrak{M}_n(R)$. En principe, il y a plusieurs façons de le faire, comme nous le verrons plus bas. Toutefois, le choix d'un faisceau convenable dépendra directement de certains critères fonctoriels.

(1.7) - Considérons le problème d'un peu plus près. Si nous voulons étudier les algèbres affines à partir de la structure de leur spectre maximal, le minimum absolu que nous pouvons espérer, c'est qu'à un morphisme d'algèbres corresponde un morphisme (dual) entre les spectres maximaux correspondants. En général, ce n'est pas le cas et c'est facile à voir. Nous pouvons y remédier de plusieurs façons.

(1.7.1) Un morphisme d'anneaux $\varphi : R \longrightarrow S$ induit une correspondance non-vide $\mathfrak{M}(S) \multimap \mathfrak{M}(R)$. En effet, si M est un idéal maximal de S, l'anneau $R/\varphi^{-1}(M)$ est non-trivial et ses idéaux maximaux (ou, de façon équivalente, ceux de R qui y correspondent, sont ceux que l'on fait correspondre à M. En plus, si $M \in \mathfrak{M}_n(S) \subset \mathfrak{M}(S)$, alors $R/\varphi^{-1}(M)$ peut être considéré comme un sous-anneau de $M_n(k)$ et n'a donc qu'un nombre fini d'idéaux maximaux. Il en résulte que la correspondance qu'on vient de définir induit pour chaque entier positif n une correspondance finie :

$$\mathfrak{M}_n(S) \multimap \Lambda_n(R)$$

qui, comme on peut le démontrer, cf.[5], est algébrique par rapport aux structures de variété algébrique sur $\mathfrak{M}_n(S)$ et sur les $\mathfrak{M}_p(R)$. Au lieu de considérer des morphismes entre spectres maximaux, nous pouvons alors étudier des correspondances entre ces espaces, afin d'en déduire des résultats algébriques. Ce point de vue donne d'une part des résultats interessants comme l'a indiqué ARTIN, cf.[3,5], mais d'autre part, comme on le sait, on ne connait pas encore très bien le comportement des faisceaux par rapport aux correspondances.

(1.7.2) On peut se restreindre aux morphismes d'anneaux <u>géométriques</u> ; rappelons (avec ARTIN et SCHELTER, cf[6]) qu'un morphisme d'anneaux $\varphi : R \longrightarrow S$ est dit géométrique si pour chaque idéal maximal $M \in \mathfrak{M}(S)$ le morphisme induit :

$$R/\varphi^{-1}(M) \longrightarrow S/M$$

est un isomorphisme. Un morphisme géométrique induit donc un morphisme

$$\begin{aligned} \mathfrak{M}(S) &\longrightarrow \mathfrak{M}(R) \\ M &\longmapsto \varphi^{-1}(M) \end{aligned}$$

que l'on prouve facilement continu pour les topologies de Zariski. Comme exemple on peut citer les extensions centrales. Autre exemple, si R est un sous-anneau d'un anneau Q et $\alpha \in R$ est inversible dans Q, alors l'anneau $R\{\alpha^{-1}\}$ engendré par α^{-1} et R donne un morphisme géométrique $R \longrightarrow R\{\alpha^{-1}\}$. Ces morphismes géométriques ont des propriétés agréables, telles que la stabilité pour les compositions et le fait que $\varphi : R \longrightarrow S$ est géométrique si et seulement si, pour chaque idéal premier $P \in \mathrm{Spec}_n(S)$ l'image inverse $\varphi^{-1}(P)$ est contenu dans $\mathrm{Spec}_n(R)$ et cela pour tout entier positif n. Rappelons d'ailleurs que $\mathrm{Spec}_n(R)$ est l'ensemble des idéaux premiers P de R tels que l'anneau R/P possède un anneau de fraction simple et de dimensin n^2 sur son centre ; nous y reviendrons plus loin. Bien que l'emploi de ces morphismes ait donné des résultats convaincants, tels qu'une version non-commutative du Théorème Principal de Zariski dû à ARTIN et SCHELTER, ils ne semblent pas réellement intéressant du point de vue faisceautique.

(1.7.3) On peut se restreindre aux extensions d'anneaux $\varphi : R \longrightarrow S$, où, rappelons-le, le morphisme φ est appelé une <u>extension</u> (au sens de PROCESI, cf[30,31]) si S est engendré en tant qu'anneau par $\varphi(R)$ et le centralisateur

$$Z_R(S) = \{ s \in S \; ; \; \varphi(r)\, s = s\, \varphi(r) \quad \text{quel que soit} \quad r \in R \}$$

Ces morphismes ont des propriétés proches de celles rencontrées pour les morphismes géométriques et certains morphismes géométriques intéressants, tels que les extensions centrales, en sont des cas particuliers. En particulier, une extension $\varphi : R \longrightarrow S$ induit un morphisme continu

$$a_\varphi : \begin{array}{l} \mathrm{Spec}(S) \longrightarrow \mathrm{Spec}(R) \\ \quad P \longmapsto \varphi^{-1}(P) \end{array}$$

et, si R et S sont affines et à identité polynomiale, on obtient par restriction un morphisme continu :

$$a_\varphi : \mathcal{R}(S) \longrightarrow \mathcal{R}(R) \quad .$$

En effet, si M est un idéal maximal de S, on obtient une extension :

$$\pi : k \longrightarrow S \longrightarrow S/M \; .$$

Si on applique à π le résultat bien connu de PROCESI [30,31] qui dit que si $R \subset S = R\{a_1,\ldots,a_n\}$ est une extension de type fini entre anneaux premiers à identité polynomiale, où R est semi-simple et S est simple, alors R est simple et S est de dimension finie sur le centre de R, on obtient que S/M est de dimension finie sur k. Mais alors son centre Z(S/M) l'est aussi, et celui-ci est un corps, comme S/M est simple et à identité polynomiale. On en déduit que Z(S/M) = k, puisque k est algébriquement clos par hypothèse et il en résulte que $S/M = M_n(k)$ pour un entier positif n. Comme $R/\varphi^{-1}(M)$ est un anneau premier à identité polynomiale on sait, d'après le théorème de Posner, que $R/\varphi^{-1}(M)$ possède un anneau de fractions classique $Q(R/\varphi^{-1}(M))$ et on obtient un diagramme commutatif :

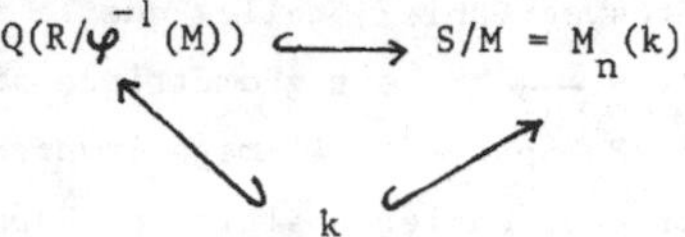

On obtient que le centre de l'anneau central simple $Q(R/\varphi^{-1}(M))$ est le corps k, d'où $Q(R/\varphi^{-1}(M)) = M_p(k)$ pour un entier positif p (qui divise n) et le centre de $R/\varphi^{-1}(M) \subset Q(R/\varphi^{-1}(M))$ est le corps k. Comme R est un anneau à identité polynomiale, on voit que $R/\varphi^{-1}(M)$ est simple et que $\varphi^{-1}(M)$ est maximal.

Bien qu'on pourrait essayer de formuler (et de prouver !) des résultats pour des morphismes plus généraux que les extensions, le point essentiel et de remarquer que, côté faisceaux, les extensions donnent des résultats très satisfaisants, comme nous le verrons plus loin.

(1.8) - Donnons une application des points de vue d'ARTIN et SCHELTER. L'outil essentiel dans leur attaque de problèmes géométriques est la notion de courbe. Soit D un anneau de Dedekind, de type fini sur k et soit A un D-ordre dans l'anneau des matrices $M_n(K)$ où K est le corps des fractions de D. On appelle $\Omega(A)$ une courbe

Exemple. Soit $A = \begin{pmatrix} k[X] & (X) \\ (X) & k[X] \end{pmatrix}$. On calcule facilement son spectre maximal $\Omega(A)$ qui consiste en les idéaux fA où $f \neq X$ est irréductible dans k[X], c'est-à-dire $f = X-a$ avec $a \neq 0$ et les idéaux

$$P = \begin{pmatrix} (X) & (X) \\ (X) & k[X] \end{pmatrix} \quad , \quad Q = \begin{pmatrix} k[X] & (X) \\ (X) & (X) \end{pmatrix}$$

Nous voyons donc que $\Omega(A)$ peut s'appliquer dans $\Omega(D)$, avec $D = k[X]$ dans notre exemple, où $\Omega_2(A)$ correspond homéomorphiquement à l'ensemble ouvert de $\Omega(D) = \mathbb{A}^1_k$, qui consiste des points p avec la propriété :

$$A \otimes_k k(p) = M_2(k).$$

Les autres points correspondent aux idéaux maximaux de $A \otimes_k k(p_o)$ où $p_o = (X) \subset k[X]$. Il est évident qu'on peut généraliser cette description à une courbe arbitraire.

(1.9) - Appliquons maintenant ces idées à l'étude de la notion d'intégralité. Rappelons qu'on appelle un morphisme d'anneaux $\varphi : R \longrightarrow S$ <u>intégral</u> si chaque élément $s \in S$ satisfait à une relation :

$$s^n = \sum m_i$$

où chaque m_i est un mot composé de s et d'éléments de R et de degré en s strictement inférieur à n. Le problème est que, par opposition au cas commutatif, on ne sait pas très bien à quoi pourrait servir cette notion. En plus, il n'est pas clair, par exemple, pourquoi le composé de deux morphismes intégraux serait intégral lui-aussi. Ceci est en rapport avec le fait que, dire que S est un R-module fini, n'a rien à voir avec l'intégralité de S sur R dans le cas non-commutatif. Pour des anneaux à identité polynomiale la situation est beaucoup plus sympathique. Définissons d'abord un morphisme fini comme un morphisme d'anneaux $\varphi : R \longrightarrow S$ tel que le centre $Z(\bar{S})$ de tout quotient $\bar{S}$ de S qui est premier, satisfait à une identité polynomiale et qui est de dimension de Krull égal à 1, est intégral sur R.

(1.10) - Quelques propriétés (cf. [5,7]).

(1.10.1) Un morphisme intégral est fini ;

(1.10.2) $\varphi : R \longrightarrow S$ est fini si et seulement si la correspondance ${}^a\varphi : \Omega(S) \longrightarrow \Omega(R)$ est propre dans le sens que le critère valuatif en géométrie algébrique commutative est satisfait, si on le traduit en termes de courbes non-commutatives et on l'applique à la correspondance ${}^a\varphi$;

(1.10.3) Si $\varphi : R \longrightarrow S$ est un morphisme fini entre algèbres affines, où S est à identité polynômiale, alors φ est intégral.

Soient maintenant R, S et T des algèbres affines et supposons que S et T sont à identité polynomiale. Si $\varphi : R \longrightarrow S$ et $\psi : S \longrightarrow T$ sont des morphismes intégraux, alors $\psi\varphi : R \longrightarrow T$ l'est aussi. En effet, comme S et T sont des anneaux à identités polynomiales, on voit que φ et ψ sont finis, donc les correspondances ${}^a\varphi : \Omega(S) \longrightarrow \Omega(R)$ et ${}^a\psi : \Omega(T) \longrightarrow \Omega(S)$ sont propres. Evidemment le composé ${}^a(\psi\varphi) : \Omega(T) \longrightarrow \Omega(R)$ l'est aussi, donc $\psi\varphi$ est un morphisme fini, qui est intégral puisque T satisfait à une identité polynomiale.

D'autres résultats de stabilité pour les morphismes intégraux se déduisent de la même façon, cf. [5,7].

(1.11) - Dans leur papier récent [6] ARTIN et SCHELTER mettent sur certains "schémas" non commutatifs un faisceau d'anneaux. On part d'un schéma (commutatif)

Z de faisceau structurel $\underline{\mathcal{O}}$ sur lequel est défini un faisceau quasi-cohérent d'algèbres d'Azumaya $\underline{\mathcal{A}}$. En utilisant des techniques essentiellement dues à GROTHENDIECK [18] on définit un espace topologique $\underline{\mathcal{S}pec}$ ($\underline{\mathcal{A}}$) sur Z comme suit. Si $U \subset Z$ est un ouvert affine, $\underline{\mathcal{A}}(U)$ est une $\underline{\mathcal{O}}(U)$-algèbre et $\underline{\mathcal{S}pec}$ ($\underline{\mathcal{A}}$) est le recollement des $\underline{\mathcal{S}pec}$ ($\underline{\mathcal{A}}(U)$). L'espace $\underline{\mathcal{S}pec}$ ($\underline{\mathcal{A}}$) est alors équipé d'un faisceau d'anneaux défini sur certains ouverts, c'est-à-dire ceux qui sont image inverse d'ouverts affines de Z. Nous verrons plus loin comment on peut étendre ce faisceau à tout ouvert, répondant ainsi à une question posée dans [6]. Le cas le plus important, auquel on peut appliquer cette construction est le schéma $Z = \mathbb{P}^n_k$, l'espace projectif de dimension n sur k. On appelle espace projectif de dimension n sur un anneau R (qui n'est pas nécessairement commutatif) l'espace $\mathbb{P}^n_R = \underline{\mathcal{S}pec}(R \otimes_k \underline{\mathcal{O}}_{\mathbb{P}^n})$ équipé de son faisceau structurel $R \otimes_k \underline{\mathcal{O}}_{\mathbb{P}^n}$ - ce qui, en fait, n'est défini que sur $\mathbb{P}^n_k$ lui-même. Un "schéma" Y sur $\mathbb{P}^n_R$ est un morphisme $Y \longrightarrow \mathbb{P}^n_R$ de schémas sur $\mathbb{P}^n_k$. De tels schémas sont définis par des k-morphismes $R \longrightarrow \underline{\mathcal{O}}_Y$ et on peut démontrer qu'ils sont propres sur Spec(R) et ont une théorie cohomologique raisonnable. Nous indiquerons plus loin une construction plus intrinsèque de ces schémas, en employant nos techniques de localisation.

Insistons sur le fait que ces considérations ont permis à ARTIN et SCHELTER de démontrer une version non commutative du Théorème Principal de Zariski, cf. loc. cit.

2 - Les représentations d'un anneau non-commutatif, d'après Procesi.

(2.1) - Une extension centrale $\alpha : R \longrightarrow M_n(K)$, où R est un anneau arbitraire et K un corps commutatif, est parfois appelée représentation absolument irréductible de R. Deux représentations absolument irréductibles $\alpha : R \longrightarrow M_n(K)$ et $\beta : R \longrightarrow M_m(L)$ sont équivalentes si $m = n$ et s'il existe un corps F contenant K et L et un F-automorphisme ψ de $M_n(F)$ qui rend le diagramme suivant commutatif :

$$\begin{array}{ccccc} & \overset{\alpha}{\nearrow} & M_n(K) & \longrightarrow & M_n(F) \\ R & & & & \downarrow \psi \\ & \underset{\beta}{\searrow} & M_n(L) & \longrightarrow & M_n(F) \end{array}$$

PROCESI a démontré dans [31] que le noyau d'une représentation absolument irréductible est un idéal premier et que deux représentations irréductibles α et β sont équivalentes si et seulement si $\mathrm{Ker}\,\alpha = \mathrm{Ker}\,\beta$. Cela nous donne une correspondance biunivoque entre les classes d'équivalence de représentations absolument irréductibles de R et les idéaux premiers P de R tels que R/P soit un anneau à identité polynomiale. Comme on l'a indiqué plus haut, on dénote par $\mathrm{Spec}_n(R)$ l'ensemble des idéaux premiers P de R qui correspondent aux représentations absolument irréductibles de degré n et par $\Sigma_n(R)$ l'union $\bigcup_{p\leq n} \mathrm{Spec}_p(R)$.

(2.2) - De façon plus générale, fixons un anneau commutatif K et considérons le foncteur M_n qui à chaque K-algèbre S associe l'anneau de matrices $M_n(S)$. Un des thèmes principaux de la théorie des représentations moderne (et classique !) est d'étudier la représentabilité du foncteur M_n. En d'autres termes, nous voulons construire un adjoint à gauche V_n de M_n, c'est-à-dire un foncteur dans la catégorie des K-algèbres ayant la propriété que pour chaque paire de K-algèbres, R et S, on a une bijection fonctorielle :

$$\mathrm{Hom}_K(R,M_n(S)) = \mathrm{Hom}_K(V_n(R),S).$$

On voit facilement qu'un tel adjoint existe. Toutefois, il est évident que l'ensemble $\mathrm{Hom}_K(R,M_n(S))$ est bien trop grand pour servir à une étude propre. Ce qui nous mène à la définition d'équivalence de morphismes $\varphi_1, \varphi_2 : R \longrightarrow M_n(S)$: ils sont équivalents s'il existe un K-automorphisme ψ de $M_n(S)$ tel que $\psi\varphi_1 = \varphi_2$. Nous cherchons à représenter le foncteur $S \leadsto \mathrm{Hom}_K(R,M_n(S))/\rho$ où ρ est la relation d'équivalence qu'on vient de définir ; mais même dans le cas commutatif, on voit qu'une représentation de ce foncteur dans la catégorie des K-algèbres (ou, dualement, dans le catégorie des K-schémas) ne peut exister en général. Même si l'on essaye de représenter ce foncteur par un ouvert convenable de $\mathrm{Spec}(V_n(R))$ stable sous l'action de $\mathrm{Aut}_K(M_n(S))$ sur lequel ce quotient existe, PROCESI a démontré qu'on est forcé de façon naturelle de considérer des algèbres d'Azumaya de la façon suivante. Si A est une K-algbère commutative et B une algèbre d'Azumaya de rang constant égal à n^2 sur A, on dit qu'un morphisme d'algèbres $\varphi : R \longrightarrow B$ est une <u>représentation absolument irréductible</u> de degré n sur A si $\varphi(R)A = B$, c'est-à-dire si φ est une extension centrale comme auparavant, deux représentations $\varphi_1 : R \longrightarrow B_1$ et $\varphi_2 : R \longrightarrow B_2$ sont dites équivalentes s'il existe un isomorphisme $\psi: B_1 \longrightarrow B_2$ de A-algèbres tel que $\psi\varphi_1 = \varphi_2$.

(2.3) - Définissons $Q_n(R,A)$ comme consistant de toutes les classes d'équivalence de représentations absolument irréductibles de R qui ont degré n sur A. Il est clair que Q_n est un foncteur covariant en A et contravariant en R si on se limite aux extensions centrales. En passant par une construction assez élaborée, PROCESI a démontré que pour R fixé, le foncteur Q_n est représentable par un ouvert d'un schéma affine, qui est le quotient d'un ouvert de $Spec(V_n(R))$ bien choisi, sous l'action du groupe algébrique $\mathcal{A}_n$ défini par $\mathcal{A}_n(S) = Aut_S(M_n(S))$ (notation fonctorielle). De manière plus spécifique, nous voulons que U_R soit un ouvert de $Spec(V_n(R))$ muni d'un faisceau d'algèbres d'Azumaya, tel que les morphismes d'espaces annelés $Spec(A) \longrightarrow U_R$ correspondent bijectivement aux éléments de $Q_n(R,A)$.

(2.4) - Une autre construction, beaucoup plus directe et intrinsèque peut être donnée en utilisant certains résultats dus à FORMANEK [11, 31]. Si R satisfait aux identités des matrices de dimension n, on définit le centre de Formanek $F(R)$ de R comme le sous-anneau de R obtenu par évaluation de tous les polynômes centraux, sans terme constant. On sait que de tels polynômes existent (en abondance) comme l'ont démontré les constructions explicites de FORMANEK, RAZMYZLOW et d'autres. On démontre assez facilement que le centre de Formanek se comporte fonctoriellement et possède des propriétés importantes, dont les principales sont les suivantes :

(2.4.1) une représentation $\varphi : R \longrightarrow M_n(K)$ est absolument irréductible si et seulement si $\varphi(F(R)) \neq 0$.

(2.4.2) R est une algèbre d'Azumaya de rang n^2 sur son centre si et seulement si $F(R)R = R$.

(2.4.3) $\sum_{n-1}(R) = V(F(R))$.

(2.5) - Soit R une A-algèbre, où A est un anneau commutatif et écrivons $\Gamma(R) = A + F(R)$, alors pour chaque $\alpha \in F(R)$ on peut démontrer que $R \otimes_{\Gamma(R)} \Gamma(R)[\frac{1}{\alpha}]$ est une algèbre d'Azumaya. Cela permet de définir de façon canonique un faisceau d'algèbres d'Azumaya sur $\tilde{U}_R = Spec(\Gamma(R)) - V(F(R))$ et l'espace annelé obtenu de cette façon est isomorphe à l'espace U_R qu'on vient de mentionner, si on met sur celui-ci son faisceau d'algèbres d'Azumaya canonique. Il en découle que l'espace $\tilde{U}_R$ est homéomorphe à $Spec_n(R)$ et que la fibre en P du faisceau canonique est exactement R_P, la localisation centrale en P.

(2.6) - Nous constatons donc qu'on peut définir un foncteur contravariant en associant à chaque R l'espace $\tilde{U}_R$ annelé par son faisceau d'algèbres d'Azumaya locales, du moins, si on veille à se restreindre aux extensions centrales. Toutefois, bien qu'il possède des propriétés très agréables, cf. [31], ce foncteur n'a pas d'adjoint en général et n'est donc que peu pratique du point de vue géométrique.

Néanmoins l'étude de ce foncteur a permis à PROCESI de formuler et de démontrer des résultats essentiels en géométrie algébrique non-commutative. Citons en guise d'exemple le théorème suivant :

(2.6.1) Si l'anneau premier R est une algèbre à identité polynomiale de type fini sur un corps K et si Z dénote le centre de l'anneau simple de fractions de R, alors :

$$\dim R = \deg \mathrm{tr}_K Z$$

Si K se réduit au corps algébriquement clos k et donc que R est une algèbre affine, (2.6.1) dit simplement que la dimension topologique de la variété algébrique irréductible $\mathcal{N}(R)$ est le degré de transcendance (sur k) du centre de "l'anneau simple de fonctions" de $\mathcal{N}(R)$. Il est évident que ceci ne fait que généraliser la théorie de la dimension des variétés algébriques bien connue en géométrie algébrique commutative.

(2.7) - Dans la démonstration de (2.6.1) un emploi fondamental a été fait du Théorème des Zéros de Hilbert non-commutatif (dû à PROCESI et AMITSUR [1, 30, 31]) qui dit que chaque algèbre R à identité polynomiale et de type fini sur un corps est une algèbre de Hilbert, c'est-à-dire que chaque idéal primitif de R (= maximal, si R possède une identité polynomiale) est de co-dimension finie et que chaque idéal premier de R est l'intersection des idéaux primitifs qui le contient. Géométriquement, cela dit que pour une telle algèbre l'ensemble des points fermés d'un sous-ensemble fermé V de $\mathrm{Spec}(R)$ est dense dans V. Bien qu'on ne les a appelé qu"exemples", ces résultats sont bien sûr d'une grande profondeur en toute géométrie non-commutative, puisque toute la théorie des variétés algébriques non-commutatives repose sur eux.

(2.8) - Autre exemple : comme toute algèbre affine à identité polynomiale est un quotient d'une algèbre de matrices génériques $k\{\xi_1^{(n)},\ldots,\xi_m^{(n)}\}$, il est essentiel de bien connaître les variétés :

$$\mathbb{A}_n^m = \mathcal{N}(k\{\xi_1^{(n)},\ldots,\xi_m^{(n)}\})$$

dans lesquelles toute variété algébrique non-commutative s'injecte. En employant les techniques citées plus haut PROCESI a démontré dans [29, 31] que :

$$\dim \mathbb{A}_n^m = \dim k\left\{\xi_1^{(n)},\dots,\xi_m^{(n)}\right\} =$$
$$= \deg.\mathrm{tr}_k\, Q(Z(k\left\{\xi_1^{(n)},\dots,\xi_m^{(n)}\right\}))$$
$$= mn^2 - (n^2-1).$$

Notons que $Q(Z(k\left\{\xi_1^{(n)},\dots,\xi_m^{(n)}\right\})) = Z(Q(k\left\{\xi_1^{(n)},\dots,\xi_m^{(n)}\right\}))$ $= Z(k\langle\xi_1^{(n)},\dots,\xi_m^{(n)}\rangle)$, où $k\langle\xi_i^{(n)}\rangle$ est le corps gauche des matrices génériques de dimension n. Le calcul explicite d'une base de transcendance de $Z(k\langle\xi_1^{(n)},\ \xi_2^{(n)}\rangle)$ -d'où toute information pour $m \geq 2$ est aisément déduite – a été fait par PROCESI pour $n=2$ [29, 31] et par FORMANEK pour $n = 3,4$ [13, 14]. Les techniques utilisées par FORMANEK dans ces deux derniers cas offrent l'espoir d'être applicables pour $n \geq 4$.

3 - La notion de variété en géométrie algébrique non-commutative.

(3.1) - Afin d'étudier les algèbres affines du point de vue géométrique, nous voulons mettre sur leur spectre maximal (ou premier, cela revient au même) un faisceau d'anneaux. Dans le cas commutatif des techniques de localisation permettent de construire de façon canonique un faisceau d'anneaux locaux sur ce spectre. En général ceci n'est pas possible. En effet, la situation idéale qu'on voudrait atteindre est celle où l'anneau R, que l'on suppose en particulier premier, noethérien et satisfaisant à une identité polynomiale, s'injecte pour chaque idéal premier P dans un anneau local Q à idéal maximal M, ayant les mêmes propriétés que R et tel que $Q \cap M = P$. Mais alors, un résultat bien connu de JATEGAONKAR [21] dit que :

$$(0) = \bigcap_{n=0}^{\infty} J(Q)^n = \bigcap_{n=0}^{\infty} M^n.$$

Comme $P \subset M$, on voit que $\bigcap P^n = 0$ et cela pour tout idéal premier P ce qui n'est pas le cas en général. Pour le voir, il suffit de prendre :

$$R = \begin{pmatrix} k[X] & (X) \\ k[X] & k[X] \end{pmatrix}, \qquad P = \begin{pmatrix} (X) & (X) \\ k[X] & k[X] \end{pmatrix}$$

et de remarquer que P est idempotent !

(3.2) Nous ne pouvons donc pas espérer obtenir un faisceau d'anneaux locaux sur $\mathcal{N}(R)$. Nous verrons plus loin comment remédier à cet inconvénient. Bien qu'on puisse obtenir des résultats beaucoup plus généraux que ceux qui suivent, par hypothèse R sera toujours une algèbre première, affine et à identité polynomiale que l'on choisira noethérienne à gauche. On dénotera par X_R

son spectre maximal $\mathcal{M}(R)$ et plus généralement par X_I l'ouvert de Zariski des idéaux maximaux P qui ne contiennent pas l'idéal (bilatère) I de R. Il est clair que X_I ne dépend que du radical rad I de I. A chaque idéal premier P de R on peut associer un foncteur noyau symétrique idempotent σ_{R-P} défini pour tout R-module à gauche M par :

$$\sigma_{R-P} M = \{ m \in M \text{ ; il existe } s \in P, \; sRm = 0 \} .$$

Rappelons qu'un <u>foncteur noyau idempotent</u> σ est un foncteur exact à gauche dans la catégorie des R-modules à gauche qui possède la propriété que $\sigma(M/\sigma M) = 0$ pour tout M. Un tel foncteur est caractérisé par son <u>filtre topologisant</u> $\mathcal{L}(\sigma)$ qui consiste des idéaux à gauche L de R tels que $\sigma(R/L) = R/L$. On dit que σ est <u>symétrique</u> si $\mathcal{L}(\sigma)$ possède une base de filtre consistant d'idéaux bilatères. Le foncteur noyau σ_{R-P} qu'on vient de définir a comme filtre :

$$\mathcal{L}(R-P) = \{ L < R \text{ ; il existe } s \notin P, \; Rs\,R \subset L \}$$

(3.3) - La théorie de la localisation par rapport à un foncteur noyau idempotent σ a été discutée amplement [15, 16, 17, 23, 41]. Pour chaque module à gauche M on trouve ainsi un module de quotients $Q_\sigma(M)$, qui est un module à gauche sur l'anneau de quotients $Q_\sigma(R)$. Dans le cas que nous étudions, on trouve ainsi un anneau de quotients :

$$Q_{R-P}(R) = \varinjlim_{L \in \mathcal{L}(R-P)} \mathrm{Hom}_R(L,R) ,$$

qui, comme nous l'avons indiqué plus haut, n'est malheureusement pas un anneau local en général.

(3.4) - De façon analogue, on associe à chaque idéal de R un foncteur noyau idempotent σ_I défini par :

$$\sigma_I M = \{ m \in M \text{ ; il existe } n \in \mathbb{N}, \; I^n m = 0 \}$$

ou par son filtre topologisant :

$$\mathcal{L}(I) = \{ L < R \text{ ; il existe } n \in \mathbb{N}, \; I^n \subset L \} .$$

Notons que ces définitions ne sont valables que dans le cas noethérien, la définition générale se formulant en termes de radicaux. Comme l'ensemble $\{I^n \text{ ; } n \in \mathbb{N}\}$ est une base de filtre pour $\mathcal{L}(I)$, on voit que l'anneau de quotients de R par rapport à I peut être donné par :

$$Q_I(R) = \varinjlim_{n \in \mathbb{N}} \mathrm{Hom}_R(I^n,R) .$$

Vu que R est un anneau premier à identité polynomiale on sait, d'après le théorème de Posner, que R possède un anneau simple de fractions $Q(R)$ et il est facile de voir que $Q_I(R) \subset Q(R)$ est l'ensemble des $q \in Q(R)$ tels qu'il existe un entier positif n pour lesquels $I^n q \subset R$. On a une propriété analogue pour $Q_{R-P}(R)$.

(3.5) - Si on associe à chaque ouvert de Zariski X_I l'anneau de quotients $Q_I(R)$, cela définit (avec des opérations de restriction évidentes) un faisceau d'anneaux $\underline{\mathcal{O}}_R$ sur X_R. Ce faisceau possède les propriétés suivantes :

(3.5.1) $\Gamma(X_R, \underline{\mathcal{O}}_R) = R$;

(3.5.2) $\underline{\mathcal{O}}_{R,P} = \varinjlim_{P \in X_I} \Gamma(X_I, \underline{\mathcal{O}}_R) = Q_{R-P}(R)$.

Remarquons qu'on peut procéder de la même façon pour les modules à gauche M afin de construire des faisceaux de $\underline{\mathcal{O}}_R$-modules $\underline{\mathcal{O}}_M$ sur X_R.

(3.6) - Soit $R = \begin{pmatrix} \mathbb{C}[X] & (X) \\ (X) & \mathbb{C}[X] \end{pmatrix}$, et écrivons A (resp. I) pour $\mathbb{C}[X]$ (resp. (X)), alors $\Omega_2(R)$ consiste en les idéaux maximaux $R(X-a)$ où $0 \neq a \in \mathbb{C}$ et $\Omega_1(R) = \{P,Q\}$ où :

$$P = \begin{pmatrix} I & I \\ I & A \end{pmatrix} \qquad Q = \begin{pmatrix} A & I \\ I & I \end{pmatrix}$$

La localisation de R en $R(X-a)$ n'est autre que la localisation centrale en $(X-a)$. Les idéaux exceptionnels P et Q donnent :

$$Q_{R-P}(R) = \begin{pmatrix} A_p & I_p \\ K(A_p:I) & K(I_p:I) \end{pmatrix} \qquad Q_{R-Q}(R) = \begin{pmatrix} K(I_p:I) & K(A_p:I) \\ I_p & A_p \end{pmatrix}$$

où $K = \mathbb{C}(X)$, le corps des fractions de $\mathbb{C}[X]$ et $K(U:V) = \{ x \in V \ ; \ Vx \subset U \}$. On constate que $Q_{R-P}(R)$ et $Q_{R-Q}(R)$ ne sont pas des anneaux locaux. Cela est dû principalement au fait que la localisation associée à σ_{R-P} et à σ_{R-Q} n'est pas parfaite.

(3.7) - On dit qu'on foncteur noyau idempotent σ induit une <u>localisation parfaite</u> si Q_σ est un foncteur exact, qui commute aux sommes directes ; de façon équivalente, si pour tout R-module à gauche M on a $Q_\sigma(M) = Q_\sigma(R) \otimes_R M$. Si la topologie de Zariski sur X_R possède une base d'ouverts X_I tels que σ_I induit une localisation parfaite, on constate que chaque σ_{R-P} a la même propriété. Ainsi, si R est une algèbre d'Azumaya ou, plus généralement, un anneau Zariski central (cf. [42 ; 48]), $\underline{\mathcal{O}}_R$ est un faisceau d'anneaux locaux. Cela permet de traiter des exemples comme les algèbres de polynômes gauches $\mathbb{C}[X,\tau]$ où τ est la conjugaison canonique sur $\mathbb{C}$. Remarquons que ces algèbres ne sont pas des algèbres d'Azumaya en général.

(3.8) - Bien que le faisceau $\underline{\mathcal{O}}_R$ possède des propriétés agréables, il ne se comporte pas fonctoriellement. Plus concrètement, nous voulons associer à chaque R un faisceau $\underline{\mathcal{O}}_R$ ayant la propriété de définir pour chaque extension $\varphi: R \longrightarrow S$ un morphisme d'espaces annelés $\Phi : (X_S,\underline{\mathcal{O}}_S) \longrightarrow (X_R,\underline{\mathcal{O}}_R)$, et pour le faisceau qu'on vient de construire, ce n'est pas le cas, sauf dans des cas spéciaux comme les exemples où φ est injectif ou épimorphe et fidèlement plat cf. [48]. C'est à ce point là qu'interviennent de façon essentielle les bimodules.

(3.9) - Dans [4] ARTIN définit un R-bimodule comme un R-module bilatère M qui est engendré comme R-module (à gauche ou à droite) par son centralisateur $Z_R(M) = \{m \in M \text{ ; quel que soit } r \in R \quad rm = mr\}$. Ainsi un morphisme d'anneaux $\varphi: R \longrightarrow S$ est une extension exactement si S est un R-bimodule pour la structure de R-module induite par φ. La catégorie des R-bimodules, dénotée par $\underline{\underline{bi}}(R)$ est une sous-catégorie pleine de la catégorie des R-modules bilatères, $\underline{\underline{b}}(R)$, qui n'est cependant pas une catégorie abélienne en général. D'autre part, la catégorie $\underline{\underline{b}}(R)$ est une catégorie de Grothendieck, ce qui permet d'y faire de la localisation. On appellera <u>foncteur noyau idempotent</u> dans $\underline{\underline{bi}}(R)$ un sous foncteur σ de l'inclusion $i : \underline{\underline{bi}}(R) \hookrightarrow \underline{\underline{b}}(R)$, qui est exact à gauche et tel que pour tout R-bimodule M l'on ait $\sigma(M/\sigma M) = 0$. En particulier, les foncteurs noyaux symétriques σ_{R-P} et σ_I qu'on a définis plus haut induisent des foncteurs noyaux idempotents dans $\underline{\underline{bi}}(R)$ par restriction. Bien que $\underline{\underline{bi}}(R)$ ne soit pas abélienne, on peut y établir des résultats de localisation par rapport à ces foncteurs noyaux, grâce à l'existence d'un adjoint à droite $bi : \underline{\underline{b}}(R) \longrightarrow \underline{\underline{bi}}(R)$ de l'inclusion canonique i. Ce foncteur associe à tout R-module bilatère M le R-bimodule $bi\, M = RZ_R(M)$. On a établi dans [43, 47] une théorie de la localisation "relative" dans $\underline{\underline{bi}}(R)$ et d'autre catégories, qui dépend directement de l'existence d'un tel adjoint.

(3.10) - On définit ainsi pour chaque foncteur noyau idempotent σ dans $\underline{\underline{bi}}(R)$ et chaque R-bimodule M un R-bimodule de quotients $Q_\sigma^{bi}(M)$ que l'on construit approximativement comme dans une catégorie de Grothendieck. D'autre part, si σ est induit par un foncteur noyau idempotent dans $\underline{\underline{b}}(R)$, on peut démontrer qu'on a un isomorphisme canonique :

$$Q^{bi}_{\sigma}(M) = bi\ Q^{b}_{\sigma}(M).$$

où Q^{b}_{σ} est la localisation dans $\underline{\underline{b}}(R)$. Si σ est un foncteur noyau idempotent dans R-mod., on voit facilement que σ induit un foncteur noyau dans $\underline{\underline{b}}(R)$ et que la localisation à gauche $Q_{\sigma}(M)$ d'un R-module bilatère M par rapport à σ dans R-mod est muni canoniquement d'une structure de R-module bilatère. Muni de cette structure, $Q_{\sigma}(M)$ coïncide avec le localisé $Q^{b}_{\sigma}(M)$ de M dans $\underline{\underline{b}}(R)$. En guise de conclusion, cela nous permet de considérer les R-bimodules $Q^{bi}_{R-P}(M) = bi\ Q_{R-P}(M)$ et $Q^{bi}_{I}(M) = bi\ Q_{I}(M)$

(3.11) - Comme nous l'avons fait plus haut, nous associons alors à l'ouvert X_I de X_R l'anneau de quotients $Q^{bi}_{I}(R)$ et on obtient de façon évidente un préfaisceau d'anneaux séparé , qui n'est pas un faisceau, en général. Appelons $\underline{\mathcal{O}}^{bi}_{R}$ le faisceau qui y est associé, alors on démontre que $\Gamma(X_R, \underline{\mathcal{O}}^{bi}_{R}) = R$ et que $\underline{\mathcal{O}}^{bi}_{R,P} = Q^{bi}_{R-P}(R)$ pour tout idéal maximal P de R. En plus, on peut démontrer [43,51] que le couple $(X_R, \underline{\mathcal{O}}^{bi}_{R})$ se comporte fonctoriellement par rapport à R, si on se restreint aux extensions d'anneaux bien-sûr.

(3.12) - En géométrie algébrique commutative on fait un usage courant du fait que Spec(R), où $\Omega(R)$, est un espace localement annelé, c'est-à-dire les fibres du faisceau structurel sont des anneaux locaux. Dans la théorie non-commutative on a vu que ce fait particulier n'est pas réalisable, donc nous aurons à faire appel à des considérations différentes. Reconsidérons donc la théorie commutative. Nous partons de la suite exacte :

$$0 \longrightarrow \underline{\mathfrak{m}}_x \longrightarrow \underline{\mathcal{O}}_x \longrightarrow \mathbb{k}(x) \longrightarrow 0$$

où $\underline{\mathcal{O}}_x$ est la fibre en x du faisceau structurel $\underline{\mathcal{O}}_X$ sur le schéma X et $\underline{\mathfrak{m}}_x$ est son unique idéal maximal ; $\mathbb{k}(x)$ est le corps de fonctions en x ($\mathbb{k}(x) = k$ pour les variétés algébriques sur k algébriquement clos). Si $x = P \in \mathrm{Spec}(R)$ nous savons que $\underline{\mathcal{O}}_x = R_P$, $\underline{\mathfrak{m}}_x = PR_P$ et $\mathbb{k}(x) = (R/P)_P = R_P/P_P$. On peut retrouver P comme $j_P^{-1}(\underline{\mathfrak{m}}_x)$ où $j_P : R \longrightarrow \underline{\mathcal{O}}_x$ est le morphisme localisant qui coïncide avec le morphisme :

$$j_x : R = \Gamma(X_R, \underline{\mathcal{O}}_R) \longrightarrow \varinjlim_{x \in X_I} (X_I, \underline{\mathcal{O}}_R) = \underline{\mathcal{O}}_x$$

mais aussi comme $P = j_P^{-1}$ (ker π_x). Nous voyons que ce n'est pas le fait que $\underline{\mathcal{O}}_x$ sont local qui est essentiel, mais la connaissance du morphisme $\pi_x : \underline{\mathcal{O}}_x \longrightarrow \mathbb{k}(x)$. Rappelons maintenant un lemme bien connu, qui es dû à GROTHENDIECK [19]. Si $(X, \underline{\mathcal{O}}_X)$ est un espace annelé et $\{F_x ; x \in X\}$ est une famille telle que F_x soit un $\underline{\mathcal{O}}_x$-module (disons à gauche) pour tout $x \in X$, alors il existe un faisceau de $\underline{\mathcal{O}}_X$-module $\underline{\mathcal{F}}$, tel que pour tout autre faisceau de $\underline{\mathcal{O}}_X$-modules $\underline{\mathcal{G}}$ nous obtenons une bijection :

$$\mathrm{Hom}_{\underline{\mathcal{O}}_X}(\underline{\mathcal{G}}, \underline{\mathcal{F}}) = \prod_{x \in X} \mathrm{Hom}_{\underline{\mathcal{O}}_x}(\underline{\mathcal{G}}_x, F_x)$$

$$g \longmapsto (g_x)_{x \in X} .$$

Bien-sûr $\underline{\mathcal{F}}$ est défini par $\Gamma(U, \underline{\mathcal{F}}) = \prod_{x \in U} F_x$ pour tout ouvert U de X. Notons aussi que la même construction peut être modifié trivialement pour d'autres structures que celle de module. Cette remarque permet de globaliser les morphismes $\pi_x : \underline{\mathcal{O}}_x \longrightarrow \mathbb{k}(x)$ en un seul, en définissant $\underline{\mathcal{K}}_X$ comme le faisceau de $\underline{\mathcal{O}}_X$-algèbres défini par la famille $\{\mathbb{k}(x) ; x \in X\}$ et en associant à la famille :

$$(\pi_x)_{x \in X} \in \prod \mathrm{Hom}_{\underline{\mathcal{O}}_x}(\underline{\mathcal{O}}_x, \mathbb{k}(x))$$

le morphisme correspondant :

$$\pi_X \in \mathrm{Hom}_{\underline{\mathcal{O}}_X}(\underline{\mathcal{O}}_X, \underline{\mathcal{K}}_X).$$

(3.13) - Cela nous mène à la définition d'un espace géométrique qui est un système $(X, \underline{\mathcal{O}}_X, \underline{\mathcal{K}}_X, \pi_X)$ où $(X, \underline{\mathcal{O}}_X)$ et $(X, \underline{\mathcal{K}}_X)$ sont des espaces annelés, et où $\pi_X : \underline{\mathcal{O}}_X \longrightarrow \underline{\mathcal{K}}_X$ est un morphisme de faisceau d'anneaux, du faisceau structurel $\underline{\mathcal{O}}_X$ dans le faisceau simple $\underline{\mathcal{K}}_X$. Si R est une algèbre affine etc... sur k, on peut y associer un espace géométrique $(\underline{X}_R, \underline{\mathcal{O}}_R^{bi}, \underline{\mathcal{K}}_R, \pi_R)$ de la façon suivante : $X_R = \Omega(R)$, $\underline{\mathcal{O}}_R^{bi}$ est le faisceau structurel qu'on vient de définir et $\underline{\mathcal{K}}_R$ et π_R sont obtenus à partir de la famille de morphismes :

$$\pi_{R,P} : Q_{R-P}^{bi}(R) \longrightarrow Q_{R-P}^{bi}(R/P) ,$$

où $Q_{R-P}^{bi}(R/P) = Q(R/P)$, comme on le voit facilement en utilisant le théorème de Posner.

(3.14) - A partir de cette situation, tout le formalisme de la géométrie algébrique se déroule approximativement comme dans le cas commutatif. On définit d'abord une notion de morphismes d'espaces géométriques, traduisant la notion de morphismes d'espaces localement annelés en termes du second faisceau structural $\underline{\mathcal{H}}_X$. On peut alors parler d'espaces géométriques isomorphes et on appelle variété algébrique (sur k) un espace géométrique qui possède un recouvrement d'ouverts mutuellement non-disjoints tel que pour chaque ouvert de ce recouvrement l'espace géométrique induit est isomorphe à un espace géométrique de la forme $(\mathcal{N}(R), \underline{\mathcal{O}}^{bi}_R, \underline{\mathcal{H}}_R, \pi_R)$ où R est une k-algèbre première, affine et satisfaisant à une identité polynomiale. Avec cette définition il est clair que nos variétés algébriques sont compactes et irréductibles.

(3.15) - <u>Quelques remarques sur le formalisme des variétés algébriques non-commutatives.</u>

(3.15.1) Il est bien connu que dans le cas commutatif chaque idéal P (éventuellement premier, si on veut se restreindre aux variétés irréductibles) définit une immersion fermée de variétés $\mathcal{N}(R/P) \longrightarrow \mathcal{N}(R)$ et inversement, chaque immersion fermée est essentiellement de cette forme. On voit ainsi qu'un morphisme de variétés algébriques (commutatives !) $f : Y \longrightarrow X$ est une immersion fermée si et seulement si pour chaque ouvert affine $U \subset X$, l'image inverse $f^{-1}(U)$ est affine et le morphisme :

$$\Gamma(U, \underline{\mathcal{O}}_X) \longrightarrow \Gamma(f^{-1}(U), \underline{\mathcal{O}}_X)$$

est surjectif.

La différence la plus importante entre le cas commutatif et le cas non-commutatif est, qu'avec les définitions qu'on vient de donner, ces résultats ne sont plus valables dans la situation non-commutative. En effet, à une projection arbitraire :

$$\pi : R \longrightarrow R/P$$

on ne peut plus associer (en général) une immersion fermée :

$$a_{\pi} : \mathcal{N}(R/P) \longrightarrow \mathcal{N}(R),$$

car cela impliquerait que pour chaque $Q \in \mathcal{N}(R)$ le morphisme associé :

$$Q^{bi}_{R-Q}(\pi) : Q^{bi}_{R-Q}(R) \longrightarrow Q^{bi}_{R-Q}(R/P)$$

serait surjectif, et il est bien connu que si σ_{R-Q} n'induit pas une localisation parfaite dans $\underline{\underline{bi}}(R)$, ce n'est pas nécessairement vrai. D'autre part, il est naturel d'espérer que seuls les $\mathcal{R}(R/P)$ seraient des sous-variétés algébriques de $\mathcal{R}(R)$, et il est clair que si l'on traduit la notion d'immersion fermée de schémas ou de variétés algébriques commutatives en termes d'espaces géométriques (en utilisant le second faisceau structural $\underline{\mathcal{K}}_X$), c'est exactement cela que l'on obtiendra. Si on utilise cette notion de sous-variété (ou dimension fermée) de façon conséquente, on obtient un formalisme qui ressemble de très près au formalisme commutatif. Ainsi les $\Sigma_n(R) \subset \mathcal{R}(R)$ sont des immersions fermées d'espaces géométriques pour tout n et on peut démontrer que chaque sous-ensemble fermé et irréductible d'une variété algébrique peut être considéré essentiellement de façon unique comme une sous-variété fermée.

(3.15.2) Si I est un idéal de R, alors $X_I = \{P \in \mathcal{R}(R) \; ; \; I \not\subset P\}$ est appelé un ouvert <u>géométrique</u> de X_R si et seulement si σ_I induit une localisation parfaite par R-bimodules. L'ouvert X_I muni du faisceau structural induit est alors une sous-variété ouverte de X_R. Plus généralement, une <u>sous-variété ouverte</u> d'une variété algébrique X est un espace géométrique $(Y, \underline{\mathcal{O}}_Y, \underline{\mathcal{K}}_Y, \pi_Y)$, où Y est un sous-ensemble ouvert de X et où les faisceaux structurels sont induits par ceux de X et tel que Y soit muni de cette façon d'une structure de variété algébrique sur k. Dans le cas commutatif, chaque <u>ouvert principal</u> X_f, $f \in R$, produit une sous-variété ouverte de la variété $\mathcal{R}(R)$. Comme les X_f forment une base de la topologie de Zariski sur $\mathcal{R}(R)$, on voit ainsi que chaque ensemble ouvert d'une variété algébrique commutative est muni d'une structure de variété ouverte. Dans le cas non-commutatif ce phénomène se présente souvent, mais n'est pas valable en général. Parmi les anneaux qui ont une base d'ouverts géométriques, il nous faut mentionner les anneaux Zariski centraux, cf. [42,48]. Du point de vue géométrique, les anneaux les plus interessants sont ceux qui possèdent cette propriété d'avoir une base d'ouverts géométriques : leur faisceau structural est un faisceau d'anneaux locaux et les projections canoniques $\pi : R \longrightarrow R/P$ induisent pour chaque idéal premier Q de R une application surjective $Q^{bi}_{R-P}(\pi) : Q^{bi}_{R-Q}(R) \longrightarrow Q^{bi}_{R-Q}(R/P)$, c'est-à-dire les problèmes mentionnés en (3.15.1) ne se produisent pas.

(3.15.3) Regardons maintenant la notion de produit de variétés algébriques. Par définition, un <u>produit</u> de deux variétés X et Y serait un objet $X \times Y$, lui-même une variété algébrique, possèdent la propriété, que pour toute variété algébrique Z, l'on ait une bijection :

$$\mathrm{Hom}(Z,X)\times \mathrm{Hom}(Z,Y) = \mathrm{Hom}(Z,X\times Y) .$$

Bien que cela paraisse gênant, des produits dans ce sens n'existent pas en général, même si X et Y sont affines. En effet, choisissons $X = \mathcal{R}(R)$, $Y = \mathcal{R}(S)$ alors on vérifie, comme dans le cas commutatif que le seul candidat possible pour $X \times Y$ est $\mathcal{R}(R \otimes_k S)$. Notons d'après BERGMAN [8] que le produit de deux algèbres premières sur un corps algébriquement clos est premier lui aussi. En plus REGEV a démontré dans [33] que le produit tensoriel de deux algèbres à identité polynomiale sur un corps est à identité polynomiale. On vérifie alors aisément que $\mathcal{R}(R \otimes_k S)$ définit bien une variété algébrique affine. Mais comme $R \otimes_k S$ n'est pas un coproduit dans la catégorie des k-algèbres et extensions, il est évident que $\mathcal{R}(R \otimes_k S)$ ne peut être un produit de $\mathcal{R}(R)$ et de $\mathcal{R}(S)$ en général. C'est pourquoi on définit le <u>produit géométrique</u> de deux variétés algébriques X et Y comme la variété algébrique $X \hat{\times} Y$ possédant la propriété suivante. Si Z est une variété algébrique et si $\mathrm{Hom}(Z ; X,Y)$ est l'ensemble des couples :

$$(\varphi,\psi) \in \mathrm{Hom}(Z,X) \times \mathrm{Hom}(Z,Y)$$

tels que les morphismes $\Gamma(\varphi) \in \mathrm{Hom}(\Gamma(X), \Gamma(Z))$ et $\Gamma(\psi) \in \mathrm{Hom}(\Gamma(Y), \Gamma(Z))$ commutent dans le sens suivant : quels que soit $(x,y) \in \Gamma(X) \times \Gamma(Y)$ on a :

$$\Gamma(\varphi)(x)\,\Gamma(\psi)(y) = \Gamma(\psi)(y)\,\Gamma(\varphi)(x),$$

alors $\mathrm{Hom}(Z,X \hat{\times} Y) = \mathrm{Hom}(Z ; X,Y)$. Comme $X \hat{\times} Y$ est défini par une propriété universelle, deux produits géométriques de X et Y sont isomorphes. A partir de là, on démontre aisément que $\mathcal{R}(R \otimes_k S)$ est le produit géométrique de $\mathcal{R}(R)$ et de $\mathcal{R}(S)$. Le formalisme des produits (resp. produits fibrés) en géométrie algébrique commutative reste valable, si on considère les produits géométriques des variétés non-commutatives. Il est facile de voir, par exemple, que si $(M,N) \in \mathcal{R}(R) \times \mathcal{R}(S) = \mathcal{R}(R \otimes_k S)$, alors l'anneau de fonctions au point (M,N) du produit ensembliste $\mathcal{R}(R) \times \mathcal{R}(S)$ est exactement $\mathbb{k}_{R \otimes S}(M,N) = \mathbb{k}_R(M) \otimes_k \mathbb{k}_S(N)$ où $\mathbb{k}_R(M)$ (resp. $\mathbb{k}_S(N)$) est l'anneau des fonctions de M (resp. N) par rapport à R (resp. S). Celà permet de démontrer que si les localisations par rapport à M et N sont parfaites, alors la fibre du faisceau structural sur $\mathcal{R}(R \otimes_k S)$ au point (M,N) est exactement la localisation de $Q^{bi}_{R-M}(R) \otimes_k Q^{bi}_{S-N}(S)$ par rapport à $Q^{bi}_{R-M}(M) \otimes_k Q^{bi}_{S-N}(S) + Q^{bi}_{R-M}(R) \otimes_k Q^{bi}_{S-N}(N)$, c'est-à-dire que :

$$\underline{\mathcal{O}}_{(x,y)} = Q^{bi}_{\mathfrak{M}}\,(\underline{\mathcal{O}}_x \otimes_k \underline{\mathcal{O}}_y)$$

où

$$\underline{\mathfrak{M}} = \underline{\mathcal{O}}_x \otimes \underline{\mathfrak{M}}_y + \underline{\mathfrak{M}}_x \otimes \underline{\mathcal{O}}_y .$$

Si une des composantes d'un produit géométrique est une variété ordinaire (c'est-à-dire commutative) on peut démontrer que ce produit géométrique est un produit véritable dans la catégorie des k-variétés algébriques. Celà nous permet d'étudier des exemples de variétés typiques, tels que les variétés affines sur un anneau R (que l'on considère toujours affine, etc..., bien-sûr !), définies par :

$$\mathbb{A}^n_R = \mathbb{A}^n_k \times \mathcal{X}(R) = \mathcal{X}(k[X_1,\ldots,X_n] \otimes_k R) = \mathcal{X}(R[X_1,\ldots,X_n]) .$$

Nous verrons plus loin de les variétés projectives donnent d'autres applications de cette construction. Pour plus de détails, cf. [43,44].

(3.16) - Avant de définir les variétés projectives, notons que dans beaucoup de cas les constructions de PROCESI qu'on a rappelé plus haut sont des cas spéciaux des nôtres. En effet, on a défini le centre de FORMANEK F(R) d'un anneau R satisfaisant aux identités des algèbres de matrices de dimension n par évaluation de tous les polynômes centraux sans terme constant. Dans [35] ROWEN considère l'anneau G(R) = F(R)R et il démontre que $\mathrm{Spec}_M(R) = X_{G(R)}$ (et de façon analogue pour $\mathcal{X}_n(R)$!). En plus, il prouve essentiellement qu'un anneau semi-premier à identité polynomiale est birationnel sur son centre, c'est-à-dire qu'on peut trouver des ouverts $X_I \subset \mathrm{Spec}(R)$ et $Y_J \subset \mathrm{Spec}(C)$, où C est le centre de R, tels que X_I et Y_J sont homéomorphes sous la correspondance $P \longrightarrow P \cap C = p$. On peut démontrer qu'il suffit de choisir (I,J) = (G(R),F(R)). Ainsi on déduit des résultats généraux dus à VAN OYSTAEYEN [34] que pour chaque idéal premier P d'un tel anneau R avec la propriété $G(R) \not\subset P$, le foncteur noyau σ_{R-P} induit une localisation parfaite dans $\underline{\underline{\mathrm{bi}}}(R)$ et que $Q^{bi}_{R-P}(R)$ est une algèbre d'Azumaya de rang constant sur son centre. Ce rang est exactement n^2, où n est le degré P.I de R, cf. PROCESI [31] . Les remarques précédentes prouvent un peu plus : en effet, on peut en déduire que σ_{R-P} est induit par le foncteur noyau central σ_{C-p} où $p = P \cap C$, et on trouve que $Q^{bi}_{R-P}(R) = R_p$, la localisation centrale par rapport à p. On retrouve donc le résultat bien connu de SMALL, qui dit que Q(R/P) = Rp /Pp pour tout $P \in \mathrm{Spec}_n(R)$, cf. [38] .

(3.17) - Soit maintenant U_R l'ouvert $\mathcal{X}_n(R)$ de $\mathcal{X}(R)$ muni du faisceau $\underline{\mathcal{F}} = \underline{\mathcal{O}}^{bi}_R|_{\mathcal{X}_n(R)}$ induit par $\underline{\mathcal{O}}^{bi}_R$. Par sa construction et les remarques précédentes U_R est donc un espace localement annelé en algèbre d'Azumaya de rang n^2. On a vu que sous quelques hypothèses supplémentaires les morphismes d'espaces géométriques $\mathcal{X}(A) \longrightarrow \mathcal{X}(R)$ sont classés par les extensions $R \longrightarrow A$. Comme R est une algèbre affine par hypothèse, il est clair que l'étude de $\mathcal{X}_n(R)$ revient à celle de $\mathrm{Spec}_n(R)$, puique R est alors une algèbre de Hilbert, c'est-à-dire on se retrouve dans la situation de PROCESI (cf. (2.3)). Ainsi, soit $\mu : \mathcal{X}(A) \longrightarrow U_R$ un morphisme d'espaces localement annelés, alors $\mu^* \underline{\mathcal{F}}$ est un faisceau cohérent d'algèbres d'Azumaya sur $\mathcal{X}(A)$ (ou sur Spec(A), si on veut). Comme Spec(A) définit un schéma commutatif, il existe une algèbre d'Azumaya B sur A tèl que $\mu^* \underline{\mathcal{F}}$ soit le faisceau canonique associé avec B de la façon habituelle. En plus, par la définition de B, on trouve un morphisme $\Gamma(\underline{\mathcal{F}}) = \Gamma(U_R, \underline{\mathcal{F}}) \longrightarrow B$, qui par composition avec le morphisme canonique $R \longrightarrow \Gamma(\underline{\mathcal{F}})$ produit une représentation $\varphi : R \longrightarrow B$. En plus, puisque ce morphisme est obtenu à partir d'un morphisme $\mu : \mathcal{X}(A) \longrightarrow U_R$, on voit que φ factorise par A de la façon suivante :

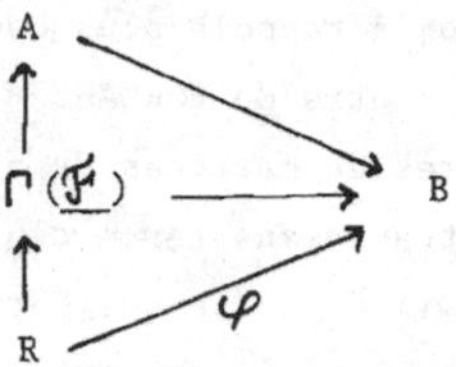

ce qui prouve finalement que $B = A\varphi(R)$, comme $R \longrightarrow \Gamma(\underline{\mathcal{F}})$ est une extension centrale. Ainsi, si R est un anneau affine semi premier à identité polynomiale qui satisfait aux identités des algèbres de matrices de dimension n et si $\underline{\mathcal{U}}_R$ est l'espace localement annelé obtenu sur $\mathcal{X}_n(R) = X_{G(R)}$ par induction de $\mathcal{X}(R)$, alors l'ensemble $Q_n(R,A)$ correspond bijectivement aux morphismes $\underline{\mathcal{X}}(A) \longrightarrow \underline{\mathcal{U}}_R$. Notons que pour une algèbre d'Azumaya R ayant les mêmes propriétés, on trouve $\underline{\mathcal{U}}_R = \underline{\mathcal{X}}(R)$, et plus généralement, si $\sigma_{G(R)}$ induit une localisation parfaite dans $\underline{\underline{\mathrm{bi}}}(R)$, alors $\underline{\mathcal{U}}_R$ est une variété affine, puisque dans ce cas :

$$\underline{\mathcal{U}}_R = \underline{\mathcal{X}}(Q^{bi}_{G(R)}(R)) \ .$$

Notons en conclusion que l'hypothèse d'affinité sur R peut être éliminée dans les constructions qu'on vient de donner, en premier lieu comme conséquence du fait qu'on ne s'interesse essentiellement qu'à la partie de degré n du spectre. Pour plus de détails, le lecteur est prié de se renvoyer à [43,51].

4 - Quelques remarques sur les variétés projectives non-commutatives.

(4.1) - En guise de conclusion, rappelons la construction et quelques propriétés des variétés projectives non-commutatives. Les remarques qui suivent retracent essentiellement une partie du contenu de [44]; pour un exposé plus complet, nous renvoyons à [43]. Nous partons d'un anneau R gradué positivement et noethérien à gauche. Un foncteur noyau idempotent σ dans R-mod est dit gradué si son filtre $\mathcal{L}(\sigma)$ possède une base de filtre qui consiste d'idéaux à gauche gradués. Si σ est un foncteur noyau idempotent gradué, alors pour tout R-module à gauche M on définit :

$$Q^+_\sigma(M)_m = \{x \in Q_\sigma(M), \text{ il existe } I \in \mathcal{L}(\sigma),$$

tel que quel que soit $n \in \mathbb{N}$, $I_n x \subset j_\sigma(M)_{n+m}\}$ où $j_\sigma : M \to Q_\sigma(M)$ est le morphisme localisant canonique. Le module des quotients gradué de M par rapport à σ est défini par :

$$Q^+_\sigma(M) = \bigoplus_{m \in \mathbb{Z}} Q^+_\sigma(M)_m \ .$$

On voit aisément que Q^+_σ peut être considéré comme une localisation dans la catégorie de Grothendieck R-gr des R-modules à gauche gradués, qui correspond à un foncteur noyau σ^g dans R-gr, associé à σ de façon canonique. Il en découle que $Q^+_\sigma(R)$ est un anneau gradué et que $Q^+_\sigma(M)$ est un $Q^+_\sigma(R)$-module à gauche gradué pour tout $M \in$ R-gr.

(4.2) - Si c est un élément central homogène de R et $S = \{1, c, c^2, \ldots\}$, on peut associer à S un foncteur noyau σ de façon bien connue et on peut démontrer que σ induit une localisation parfaite. On en déduit facilement qu'il y a correspondance biunivoque entre les idéaux premiers gradués de R qui ne contiennent pas l'élément c et les idéaux premiers gradués de $Q^+_\sigma(R)$. En plus Q_σ est une localisation centrale et le centre $Z(Q_\sigma(R))$ de $Q_\sigma(R)$ se réduit au localisé gradué commutatif C^+_c du centre C de R par rapport à l'élément homogène c.

(4.3) - Soit Proj(R) l'ensemble des idéaux premiers gradués de R tels que $P \not\supset R_+ = \bigoplus_{n > 0} R_n$. On met sur Proj(R) la topologie induite par la topologie de Zariski de Spec(R), c'est-à-dire ses sous-ensembles ouverts sont les :

$$X_+(I) = \{P \in \text{Proj}(R) \ ; \ I \not\subset P\}$$

où I parcourt l'ensemble des idéaux de R. Il est clair que $X_+(I)$ ne dépend que de la partie graduée strictement positive de I.

(4.4) - Nous allons construire un faisceau d'anneaux canoniques sur Proj(R) muni de cette topologie de la façon suivante. A chaque ouvert $X_+(I)$ de Proj(R) - où on peut supposer que $I = I_+$ - correspond un foncteur noyau idempotent σ_I^+ décrit par son filtre $\mathcal{L}(\sigma_I^+)$ qui consiste des R-idéaux à gauche L qui contiennent un R-idéal gradué J tel que rad $J \supset I_+$. On constate, en effet, que $X_+(I)$ et σ_I^+ ne dépendent que du radical et de la partie positive de I, et il en découle donc que les ouverts $X_+(I)$ et les foncteurs noyaux σ_I^+ se correspondent biunivoquement. Si l'on associe à l'ouvert $X_+(I)$ l'anneau des quotients $Q_I^+(R)$ on obtient de façon évidente un préfaisceau d'anneaux gradués $\underline{Q}_R^+$ pour la topologie de Zariski sur Proj(R). Comme R est noethérien à gauche par hypothèse, chaque ouvert de Proj(R) est quasi-compact, ce qui permet de démontrer que le préfaisceau $\underline{Q}_R^+$ est séparé. Il suit que $\underline{Q}_R^+$ s'injecte dans son faisceau associé $L\,\underline{Q}_R^+$ qui est un faisceau d'anneaux gradués. Rappelons quelques propriétés de ce faisceau :

(4.4.1) pour chaque $P \in \mathrm{Proj}(R)$, on a $\sigma_{R-P}^+ = \sup\{\sigma_I^+ \,;\, P \in X_+(I)\}$, où σ_{R-P}^+ est défini par son filtre $\mathcal{L}(\sigma_{R-P}^+)$ qui est engendré par les idéaux à gauche gradué contenus dans $\mathcal{L}(R-P)$;

(4.4.2) la fibre en P de $\underline{Q}_R^+$ et de $L\,\underline{Q}_R^+$ est exactement $Q_{R-P}^+(R)$;

(4.4.3) si R est un anneau premier, on constate que $\underline{Q}_R^+ = L\,\underline{Q}_R^+$, c'est-à-dire que $\underline{Q}_R^+$ est lui même un faisceau.

On procède maintenant comme dans le cas commutatif. En effet, si $\underline{\mathcal{R}}$ est un préfaisceau d'anneuax gradués sur un espace topologique X, alors le sous-préfaisceau $\underline{\mathcal{R}}_o$ de $\underline{\mathcal{R}}$ défini par $\underline{\mathcal{R}}_o(U) = (\underline{\mathcal{R}}(U))_o$ est un préfaisceau lui même ; de plus si $\underline{\mathcal{R}}$ est un faisceau, $\underline{\mathcal{R}}_o$ en est un aussi. De ce fait, on voit que $(L\,\underline{Q}_R^+)_o$ est un faisceau d'anneaux sur Proj(R) qu'on pourrait appeler structural. Si R est un anneau de Zariski central, on peut démontrer que l'espace Proj(R) muni de ce faisceau structural est un "schéma".

(4.5) - Nous allons modifier un peu nos techniques afin de les adapter à la situation des anneaux à identité polynomiale. Nous supposerons dans ce qui suit que R est un anneau à identité polynomiale, qui est gradué positivement et noethérien à gauche. Nous imposons en plus, les conditions suivantes :

(4.5.1) $R_+ \subset \mathrm{rad}\, RC_+$; cette condition dit essentiellement que Proj(R) peut être recouvert par des ouverts $X_+(Rc)$, où c est un élément homogène de C_+ et que l'application $P \longrightarrow P \cap C$ définit un morphisme $\mathrm{Proj}(R) \longrightarrow \mathrm{Proj}(C)$.

(4.5.2) C est engendré par C_1 comme C_o-algèbre. Notons qu'il aurait suffit de postuler que, pour tout élément de degré t dans C_+, il y a correspondance biunivoque (et donc un homéomorphisme) entre les espaces $\mathrm{Spec}(Q_c^+(R))$ et $\mathrm{Spec}(Q_c^+(R)^{(t)})$, où pour tout anneau gradué S on définit $S^{(t)} = \bigoplus_{i \in \mathbb{Z}} S_{it}$. Nous avons préféré l'hypothèse un peu plus forte (4.5.2), cela simplifiant considérablement la plupart des démonstrations. En plus les constructions de ARTIN et de SCHELTER qu'on a citées plus haut s'appliquent notamment à des exemples qui satisfont à (4.5.1) et (4.5.2) comme nous le remarquerons plus loin.

(4.6) - Nos hypothèses impliquent que $\mathrm{Proj}(R)$ peut être recouvert par des ouverts $X_+(Rc)$ où c est homogène, central et de degré 1. Comme on l'a vu plus haut, le foncteur noyau idempotent gradué σ_{Rc}^+ induit une localisation parfaite et Q_{Rc}^+ n'est autre que la localisation centrale Q_c^+. En utilisant ceci on peut prouver que les espaces $X_+(Rc)$ et $\mathrm{Spec}(Q_c^+(R)_o)$ sont homéomorphes, si le premier est muni de la topologie induite par $\mathrm{Proj}(R)$ et le second de sa topologie de Zariski. En plus, si I est un idéal gradué de R et $I' = (Q_c^+(R)I)_o$, alors les anneaux $Q_{I'}(Q_c^+(R)_o)$ et $Q_{Ic}^+(R)_o$ s'identifient canoniquement. Notons maintenant que $Q_{Ic}^+(R)_o$ n'est autre que l'anneau des sections de $X_+(Ic)$ du préfaisceau $(\underline{Q}_R^+)_o$ sur $\mathrm{Proj}(R)$ et $Q_{I'}(Q_c^+(R)_o)$ est l'anneau de sectionsdu faisceau structural sur $\mathrm{Spec}(Q_c^+(R)_o) = Y$ sur l'ouvert $Y_{I'}$, qui correspond à $X_+(Ic)$ sous l'homéomorphisme qu'on a mentionné plus haut.

(4.7) - On obtient le préfaisceau $\underline{Q}_R^o$ sur $\mathrm{Proj}(R)$ en associant à chaque sous-ensemble ouvert $X_+(I)$ l'anneau $\mathrm{bi}(Q_I^+(R)_o)$, et on voit facilement que $\underline{Q}_R^o$ est un faisceau d'anneaux séparé qui possède la propriété que pour chaque ouvert $X_+(Rc)$, où c est homogène, central et de degré 1, on a $\mathrm{bi}(Q_{Rc}^+(R)_o) = Q_{Rc}^+(R)_o$. Nous définissons $\underline{\mathcal{O}}_R^+$ comme le faisceau associé au préfaisceau $\underline{Q}_R^o$. La restriction de $\underline{Q}_R^o$ à l'ouvert $X_+(Rc)$ est exactement le préfaisceau structural $\underline{\mathcal{O}}^{bi}_{Q_c^+(R)_o}$ sur $\mathrm{Spec}(Q_c^+(R)_o)$. Rappelons maintenant un résultat utile, mais tout-à-fait évident, sur les faisceaux ; soit $\underline{\mathcal{P}}$ un préfaisceau sur un espace topologique X et soit $\{X_\lambda\ ;\ \lambda \in \Lambda\}$ un recouvrement ouvert de X. Soit $\underline{\mathcal{P}}_\lambda = \underline{\mathcal{P}}|X_\lambda$, le préfaisceau induit sur X_λ et soient $\underline{\mathcal{S}}$ (resp. $\underline{\mathcal{S}}_\lambda$) les faisceaux sur X (resp. X_λ) associés à $\underline{\mathcal{P}}$ (resp. $\underline{\mathcal{P}}_\lambda$)

alors $\underline{\mathcal{S}} = \underline{\mathcal{S}}_{|X_\lambda}$. En appliquant cette remarque, on obtient que la restriction de $\underline{\mathcal{O}}^+_R$ à $X_+(Rc)$ est le "bon" faisceau structurel sur $\mathrm{Spec}(Q^+_c(R)_o)$ lorsque $X_+(Rc)$ et ce dernier espace topologique sont identifiés. Modulo quelques considérations très évidentes, cela prouve donc qu'on peut munir Proj(R) d'une structure de "schéma non-commutatif" (notion non-définie dans le présent texte). Si R est une algèbre affine, on peut se restreindre aux idéaux maximaux. Le sous-espace P(R) de Proj(R) qui consiste en les points fermés de Proj(R) et ainsi doté d'une structure de variété algébrique non commutative. Pour des exemples concrets, le lecteur est prié de se rapporter à [43, 44]. Notons seulement que si R est une algèbre d'Azumaya graduée, alors Proj(R) et Proj(C) sont homéomorphes, tandis que si $\mathcal{O}$ est le faisceau structural sur Proj(C), alors $\underline{\mathcal{O}}^+_R = \underline{\mathcal{O}} \otimes_C R$.

(4.8) - On peut démontrer (cf. loc. cit.) que le "schéma" Proj(R) (resp. la variété projective P(R)) se comporte fonctoriellement comme son analogue commutatif. Nous n'entrons pas dans les détails, mais citons néanmoins l'exemple suivant. Soient R et S deux anneaux premiers affines qui sont gradués positivement, noethériens à gauche et qui possèdent une identité polynomiale.Comme avant, on suppose que les conditions (4.5.1) et (4.5.2) sont satisfaites. Si $\varphi : R \longrightarrow S$ est une extension graduée, ceci implique en particulier que le centre de R s'applique dans le centre de S et que l'image inverse $\varphi^{-1}(P)$ d'un idéal premier (gradué) de S est un idéal premier (gradué) de R. Notons $Y = \mathrm{Proj}(R)$, $X = \mathrm{Proj}(S)$ et $U_\varphi = X_+(\varphi(R_+)) \subset X$, alors φ induit un morphisme continu ${}^a\varphi : U_\varphi \longrightarrow Y$ en appliquant P sur $\varphi^{-1}(P)$. Si $\underline{\mathcal{U}}_\varphi$ est le sous-schéma ouvert de $\underline{\mathcal{P}roj}(S)$ induit sur l'ouvert U_φ, alors φ induit un morphisme de "schémas non commutatifs" (ou de variétés algébriques non commutatives, cela revient au même dans notre contexte) :

$$\underline{\mathcal{P}roj}(\varphi) : \underline{\mathcal{U}}_\varphi \longrightarrow \underline{\mathcal{P}roj}(R).$$

(4.9) - Montrons maintenant comment la construction de ARTIN et de SCHELTER de l'espace projectif sur un anneau R, se réduit à nos techniques. Pour la simplicité, soit R un anneau premier, noethérien à gauche, affine et à identité polynomiale sur k. Comme dans (1.11), soit $Z = \mathbb{P}^n_k = \mathbb{P}^n$ l'espace projectif de dimension n sur k, couvert par les n+1 espaces affines : $U_j = \mathrm{Spec}(k[u_{oj},\ldots,u_{nj}])$, où $u_{ij} = u_i/u_j$ et où les u_j (j=0,...,n) dénotent les coordonnées canoniques. Si nous dénotons par T_j l'anneau $k[u_{oj},\ldots,u_{nj}]$, il suit que $\mathbb{P}^n$ est l'union des spectres obtenus à partir des diagrammes d'anneaux et de localisations suivants :

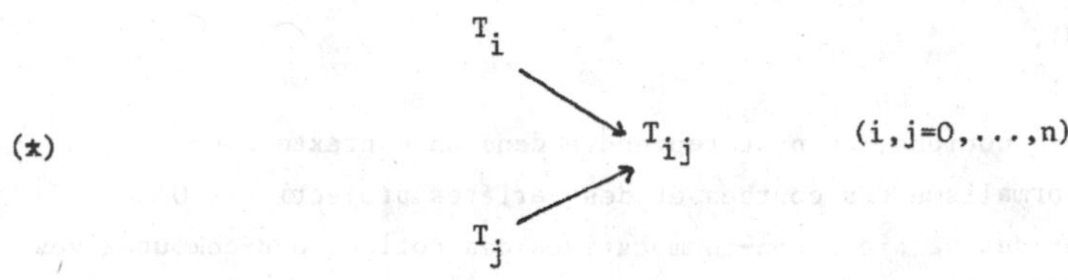

où $T_{ij} = T_i[u_{ji}^{-1}] = T_j[u_{ij}^{-1}] = T_{ji}$.

Si $\mathcal{A}$ est un faisceau d'algèbres quasi-cohérent sur $\mathbb{P}^n$, alors $\Gamma(U_i,\mathcal{A}) = A_i$ est une T_i-algèbre et nous obtenons des diagrammes :

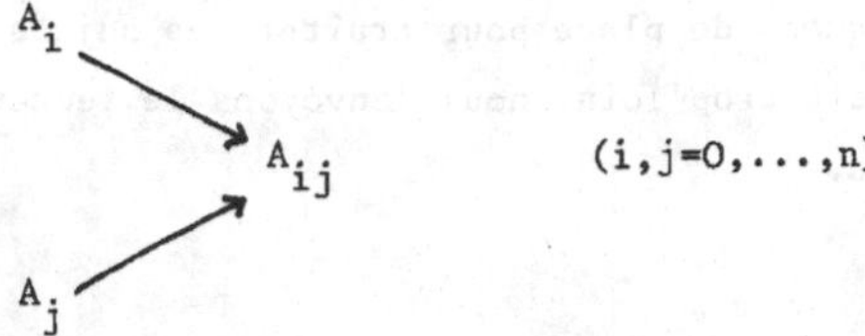

dont les morphismes sont des localisations centrales induites par les morphismes des diagrammes (*) par produit tensoriel. L'espace $\underline{Spec}(\mathcal{A})$ est l'union des $Spec(A_i)$ identifié le long de leurs sous-ensembles ouverts $Spec(A_{ij})$.

(4.10) - L'espace projectif de dimension n sur R est défini comme $\mathbb{P}^n_R = \underline{Spec}(R \otimes_k \mathcal{O}_{\mathbb{P}^n})$ et les anneaux A_i qui le définissent sont les :

$$R \otimes_k T_i = R[u_{oi},\dots,u_{ni}]$$

Soit $\gamma : \mathbb{P}^n_R \longrightarrow \mathbb{P}^n$ le morphisme structurel canonique, alors on a démontré dans loc. cit. le résultat suivant :

(4.10.1) Soit R un anneau premier noethérien qui est affine et satisfait une identité polynomiale sur k, soit $S = R[y_o,\dots,y_n]$ et $\mathcal{A} = R \otimes_k \underline{\mathcal{O}}_{\mathbb{P}^n}$, alors :

(4.10.1.1) $\mathbb{P}^n_R = Proj(S)$;

(4.10.1.2) pour tout ouvert U_j de $\mathbb{P}^n$ les faisceaux $\mathcal{A}$ et $\underline{\mathcal{O}}^+_S$ coïncident sur $\gamma^{-1}(U_j)$ c'est-à-dire $\Gamma(U_j,\mathcal{A}) = \Gamma(\gamma^{-1}(U_j), \underline{\mathcal{O}}^+_S)$.

Le fait que $\mathcal{A}$ et $\gamma_* \underline{\mathcal{O}}^+_S$ coïncident sur <u>tout</u> ouvert affine U de $\mathbb{P}^n_R$, c'est-à-dire que $\Gamma(U,\mathcal{A}) = \Gamma(\gamma^{-1}(U), \underline{\mathcal{O}}^+_S)$ démontre que $\underline{\mathcal{O}}^+_S = \underline{\mathcal{O}}^+_{R[y_o,\dots,y_n]}$ est une extension de $\mathcal{A}$ à tout ouvert de

$\mathbb{P}^n_R = \mathrm{Proj}(R[y_o,\ldots,y_n])$.

(4.11) - A partir de ces notions, on peut reprendre dans un contexte non-commutatif "tout" le formalisme des courbes et des variétés projectives. On peut définir et étudier des versions non-commutatives des notions non-commutatives des notions de morphismes propres et projectives, on peut établir une théorie cohomologique de ces variétés en utilisant une notion adaptée de faisceau cohérent, pour laquelle des principes généraux, tels que le théorème de Serre, ou la théorie de la dualité cohomologique restent valables, ce qui doit nous mener à des résultats tels qu'un théorème de Riemann-Roch non-commutatif, cf[43], etc... Comme nous manquons de place pour traiter ces sujets ici et que de toute façon cela nous mènerait trop loin, nous renvoyons le lecteur à [43] et à des papiers en préparation.

Bibliographie

[1] Amitsur S.A., Procesi C., Jacobson rings and Hilbert algebras with polynomial identities, Annali Math. Pura Applicata, 71, p.61-72 (1960).

[2] Amitsur S.A., Small L, Prime P.I. rings (to appear);

[3] Artin M., Integral homomorphisms of PI rings dans "Noetherian rings and rings with polynomial identities", Proceedings of a conference held at the University of Durham, 23-31 July, 1979.

[4] Artin M., On Azumaya algebras and finite dimensional representations of rings, J. Algebra, 11, p.532-563 (1969).

[5] Artin M., Specialization of representations of rings, Symp. on Alg. Geometry, Kyoto, p. 237-247 (1977).

[6] Artin M., Schelter W., A version of Zariski's main theorem for polynomial identity rings, Amer. J. math., 101, p.301-330 (1979).

[7] Artin M., Schelter W., Integral ring homomorphisms (to appear).

[8] Bergman G., Zero-divisors in Tensor products, the Kent state conference on Non commutative rings, Lect. Notes in Mathematics 545, Springer Verlag, Berlin (1977).

[9] Bergman G., Small L.W., PI degrees and prime ideals, J. Algebra, 33, p.435-462 (1975).

[10] Bourbaki N., Algèbre Commutative, Eléments de Mathématique 27, 28, 30, 31, Hermann, Paris (1961-1965).

[11] Formanek E., Central polynomials for matrix rings J. Algebra, 23, p.129-133 (1972).

[12] Formanek E., Noetherian PI rings, Comm. in Algebra, 1, p.79-86 (1974)

[13] Formanek E., The center of the ring of 3 X 3 generic matrices, Linear and Mult. Algebra, 7, pp 203-212 (1979).

[14] Formanek E., The Center of the ring of 4 X 4 generic matrices, J of Algebra 62, 304-319, (1980).

[15] Gabriel P., Des Catégories abéliennes, Bull. Soc. Math. France, 90, p.323-448 (1962).

[16] Golan J., Localization of Noncommutative Rings, Marcel Dekker, New-York (1975).

[17] Goldman O., Rings and Modules of Quotients, J. Algebra, 13, p.10-47 (1969).

[18] Grothendieck A., Eléments de Géométrie Algébrique, Publ. Math. Inst. Hautes Etudes Sci., (1960).

[19] Grothendieck A., sur Quelques Points d'Algèbre Homologique, Tohoku Math. J., p.119- 221 (1958).

[20] Jacobson N., PI-algebras, Lect. Notes in Mathematics 441, Springer-Verlag, Berlin (1975).

[21] Jategaonkar A.V., Jacobson's Conjecture and Modules over fully bounded Noetherian Rings, J. Algebra, 30, p.103-121, (1974).

[22] Jategaonkar A.V., Principal Ideal Theorem for Noetherian. P.I. Rings, J. Algebra, 35, p.17-22 (1975).

[23] Lambek J., Torsion Theories, Additive Semantics and Rings of Quotients, Lect. Notes in Mathematics 197, Springer-Verlag, Berlin (1971).

[24] Mumford D., Introduction to Algebraic Geometry, Preliminary version of first 3 chapters.

[25] Murdoch D.C., Van Oystaeyen F., Noncommutative localization ans scheaves, J. Algebra, 35, p.500-515 (1975).

[26] Nastasescu C., Van Oystaeyen F., Graded and Filtered Rings and Modules, Lect. Notes in Mathematics, Springer-Verlag, Berlin (1979).

[27] Nauwelaerts E., Localization of P.I. algebras, L.U.C., preprint, 1977.

[28] Nauwelaerts E., Zariski extension of rings, ph. d. Thesis, University of Antwerp, U.I.A., (1979).

[29] Procesi C., Noncommutative affine rings, Att. Acc. Naz. Lincei, s. VIII, v. VIII f.6., p.230-255 (1967).

[30] Procesi C., Non commutative Jacobson rings, Annali Sci. Norm. Sup. Pisa, v. XXI, f.II, p.381-390 (1967).

[31] Procesi C., Rings with Polynomial Identities, Marcel Dekker, New-York (1973).

[32] Procesi C., Sugli Anelli zero dimensionali con identita polinomiale, Rend. Circolo Mat. Palermo, s. II, T. XVII, p.5-11 (1968).

[33] Regev A., Existence of identities in $A \otimes B$, Israel J. of Math., 11, p.131-152 (1972).

[34] Rowen L., Some results on the Center of a Ring with Polynomial Identity, Bull. Amer. Math. Soc, 79, p.219-223 (1973).

[35] Rowen L., On rings with central polynomials, J. Algebra, 31, p.393-426 (1974).

[36] Schelter W., Integral extensions of rings satisfying a polynomial identity, J. Algebra, 40, p.245-257 (1976).

[37] Schelter W., Pare R., Finite extensions are integral (to appear).

[38] Small L., Localization of P.I. Rings, J. Algebra, 18, p.269-270 (1971).

[39] Van Oystaeyen F., Birational ring homomorphisms and ring extensions in "Proc. Ring Theory Antwerp 1978", M. Dekker, New-York (1979).

[40] Van Oystaeyen F., On Graded Rings and Modules of Quotients, Comm. in Algebra, 18, p.1923-1959 (1978).

[41] Van Oystaeyen F. Prime Spectra in Non-commutative Algebra, Lect. Notes in Mathematics 444, Springer Verlag, Berlin (1975).

[42] Van Oystaeyen F., Zariski Central Rings, Comm. in Algebra, 6, p.799-821 (1978).

[43] Van Oystaeyen F., Verschoren A., Introduction to Non commutative Algebraic Geometry, (monograph, to appear).

[44] Van Oystaeyen F., Verschoren A., Graded P.I. Rings and Noncommutative Projective Schemes (to appear).

[45] Van Oystaeyen F., Verschoren A., Localization of Presheaves of Modules, Indag. Math., 79, p.335-348 (1976).

[46] Van Oystaeyen F., Verschoren A., Reflectors and Localisation-Application to sheaf Theory, Marcel Dekker, New-York (1978).

[47] Van Oystaeyen F., Verschoren A., Relative localization, Bimodules and Semiprime P.I. Rings, Comm. in Algebra, 7, p.955-988 (1979).

[48] Verschoren A., Les Extensions et les Schémas Noncommutatifs (to appear in Publi. Math. Debrecen).

[49] Verschoren A., Localization and the Gabriel Popescu Embedding, Comm. in Algebra, 6, p.1563-1587 (1978)

[50] Verschoren A., Perfect Localizations and Torsion-free Extensions, Comm. in Algebra , 8, 839-860, 1980.

[51] Verschoren A., Some Ideas in Noncommutative Algebraic Geometry, ph. d. Thesis, University of Antwerp, U.I.A. (1979).

Sur la formule de Molien dans certaines algèbres enveloppantes.

J . AlEV

Introduction :

Dans ce travail on étudie un analogue de la formule de Molien dans certaines algèbres enveloppantes. Plus précisément, si G est un groupe fini d'automorphismes d'une algèbre de Lie nilpotente $\mathfrak{g}$, G s'étend à l'algèbre enveloppante $U(\mathfrak{g})$. Si $\mathfrak{g}$ est abélienne $U(\mathfrak{g})^G$ est un anneau gradué dont la série de Hilbert est explicitement donnée par la formule de Molien en termes de l'action de G . Dans le cas nilpotent général nous introduisons une graduation dans $U(\mathfrak{g})$ de façon à retrouver la formule de Molien pour $U(\mathfrak{g})^G$.

§1 Algèbres de Lie nilpotentes graduées

1.1 Soit $\mathfrak{g}$ une algèbre de Lie nilpotente de dimension finie p sur un corps K , de caractéristique 0 . Considérons la suite centrale descendante de $\mathfrak{g}$.

(1.1.1.) $$\mathfrak{g} = \mathfrak{g}^1 \supset \mathfrak{g}^2 \supset \dots \supset \mathfrak{g}^q \supset \mathfrak{g}^{q+1} = (0).$$

où $\mathfrak{g}^{i+1} = [\mathfrak{g}^i, \mathfrak{g}]$ pour $1 \leq i \leq q$ et $q+1$ est l'indice de nilpotence de $\mathfrak{g}$. On définit sur $\mathfrak{g}$ une fonction "hauteur" de la façon suivante :

$$h(x) = \sup\{i \mid x \in \mathfrak{g}^i\} \quad \text{si} \quad x \neq 0 , \quad h(0) = +\infty .$$

1.2 Lemme : On a

(i) $h(x+y) \geqslant \inf\{h(x), h(y)\}$

(ii) $h(x+y) = \inf\{h(x), h(y)\}$ si $h(x) \neq h(y)$

(iii) $h([x,y]) \geqslant h(y)+h(x)$

Preuve :

(i) résulte du fait que : $x+y \in \mathfrak{g}^{\inf\{h(x),h(y)\}}$.

(ii) Supposons que $h(x) > h(y)$.

Si $x+y \in \mathfrak{g}^{\inf\{h(x),h(y)\}+1}$, alors $y \in \mathfrak{g}^{h(y)+1}$,

ce qui est contradictoire avec la définition de $h(y)$.

(iii) résulte de l'inclusion :

$$[\mathfrak{g}^i, \mathfrak{g}^j] \subset \mathfrak{g}^{i+j} .$$

1.3 Considérons l'algèbre de Lie $gr\,\mathfrak{g}$ associée à $\mathfrak{g}$:

$$gr\,\mathfrak{g} = \frac{\mathfrak{g}^1}{\mathfrak{g}^2} \oplus \frac{\mathfrak{g}^2}{\mathfrak{g}^3} \oplus \ldots \oplus \mathfrak{g}^q$$

où le crochet est défini par :

$$[x+\mathfrak{g}^{i+1}, y+\mathfrak{g}^{j+i}] = [x,y] + \mathfrak{g}^{i+j+1}.$$

$gr\,\mathfrak{g}$ est une algèbre de Lie nilpotente de même dimension que $\mathfrak{g}$ et on a :

(1.3.1.) $$(gr\,\mathfrak{g})^i = \frac{\mathfrak{g}^i}{\mathfrak{g}^{i+1}} \oplus \ldots \oplus \mathfrak{g}^q \qquad 1 \leq i \leq q.$$

(1.3.2.) Soit $e_p, e_{p-1}, \ldots, e_2, e_1$ une base adaptée à la suite (1.1.1.). Considérons l'application $\varphi : \mathfrak{g} \longrightarrow gr\,\mathfrak{g}$ définie par $e_i \longmapsto \bar{e}_i$ et prolongée à $\mathfrak{g}$ par linéarité. ($\bar{e}_i$ désigne la forme initiale de e_i dans $gr\,\mathfrak{g}$).

1.4 <u>Lemme : L'application φ est un isomorphisme d'algèbres de Lie si et seulement si on a la propriété suivante : quels que soient</u> $i,j \in \{1,2,\ldots,p\}$, $h(e_k) = h(e_i)+h(e_j)$ <u>pour les</u> e_k <u>apparaissant dans les relations de structure et tels que</u> : $\lambda_{ijk} \neq 0$.

$$[e_i,e_j] = \sum_{k > \sup\{i,j\}} \lambda_{ijk} e_k \quad \underline{\text{quand}} \quad [e_i,e_j] \neq 0.$$

<u>Preuve</u> : Supposons que φ soit un isomorphisme d'algèbres de Lie. Alors, on a :

$\varphi([e_i,e_j]) = [\bar{e}_i,\bar{e}_j]$. Supposons que $[e_i,e_j] \neq 0$.

Mais, $\varphi([e_i,e_j]) = \varphi(\sum_k \lambda_{ijk} e_k) = \sum_k \lambda_{ijk} \bar{e}_k$ et

$[\bar{e}_i,\bar{e}_j] = [e_i,e_j] + \mathcal{G}^{h(e_i)+h(e_j)+1}$ qui est un élément homogène de degré $h(e_i)+h(e_j)$ dans $gr\,\mathcal{G}$. Il s'ensuit que tous les e_k sont homogènes de degré $h(e_i)+h(e_j)$ dans $gr\,\mathcal{G}$. Réciproquement, si $x = \sum \alpha_i e_i$ et $y = \sum \beta_j e_j$ on a :

$$\varphi([x,y]) = \varphi(\sum \alpha_i \beta_j [e_i,e_j]) = \sum \alpha_i \beta_j \varphi([e_i,e_j])$$

$$= \sum_{i,j,k} \alpha_i \beta_j \lambda_{ijk} \bar{e}_{k,i,j}$$

tandis que :

$$[\varphi(x), \varphi(y)] = [\sum \alpha \bar{e}_i, \sum \beta_j \bar{e}_j] = \sum \alpha_i \beta_j [\bar{e}_i,\bar{e}_j]$$

$$= \alpha_i \beta_j \, ([e_i,e_j] + \mathcal{G}^{h(e_i)+h(e_j)+1})$$

$$= \sum_{i,j} \alpha_i \beta_j \, (\sum_k \lambda_{ijk} e_{k,i,j} + \mathcal{G}^{h(e_i)+h(e_j)+1})$$

$$= \sum_{i,j,k} \alpha_i \beta_j \lambda_{ijk} \bar{e}_{k,i,j}$$

1.5 <u>Définition</u> : Une base adaptée qui vérifie la propriété du Lemme 1.4 sera dite <u>base spéciale</u> et une algèbre de Lie nilpotente qui admet une base adaptée spéciale est dite <u>graduée</u>. On a alors : $\mathfrak{g} \simeq \mathrm{gr}\,\mathfrak{g}$.

<u>Exemples</u> : 1) $\mathrm{gr}\,\mathfrak{g}$ est graduée pour toute algèbre de Lie nilpotente $\mathfrak{g}$. En effet, si $e_p, e_{p-1}, \ldots, e_2, e_1$ est une base adaptée de $\mathfrak{g}$, la base de $\mathrm{gr}\,\mathfrak{g}$ formée par les formes initiales $\bar{e}_i$, $1 \leq i \leq p$, est spéciale. On a en particulier : $\mathrm{gr}(\mathrm{gr}\,\mathfrak{g}) \simeq \mathrm{gr}\,\mathfrak{g}$ pour toute algèbre de Lie nilpotente $\mathfrak{g}$ de dimension finie. Avec les notations du [1], $\mathfrak{g}_3$ et $\mathfrak{g}_{5,5}$ sont graduées, tandis que $\mathrm{gr}\,\mathfrak{g}_{5,6} \simeq \mathfrak{g}_{5,5}$.

2) On dit que $\mathfrak{g}$ est <u>filiforme</u> si $\dim \mathfrak{g}^1/\mathfrak{g}^2 = 2$, $\dim \mathfrak{g}^i/\mathfrak{g}^{i+1} = 1$, pour $2 \leq i \leq q$. Dans [6], on caractérise les algèbres de Lie filiformes graduées de toute dimension.

3) Dans [4], on définit 3 familles infinies d'algèbres de Lie nilpotentes non filiformes par les crochets suivants :

$$\mathfrak{g}_{m,\varepsilon} = \bigoplus_{i=1}^{3m+\varepsilon} K e_i \quad \text{où } \varepsilon = 0, 1, -1 \text{ et } 2 \leq m < \infty \text{ muni}$$

des crochets : $[e_{3i-1}, e_{3j+1}] = e_{3(i+j)}$;

$[e_{3i}, e_{3j+1}] = 2e_{3(i+j)+1}$ et $[e_{3i}, e_{3j-1}] = -2e_{3(i+j)-1}$.

§2 Graduations dans l'algèbre enveloppante d'une algèbre de Lie nilpotente graduée.

Soit $\mathfrak{g}$ une algèbre de Lie nilpotente graduée de dimension p sur K et $e_p, e_{p-1}, \ldots, e_2, e_1$ une base spéciale de $\mathfrak{g}$.

2.1 Définition : On appelle monôme en les e_i une expression du type $e_{i_1}^{t_1} e_{i_2}^{t_2} \ldots e_{i_s}^{t_s}$ où les $t_i \in \mathbb{N}$ et où les i_j ne sont pas tous distincts. C'est un élément de l'idéal d'augmentation de $U(\mathfrak{g})$, l'algèbre enveloppante de $\mathfrak{g}$.

On appelle monôme standard un monôme dans lequel on a :

$$i_1 \geqslant i_2 \geqslant \ldots \geqslant i_s$$

On peut attacher trois entiers au monôme

$$m = e_{i_1}^{t_1} e_{i_2}^{t_2} \ldots e_{i_s}^{t_s} .$$

1) Le degré total de m, noté $d(m)$ et donné par :

$$d(m) = \sum_{j=1}^{s} t_j .$$

2) La hauteur de m, notée $h(m)$ et notée par :

$$h(m) = \sum_{j=1}^{s} h(e_{i_j}) t_j \quad \text{où} \quad h(e_{i_j})$$

désigne la hauteur définie dans 1.1.

3) Un entier noté $v(m)$ mesurant la déviation de m par rapport au caractère d'être standard. C'est par le procédé de redressement intervenant dans le théorème de Poincaré-Birkoff-Witt qu'on le définit :

Soit $m = e_{i_1}^{t_1} e_{i_2}^{t_2} \ldots e_{i_s}^{t_s}$ le monôme considéré.

Si e_1 apparaît quelque part dans m, on appellera inversion de m en <u>cet</u> e_1 le nombre de facteurs différents de e_1 qui se trouvent à la droite de cet e_1. Faisons passer tous les e_1 apparaissant dans m à l'extrême droite. On obtient alors un monôme

$\tilde{m}_1 = e_{i_1}^{t_1} e_{i_2}^{t_2} \ldots e_{i_k}^{t_k} e_1^{r_1}$ avec

$i_1, i_2, \ldots, i_k \in \{2, \ldots, p\}$. Notons par $v_1(m)$ le nombre total d'inversions faites avec des e_α différents de e_1 pour passer de m à $\tilde{m}_1$. On peut recommencer la même opération sur le monôme $m_1 = e_{i_1}^{t_1} \ldots e_{i_k}^{t_k}$ avec e_2. Le procédé se termine quand on arrive à un monôme standard. On peut alors définir :

$$v(m) = \sum_{k=1}^{p} v_k(m) \quad .$$

($v_k(m) = 0$ si e_k n'intervient pas dans m).

2.2 <u>Proposition</u> : <u>Tout monôme</u> m <u>de hauteur</u> $h(m)$ <u>est combinaison linéaire de monômes standard de hauteur</u> $h(m)$.

<u>Preuve</u> : La première partie de l'assertion est le théorème de Poincaré-Birkoff-Witt appliqué à la base $e_p, e_{p-1}, \ldots, e_1$ La proposition donne l'information plus précise : à savoir que la hauteur des monômes standards qui apparaissent est la même que celle de m. La preuve se fait par récurrence sur $d(m)+v(m)$.

Si $d(m)+v(m) = 1$; alors $m = e_j$ et c'est clair.

Supposons $d(m)+v(m)>1$; Soit j_o le plus petit entier tel que $v_{j_o}(m) \neq 0$. On a alors :

(2.2.1.)

$$\ldots e_{j_o}^{s-1}\, e_{j_o}\, e_k\, e_k^{t-1} \ldots = \ldots e_{j_o}^{s-1}\, e_k e_{j_o}\, e_k^{t-1} \ldots + \ldots e_{j_o}^{s-1} [e_{j_o}, e_k]\, e_k^{t-1} \ldots$$

avec $k > j_o$

D'après le choix de la base :

$$[e_{j_o}, e_k] = \sum \lambda_r e_r \quad \text{avec } h(e_r) = h(e_{j_o}) + h(e_k).$$

On a alors :

a) $v(\ldots e_{j_o}^{s-1}\, e_k\, e_{j_o}\, e_k^{t-1} \ldots) = v(m)-1$

b) $\ldots e_{j_o}^{s-1} \left[e_{j_o}, e_k\right] e_k^{t-1} \ldots = \sum_r \lambda_r \, (\ldots e_{j_o}^{s-1}\, e_r\, e_k^{t-1} \ldots)$

Les monômes apparaissant dans cette somme sont de degré total $d(m)-1$.

c) La hauteur de tout monôme apparaissant dans la combinaison linéaire de b) est $h(m)$.

On peut donc appliquer l'hypothèse de récurrence à chacun des termes qui se trouvent dans le membre droit de (2.2.1.) et ainsi la proposition est démontrée.

La proposition 2.2 implique que l'algèbre enveloppante d'une algèbre de Lie nilpotente graduée est graduée par la la hauteur. Remarquons que cette graduation dépend de la base spéciale choisie comme le montre l'exemple de $\mathcal{G}_3$, définie par : $[e_1, e_2] = e_3$. En effet, e_3, e_2, e_1 et e_3, e_2+e_3, e_1+e_3 sont des bases spéciales, mais elles définissent des graduations distinctes de $U(\mathcal{G}_3)$.

Par contre, si $H_n^{\mathcal{B}}$ désigne l'espace des éléments homogènes de degré n dans la graduation définie par la base spéciale $\mathcal{B}$, la proposition 2.2 implique que $\dim_k H_n^{\mathcal{B}}$ ne dépend pas de $\mathcal{B}$ et que par conséquent la série de Hilbert :

$$H(U(\mathcal{G}),T) = \sum_{n\geq 0} \dim_k H_n^{\mathcal{B}} T^n = \frac{1}{\prod_{i=1}^{p} (1-T^{h(e_i)})}$$

est un invariant de $U(\mathcal{G})$.

Soient $\mathcal{B} = (e_p,\ldots,e_1)$ et $\mathcal{B}' = (f_p,\ldots,f_1)$ deux bases spéciales de l'algèbre de Lie nilpotente graduée $\mathcal{G}$. On a alors :

$$\mathcal{G} = \mathcal{G}^q \oplus \mathcal{G}_{q-1} \oplus \ldots \oplus \mathcal{G}_1 \quad \text{et} \quad \mathcal{G} = \mathcal{G}^q \oplus \mathcal{H}_{q-1} \oplus \ldots \oplus \mathcal{H}_1$$

où $\mathcal{G}_i$ (resp. $\mathcal{H}_i$) est le sous-espace engendré par les e_α (resp. f_β) de hauteur i. Il est facile de voir que les bases $\mathcal{B}$ et $\mathcal{B}'$ définissent la même graduation dans $U(\mathcal{G})$ si et seulement si $\mathcal{G}_i = \mathcal{H}_i$ pour tout $1 \leq i \leq q-1$.

Exemple : Ce cas se présente quand $\mathcal{B}$ et $\mathcal{B}'$ sont deux bases de $gr\,\mathcal{G}$ provenant de deux bases adaptées de $\mathcal{G}$ où $\mathcal{G}$ est une algèbre de Lie nilpotente quelconque. La graduation de $U(gr\,\mathcal{G})$ définies par ces bases spéciales de $gr\,\mathcal{G}$ sera dite principale.

§3 Groupes d'automorphismes finis d'une algèbre de Lie nilpotente.

Soit G un groupe d'automorphismes fini d'une algèbre de Lie nilpotente $\mathcal{G}$ de dimension finie p sur un corps K de caractéristique nulle. G définit naturellement un groupe d'automorphismes $\overline{G}$ de $gr\,\mathcal{G}$.

3.1 Lemme : $\overline{G} \simeq G$.

Preuve : Considérons l'application $\psi : G \longrightarrow \overline{G}$.

Soit $g \in G$ t.q. $\overline{g} = id_{gr\mathcal{G}}$. Alors :

$$\forall x \in \mathcal{G}^i \ , \quad x-g(x) \in \mathcal{G}^{i+1} \quad \text{pour} \quad 1 \leq i \leq q.$$

Montrons par récurrence que :

$$(g-1)^i \Big|_{\mathcal{G}^{q+1-i}} = 0 \quad \text{pour} \quad 1 \leq i \leq q.$$

$$i = 1, \qquad g \Big|_{\mathcal{G}^q} = id_{\mathcal{G}^q} .$$

$$i > 1, \qquad \overline{g} \Big|_{\frac{\mathcal{G}^{q+1-i}}{\mathcal{G}^{q+1-(i-1)}}} = id_{\frac{\mathcal{G}^{q+1-i}}{\mathcal{G}^{q+1-(i-1)}}}$$

et par conséquent,

$(g-1)(\mathcal{G}^{q+1-i}) \subset \mathcal{G}^{q+1-(i-1)}$. Il s'ensuit par récurrence que $(g-1)^i \Big|_{\mathcal{G}^{q+1-i}} = 0$.

On a donc $g = 1+n$ avec n nilpotent d'ordre q.

Comme g est d'ordre fini cela implique $n = 0$ en caractéristique nulle.

Remarque : La conclusion du lemme tombe en défaut en caractéristique positive.

En effet, si car $K = 2$, considérons l'automorphisme σ de $\mathcal{G}_3$ défini par :

$$X \longrightarrow X+Z$$

$$Y \longrightarrow Y+Z$$

$$Z \longrightarrow Z$$

$\overline{\sigma} = id$, tandis que σ est d'ordre fini et $\sigma \neq id$.

3.2 <u>Lemme</u> : <u>Pour tout $\bar{g} \in \bar{G}$, il existe une base spéciale de</u> $gr\mathfrak{g}$ qui définit la graduation principale de $U(gr\mathfrak{g})$ dans laquelle $\bar{g}$ se diagonalise. (K algébriquement clos).

<u>Preuve</u> : Il suffit de remarquer que $\bar{g}$ conserve $\frac{\mathfrak{g}^i}{\mathfrak{g}^{i+1}}$ pour $1 \leqslant i \leqslant q$ et qu'on peut diagonaliser $\bar{g}$ dans chacun des sous-espaces $\frac{\mathfrak{g}^i}{\mathfrak{g}^{i+1}}$ puisque $\bar{g}$ vérifie le polynôme $X^{|G|} - 1$. La base de $gr\mathfrak{g}$ obtenue par juxtaposition est spéciale et définit la graduation principale de $U(gr\mathfrak{g})$.

3.3 D'après le lemme 3.2 , $\bar{G}$ conserve la graduation principale de $U(gr\mathfrak{g})$. Il est alors possible de considérer la série de Hilbert de $U(gr\mathfrak{g})^G$ qui est égale à $\sum_{n \geqslant 0} \dim_k(H_n^G)\, T^n$ où H_n désigne l'ensemble des éléments homogènes de degré n de $U(gr\mathfrak{g})$ dans la graduation principale.

3.4 <u>Théorème</u> : <u>On a la formule</u> :

$$H(U(g_r\mathfrak{g})^{\bar{G}},T) = \frac{1}{|G|} \sum_{g \in G} \frac{1}{(1-\xi_1(\bar{g})T^{h(\xi_1(\bar{g}))})\dots(1-\xi_p(\bar{g})T^{h(\xi_p(\bar{g}))})}$$

<u>où</u> $\xi_i(g)$, $1 \leqslant i \leqslant p$ <u>désigne les</u> p <u>valeurs propres distinctes de</u> $\bar{g}$ <u>et</u> $h(\xi_i(g))$, $1 \leqslant i \leqslant p$, <u>désigne la hauteur dans</u> $g_r\mathfrak{g}$ <u>d'un vecteur propre relatif à la</u>

valeur propre $\xi_i(g)$.

La preuve se fait à l'aide du lemme suivant :

3.5 Lemme : Si $(\bar{g},H_n)$ désigne l'application linéaire définie par $\bar{g}$ dans H_n, on a :

$$\mathrm{tr}(\bar{g},H_n) = \sum_{\sum_{k=1}^{p} h(\xi_k(\bar{g}))i_k=n} \xi_1^{i_1}(\bar{g})\xi_2^{i_2}(\bar{g}) \dots \xi_p^{i_p}(\bar{g}).$$

Preuve : Le lemme 3.2 fournit une base spéciale $x_p(\bar{g}), \dots, x_1(\bar{g})$ de $g \wedge \mathcal{G}$ qui définit la graduation principale de $U(g\wedge\mathcal{G})$ et dans laquelle $\bar{g}$ se diagonalise. Il est alors clair que les monômes standards en les x_i de hauteur n forment une base de H_n dans laquelle $(\bar{g},H_n)$ se diagonalise avec les valeurs propres données par $\xi_1^{i_1}(\bar{g}) \dots \xi_p^{i_p}(\bar{g}) \in K$ telles que $h(\xi_1(\bar{g}))i_1 + \dots + h(\xi_p(\bar{g}))i_p = n$.

Preuve du théorème : Remarquons d'abord que :

$\dim H_n^G = \frac{1}{|G|}\mathrm{tr}\,(\sum_{g\in G}(\bar{g},H_n))$. En effet, l'application $\frac{1}{|G|}\sum_{g\in G}(\bar{g},H_n)$ réalise la projection de H_n sur H_n^G.

On a alors :

$$H(U(g\wedge\mathcal{G})^{\bar{G}},T) = \sum_{n\geq 0}(\dim H_n^G)\,T^n =$$

$$\sum_{n\geq 0}\frac{1}{|G|}\sum_{g\in G}\mathrm{tr}(\bar{g},H_n)\,T^n = \frac{1}{|G|}\sum_{g\in G}\sum_{n\geq 0}\mathrm{tr}(\bar{g},H_n)\,T^n =$$

$$= \frac{1}{|\bar{G}|} \sum_{g \in G} \sum_{n \geq 0} \sum_{\sum_{k=1}^{p} h(\xi_k(\bar{g}))\, i_k = n} \xi_1^{i_1}(\bar{g}) \cdots \xi_p^{i_p}(\bar{g})\, T^{n} =$$

$$\frac{1}{|G|} \sum_{g \in G} \sum_{n \geq 0} \sum_{\sum_{k=1}^{p} h(\xi_k(\bar{g}))\, i_k = n} (\xi_1^{i_1}(\bar{g})\, T^{i_1 h(\xi_1(\bar{g}))}) \cdots (\xi_p^{i_p}(\bar{g})\, T^{i_p h(\xi_p(\bar{g}))})$$

$$= \frac{1}{|G|} \sum_{g \in G} \frac{1}{(1 - \xi_1(\bar{g})\, T^{h(\xi_1(\bar{g}))}) \cdots (1 - \xi_p(\bar{g})\, T^{h(\xi_p(\bar{g}))})} .$$

3.6 Remarque : Si $\sigma(\bar{g})$ désigne l'endomorphisme linéaire de $g \cap \mathscr{G}$ défini par $\sigma(\bar{g})(\chi_i(\bar{g})) = T^{h(\chi_i(\bar{g}))} \chi_{i(\bar{g})}$,

l'expression $\prod_{i=1}^{p} \frac{1}{1 - \xi_i(\bar{g})\, T^{h(\xi_i(\bar{g}))}}$ devient

$\frac{1}{\det(1 - \bar{g}\sigma(\bar{g}))}$. Remarquons d'autre part, que pour tout $\bar{g} \in \bar{G}$, $\sigma(\bar{g})$ n'est autre que la multiplication par T^{i} dans $\frac{\mathscr{G}^{i}}{\mathscr{G}^{i+1}}$ qui ne dépend pas de $\bar{g}$. La formule du théorème 3.4 prend donc la forme plus habituelle de la fromule de Molien :

$$H(u(g \cap \mathscr{G})^G, T) = \frac{1}{|G|} \sum_{g \in G} \frac{1}{\det(1 - \bar{g}\sigma)} .$$

§4 L'étude d'un exemple

a) Considérons l'anneau de polynômes commutatif $K[X,Y,Z]$

et G le groupe $\{1,\tau\}$ où τ est défini par :

$$\tau(X) = Y \quad ; \quad \tau(Y) = X \quad ; \quad \tau(Z) = -Z.$$

La formule classique de Molien donne

$$H(K[X,Y,Z]^G, T) = \frac{1}{2}\left[\frac{1}{\det(1-\tau T)} + \frac{1}{\det(1-T)}\right] = \frac{1+T^2}{(1+T)^2\,(1-T)^3}$$

b) Considérons $U(\mathcal{G}_3) = K[X,Y,Z]$ avec $XY-YX = Z$ et

le même automorphisme que dans a). On obtient alors :

$$H(U(\mathcal{G}_3)^G, T) = \frac{1}{2}\left[\frac{1}{\det(1-\tau\sigma)} + \frac{1}{\det(1-\sigma)}\right]$$

où σ est l'application définie par $\sigma(X)=TX, \sigma(Y)=TY$

et $\sigma(Z) = T^2 Z$. Il vient :

$$H(U(\mathcal{G}_3)^G, T) = \frac{1}{2}\left[\frac{1}{(1+T^2)(1-T^2)} + \frac{1}{(1-T)^2(1-T^2)}\right] =$$

$$= \frac{1-T+T^2}{(1+T^2)(1+T)(1-T)^3} .$$

Références

(1) J. Dixmier, Sur les représentations unitaires des groupes de Lie nilpotents. III, Canadian Journal of Math, 10, 1958, p. 321-348.

(2) J. Dixmier, Algèbres enveloppantes, Gauthier-Villars, 1974.

(3) N. Jacobson, Lie Algebras, Interscience Publishers.

(4) L.J. Santha Roubane thèse de 3e cycle, Paris 1979.

(5) R.P. Stanley, Invariants of finite groups and their applications to com binatorics, Bulletin of the American Math. Soc., Vol.1, N°3, May 1979.

(6) M. Vergne, Cohomologie des algèbres de Lie nilpotentes, Bull. Soc. Math. France, 98, 1970, 81-116.

SEMIPRIME IDEALS IN RINGS WITH FINITE GROUP ACTIONS REVISITED

by

Joe W. FISHER

Résumé. The connections between the prime ideal structure of a ring and that of its subring fixed by the action of a finite group of automorphisms is studied. Such a study was begun by Osterburg and the author in a paper with the same title. Recent work by Lorenz and Passman has added new impetus to the study. We will look at their work together with its ramifications.

Acknowledgment. I would like to thank Professor M.P. Malliavin for inviting me to participate in this seminare. Many thanks to Professor G. Renault for inviting me to spend the spring quarter visiting at the University of Poitiers where this paper was written. I also would like to acknowledge partial support from NSF grant No. MCS -78-00904 and from the Taft Committee at the University of Cincinnati.

Let R be an associative ring with unity and let G be a finite group of automorphisms acting on R. Throughout this paper we will assume that G is finite and the order of G is invertible in R. By r^g we will mean the image of r under g in G. The fixed ring of R is $R^G = \{r \in R : r^g = r$ for all g in $G\}$. Also '$\subset$' denotes proper inclusion.

Our objective is to study the connections between the prime ideal structures of R and R^G. The following example shows that prime ideals of R do not contract to prime ideals of R^G.

Example. Let F be a field of characteristic not 2 and $R = \mathrm{Mat}_2(F)$. Then $g : \begin{bmatrix} a & b \\ c & d \end{bmatrix} \to \begin{bmatrix} a & -b \\ -c & d \end{bmatrix}$ is an automorphism of order 2 and $R^G \cong F \oplus F$. Hence, the prime ideal zero in R contracts to zero in R^G which is not prime. However, we note that this prime in R does contract to a semiprime ideal in R^G.

Theorem (Bergman-Isaacs [1])

(1) If R is semiprime, then R^G is semiprime.

(2) If R is semiprime and L is a G-invariant ideal of R with $L \cap R^G = 0$, then $L = 0$.

If Q is a prime ideal of R, then $\cap Q^g = \cap\{Q^g : g \in G\}$ is a G-invariant semi-prime ideal of R and $Q \cap R^G = (\cap Q^g) \cap R^G$. Set $B = \cap Q^g$. Then $(R/B)^G = (R^G + B)/B \cong R^G/(B \cap R^G)$. Hence $Q \cap R^G$ is semiprime by Bergman-Isaacs' theorem. Similarly any G-invariant semiprime ideal of R contracts to semiprime ideal of R^G.

Theorem. (Fisher-Osterburg [2]) (Incomparability) If $Q_1 \supset Q_2$ are prime ideals of R, then $\cap Q_1^g \supset \cap Q_2^g$ and $Q_1 \cap R^G \supset Q_2 \cap R^G$.

Proof. If $\cap Q_1^g = \cap Q_2^g$, then $Q_2 \supset Q_1^{g_1} \supset Q_2^{g_1} \supset Q_1^{g_2} \supset Q_2^{g_2} \supset \dots$ for some $g_1, g_2, \dots$ in G. By the pigeon-hole principle, there exists g_i in G such that $Q_2^{g_i} \supset \dots \supset Q_1^{g_i} \supset Q_2^{g_i}$. Hence $Q_1 = Q_2$.

If $Q_1 \cap R^G = Q_2 \cap R^G$, then pass modulo $\cap Q_2^g$ to get $\overline{\cap Q_1^g} \cap \overline{R}^G = 0$. By Bergman-Isaacs' theorem, $\overline{\cap Q_1^g} = \overline{0}$.

Theorem. (Fisher-Osterburg [2]). If R^G satisfies the ACC on semiprime ideals, then R satisfies the ACC on semiprime ideals.

Proof. From the previous theorem, we immediately get the ACC on prime ideals of R. Then we showed that every G-invariant semiprime ideal of R was a finite intersection of primes. From this we showed that each semiprime ideal of R was a finite intersection of primes. Then an application of the König graph theorem completed the proof.

Later the author proved in [3] that if R satisfies any "decent" chain condition on G-invariant ideals, then R would satisfy the same chain condition on all ideals. In particular, ACC (DCC) on G-invariant semiprime ideals implies ACC (DCC) on all semiprime ideals. Hence the following theorem.

Theorem. (Fisher [3]). If R^G satisfies DCC on semiprime ideals, then R satisfies DCC on semiprime ideals.

At that point in time there were many nagging questions remaining concerning semiprime ideals. Do the converses of the previous two theorems hold ? Do the prime ranks of R and R^G coincide ? Is every contraction of a prime ideal in R^G a finite intersection of primes of R^G ? All of these seemed formitable problems considering that "Lying Over" failed [2]. (In the above example there is no prime in R lying over the prime $F \oplus 0$ in R^G). In [2] we answered these questions when R^G is contained in the center of R ; however, this was little solace.

Lorenz and Passman [4] found the key to unlock this mystery in the theory of skew group rings, i.e. crossed products with trivial factor sets. Actually most of what they did holds for arbitrary crossed products, but for the applications here we will only need skew group rings.

The skew group ring $R*G$ is the free left R-module $\sum\{Rg:g\in G\}$ with basis G. Multiplication comes from that of G together with $rg = gr^g$ for r in R and g in G. Since we are assuming that $|G|^{-1}$ in R, we have an isomorphic copy of R^G sitting in $R*G$ as an idempotent generated subring. More precisely, set $e = |G|^{-1}\sum\{g:g\in G\}$. Then $e^2 = e$ and $e(R*G)e = eR^G = R^Ge \cong R^G$.

As is well-known for such idempotent generated subrings, the contraction map, ϕ, gives a one to one correspondence between prime (primitive) ideals of $R*G$, not containing e, and prime (primitive) ideals of $R^G \cong e(R*G)e$. This essentially reduces the problem of studying the prime (primitive) ideal structures of R and R^G to that of studying the prime (primitive) ideal structures of R and $R*G$. The first result in this direction was the following.

Theorem. (Fisher-Montgomery [5]). If R is semiprime (semiprimitive), then $R*G$ is semiprime (semiprimitive).

For the prime ideal structures, let P be a prime ideal of $R*G$. Then $P\cap R$ is a G-invariant ideal of R and there exists a prime ideal Q of R, unique up to G-conjugacy, such that $P\cap R = \cap Q^g$ [4, lemmas 1.1, 3.1]. Hence $P\cap R$ is <u>G-prime</u> in the sense that if A and B are G-invariant ideals of R with $AB\subset(P\cap R)$, then $A\subset(P\cap R)$ or $B\subset(P\cap R)$.

Theorem. (Lorenz-Passman [4]).

(i) If Q is a prime ideal of R, then there exists finitely many minimal primes $P_1,\dots,P_n$ over $(\cap Q^g) * G$ with $P_i \cap R = \cap Q^g$ for each i. If $J = P_1 \cap P_2 \cap \dots \cap P_n$, then $J^{|G|} \subset (\cap Q^g) * G$.

(ii) If P is a prime ideal of $R * G$, then there exists finitely many minimal primes $P = P_1, P_2, \dots, P_n$ over $(P \cap R) * G$ with $P_i \cap R = P \cap R$ for each i. Again $J^{|G|} \subset (P \cap R) * G$.

Since we are assuming $|G|^{-1}$ in R, in our situation, we actually have $(\cap Q^g) * G = P_1 \cap P_2 \cap \dots \cap P_n = (P \cap R) * G$ by Fisher-Montgomery's theorem.

Corollary. (Going Down). If $A_1 \supset A_2$ are G-prime ideals of R and P_1 is a prime ideal of $R * G$ with $P_1 \cap R = A_1$, then there exists a prime ideal P_2 of $R * G$ with $P_1 \supset P_2$ and $P_2 \cap R = A_2$.

Theorem. (Lorenz-Passman [4]) (Incomparability). If $P \subset I$ are ideals of $R * G$ with P prime, then $P \cap R \subset I \cap R$.

Since we are mainly interested in primes we have only stated these theorems for primes ; however, we should mention that they also hold for primitive ideals.

It should also be noted that these theorems are easy to state but, they are long and difficult to prove. By using them it is now easy to prove the following theorem.

Theorem. (Lorenz-Passman [4]). The prime (primitive) rank of R is equal to the prime (primitive) rank of R^G.

Proof. If $Q_1 \subset Q_2 \subset \dots \subset Q_n$ is a chain of primes in R, then $\cap Q_1^g \subset \cap Q_2^g \subset \dots \subset \cap Q_n^g$ by

"Incomparability". Then there exists a minimal prime P_n over $(\cap Q_n^g) * G$ with $P_n \cap R = \cap Q_n^g$ and $e \notin P_n$. By using "Going Down" repeatedly we get $P_1 \subset P_2 \subset \ldots \subset P_n$ with $P_i \cap R = \cap Q_i^g$. Consequently $P_1^\phi \subset P_2^\phi \subset \ldots \subset P_n^\phi$ in R^G. Therefore rank $R \leq$ rank R^G.

Conversely, if $P_1^\phi \subset P_2^\phi \subset \ldots \subset P_n^\phi$ is a chain of primes in R^G, then $P_1 \subset P_2 \subset \ldots \subset P_n$ is a chain of primes in R. By "Incomparability" $P_1 \cap R \subset P_2 \subset R \subset \ldots \subset P_n \cap R$. There exists primes $Q_1, \ldots, Q_n$ in R such that $P_i \cap R = \cap Q_i^g$. By taking appropriate translates we may assume that $Q_1 \subset Q_2 \subset \ldots \subset Q_n$ and these inclusions must be proper. Hence rank $R \geq$ rank R^G.

APPLICATIONS OF LORENZ AND PASSMANS WORK

The first application yields a converse to our earlier theorem on the ACC for semiprime ideals.

Theorem. If R satisfies the ACC on semiprime ideals, then R^G satisfies the ACC on semiprime ideals.

Proof. First, we claim that every semiprime ideal of $R * G$ is a finite intersection of prime ideals of $R * G$. Assume to the contrary that Q is a semiprime ideal of $R * G$ which is not a finite intersection of primes. Amongst such Q, we may choose one with $Q \cap R$ maximal, because R satisfies the ACC on semiprime ideals.

By hypothesis $Q \cap R$ is a finite intersection of prime ideals. Hence there exists primes $P_1, \ldots, P_n$ in R with $Q \cap R = \bigcap_{i=1}^{n} (\cap P_i^g)$. Then

$(Q \cap R) * G = \bigcap_{i=1}^{n} ((\cap P_i^g) * G)$. By Lorenz-Passman there exists finitely many minimal primes over each $\cap P_i^g * G$. So there exists finitely many minimal primes over $(Q \cap R) * G$, say $T_1, T_2, \ldots, T_k$.

Every minimal prime over Q contains some T_i. Let S_i denote the intersection of all those minimal primes over Q which contain T_i. Since Q is not a finite intersection of minimal primes, some S_i, say S_1, must not be a finite intersection of minimal primes. Consequently, $T_1 \neq S_1$. We have $(Q \cap R) \subset (S_1 \cap R)$; for otherwise, $((Q \cap R) * G) \subseteq T_1 \subset S_1$ and they all contract to $Q \cap R$. This contradicts Lorenz-Passman's "Incomparability". Hence we have a contradiction to the choice of Q because $Q \subset S_1$, $(Q \cap R) \subset (S_1 \cap R)$ and S_1 is not a finite intersection of primes.

Second, it is immediate from the hypothesis and "Incomparability" that $R * G$ satisfies the ACC on prime ideals.

Third, we claim that every semiprime ideal of R^G is a <u>finite</u> intersection of prime ideals of R^G. Let S be a semiprime ideal of R^G. Then $\phi^{-1}(S)$ is a semiprime ideal of $R * G$ and $\phi^{-1}(S)$ is a finite intersection of primes of $R * G$. The contractions of these primes yield finitely many primes of R^G which intersect in S.

Fourth, R^G satisfies the ACC on primes because $R * G$ does and ϕ is a one to one correspondence.

Finally, an application of the König graph theorem [6, Theorem 7.7] establishes that R^G satisfies the ACC on semiprime ideals.

<u>Question.</u> Does there exists a converse to our earlier theorem on the DCC for semiprime ideals, i.e., if R satisfies the DCC on semiprime ideals, then does R^G ?

As we showed earlier "Lying Over" in the usual sense fails for the pair $R^G \subsetneq R$. However, if we modify the notion of "Lying Over" slightly, we not only get "Lying Over" for the pair $R^G \subsetneq R$, but also "Going Up" and "Going Down". The idea for such an approach comes from a paper of S. Montgomery [7]. First a lemma.

Lemma. Let Q be a prime ideal of R. Then $Q \cap R^G$ is a finite intersection of primes of R^G. More precisely, if $P_1, P_2, \ldots, P_n$ are the minimal primes in $R * G$ intersecting in $(\cap Q^g) * G$ then $Q \cap R^G = \bigcap_{i=1}^{n} P_i^\phi$.

Proof. One must merely follow the isomorphisms $R \cong R.1 \subset R * G$ and $R^G \cong R^G e = e(R * G)e$ to see that it works out as claimed.

Theorem. (Lying Over). For each prime ideal P in R^G, there exists a prime ideal Q in R, unique up to G-conjugacy, such that P is minimal over $Q \cap R^G$.

Proof. Let P_1 be the prime ideal in $R * G$ such that $P_1^\phi = P$. By Lorenz-Passman, there exists a prime ideal Q in R, unique up to G-conjugacy, such that $P_1 \cap R = \cap Q^g$ and finitely many minimal primes $P_1, P_2, \ldots, P_n$ in $R * G$ intersecting in $(\cap Q^g) * G$. By the above lemma $Q \cap R^G = P \cap P_2^\phi \cap \ldots \cap P_n^\phi$. We claim that P is minimal over $Q \cap R^G$. Suppose that $P \supseteq T \supseteq Q \cap R^G$ with T a minimal prime over $Q \cap R^G$. Then either $T \supseteq P$ or $T \supseteq P_i^\phi$ for some i with $2 \le i \le n$. If $T \supseteq P$, then P is minimal. If $P_1^\phi \supseteq T \supseteq P_i^\phi$, then $P_1 \supseteq \phi^{-1}(T) \supseteq P_i$ which contradicts the minimality of P_1 over $(\cap Q^g) * G$.

Theorem. (Going Up). If $P_2 \supset P_1$ are prime ideals in R^G and Q_1 is a prime ideal in R such that P_1 is minimal over $Q_1 \cap R^G$, then there exists a prime ideal Q_2 in R with $Q_2 \supset Q_1$ and P_2 minimal over $Q_2 \cap R^G$.

Proof. From the lemma we have finitely many minimal primes $T_1, T_2, \ldots, T_n$ in $R * G$ intersecting in $(\cap Q_1^g) * G$ and $(Q_1 \cap R^G) = \bigcap_{i=1}^{n} T_i^{\phi}$. Since P_1 is minimal over $Q_1 \cap R^G$, some T_i^{ϕ}, say T_1^{ϕ}, is equal to P_1. Let S_1 be a prime ideal in $R * G$ with $S_1^{\phi} = P_2$. Then $S_1 \supset T_1$ and by "Incomparability" $(S_1 \cap R) \supset (T_1 \cap R) = \cap Q_1^g$. We know that there is a prime ideal Q_2 in R such that $(S_1 \cap R) = \cap Q_2^g$ and we may take $Q_2 \supset Q_1$ since some conjugate contains Q_1. As in the above proof, P_2 will be minimal over $Q_2 \cap R^G$.

Theorem. (Going Down). If $P_2 \supset P_1$ are prime ideals in R^G and Q_2 is a prime ideal in R such that P_2 is minimal over $Q_2 \cap R^G$, then there exists a prime ideal Q_1 in R with $Q_2 \supset Q_1$ and P_1 minimal over $Q_1 \cap R^G$.

Proof. The technique of proof is similar to that used in proving "Going Up".

We will use these notions of "Lying Over" and "Going Up" in proving that Lorenz and Passman's theorem on prime rank is still valid if the notion of prime rank is extended to arbitrary ordinals. We define the <u>classical Krull dimension of R, cl. Krull dim R</u>, as follows : Let $\mathcal{P}_0(R)$ denote the set of maximal ideals of R and if $\alpha > 0$ is an ordinal let $\mathcal{P}_\alpha(R) = \{P \in \operatorname{Spec} R : P \notin \mathcal{P}_\beta(R)$ for any $\beta < \alpha$ and for all $Q \in \operatorname{Spec} R$, $Q \supset P$ implies $Q \in \mathcal{P}_\beta(R)$ for some $\beta < \alpha\}$. Then the cl. Krull dim R, if it exists, is the first α such that $\operatorname{Spec} R = \cup\{\mathcal{P}_\beta(R) : \beta \le \alpha\}$. See [6, p. 48].

Theorem. (i) If R has classical Krull dimension, then R^G has classical Krull dimension and cl. Krull dim R^G = cl. Krull dim R.

(ii) If R^G has classical Krull dimension, then R has classical Krull dimension and cl. Krull dim R = cl. Krull dim R^G.

Proof. (i) Suppose that cl. Krull dim $R = \alpha$. We use induction on α. If $\alpha = 0$, then the result follows from Lorenz-Passman's prime rank theorem. Assume that the result is true for every $\beta < \alpha$. Let $P \in \operatorname{Spec} R^G$. We want to show that $P \in \mathcal{P}_\beta(R^G)$. for some $\beta \leq \alpha$. Suppose that $P \notin \mathcal{P}_\beta(R^G)$ for any $\beta < \alpha$. Let $P_o \in \operatorname{Spec} R^G$ with $P_o \supset P$. By "Lying Over" and "Going Up" there exists $Q_o \supset Q$ in Spec R with P_o minimal over $Q_o \cap R^G$ and P minimal over $Q \cap R^G$. Since cl. Krull dim $R = \alpha$ and $Q_o^g \supset Q^g$ for each $g \in G$, we have $\{Q_o^g : g \in G\} \subseteq \mathcal{P}_\beta(R)$ for some $\beta < \alpha$.
Hence cl. Krull dim $(R/\cap Q_o^g) = \beta$ for $\beta < \alpha$. By the induction hypothesis cl. Krull dim$(R^G/Q_o \cap R^G) = \beta$. Thence $P_o \in \mathcal{P}_\sigma(R^G)$ for some $\sigma \leq \beta < \alpha$. Now by definition $P \in \mathcal{P}_\alpha(R^G)$. Whence cl. Krull dim $R^G = \alpha$.

(ii) Suppose that cl. Krull dim $R^G = \alpha$. Again we use induction and the $\alpha = 0$ case is provided by Lorenz-Passman. Assume that the result is true for all $\beta < \alpha$. If $Q \in \operatorname{Spec} R$, then we need to show that $Q \in \mathcal{P}_\beta(R)$ for some $\beta \leq \alpha$. Suppose that $Q \notin \mathcal{P}_\beta(R)$ for any $\beta < \alpha$. Take $Q_o \in \operatorname{Spec} R$ with $Q_o \supset Q$. By "Incomparability" $\cap Q_o^g \supset \cap Q^g$ and $Q_o \cap R^G \supset Q \cap R^G$. By Lorenz-Passman's theorem, we have that $(\cap Q^g) * G = P_1 \cap P_2 \cap \ldots \cap P_n$ for $P_i \in \operatorname{Spec} R$ and each $P_i \cap R = \cap Q^g$ and $(\cap Q_o^g) * G = T_1 \cap T_2 \cap \ldots \cap T_m$ for $T_j \in \operatorname{Spec} R$ and each $T_j \cap R = \cap Q_o^g$. Hence each $T_j \supset P_i$ for some $1 \leq i \leq n$.

We claim that cl. Krull dim $(R^G/Q_o \cap R^G) < \alpha$. If V is a prime in

$R^G/(Q_o \cap R^G)$, then $V \supseteq T_j^\phi$ for some $1 \leq j \leq m$ because $\cap T_j^\phi = Q_o \cap R^G$ and $V \supseteq T_j^\phi \supset P_i^\phi$ for some i. Since cl. Krull dim $R^G = \alpha$, each $T_j^\phi \in \mathcal{P}_\rho(R^G)$ for some $\rho < \alpha$. Since $\{T_j^\phi\}$ is finite, it follows that cl. Krull dim $(R^G/Q_o \cap R^G) < \alpha$.

By the induction hypothesis cl. Krull dim $(R/\cap Q_o^g) < \alpha$. Therefore $Q_o \in \mathcal{P}_\beta(R)$ for some $\beta < \alpha$. By definition $Q \in \mathcal{P}_\alpha(R)$ and so cl. Krull dim $R = \alpha$.

<u>Question.</u> One question which arises from this proof is the following : If R has classical Krull dimension and Q is a prime ideal in R, then do the finitely many minimal primes over $Q \cap R^G$ all lie in the <u>same</u> $\mathcal{P}_\alpha(R^G)$ for some appropriate α ?

Recall that a ring is said to be <u>Jacobson</u> if every prime factor ring is semiprimitive. In 1976 Armendariz and Lorenz [10] proved that if R^G is Jacobson, then R is Jacobson. Susan Montgomery [7] has noted that the converse of this result follows from Lorenz-Passman's work.

Theorem. If R is Jacobson, then R^G is Jacobson.

Proof. Again by exploiting ϕ, it is sufficient to prove that $R * G$ is Jacobson. Let P be a prime ideal of $R * G$. Then $P \cap R$ is semiprime and hence semiprimitive. Therefore by Fisher-Montgomery's theorem $R * G/(P \cap R) * G \cong (R/P \cap R) * G$ is semiprimitive. From Lorenz-Passman's work P is one of finitely many minimal primes intersecting in $(P \cap R) * G$. From $(P \cap R) * G$ being an intersection of primitives, it is easy to see that P is an intersection of primitives. Hence $R * G$ is Jacobson.

We will need the following theorem in a moment.

Theorem. (Lorenz-Passman [8]). (Going Up). Let $A_1 \subset A_2$ be G-primes of R and P_1 a prime in $R*G$ with $P_1 \cap R = A_1$. Then there exists a prime ideal P_2 in $R*G$ with $P_1 \subset P_2$ and $P_2 \cap R = A_2$.

We conclude this paper by proving one final theorem relating the primitive ideal structure of R^G to that of R. Recall that the set of all (left) primitive ideals of R comes equipped with the Jacobson topology. Here a collection of primitive ideals is closed if it is precisely the set of all the primitive ideals lying over some ideal of R. We shall denote this space by Priv R. Priv R is a Baire space if the countable intersection of dense open sets is always dense.

We define a Kaplansky ring to be a Jacobson ring in which the primitive ideal space of every homorphic image of R is a Baire space. See [11].

Theorem. If R is a left Noetherian Kaplansky ring, then R^G is a Kaplansky ring.

Proof. First we note that R^G is Jacobson by the previous theorem and left Noetherian by well-knownness. Hence, by [11, lemmas 2 and 4] it suffices to prove the following assertion. If P^ϕ is a prime ideal of R^G and $\{P_i^\phi : i \in I\}$ is a countable collection of prime ideals of R^G which property contain P^ϕ, then there exists a primitive ideal $Q^\phi \supset P^\phi$ which does not contain any of the ideals P_i^ϕ. Again by exploiting ϕ, it is sufficient to prove that $R*G$ satisfies this assertion. This is precisely what Lorenz and Passman do in [8, Theorem 6]. We will reproduce it here for the sake of completeness. By "Incomparability" each $P_i \cap R$ is a G-prime ideal of R property containing $P \cap R$. By assumption $\mathrm{Priv}(R/P \cap R)$ is a Baire space. Moreover $P \cap R$ is semiprime and hence semiprimitive. By a slight generalization of [11, Lemma 3], there exists a primitive ideal $T \supset P \cap R$ of R which does not contain any of the ideals $P_i \cap R$. The "Going Up" theorem applied to P and $\cap T^g$ yields the existence of a prime ideal Q of S with $Q \supset P$ and $Q \cap R = \cap T^g$. It follows from [4, Lemma 4.1. (ii)] that Q is primitive. Since Q clearly does not contain any of the P_i, the result follows.

Question. If R^G is a left Noetherian Kaplansky ring, then is R a Kaplansky ring ?

Even through R is not necessarily left Noetherian, there are only finitely many primes minimal over any ideal of R by Fisher-Osterburg [2, Theorem 2.10]. This is precisely the ingredient that Farkas needs for [11, Lemmas 2 and 4]. Hence again, it would suffice to take a countable collection $\{Q_i : i \in I\}$ of prime ideals of R which property contain a given prime ideal Q and produce a primitive ideal which does not contain any of the ideals Q_i.

REFERENCES

1. Bergman, G.M. and I.M. Isaacs. Rings with fixed point free group actions, Proc. London Math. Soc., 27 (1973), 69-73.

2. Fisher, J.W. and J. Osterburg. Semiprime ideals in rings with finite group actions., J. Algebra, 50 (1978), 488-502.

3. Fisher, J.W. Chain conditions for modular lattices with finite group actions, Can. J. Math., 31 (1979), 558-564.

4. Lorenz, M. and D.S. Passman. Prime ideals in crossed products of finite groups., Israel J. Math., 33 (1979), 89-132.

5. Fisher, J.W., and S. Montgomery. Semiprime skew group rings, J. Algebra, 52 (1978), 241-247.

6. Gordon, R., and J.C. Robson. Krull dimension, Memoirs Amer. Math. Soc., 133 (1973).

7. Montgomery S. Prime ideals in fixed rings, preprint.

8. Lorenz, M., and D.S. Passman. Integrality and normalizing extensions of rings, J. Algebra, (to appear).

9. Lorenz, M., and D.S. Passman. Observations on crossed products and fixed rings, Comm. in Algebra, to appear.

10. Lorenz, M. Primitive ideals in crossed products and rings with finite group actions, Math Z., 158 (1978), 285-294.

11. Farkas, D.R. Baire category and Laurent extensions, Can. J. Math., 31 (1979), 824-830

University of Cincinnati
Cincinnati, Ohio 45221
U.S.A.

Université de Poitiers
Poitiers, France

21 Avril 1980

LES MODULES ARTINIENS ET LEURS ENVELOPPES QUASI-INJECTIVES

par

François COUCHOT

Nous nous intéressons d'abord aux modules artiniens coirréductibles sur un anneau commutatif. Nous démontrons que ces modules sont quasi-injectifs, que leur anneau d'endomorphismes est un anneau commutatif local nœthérien complet pour la topologie définie par l'idéal maximal, et que ces modules sont injectifs sur leur anneau d'endomorphismes. Ceci nous permet d'affirmer que sur un anneau local nœthérien et complet, tout module artinien coirréductible est injectif modulo son annulateur ; mais ceci n'est pas toujours vérifié avec des hypothèses plus faibles comme le montrent les exemples 1 et 2.

Dans le paragraphe 2, nous établissons des résultats similaires avec des modules unisériels (l'ensemble des sous-modules est totalement ordonné pour l'inclusion) linéairement compacts (pour la topologie discrète) et extensions essentielles d'un module simple.

Enfin dans le paragraphe 3, nous montrons que si M est un A-module artinien, son enveloppe quasi-injective $\overline{M}$ est un module artinien. Nous montrons en outre qu'il existe un anneau semi-local $\hat{A}$ nœthérien et complet pour la topologie définie par son radical de Jacobson, tel que M et $\overline{M}$ aient une structure de $\hat{A}$-modules qui coïncident avec leur structure de A-modules, et tel que $\overline{M}$ soit l'enveloppe injective de M sur $\hat{A}$. On retrouve ainsi un résultat de (1) sur la structure des modules artiniens.

Nous donnons un exemple de module M de longueur finie et coirréductible sur un anneau non-commutatif, tel que M ne soit pas quasi-injectif et dont l'enveloppe quasi-injective n'est pas un module artinien.

Tous les anneaux et modules considérés sont unitaires.

1. <u>Quasi-injectivité des modules artiniens coirréductibles</u> :

Dans tout ce paragraphe, sauf pour la proposition 1.8, l'anneau considéré est commutatif.

<u>Théorème 1.1.</u> : <u>Soit</u> M <u>un A-module artinien indécomposable. Alors les les conditions suivantes sont équivalentes</u> :

1°) M <u>est un module coirréductible</u>.

2°) M <u>est un module quasi-injectif</u>.

<u>Démonstration de 2 ⟹ 1</u> : Si M n'est pas coirréductible, on a $E = E_A(M) = E_1 \oplus E_2$. Comme d'après (8), M est stable pour tout endomorphisme de E, on en déduit que $M = (M \cap E_1) \oplus (M \cap E_2)$ avec $M \cap E_i \neq \{0\}$ pour i = 1 et 2. Donc contradiction.

Avant de montrer 1) ⟹ 2), nous allons établir les résultats suivants :

<u>Proposition 1.2.</u> : <u>Soit</u> M <u>un module coirréductible de longueur finie</u>. <u>Alors on a</u> :

1°) M <u>est un module injectif modulo sur annulateur</u>.

2°) <u>On a</u> $\mathrm{End}_A M = {}^{A}/_{\mathrm{ann}\, M}$.

Pour démontrer cette proposition, nous utilisons le lemme classique suivant :

<u>Lemme 1.3.</u> : <u>Soient</u> A <u>un anneau local d'idéal maximal</u> $\mathfrak{m}$, E <u>l'enveloppe injective de</u> $A/\mathfrak{m}$. <u>Alors</u> :

1°) <u>Pour tout A-module</u> M <u>de longueur finie</u>, M <u>et</u> $\mathrm{Hom}_A(M,E)$ <u>sont de même longueur</u>.

2°) <u>L'homomorphisme canonique de</u> M <u>dans</u> $\mathrm{Hom}_A(\mathrm{Hom}_A(M,E),E)$ <u>est un isomorphisme si</u> M <u>est de longueur finie</u>.

3°) <u>Si</u> A <u>est un anneau artinien, on a</u> $A = \mathrm{End}_A E$, <u>et</u> long A = long E.

<u>Démonstration</u> :

1°) se démontre par récurrence sur la longueur de M.

2°) l'homomorphisme canonique est injectif puisque E est un cogénérateur injectif, et on conclut d'après 1) en constatant que les deux modules sont de même longueur.

3°) On applique 2) en prenant $M = A$.

Démonstration de la proposition 1.2. : Soit M un A-module coirréductible de longueur finie. Alors $A/_{\text{ann } M}$ s'identifie à un sous-module de M^n. On peut donc supposer M fidèle et A local artinien. (A, étant artinien, est un produit fini d'anneaux locaux. Comme M est indécomposable, on en déduit que A est local). Soit alors $E = E_A(M)$. Alors E est isomorphe à l'enveloppe injective du corps résiduel de A. D'après le lemme 1.3., on a $0 = \text{ann}_A M = \text{Hom}_A(E/M,E)$. On en déduit que $M = E$.

Proposition 1.4. : Soient M un module artinien, $M_1 = \text{Soc}(M)$, et si $n > 1$, soit M_n l'image réciproque de $\text{Soc}(M/_{M_{n-1}})$ par l'épimorphisme canonique de M sur $M/_{M_{n-1}}$. Alors on a :

1°) $M = \bigcup_{n \in \mathbb{N}} M_n$

2°) si N est un sous-module de M, on a $N_n = M_n \cap N$

3°) Pour tout homomorphisme f de M dans un module artinien N, on a $f(M_n) \subseteq N_n \quad \forall n \geq 1$.

Remarque 1.5. : La suite croissante des sous-modules $(M_n)_{n \geq 1}$ est appelée la suite de Lœwy de M. S'il existe n tel que $M = M_n$, alors n est appelé l'élévation du module M.

Démonstration de la proposition 1.4. :

1°) Soit $x \in M$, $x \neq 0$. Alors le sous-module Ax est un module nœthérien. La suite des sous-modules $(Ax \cap M_n)$ est donc stationnaire. Donc $\exists n$ tel que $Ax \cap M_n = Ax \cap M_{n+1}$. Si $x \notin M_n$, alors $\exists b$ dans A tel que $bx \in M_{n+1}$ et $bx \notin M_n$ puisque $M_{n+1}/_{M_n} = \text{Soc}(M/_{M_n})$. Mais puisque $Ax \cap M_n = Ax \cap M_{n+1}$, on a aussi $bx \in M_n$. Donc contradiction et $x \in M_n$.

2°) On sait que $N_1 = \text{Soc } N = M_1 \cap N$. Supposons que $N_n = M_n \cap N$

Alors $\text{Soc}\left(\frac{N+M_n}{M_n}\right) = \left(\frac{N+M_n}{M_n}\right) \cap \left(\frac{M_{n+1}}{M_n}\right) = \frac{(N+M_n) \cap M_{n+1}}{M_n}$. Or $\frac{N \cap M_{n+1}}{N_n} = \frac{N \cap M_{n+1}}{N \cap M_n}$

est isomorphe à $\frac{(N \cap M_{n+1}) + M_n}{M_n} = \frac{(N+M_n) \cap M_{n+1}}{M_n}$. On a donc $\text{Soc}(N/_{N_n}) = \frac{N \cap M_{n+1}}{N_n}$

d'où $N_{n+1} = N \cap M_{n+1}$.

3°) se fait par récurrence sur n.

Fin de la démonstration du théorème 1.1.

1) $\Longrightarrow$ 2). Soient M un module artinien coirréductible, N un sous-module de M, $(M_n)_{n\geq 1}$ et $(N_n)_{n\geq 1}$ les suites de Lœwy de M et N respectivement, et $f : N \longrightarrow M$ un homomorphisme. Alors $\forall n \geq 1$ on a $f(N_n) \subsetneq M_n$. Soit $f_n = f|_{N_n}$.

Soit g_1 un prolongement de f_1 à M_1. Supposons construit g_n un prolongement de f_n à M_n. Considérons l'homomorphisme $h_{n+1} : N_{n+1} + M_n \longrightarrow M_{n+1}$ défini par : $h_{n+1}(x+y) = f(x) + g_n(y)$ où $x \in N_{n+1}$ et $y \in M_n$. Alors h_{n+1} est bien défini car $M_n \cap N_{n+1} = N_n$. Puisque M_{n+1} est quasi-injectif, on peut prolonger h_{n+1} en un homomorphisme g_{n+1} de M_{n+1} dans M_{n+1}. On définit ainsi un prolongement g de f à M par récurrence sur n.

Théorème 1.6. : Soient M un module artinien coirréductible, et $H = \mathrm{End}_A M$. Alors :

1) M est un H-module injectif et cogénérateur.

2) H est un anneau commutatif local nœthérien complet pour la topologie J-adique, où J est l'idéal maximal de H. Si $(M_n)_{n\geq 1}$ est la suite de Lœwy de M, on a $H = \varprojlim A/\mathrm{ann}\, M_n$.

Remarque : Dans (7) ce résultat est démontré en partie en prenant comme hypothèse M artinien indécomposable et quasi-injectif. Nous donnons ici une autre démonstration qui peut aussi s'appliquer à certains modules linéairement compacts comme nous le verrons dans le paragraphe 2.

Avant de démontrer le théorème 1.6, rappelons quelques définitions et résultats.

Définitions 1.7. : Soient A un anneau (non nécessairement commutatif), M un A-module à gauche. On dit que M est f_p-injectif (ou absolument pur) si pour tout A-module à gauche F de présentation finie, on a $\mathrm{Ext}^1_A(F,M) = 0$.

On dit que M est linéairement compact (pour la topologie discrète) (resp. semi-compact) si toute famille filtrante décroissante $(x_i + M_i)_{i\in I}$ où pour tout i, $x_i \in M$ et M_i est un sous-module de M (resp. l'annulateur dans M d'un idéal à gauche de A) a une intersection non vide.

Alors M est un module injectif si et seulement si il est f_p-injectif et semi-compact (voir (11) propositions 2 et 3).

Si E est un A-module à gauche, alors la E-topologie de M est la topologie linéaire sur M définie en prenant pour système fondamental de voisinages de zéro les noyaux des homomorphismes de M dans un produit fini de modules isomorphes à E.

Proposition 1.8. : Soient A un anneau (non nécessairement commutatif), M un A-module à gauche et $H = \mathrm{End}_A M$. On suppose que M est quasi-injectif et que tout module quotient d'un produit de modules isomorphes à M est séparé pour la M-topologie.

Alors on a :

1) Pour tout H-module à gauche F de présentation finie (resp. de type fini) l'homomorphisme canonique de F dans $\mathrm{Hom}_A(\mathrm{Hom}_H(F,M),M)$ est un isomorphisme (resp. un épimorphisme).

2) M est un H-module à gauche f_p-injectif.

3) M est un H-module à gauche injectif si et seulement si M est un A-module linéairement compact.

4) M est un H-module à gauche cogénérateur et injectif si et seulement si M est un A-module linéairement compact et extension essentielle d'un socle de longueur finie.

Démonstration :

1) On vérifie d'abord qu'on a bien un isomorphisme si F est un H-module libre de type fini. On utilise ensuite une présentation de F et le fait que M soit quasi-injectif pour conclure. (C'est-à-dire que M est M^n-injectif $\forall n$).

2) Soient F un H-module à gauche de présentation finie, $0 \longrightarrow K \xrightarrow{u} L \xrightarrow{v} F \longrightarrow 0$, une suite exacte où L est un H-module à gauche libre de type fini et où K est par conséquent de type fini. Notons $T = \mathrm{Hom}_H(-,M)$, $S = \mathrm{Hom}_A(-,M)$ et φ_F l'homomorphisme canonique de F dans $ST(F)$.

On a alors la suite exacte suivante :

$$0 \longrightarrow T(F) \xrightarrow{T(v)} T(L) \xrightarrow{T(u)} T(K) \xrightarrow{w} \mathrm{Ext}^1_H(F,M) \longrightarrow 0$$

On en déduit le diagramme commutatif suivant :

$$\begin{array}{ccccccccc}
0 & \longrightarrow & K & \xrightarrow{u} & L & \xrightarrow{v} & F & \longrightarrow & 0 \\
 & & \downarrow{\varphi_K} & & \downarrow{\varphi_L} & & \downarrow{\varphi_F} & & \\
0 \longrightarrow S(\mathrm{Ext}^1_H(F,M)) & \xrightarrow{S(\omega)} & ST(K) & \xrightarrow{ST(u)} & ST(L) & \xrightarrow{ST(v)} & ST(F) & \longrightarrow & 0
\end{array}$$

où la dernière ligne est exacte puisque M est quasi-injectif que $T(L)$ est isomorphe à un produit fini de modules isomorphes à M et que $T(K)$ est isomorphe à un sous-module d'un produit fini de modules isomorphes à M. (si M est quasi-injectif, alors M est M^n-injectif $\forall n$).

Alors $ST(u) \circ \varphi_K = \varphi_L \circ u$ est injectif et par conséquent φ_K est un isomorphisme. On en déduit que $ST(u)$ est injectif et que $S(\mathrm{Ext}^1_H(F,M)) = 0$. Or $\mathrm{Ext}^1_H(F,M)$ est isomorphe à un sous-module d'un module-quotient d'un produit fini de modules isomorphes à M, et par conséquent, $\mathrm{Ext}^1_H(F,M)$ est séparé pour la M-topologie. D'où $\mathrm{Ext}^1_H(F,M) = 0$.

3) Soit N un sous-A-module de M. Puisque M/N est séparé pour la M-topologie, l'homomorphisme canonique de N dans $TS(N)$ est un isomorphisme. Par conséquent tout sous A-module de M est l'annulateur d'un idéal à gauche de H. On en déduit donc 3) d'après 1.7.

4) Supposons d'abord que M soit un A-module linéairement compact et extension essentielle d'un socle de longueur finie. Alors M est un H-module à gauche injectif et il suffit de montrer que pour tout idéal à gauche J de H, on a $\mathrm{Hom}_H(H/J, M) \neq 0$. Or $J = \bigcup_{\lambda \in \Lambda} J_\lambda$, où $(J_\lambda)_{\lambda \in \Lambda}$ est la famille des idéaux à gauche de type fini de H inclus dans J. On a $\mathrm{ann}_M J = \bigcap_{\lambda \in \Lambda} \mathrm{ann}_M J_\lambda$. D'après (13) (proposition 1*), pour avoir $\mathrm{ann}_M J \neq \{0\}$, il suffit de montrer que $\forall \lambda \in \Lambda$, $\mathrm{ann}_M J_\lambda \neq \{0\}$. Or d'après 1) on a $J_\lambda = \mathrm{ann}_H(\mathrm{ann}_M J_\lambda)$.

Réciproquement, soit $(N_\lambda)_{\lambda \in \Lambda}$ une famille filtrante décroissante de sous-A-modules non nuls de M.

On a $\sum_{\lambda \in \Lambda} \mathrm{ann}_H N_\lambda \subseteq \mathrm{ann}_H(\bigcap_{\lambda \in \Lambda} N_\lambda)$. Or d'après 3), on a :

$$\bigcap_{\lambda \in \Lambda} N_\lambda = \mathrm{ann}_M(\mathrm{ann}_H(\bigcap_{\lambda \in \Lambda} N_\lambda)) \subseteq \mathrm{ann}_M(\sum_{\lambda \in \Lambda} \mathrm{ann}_H N_\lambda) = \bigcap_{\lambda \in \Lambda} \mathrm{ann}_M(\mathrm{ann}_H N_\lambda)$$

$$= \bigcap_{\lambda \in \Lambda} N_\lambda .$$

Puisque M est cogénérateur injectif sur H on en déduit que $\sum_{\lambda \in \Lambda} \mathrm{ann}_H N_\lambda = \mathrm{ann}_H(\bigcap_{\lambda \in \Lambda} N_\lambda)$. Puisque $N_\lambda \neq 0$ on a $\sum_{\lambda \in \Lambda} \mathrm{ann}_H N_\lambda \neq H$.

On en déduit que $\bigcap_{\lambda \in \Lambda} N_\lambda \neq \{0\}$ et on applique (13) (proposition 1^*) pour conclure.

Proposition 1.9. : Soient A un anneau commutatif, M un A-module linéairement compact et extension essentielle d'un module simple S, $\mathfrak{m}$ l'idéal maximal de A annulateur de S. Alors M est un $A_{\mathfrak{m}}$-module.

Démonstration : Soit $s \in A$, $s \notin \mathfrak{m}$. Soit $\hat{s}$ la multiplication par s dans M. Puisque $s \notin \mathfrak{m}$, $\ker \hat{s} \cap S = \{0\}$ et donc $\hat{s}$ est injective. Soit $x \in M$. Alors $A/_{\mathrm{ann}(x)}$ est un anneau linéairement compact local (voir (2) exercice 21 c, p. 112). La multiplication par s dans Ax est donc bijective et il existe $y \in Ax$ tel que : $x = sy$.

Démonstration du théorème 1.6. : D'après la proposition 1.9., on peut supposer A local. Soient E l'enveloppe injective de M, N un sous-module de M^n, n entier $\geqslant 1$. Alors $\forall x \in M^n$, $x \notin N$, $\exists f : M^n \longrightarrow E$ tel que $f(x) \neq 0$ et $f(N) = \{0\}$. Comme M est quasi-injectif, $f(M^n) \subseteq M$. Par conséquent M^n/N est séparé pour la M-topologie. On applique la proposition 1.8.

Soit $(M_n)_{n \in \mathbb{N}}$ la suite de Lœwy de M. Alors pour $h \in H$, $h(M_n) \subseteq M_n$. Comme $\mathrm{End}_A M_n = A/_{\mathrm{ann}\, M_n}$, on a $H = \varprojlim A/_{\mathrm{ann}\, M_n}$. Et M étant un H-module artinien, comme $H = \mathrm{End}_H M$, on en déduit le reste de 2, d'après (10).

Corollaire 1.10. : Soit A un anneau local nœthérien d'idéal maximal $\mathfrak{m}$. Alors si A est complet pour la topologie $\mathfrak{m}$-adique, tout A-module M artinien et coirréductible est injectif modulo son annulateur et on a $\mathrm{End}_A M = A/_{\mathrm{ann}\, M}$.

Démonstration : Puisque A est complet pour la topologie $\mathfrak{m}$-adique, A est linéairement compact et donc $A/_{\mathrm{ann}\, M} = \varprojlim A/_{\mathrm{ann}\, M_n} = \mathrm{End}_A M$, où $(M_n)_{n \geqslant 1}$ est la suite de Lœwy de M.

Remarque 1.11. : Soit M un A-module artinien coïrréductible et fidèle, $\mathfrak{m}$ l'annulateur de son socle. Si M est un A-module injectif, alors $M \simeq E_A(A/\mathfrak{m})$ et on en déduit que $A_{\mathfrak{m}}$ est un anneau nœthérien.

Cependant on peut trouver des modules artiniens coïrréductibles et fidèles sur des anneaux locaux non nœthériens.

Exemple 1 : Soient K un corps commutatif, C l'anneau $K[X,Y]$, $\mathfrak{p}$ l'idéal maximal $CX+CY$ B l'anneau local $C_{\mathfrak{p}}$, $\mathfrak{m}$ sont idéal maximal, $\hat{B}$ le complété de B pour la topologie $\mathfrak{m}$-adique, isomorphe à $K[[X,Y]]$. Soit A le sous-anneau de $\hat{B}$ formé des séries formelles du type $Xf(X,Y)+\frac{P(Y)}{Q(Y)}$ où $f \in \hat{B}$ et où P et Q sont des polynômes de $K(Y)$ tels que $Q(0) \neq 0$. Alors A est un anneau local non nœthérien dont l'idéal maximal $\mathfrak{M}$ et $A\mathfrak{M}$ et tel que $B \subsetneq A \subseteq \hat{B}$. Voir (2) exercice 14 p. 119.

Soit M l'enveloppe injective sur B de $B/_{\mathfrak{M}}$. D'après (10) M est un B-module artinien et c'est aussi un $\hat{B}$-module. Donc M est un A-module artinien coïrréductible et fidèle (et non injectif).

On peut aussi trouver des modules artiniens coïrréductibles, fidèles et non injectifs sur des anneaux locaux nœthériens.

<u>Exemple 2</u> : Soient B l'anneau local de l'exemple 1 en supposant K algébriquement clos de caractéristiques 0 et $P(X,Y) = X(X^2+Y^2) + X^2 - Y^2$. Alors l'idéal BP est premier. On sait qu'il existe $u(X) \in K[(X)]$ tel que $1+X = (u(X))^2$. On a donc $P(X,Y) = (Xu(X) - Yu(-X))(Xu(X) + Yu(-X))$. Comme on peut prendre le premier cœfficient de $u(X)$ égal à 1, on a $Xu(X) - Yu(-X) = X-Y + \sum_{n=2}^{+\infty} a_n$, où a_n est un polynôme homogène de degré n.

On considère la suite décroissante $(\mathfrak{P}_n)_{n\geq 1}$ d'idéaux de B suivante :

$$\mathfrak{P}_1 = \mathfrak{M} = BX + BY, \quad \mathfrak{P}_2 = B(X-Y) + \mathfrak{M}^2, \quad \text{et } \forall n \geq 2$$

$$\mathfrak{P}_n = B(X-Y+\sum_{p=2}^{n-1} a_p) + \mathfrak{M}^n.$$

Soit E l'enveloppe injective sur B de $B/_{\mathfrak{M}}$. Alors E est un B-module artinien et $\mathrm{End}_B E = \hat{B} = K[(X,Y)]$ d'après (10). On considère le sous-module M de E, tel que $M = \bigcup_{n\geq 1} \mathrm{ann}_E\, \mathfrak{P}_n$.

Alors $\quad \mathrm{ann}_B M = \bigcap_{n\geq 1} \mathrm{ann}_B(\mathrm{ann}_E\, \mathfrak{P}_n) = \bigcap_{n\geq 1} \mathfrak{P}_n.$

Puisque M est quasi-injectif, M est stable par les endomorphismes de E, et donc M est un $\hat{B}$-module. On a donc $\mathrm{ann}_B M = \mathrm{ann}_{\hat{B}} M \cap B$. Comme pour tout B-module de type fini N, on a $\hat{B} \otimes_B N$ isomorphe à $\mathrm{Hom}_{\hat{B}}(\mathrm{Hom}_B(N,E),E)$, on a $\mathrm{ann}_{\hat{B}}(\mathrm{ann}_E\, \mathfrak{P}_n) = \hat{B}\, \mathfrak{P}_n$.

Par conséquent $\mathrm{ann}_{\hat{B}} M = \bigcap_{n\geq 1} \hat{B}\, \mathfrak{P}_n$. Comme $\hat{B}$ est linéairement compact, on a alors :

$$\mathrm{ann}_{\hat{B}} M = \bigcap_{n\geq 1} (\hat{B}(Xu(X) - Yu(-X)) + \hat{\mathfrak{M}}^n) = \hat{B}(Xu(X) - Yu(-X)).$$

Comme l'idéal $\hat{B}(Xu(X)-Yu(-X))$ est premier, on en déduit que $\mathrm{ann}_B M = BP$. Posons $A = {}^B/_{BP}$. Alors M est un A-module artinien coïrréductible et fidèle. Il n'est pas injectif puisqu'il n'est pas fidèle sur $\hat{A} = \hat{B}/\hat{B}P$.

2. Quasi-injectivité de certains modules linéairement compacts et coïrréductibles :

Dans tout ce paragraphe, l'anneau A considéré est commutatif.

Définition 2.1. : On dit qu'un A-module M est unisériel si l'ensemble de ses sous-modules est totalement ordonné pour l'inclusion.

On dit qu'un A-module M vérifie (P) si on a les propriétés suivantes :

a) M est linéairement compact

b) M est extension essentielle d'un module simple

c) M est un module unisériel

Avant d'établir la quasi-injectivité des modules M vérifiant (P) rappelons quelques définitions et résultats.

Définition 2.2. : On dit qu'un A-module M est prélinéairement compact si ${}^M/_N$ est linéairement compact pour tout sous-module N non nul de M.

Théorème 2.3. : Soit A un anneau unisériel. Alors les conditions suivantes sont équivalentes

1) A est un anneau auto-injectif

2) A est linéairement compact et tout élément de A est soit inversible soit un diviseur de zéro.

C'est le théorème 2.3. de (9).

Théorème 2.4. : Soient A un anneau local, $\mathfrak{m}$ sur idéal maximal, E l'enveloppe injective de $A/\mathfrak{m}$. Alors les conditions suivantes sont équivalentes :

1) A est unisériel et prélinéairement compact

2) E est unisériel

De plus, si ces conditions sont vérifiées, E est un module linéairement compact.

L'équivalence 1) $\Longleftrightarrow$ 2) est une partie du théorème principal de (5). Pour la dernière assertion voir la proposition 4.4. de (14).

Théorème 2.5. : Soit M un module vérifiant (P). Alors M est un module quasi-injectif.

Pour la démonstration, nous avons besoin du lemme suivant :

Lemme 2.6. : Soient M un module vérifiant les propriétés b) et c) de la définition 2.1., N un sous-module de M, $f : N \longrightarrow M$ un homomorphisme : Alors $\forall x \in N,\ f(x) \in Ax$.

Démonstration : Ou bien $f(x) \in Ax$, ou bien $x \in Af(x)$. Supposons que $x \in Af(x)$. Alors $\exists \lambda \in A$, tel que $x = \lambda f(x)$. Or il existe $a \in A$ tel que $ax \in \mathrm{Soc}\, M = S$, avec $ax \neq 0$. Nous avons $\mathrm{ann}\, x = \mathrm{ann}\, f(x) = I$ et A/I est un anneau unisériel. Puisque $ax \in S$, $f(ax) \in S$. Si λ appartient à l'idéal maximal annulateur de S, on obtient que $ax = \lambda f(ax) = 0$. Contradiction. Par conséquent on a : $Ax = Af(x)$.

Démonstration du théorème 2.5. : Soient N un sous-module de M, $f : N \longrightarrow M$ un homomorphisme. Considérons la famille $\mathcal{F} = \{(P,g)$ où P est un sous-module de M contenant N, $g : P \longrightarrow M$ un homomorphisme prolongeant $f\}$. On ordonne $\mathcal{F}$ de la façon suivante : $(P,g) \leq (Q,h)$ si et seulement si $P \subseteq Q$ $h|_P = g$. Alors $\mathcal{F}$ est inductive. Soit (P,g) un élément maximal de $\mathcal{F}$.

Supposons $P \neq M$. Alors $\exists x \in M, x \notin P$ et donc $P \subseteq Ax$. D'après le théorème 2.3., $A/_{\mathrm{ann}\,(x)}$ est auto-injectif, et donc Ax est un module quasi-injectif. D'après le lemme 2.6., $g(P) \subseteq Ax$. Donc on peut prolonger g à Ax, ce qui contredit le caractère maximal de P. Donc $P = M$.

Théorème 2.7. : Soient M un module vérifiant (P), et $H = \mathrm{End}_A M$. Alors :

1) H est un anneau commutatif et c'est un anneau unisériel et linéairement compact.

2) M est un H-module injectif.

Démonstration : D'après le lemme 2.6., H est un anneau commutatif. On peut reprendre le début de la démonstration du théorème 1.6. pour vérifier qu'on a bien les conditions de la proposition 1.8. et en déduire que M est un H-module injectif. Puisque M est un H-module cogénérateur injectif, en utilisant les résultats de (12), on en déduit que H est un anneau unisériel et linéairement compact.

Théorème 2.8. : Soient un A-module fidèle M vérifiant (P), et $\mathfrak{m}$ l'idéal maximal de A qui annule le socle de M. Alors les conditions suivantes sont équivalentes :

1) M est un A-module injectif

2) $A_{\mathfrak{m}}$ est un anneau unisériel et prélinéairement compact.

Démonstration :

1) ⟹ 2) En utilisant la proposition 1.9., M est alors isomorphe à $E_{A_{\mathfrak{m}}}\left({A_{\mathfrak{m}}}/{\mathfrak{m}A_{\mathfrak{m}}}\right)$ et on en déduit le résultat en utilisant le théorème principal de (5).

2) ⟹ 1) D'après le théorème principal de (5), $E_A(M)$ est unisériel. Supposons $M \neq E_A(M)$. Alors $\exists x \in E_A(M)$, $x \notin M$ tel que $M \subset Ax$. Alors M s'identifie à un idéal propre de $A/_{\mathrm{ann}(x)}$ et par conséquent $\mathrm{ann}\, M \neq \{0\}$. Contradiction.

Remarque 2.9. : Le module M de l'exemple 2 du paragraphe 1 vérifie (P) puisque pour tout $n \geq 1$ long $({}^{\mathfrak{p}^n}/_{\mathfrak{p}^{n+1}}) = 1$, et donc long $({}^{M_{n+1}}/_{M_n}) = 1$ d'après le lemme 1.3. Et M est un exemple de module fidèle vérifiant (P), qui n'est pas injectif.

Corollaire 2.10. : Soient A un anneau unisériel et prélinéairement compact, M un A-module vérifiant (P). Alors on a :

1) M est injectif modulo sur annulateur.

2) Si $\mathrm{ann}\, M \neq \{0\}$ ou si A n'est pas intègre, on a $A/_{\mathrm{ann}\, M} = \mathrm{End}_A M$.

Démonstration : D'après (5) $A/_{\mathrm{ann}\, M}$ est alors un anneau unisériel et linéairement compact. Il suffit alors d'appliquer le lemme 2.6. pour avoir le résultat.

Remarque 2.11. : Dans les paragraphes 1 et 2 nous venons d'établir que certains modules linéairement compacts extension essentielle d'un module simple sont quasi-injectifs.

Alors est-ce que tout module M linéairement compact et extension essentielle d'un module simple est quasi-injectif ?

Mais pour répondre à cette question, même lorsque M est de type fini, nous sommes amené à résoudre un problème posé à la fois par Müller dans (12) et Goblot dans (6) :

<u>Est-ce que tout anneau</u> A <u>linéairement compact et extension essentielle d'un module simple est auto-injectif</u> ?

3. <u>Enveloppes quasi-injectives des modules artiniens</u>

<u>Définitions et rappels</u> : Soient M un module, u un monomorphisme de M dans un module $\overline{M}$ quasi-injectif. On dit que $\overline{M}$ est une <u>enveloppe quasi-injective</u> de M, si pour tout monomorphisme v de M dans un module quasi-injectif N, il existe un monomorphisme $w : \overline{M} \longrightarrow N$ tel que $v = w \circ u$. Alors deux enveloppes quasi-injectives de M sont isomorphes et si E est l'enveloppe injective de M, Λ l'anneau des endomorphismes de E, alors ΛM est une enveloppe quasi-injective de M.

Voir (3) (proposition 19.7. p. 64).

Soient M un module artinien sur un anneau commutatif A, $\mathfrak{m}_1,\ldots,\mathfrak{m}_p$ les idéaux maximaux qui annulent les modules simples qui composent le socle de M. Alors on a $\operatorname{supp} M = \{\mathfrak{m}_1,\ldots,\mathfrak{m}_p\}$ d'après (1) proposition 1.2. On dit que M est un <u>module anti-primaire</u> si son support ne contient qu'un seul idéal maximal. On dit $M = M_1 \oplus \ldots \oplus M_p$ est une décomposition <u>anti-primaire réduite</u> de M si les $(M_i)_{1\leqslant i\leqslant p}$ est une famille de sous-modules anti-primaires de M tels que $\operatorname{supp} M_i \cap \operatorname{sup} M_j = \phi$ si $i \neq j$. Alors tout module artinien admet une décomposition anti-primaire réduite unique d'après (1).

On se propose de démontrer le théorème suivant :

<u>Théorème : Soient</u> M <u>un module artinien,</u> $\overline{M}$ <u>une enveloppe quasi-injective de</u> M, $(M_n)_{n\geqslant 1}$ <u>la suite de Lœwy de</u> M <u>et</u> $\hat{A} = \varprojlim_n A/_{\operatorname{ann} M_n}$. <u>Alors on a les résultats suivants</u> :

1) $\hat{A}$ <u>est un anneau nœthérien semi-local complet pour la topologie J-adique, où</u> J <u>est le radical de Jacobson de</u> $\hat{A}$. <u>Et</u> M <u>et</u> $\overline{M}$ <u>ont une structure de</u> $\hat{A}$<u>-modules qui coïncident avec leurs structures de A-modules.</u>

2) $\overline{M}$ <u>est un module artinien et on a</u> $\overline{M} = E_{\hat{A}}(M)$.

3) Si M est un module anti-primaire, $\overline{M}$ est isomorphe à N^p, où N est un A-module artinien coïrréductible et où p est la longueur du socle de M. De plus, si M est de longueur finie, $\overline{M}$ est aussi de longueur finie et on a $\text{long}\,\overline{M} = \text{long}\,(\text{Soc}\,M) \times \text{long}\,(A/_{\text{ann}\,M})$.

4) Si $M = M_1 \oplus \ldots \oplus M_p$ est une décomposition anti-primaire réduite de M, on a $\overline{M} = \overline{M}_1 \oplus \ldots \oplus \overline{M}_p$, où $\overline{M}_i$ est une enveloppe quasi-injective de M_i, pour tout i $1 \leq i \leq p$.

Démonstration : Soient $E = E_A(M)$, $\Lambda = \text{End}_A E$.

Supposons que M soit anti-primaire et de longueur finie. Puisque $\overline{M} = \Lambda M$, on a $\text{ann}\,M = \text{ann}\,\overline{M}$. Par conséquent $\overline{M} \subseteq E_{A/_{\text{ann}\,M}}(M)$. Or $A/_{\text{ann}\,M}$ est un anneau artinien local car M est de longueur finie et $\text{ann}\,M$ est $\mathfrak{M}$-primaire, où $\mathfrak{M}$ est l'idéal maximal du support de M.

Nous avons donc $E_{A/_{\text{ann}\,M}}(M) \simeq E_{A/_{\text{ann}\,M}}(A/\mathfrak{M})^p$, où p est la longueur du socle de M. D'après le lemme 1.3., on en déduit que $\overline{M} = E_{A/_{\text{ann}\,M}}(M)$, et que $\text{long}\,E_{A/_{\text{ann}\,M}}(A/\mathfrak{M}) = \text{long}\,A/_{\text{ann}\,M}$.

Si M n'est pas de longueur finie, montrons que l'on a $\overline{M} = \bigcup_{n \geq 1} \text{ann}_E(\text{ann}\,M_n)$ où (M_n) est la suite de Lœwy de M. Il est évident que $\Lambda M \subseteq \bigcup_{n \geq 1} \text{ann}_E(\text{ann}\,M_n)$. Puisque $E_A(M_n) = E$, d'après la 1ère partie de la démonstration $\text{ann}_E(\text{ann}\,M_n)$ est l'enveloppe quasi-injective de M_n et on a donc $\Lambda M_n = \text{ann}_E(\text{ann}\,M_n)$. On en conclut que $\overline{M} = \bigcup_{n \geq 1} \text{ann}_E(\text{ann}\,M_n)$.

Si $\text{supp}\,M = \{\mathfrak{M}\}$, soit $E' = E_A(A/_{\mathfrak{M}})$. Alors $E \simeq E'^p$ où p est la longueur du socle de M. Soit $N = \bigcup_{n \geq 1} \text{ann}_{E'}(\text{ann}\,M_n)$. Alors on a $\overline{M} \simeq N^p$. Posons $N_n = \text{ann}_{E'}(\text{ann}\,M_n)$. On a d'après le lemme 1.3., $A/_{\text{ann}\,M_n} = \text{End}_A N_n$ et par conséquent $\text{End}_A N = \varprojlim A/_{\text{ann}\,M_n}$. Soient $\hat{a} \in \hat{A}$ et $x \in \overline{M}$ (resp. M). Alors, il existe un n tel que $x \in \text{ann}_E(\text{ann}\,M_n) \simeq N_n^p$ (resp. $M_n \subseteq N_n^p$) et on a $x = x_1 + \ldots + x_p$ où $x_j \in N_n$, $\forall_j$. Or sur N_n la multiplication par $\hat{a}$ coïncide avec la multiplication par un élément a appartenant à A. On en déduit donc $\hat{a}x = ax$. Donc M et $\overline{M}$ ont une structure de $\hat{A}$-module qui coïncident avec leur structure de A-module.

Puisque $\overline{M}$ est stable par tout endomorphisme de E, et que tout endomorphisme de $\overline{M}$ se prolonge à E, on peut prendre $\Lambda = \text{End}_A \overline{M}$. Alors Λ est un $\hat{A}$-module libre de type fini, et on en déduit que $\overline{M} = \Lambda M$ est un $\hat{A}$-module artinien et aussi un A-module artinien. Donc N est un A-module artinien coïrréductible et par conséquent N est $\hat{A}$-cogénérateur injectif et $\hat{A}$ est un anneau local nœthérien complet.

Nous avons donc établi 1) 2) et 3) dans le cas où M est anti-primaire.

Soient $M = M_1 \oplus \ldots \oplus M_j \oplus \ldots \oplus M_p$ une décomposition anti-primaire réduite de M, $\mathfrak{M}_j$ l'idéal maximal de supp M_j $1 \leq j \leq p$, $(M_{j,n})_{n \geq 1}$ la suite de Lœwy de M_j. On a $M_n = \bigoplus_{j=1}^{p} M_{j,n}$. On a donc $\operatorname{ann} M_n = \bigcap_{j=1}^{p} \operatorname{ann} M_{j,n}$, et comme $\operatorname{ann} M_{j,n}$ est $\mathfrak{M}_j$-primaire $\forall j$, on en déduit que $A/_{\operatorname{ann} M_n} = \prod_{j=1}^{p} A/_{\operatorname{ann} M_{j,n}}$ et donc $\hat{A} = \prod_{j=1}^{p} \hat{A}_j$, où $\hat{A}_j = \varprojlim A/_{\operatorname{ann} M_{j,n}}$. Puisque pour tout j, $\overline{M_j}$ est un $\hat{A}_j$-module injectif, on en déduit que $\bigoplus_{j=1}^{p} \overline{M_j}$ est un $\hat{A}$-module injectif et par conséquent un A-module quasi-injectif. On a donc $\overline{M} \subseteq \bigoplus_{j=1}^{p} \overline{M_j}$. Mais comme $\forall j$, $M_j \subseteq M$, on a $\bigoplus_{j=1}^{p} \overline{M_j} \subseteq \overline{M}$ (on a $E = \bigoplus_{j=1}^{p} E_A(M_j)$).

Remarques :

1) Si M est un module artinien anti-primaire et si $M = M_1 \oplus M_2$, on n'a pas toujours $\overline{M} = \overline{M_1} \oplus \overline{M_2}$ comme le montre l'exemple suivant. Soient K un corps, $A = \frac{K[X,Y]}{(X^2,Y^2)}$ x et y les images respectives de X et Y dans A, M_1 le sous-module A engendré par x, M_2 le sous-module de A engendré par y. Alors on a $M_1 = \overline{M_1}$ et $M_2 = \overline{M_2}$. Mais $M_1 \oplus M_2$ étant fidèle sur $A/_{(xy)}$ on a long $(\overline{M_1 \oplus M_2}) = 6$, alors que long $(M_1 \oplus M_2) = 4$.

2) Si $(M_n)_{n \geq 1}$ est la suite de Lœwy d'un module artinien M, on a toujours l'inclusion $\overline{M_n} \subseteq (\overline{M})_n$, mais on n'a pas toujours l'égalité comme le montre l'exemple suivant.

Soient $A = \frac{k(X,Y)}{(X^2,XY,Y^3)}$, x et y les images respectives de X et Y dans A. Alors on a $A = A_3$ et long $(\operatorname{Soc} A) = 2$. Par conséquent $\overline{A} = E^2$ où $E = E_A(k)$. On a donc $(\overline{A})_2 = E_2^2$. Or $E_2 = \operatorname{ann} \mathfrak{M}^2$ où $\mathfrak{M}$ est l'idéal maximal de A. On a $A/_{\mathfrak{M}^2} = \frac{k[X,Y]}{(X^2,XY,Y^2)}$ et d'après le lemme 1.3., on a long $(\overline{A})_2 = 6$. D'autre part on a $\operatorname{ann} A_2$ engendré par x et y^2 et $A/_{\operatorname{ann} A_2} = \frac{k[X,Y]}{(X,Y^2)}$. On a donc long $\overline{A_2} = 4$ et par conséquent $\overline{A_2} \neq (\overline{A})_2$.

3) Le 1) du théorème avait déjà été démontré en partie dans (1).

4) Contre-exemple : Soient L un corps commutatif, K un sous-corps de L. On considère l'anneau A des matrices carrées d'ordre 2 suivant :

$$A = \begin{pmatrix} L & L \\ 0 & K \end{pmatrix} \quad \text{et } M \text{ le A-module à gauche } \begin{pmatrix} L \\ K \end{pmatrix}$$

Alors M est un A-module à gauche de longueur 2 et coïrréductible. On a $S = \mathrm{Soc}\, M = \begin{pmatrix} L \\ 0 \end{pmatrix}$, $\mathrm{End}_A S = L$ et $\mathrm{End}_A M = K$. Donc si $L \neq K$, M n'est pas quasi-injectif. On a $\overline{M} = \begin{pmatrix} L \\ L \end{pmatrix}$ et donc si $(L : K)$ est infini, $\overline{M}$ n'est pas un A-module artinien.

Bibliographie

[1] B. BALLET : Topologies linéaires et modules artiniens (Thèse) - J. of Algebra 41, 365-397 (1976).

[2] N. BOURBAKI : Algèbre commutative - Chap. 3.

[3] C. FAITH : Algebra II - Ring Theory (Springer).

[4] C. FAITH - Y. UTUMI : Quasi-injectives modules and their endomorphism rings - Arch. Math. 15. 166-174 (1966).

[5] D.T. GILL : Almost valuation rings - J. London. Math. Soc. (2) 4 140-146 (1971).

[6] R. GOBLOT : Sur deux classes de catégories de Grothendieck - Thèse 1971. Université des Sciences et techniques de Lille.

[7] M. HARADA : On quasi-injective modules with a chain condition over a commutative ring - Osaka J. Math. 9 421-426 (1972).

[8] R.E. JOHNSON - E.T. WONG : Quasi-injective modules and irreductible rings - J. London. Math. Soc. 36 260-268 (1961).

[9] G.B. KLATT - L.S. LEVY : Pre-self injective rings - Trans. Amer. Math. Soc. 122 407-419 (1969).

[10] E. MATLIS : Modules with descending chain conditions - Trans. Amer. Math. Soc. 97 495-508 (1960).

[11] E. MATLIS : Injective modules over Prüfer Rings - Nagoya Math. J. 15 57-69 (1959).

[12] B.J. MÜLLER : Linear compactness and Morita Duality - J. Algebra 16. 60-66 (1970).

[13] P. VAMOS : The dual of the notion of "finitely generated" - J. London Math. Soc. 49 (1968) p. 643-646.

[14] P. VAMOS : Classical rings - J. Algebra 34 114-129 (1975).

M. François COUCHOT
Département de Mathématiques
Esplanade de la Paix
Université de Caen
14032 CAEN CEDEX

Manuscrit reçu le 4 février 1980

THE REDUCED RANK IN NOETHERIAN RINGS

A.W. GOLDIE

0. Introduction

This lecture introduces the concept of reduced rank of a module over a non-commutative noetherian ring and indicates a number of applications. In effect the reduced rank of a module can be used to circumvent those difficulties which arise in module theory due to the lack of classical localisations, provided that the problem requires only an analysis of composition length of modules once the localisation has been effected

R is a ring which is right noetherian (denoted here by max-r). All right R-modules to be considered are finitely generated (f.g.) . The ring R is right fully bounded (R F B) if in any prime factor ring, every essential right ideal contains a non-zero ideal. Rings with polynomial identity are natural examples of these rings.

Let T be an ideal of R , define :

$$\mathscr{C}'(T) = \{ c \in R \ ; \ cx \in T \Rightarrow x \in T \}$$

$$'\mathscr{C}(T) = \{ c \in R \ ; \ xc \in T \Rightarrow x \in T \}$$

$$\mathscr{C}(T) = \mathscr{C}'(T) \cap {}'\mathscr{C}(T).$$

The elements of $\mathscr{C}(T)$ are said to be regular modulo T. For example, R is right fully bounded when $cR + P$ with $c \in \mathscr{C}(P)$ contains $Rd + P$ for some $d \in \mathscr{C}(P)$.

Let I be a right ideal then bound $I = \operatorname{ann} \frac{R}{I} = \{ x \in R \ ; \ Rx < I \}$.

Reduced rank of a module

Let M be a f.g. right R-module. Then rank $M = r$, provided that M contains a direct sum of r submodules ($\neq 0$) but no longer sum. It is known that any direct sum S of uniform submodules is essential as a submodule of M if and only if S has r uniform summands.

Let R be a semi-prime ring, then define :

$$\rho(M) = \text{rank } M = \text{composition length } (M \otimes_R Q),$$

where Q is the quotient ring of R . Note that Q is a semi-simple artinian ring. More generally when N is the nilpotent radical of the ring R then define :

$$\rho(M) = \rho_{R/N}(M) \text{ , provided } MN = 0$$

and generally when $MN^k = 0$, then :

$$\rho(M) = \sum_{j=o}^{k-1} \rho(MN^j / MN^{j+1})$$

The definition for the case when $N = 0$ shows that $\rho(M)$ is additive on short exact sequences ; this clearly extends to the case of general rings and shows that if we take any Loewy series of M

$$M = M_o > M_1 > \dots > M_k = 0 \quad \text{where} \quad M_j N \leqslant M_{j+1}$$

then

$$\rho(M) = \sum_{j=o}^{k-1} \rho_{R/N}(M_i / M_{i+1}) .$$

The definition of $\rho(M)$ can therefore be made in a manner independent of the choice of series for M ; it is an invariant of M . We call it the reduced rank of M .

Note that these definitions can be made under more general conditions. One must have that rank $M < \infty$ and that N exists (the lower nil radical is nilpotent) and $R_{/N}$ has a quotient ring . Thus, for example, that R and M have Krull dimension as modules or that R is a prime ring with polynomial identity and rank $M < \infty$; each suffices to give the existence of $\rho(M)$.

Note also that $\rho(M) = 0$ if and only for $m \in M$ there exists $c \in \mathscr{C}(N)$ with $mc = 0$. Then M consists of torsion elements. When M is a bimodule f.g. on the left, then $Md = 0$ for some $d \in \mathscr{C}(N)$.

. Application to Gabriel H-rings

Theorem : Let R be a ring with max-r. The following are equivalent :

(α) $\phi : \mathscr{E}(R) \to \text{Spec } R$ is bijective ;

$\phi(E) = \text{ass } E \in \text{Spec } R$, $\mathscr{E}$ is the set of indecomposable injectives of R ;

(β) R is right fully bounded ;

(γ) Cotertiary modules are isotopic ;

(δ) R is a Gabriel H-ring (For any f.g. $M = M_R$ there exists a finite subset

$m_1, \ldots, m_k \in M$ with $\operatorname{ann} M = \bigcap_{i=1}^{k} (m_i))$.

The theorem is well known and most of it is given in Stenström's book on ring theory. However $(\beta) \Longrightarrow (\delta)$ is not given there and was first proved by G. Cauchon in [2]. The reduced rank provides a simple proof of this part of the theorem.

To prove that $(\beta) \Longrightarrow (\delta)$ first observe that the f.g. module M can be taken to be cyclic, say $R_{/I}$, where $I \triangleleft_r R$. Let $I = K_1 \cap \ldots \cap K_s$, where each $R_{/K_i}$ is a uniform module and $K_i \triangleleft_r R$. It is enough to prove the result for $R_{/K_i}$, so we consider $R_{/K}$, where $K \triangleleft_r R$.

<u>Lemma</u> - <u>Let</u> R <u>be a ring with max-r and right fully bounded</u>. <u>Let</u> K <u>be a right ideal of</u> R <u>with</u> $R_{/K}$ <u>a uniform module. Take</u> bound $K = 0$. <u>Then</u> $\operatorname{Ass} R_R = \operatorname{Ass} R_{/K} = \{P\}$. (Ass $R_{/K}$ is a single prime ideal P ; Ass R_R is the repeated prime P).

We refer to stenström's book for the proof ; this is part (γ) of the theorem (in the Lesieur-Croisot terminology). To prove that $(\beta) \Longrightarrow (\delta)$ we need only the following lemma :

<u>Lemma</u> - <u>Let</u> R <u>be a ring with max-r and be</u> RFB. <u>Let</u> K <u>be a right ideal such that</u> $R_{/K}$ <u>is uniform and</u> bound $K = 0$. <u>Then for each</u> $a \in R$ <u>set</u> $K_a = \{x \in R \,;\, ax \in K\}$. <u>There exists a finite set</u> $a_1, \ldots, a_n \in R$ <u>with</u>

$$0 = K_{a_1} \cap \ldots \cap K_{a_n}$$

<u>Proof</u> - Let $\{P\} = \operatorname{Ass} R_R = \operatorname{Ass} R_{/K}$, then there exists an essential right ideal E of R with $EP = 0$. Examine $\rho(E)$ with respect to the ring $R_{/P}$. If $E = E \cap K_a$ for all $a \in R$ then $E = 0$, since $\bigcap_{a \in R} K_a = 0$. So let $E \underset{\neq}{>} E \cap K_{a_1}$. Then $\dfrac{E}{E \cap K_{a_1}} \overset{\text{inj}}{\hookrightarrow} R_{/K_{a_1}} \overset{\text{inj}}{\hookrightarrow} R_{/K}$. Now $R_{/K}$ has zero torsion with respect to $\mathscr{C}(P)$, for $xc \in K \Longrightarrow x(cR + P) \in K \Longrightarrow xT \in K$ $(P \underset{\neq}{<} T \triangleleft R)$, which contradicts that P is the associated prime. Hence

$\rho(E/_{E\cap K_{a_1}}) = 1$ and we repeat with : $E > E\cap K_{a_1} > E\cap K_{a_1}\cap K_{a_2} \supset \dots$ so by the additivity of $\rho(E)$ there exists $F = E\cap K_{a_1}\cap K_{a_2}\cap \dots \cap K_{a_n}$ such that $\rho(F/_{F\cap K_a}) = 0$ for all $a\in R$. Then $F = F\cap K_a = F\cap(\bigcap_R K_a) = 0$. Now E is essential , hence :

$$K_{a_1}\cap\dots\cap K_{a_n} = 0 .$$

2. The Principal Ideal Theorem

There are a number of theorems ; these are generalisations of the commutative case and of theorems obtained by Jategaonkar [5].

Theorem 2.1* Let R be a prime ring which is right noetherian and has polynomial identity. Let $c\in R$ be a regular element ; then any minimal prime P over $bd(cR)$ for which $c\notin \mathscr{C}(P)$ has height one (rank one). There is a least one such prime.

This theorem was proved by Jategaonkar when c is a central element. In this case $cR = bd(cR)$ and $cR \leqslant P$ so that $c\notin\mathscr{C}(P)$.

The general theorem is given in the present form because we may have $RcR = R$, so that cR itself has no minimal primes. For example $R = (Z[x])_2$ and $c = \begin{pmatrix}1 & 0\\ 0 & x\end{pmatrix}$, then $RcR = R$ but $bd(cR) = xR$ is prime of height 1.

Now let $R = \begin{pmatrix}Z & 2Z\\ Z & Z\end{pmatrix}$, and $c = \begin{pmatrix}1 & 0\\ 0 & 2\end{pmatrix}$, then

$$bd(cR) = \begin{pmatrix}2Z & 2Z\\ 2Z & 2Z\end{pmatrix} \quad\text{and for}\quad P_1 = \begin{pmatrix}2Z & 2Z\\ Z & Z\end{pmatrix}$$

we have $c\in\mathscr{C}(P)$; however observe that height P is one . For $P_2 = \begin{pmatrix}Z & 2Z\\ Z & 2Z\end{pmatrix}$

---

*See also M. Chamarie and Guy Maury C.R. 286 (1978), 609 - 611, for a principal ideal theorem using other methods. This is a special case of theorem (2.1), though I understand that a new generalisation has been obtained which does not require that the ring be noetherian.

the theorem will apply. So in a sense, we may enquire whether the theorem is really unfinished, or perhaps there exists an example for which $c \in \mathscr{C}(P)$ and height $P > 1$.

<u>Proof</u> - Let $B = \text{bound}(cR)$ and $0 \neq Q \underset{\neq}{<} P$ be a prime ideal of R ; suppose that $\delta \neq 0$ is a central element in Q. Set $c^{-k}(\delta R) = \{x \in R \; ; \; c^k x \in \delta R\}$. For large k, $c^{-k}(\delta R) = c^{-k-1}(\delta R)$, so set $c^k = a$, k large enough, and have $a^2 x \in \delta R \Rightarrow a x \in \delta R$. Also $\text{bd}(aR) \geqslant B^k$, so $\text{bd}(aR)$ and $\text{bd}(cR)$ have the same minimal primes. Now, as R-modules :

$$\frac{aR + \delta R}{aR} \simeq \frac{a^2R + a\delta R}{a^2R} \quad \text{and} \quad \frac{aR}{a^2R} \simeq \frac{a^2R + \delta R}{a^2R + a\delta R}.$$

The module $\frac{aR + \delta R}{a^2 R}$, considered as over the ring R/B^{2k}, has a reduced rank ρ and the above shows that $\rho\left(\frac{aR + \delta R}{a^2R + \delta R}\right) = 0$. Hence this factor module is torsion over the ring $R/\sqrt{B}$, where $\sqrt{B}$ = nilpotent radical of B.

Now $\mathscr{C}(\sqrt{B}) = \mathscr{C}(P_1) \cap \ldots \cap \mathscr{C}(P_k)$, where $P_1, \ldots, P_k$ are the minimal primes over B and $P = P_1$, say.

Let $ae = a^2y + \delta z$, where $e \in \mathscr{C}(\sqrt{B})$. Now $a \in \mathscr{C}(Q)$, since $c \in \mathscr{C}(Q)$ as $cR > B \not< Q$. Thus $e - ay \in Q$, hence $ay \in \mathscr{C}(P)$ and $a \in \mathscr{C}(P)$. This part is proved.

Next suppose that $c \in \mathscr{C}(P_i)$ for $i = 1, \ldots, k$; then $c \in \mathscr{C}(\sqrt{B})$.

However $\left| R/cR \right| = \left| R/B \right|$ by Cauchon's theorem above,

$$\left| R/cR \right| = \left| R/\sqrt{B} \right| > \left| R/(cR + \sqrt{B}) \right|,$$ since $c \in \mathscr{C}(\sqrt{B})$. Now R is known to be ideal invariant, being fully bounded nœtherian, and hence

$$\left| R/(cR + \sqrt{B}) \right| = \left| R/(cR + (\sqrt{B})^k) \right| \quad \text{for } k = 1, \ldots \quad \left| R/(cR + (\sqrt{B})^k) \right| = \left| R/cR \right|$$

since $(\sqrt{B})^k < cR$ for large k. The argument uses the notation $|M|$ for the Krull dimension of a right R-module and results from Krause-Lenagan-Stafford [6] .

<u>Theorem 2.2</u> - <u>Let</u> R <u>be a right noetherian ring with an invertible ideal</u> $X \neq R$. <u>If</u> P <u>is a prime ideal minimal over</u> X <u>then height</u> $P \leq 1$.

This theorem is from Chatters - Goldie - Hajarnavis - Lenagan [3] .

<u>Proof</u> - Let T be an overring of R ; it is convenient to set $T = \bigcup_{n=1}^{\infty} X^{-n}$. Taking P to be a prime ideal of R such that $X \nleq P$ then $\frac{X + P}{P}$ is an invertible ideal in $R_{/P}$ with inverse $\frac{X^{-1} + PT}{PT}$ in the ring $\frac{T}{PT}$. Accordingly we can reduce the problem to the case where R is a prime ring .

So let height $P \geq 2$ in the prime ring R and let $P \underset{\neq}{>} Q \underset{\neq}{>} 0$, where Q is a prime ideal. Let c be a regular element of R with $c \in Q$. For large n , $cR \cap X^{2n} \leq cX^n$, because the ascending chain in R :

$$cR \leq (cR \cap X)X^{-1} \leq (cR \cap X^2)X^{-2} \leq \ldots\ldots \leq R$$

must become stationary. Now P is minimal over X^n , so replace X by X^n and have $cR \cap X^2 \leq cX$; equivalently , $cR \cap (X^2 + cX) = cX$.

Let $\rho(M)$ be the reduced rank of a f.g. module M over the ring $R_{/X}$ and let $N = \sqrt{X}$ the nilpotent radical of X .

$$\rho\left(\frac{X^2 + cR}{X^2 + cX}\right) = \rho\left(\frac{X^2 + cX + cR}{X^2 + cX}\right) = \rho\left(\frac{cR}{cR \cap (X^2 + cX)}\right)$$

$$= \rho\left(\frac{cR}{cX}\right) = \rho\left(R_{/X}\right) ,$$

and

$$\rho\left(\frac{X + cR}{X^2 + cX}\right) = \rho\left(\frac{X + cR}{X}\right) + \rho\left(\frac{X}{X^2 + cX}\right)$$

$$= \rho\left(\frac{X + cR}{X}\right) + \rho\left(R_{/X + (cR)}\right) = \rho\left(R_{/X}\right) .$$

Thus $\rho\left(\frac{X + cR}{X^2 + cR}\right) = 0$. Let $x \in X$ and $d \in \mathcal{C}(N)$ with $xd \in X^2 + cR$, then

$x d \in X^2 + (Q \cap X) = X^2 + XQ$. Thus $X^{-1} x d \leqslant X + Q \leqslant P$. Now $X^{-1} x \leqslant R$ and $d \in \mathscr{C}(N) \leqslant \mathscr{C}(P)$, hence $X^{-1} x \leqslant P$. As x is any element of R, this gives $R = X^{-1} X \leqslant P$; a contradiction.

In the argument we used $\rho\left(\frac{R}{X + cR}\right) = \rho\left(\frac{X}{X^2 + cX}\right)$, which requires proof.

<u>Lemma 2.3 - Let</u> $B \leqslant A$ <u>be f.g. right</u> R - <u>sub-modules of</u> $T = \bigcup_1^\infty X^{-n}$ <u>such that</u> $AX \leqslant B$ <u>, then</u>

$$\rho_{R/X}(A/B) = \rho_{R/X}(AX/BX)$$

<u>Proof</u> - Let $N = \sqrt{X}$ as before. Evidently $XN = NX$. Suppose that $AN \leqslant B$ hence $AXN \leqslant BX$. Then is a natural lattice isomorphism :

$$C/B \longleftrightarrow CX/BX$$

between submodules of A/B and AX/BX and it is enough to show that torsion modules correspond. Let A/B be a torsion R/N - module, then so is AY/BY, where Y is either X or X^{-1}. Note $YN = NY$ and for $t \in T$, $tY \leqslant N \Leftrightarrow Yt \leqslant N$.

Let $a \in AY$, set $K = \{r \in R ; ar \in BY\} \geqslant N$. We need to prove that $K \cap \mathscr{C}(N)$ is not empty ; it is done here by proving that K/N is an essential right ideal of R/N. Let $I \triangleleft_r R$, $I > N$ and $I \cap K = N$. Let $s \in IY^{-1}$, $s \in Y^{-1}$. Now as $\in AYY^{-1} = A$ so there exists $c \in \mathscr{C}(N)$ with $asc \in B$ and $ascY \leqslant BY$. Now $scY \leqslant IY^{-1}RY \leqslant I \leqslant R$, hence $scY \leqslant I \cap K = N$, hence $Ysc \leqslant N$. Now $Ys \leqslant YIY^{-1} \leqslant R$, so as $c \in \mathscr{C}(N)$, then $Ys \leqslant N$ and $sY \leqslant N$ However s is any element of IY^{-1}, hence $I = IY^{-1}Y \leqslant N$ and it is done.

<u>Corollary - Let</u> R <u>be a right noetherian ring with a normal non-unit</u> u <u>and</u> P <u>be a prime minimal over</u> $Ru = uR$ <u>then height</u> P <u>is</u> $\leqslant 1$.

This is the original theorem of Jategaonkar [5].

3. Rings of finite right global dimension

The following theorem is due do Brown - Hajarnavis - Mac-Eachan [1].

Theorem 3.1 - *Let R be a right noetherian local ring of finite global dimension on the right and set N to be its nilpotent radical. Then $R/_N$ is a full matrix ring over a local domain and for each $n \in \mathbb{N}$ there exists $c \in \mathcal{C}(N)$ with $nc = 0$.*

The theorem is a partial generalization of the classical result that a regular local ring is an integral domain. Here one is able to prove that $R/_N$ is a prime ring and the analogy would be complete if it could be shown that $N = 0$. When R is left noetherian the result proves that $Nc = 0$ for some $c \in \mathcal{C}(N)$. Then, of course, in some special cases the theorem can be completed. For example, when R has an artinian quotient ring then c is regular and $N = 0$. Again when R has zero singular ideal then R is a prime ring. For if $N^k = 0$, $N^{k-1} \neq 0$, then $N^{k-1}(N + cR) = 0$ and $N + cR$ is right essential, so N^{k-1} is the singular ideal. But both of these cases are really special pleading, for example, in the commutative case the singular ideal is N itself. Probably the must natural such assumption is Ass R_R consists of minimal primes ; this is weaker than that of having a right artinian quotient ring.

Proof - R is known to have a unique minimal projective module P, $P = eR$, with $e^2 = e$, M be a f.g. R-module, then we obtain $\rho(M) = k\,\rho(P)$ for some $k \in \mathbf{Z}^+$, where ρ is the reduced rank. When M is projective it is a direct sum of copies of P and it is clear. So we shall prove the result by induction on the projective dimension of M. Suppose that $0 \to K \to F \to M \to 0$ is exact with F free and proj dim K < proj dim M. Then :

$$\rho(K) = k_1\ \rho(P) \qquad \text{(induction)}$$

$$\rho(F) = k_2\ \rho(P) \qquad k_1, k_2 \in \mathbf{Z}^+ .$$

Then $\rho(M) = (k_2 - k_1)\ \rho(P)$, so it is done.

Now decompose R as $R = e_1 R \oplus \dots \oplus e_n R$ with $e_i R \simeq P$, the e_i being primitive idempotents ; we prove that $e_i R/_{e_i N}$ is a uniform module. Dropping the subscript, choose $x \in eR$ with $\frac{xR + eN}{eN}$ uniform, then :

$$\rho({}^{eR}/_{xR}) + \rho(xR) = \rho(eR)$$

so either $\rho(xR) = 0$ or $\rho({}^{eR}/_{xR}) = 0$. If $\rho(xR) = 0$ then $xc = 0$ for some $c \in \mathscr{C}(N)$, hence $x \in N$ and $x \in eN$; a contradiction. Thus $\rho({}^{eR}/_{xR}) = 0$ so $ed \in xR$ for some $d \in \mathscr{C}(N)$. This means that $(xR + eN)/_{eN}$ is both uniform and essential in $eR/_{eN}$. We conclude that each $e_i R/_{e_i N}$ is a uniform module, as stated.

Let $D \in \mathrm{End}_{R/_N}(e_1 R/_{e_1 N})$; it is known that D is an integral domain. As $e_1 R \simeq e_i R$ $(1 \leqslant i \leqslant n)$ then $e_1 R/_{e_1 N} \simeq e_i R/_{e_i N}$ and :

$$R/_N \simeq e_1 R/_{e_1 N} \oplus \ldots\ldots \oplus e_n R/_{e_n N}$$

from which it follows that $R/_N \simeq \mathbb{M}_n(D)$ and D is a local integral domain, hence N is a prime ideal of R.

Finally $0 \longrightarrow N \longrightarrow R \longrightarrow R/_N \longrightarrow 0$ is exact and $\rho(R) = n\,\rho(P)$ and $\rho({}^{R}/_{N}) = n\,\rho({}^{eR}/_{eN})$ and $\rho(R) = \rho(N) + \rho({}^{R}/_{N})$, so that $\rho(N) = 0$. Then for each $n \in N$, there exists $c \in \mathscr{C}(N)$ with $nc = 0$.

If R is left noetherian, then $N^i/_{N^{i+1}}$ has a finite set of left generators : $\bar{x}_1, \ldots, \bar{x}_m$ with $\bar{x}_j(d_i R + N) = 0$ for some $d_i \in \mathscr{C}(N)$. Thus $N^i d_i < N^{i+1}$ and $N d_1 d_2 \ldots d_k = 0$; where $d_1 d_2 \ldots d_k \in \mathscr{C}(N)$ also.

REFERENCES

[1] K. A. BROWN , C.R. HAJARNAVIS, A.B. MACEACHAN
Noetherian rings of finite global dimension. Warwick Math Institute, 1978.

[2] G. CAUCHON Les T-anneaux, la condition (H) de Gabriel et ses conséquences. Comm. Algebra 4 (1976), 11-50.

[3] A.W. CHATTERS, A.W. GOLDIE, C.R. HAJARNAVIS, T.H. LENAGAN
Reduced rank in nœtherian rings. J. Algebra (to appear in 1980)

[4] A.W. GOLDIE Torsion free modules and rings. J. Algebra 1 (1964), 268-287.

[5] A.V. JATEGAONKAR Relative Krull dimension and prime ideals in right nœtherian rings. Comm. Algebra 4 (1974), 429 - 468.

[6] G. KRAUSE , T.H. LENAGAN, J.T. STAFFORD
Ideal invariance and artinian quotient rings. J. Algebra 55 (1978) 145 - 154.

[7] B. STENSTRÖM Rings of quotients. Springer-Verlag 1975.

Prime Ideals in Group Algebras of Polycyclic-by-Finite Groups : Vertices and Sources

by

MARTIN LORENZ

These notes represent a somewhat expanded version of a talk that I gave in this seminar in November 79. The results presented here are joint work with D.S. Passman.

In Section 1 we describe the machinery that has been developed to study prime ideals in $K[G]$ whith G polycyclic-by-finite. We briefly discuss Roseblade's fundamental work on group algebras of orbitally sound groups and its extension to general polycyclic-by-finite groups by Passman and the author. Although crossed products have played an important role here and some results do in fact hold in this more general setting, we will concentrate on group algebras here. Sections 2 and 3 contain previously unpublished material. The main purpose of these sections is to illustrate the notions of vertex and source for prime ideals in $K[G]$ that were introduced in [4] . Our general point of view in Section 2 is to consider the vertex of a prime P as being given and derive information about P'. In particular, we will describe the set $\mathrm{Spec}_H(K[G])$ of all prime ideals in $K[G]$ having a fixed vertex H . Section 3 is devoted to the catenarity problem.

Throughout these notes, G will always be a polycyclic-by-finite group and K will be a commutative field.

§ 1 - Preliminary results

1. A - Induced Ideals ([6], [4]). Let H be a subgroup of G and let L be an ideal of K[H]. Then we let L^G denote the unique largest ideal of K[G] contained in L K[G], that is :

$$L^G = \mathrm{ann}_{K[G]}\ (K[G] / L\, K[G]) = \cap_{g\in G}\ L^g\, K[G].$$

Any ideal of K[G] of the form $I = L^G$ will be called an induced ideal or, more precisely, induced from H. Let $\pi_H : K[G] \to K[H]$ be the projection map sending $\Sigma_{g\in G}\, k_g\, g$ to $\Sigma_{g\in H}\, k_g\, g$. Then L^G can also be characterized as the unique largest ideal of K[G] satisfying $\pi_H(I) \subseteq L$. In particular, the above definition of L^G is left-right symmetric. If H is normal in G, then the above expres- expression for L^G becomes :

(1.1) $$L^G = (\cap_{g\in G}\ L^g)\ K[G] = (L^G \cap K[H])\ K[G].$$

Thus, for an ideal of K[G], being induced from a normal subgroup H of G is the same as being controlled by H, in the usual sense. The basic result that we will need is a follows :

(1.2) Theorem ([4, Theorem 1.7]). Let N be a normal subgroup of G of finite index, let Q be a prime ideal of K[N] and let A be any subgroup of G containing the stabilizer of Q in G, that is $A \supseteq \{g\in G \mid Q^g = g^{-1}\, Q\, g = Q\}$. Then the induction map $(.)^G$ yields a 1-1 correspondence between the prime ideals T of K[A] with $T \cap K[N] = \cap_{a\in A}\ Q^a$ and the primes P of K[G] with $P \cap K[N] = \cap_{g\in G}\ Q^g$. Moreover, if $P = T^G$ as above, then T is the unique minimal covering prime of $P \cap K[A]$ with $\cap_{g\notin A}\ Q^g \not\subseteq T \cap K[N]$.

We remark that if N is normal in G of finite index and P is a prime ideal of K[G], then $P \cap K[N]$ always has the form $P \cap K[N] = \cap_{g\in G}\ Q^g$ for some prime ideal Q of K[N] which is unique up to G-conjugacy.

(1.3) Corollary Let N be a normal subgroup of G of finite index and let P be a prime ideal of $K[G]$. Write $P \cap K[N] = \cap_{g \in G} Q^g$ for some prime Q of $K[N]$. If A is a normal subgroup of G containing $\mathrm{Stab}_G(Q)$, then $P = (P \cap K[A])K[G]$.

Proof By Theorem 1.2, $P = T^G$ for some prime T of $K[A]$ and, by (1.1), $T^G = (T^G \cap K[A])K[G]$, since A is normal.

1. B - Orbitally Sound Groups (Roseblade [8]). A subgroup H of G is called orbital if $[G : \mathbb{N}_G(H)] < \infty$. An orbital subgroup H is said to be isolated orbital if and only if H is the only orbital subgroup M with $M \supseteq H$ and $[M : H] < \infty$. In general one defines, for H orbital in G,

$$i_G(H) = \langle M \mid H \subseteq M \subseteq G, \ M \text{ orbital}, \ [M : H] < \infty \rangle .$$

One can show (see [8, p. 400/401]) that $[i_G(H) : H] < \infty$ and that $i_G(H)$ is isolated orbital in G. Therefore, $i_G(H)$ is called the isolator of H in G. The definition of $i_G(H)$ makes it clear that we have :

$$(1.4) \qquad \mathbb{N}_G(H) \subseteq \mathbb{N}_G(i_G(H)) .$$

The group G is said to be orbitally sound if and only if all isolated orbital subgroups of G are normal. The following important result is due to Roseblade.

(1.5) Theorem ([8, Theorem C2]). Set $\mathrm{nio}(G) = \cap_H \mathbb{N}_G(H)$, where the intersection runs over all isolated orbital subgroups of G. Then $\mathrm{nio}(G)$ is an orbitally sound characteristic subgroup of G of finite index. Moreover, $\mathrm{nio}(G)$ contains every orbitally sound normal subgroup of G of finite index and every finite-by-nilpotent normal subgroup of G.

Using the linearity of polycyclic-by-finite groups, Wehrfritz has given an alternate proof for the existence of an orbitally sound normal subgroup of G of finite index. (See [2, § 2]) .

1. C Standard Prime Ideals ([3]). We let $\Delta = \Delta(G)$ denote the f.c.center of G, that is :

$$\Delta = \{ g \in G \mid [G : \mathbb{C}_G(g)] < \infty \} .$$

For any ideal I of $K[G]$ we set $I^\dagger = \{ g \in G \mid 1 - g \in I \}$. Thus $I^\dagger$ is the kernel of the natural map $G \to K[G]/I$ and hence is normal in G . Following Roseblade [8] , I will be called faithful if $I^\dagger = \langle 1 \rangle$ and almost faithful if $I^\dagger$ is finite. Note that any ideal I of $K[G]$ is the complete inverse image of a faithful ideal in $K[G/I^\dagger]$.

(1.6) Definition ([3]) . A prime ideal P of $K[G]$ is said to be standard if and only if $P = L^G$ for some almost faithful prime ideal L of $K[\Delta]$.

In [3, Proposition 1.4] it is shown that for any almost faithful prime L of $K[\Delta]$ the induced ideal L^G is always prime in $K[G]$. Any standard prime is in particular almost faithful. We call P virtually standard if the image of P in $K[G/P^\dagger]$ is standard. Although the defining conditions seem to be very restrictive, virtually standard primes do in fact occur quite often. Indeed, Roseblade's theorem [8, Theorem C 1] can be stated as follows :

If G is orbitally sound then all primes of $K[G]$ are virtually standard.

A converse to this will be proved in Section 2 .

1.D Vertices and Sources of Prime Ideals ([4]) For any subgroup H of G we let :

$$\nabla_G(H)$$

denote the complete inverse image in $\mathbb{N}_G(H)$ of $\Delta(\mathbb{N}_G(H)/H)$. Thus, clearly, $\nabla_G(H) \lhd \mathbb{N}_G(H)$ and $\nabla_G(H)/H$ is finite-by- abelian. Furthermore, if H is orbital then so is $\nabla_G(H)$ and if H is isolated orbital then $\nabla_G(H)/H \cong \mathbf{Z}^n$ for some n. Finally, it is not hard to see that for isolated orbital subgroups H_1 and H_2 we have :

$$(1.7) \qquad H_1 \subseteq H_2 \quad \text{implies} \quad \nabla_G(H_1) \subseteq \nabla_G(H_2)$$

([4, Lemma 3.1]). Let I be an ideal of $K[G]$ and let $N \trianglelefteq G$. Then we say that I is <u>almost faithful sub</u> N if and only if $I^\dagger \subseteq N$ and $[N : I^\dagger] < \infty$.

(1.8) <u>Theorem</u> ([4, Theorem I, II, III]). <u>Let</u> K <u>be a field and</u> let G <u>be a polycyclic-by-finite group</u>.

(i) (Existence) <u>If</u> P <u>is a prime ideal of</u> $K[G]$, <u>then there exists an isolated orbital subgroup</u> H <u>of</u> G <u>and an almost faithful sub</u> H <u>prime ideal</u> L <u>of</u> $K[\nabla_G(H)]$ <u>with</u> $P = L^G$.

(ii) (Uniqueness) <u>In the situation of part</u> (i), H <u>is unique up to conjugation in</u> G <u>and, for a given</u> H, L <u>is unique up to conjugation by</u> $N_G(H)$.

(iii) (Converse) <u>If</u> H <u>is an isolated orbital subgroup of</u> G <u>and</u> L <u>is an almost faithful sub</u> H <u>prime ideal of</u> $K[\nabla_G(H)]$, <u>then</u> L^G <u>is a prime ideal of</u> $K[G]$.

(1.9) <u>Definition</u> ([4]). Let P be a prime ideal of $K[G]$ and let H and L be as in Theorem 1.8(i),(ii). Then we call H a <u>vertex</u> of P and write :

$$H =_G \mathrm{vx}(P).$$

Furthermore, we call L a <u>source</u> of P (corresponding to the vertex H).

If P is a given prime of $K[G]$, then $\mathrm{vx}(P)$ can be obtained as follows. Write $P \cap K[\mathrm{nio}(G)] = \bigcap_{g \in G} Q^g$ for some prime ideal Q of $K[\mathrm{nio}(G)]$. (Q is unique up to G-conjugacy). Then we have :

$$(1.10) \qquad \mathrm{vx}(P) =_G i_G(Q^\dagger)$$

([4, Theorem 2.4(i)]). If $H = \mathrm{vx}(P)$ is given, then the possible sources of P, for this H , are obtained as follows. Set $A = \mathbb{N}_G(H)$. Then, by *(1.4)* and *(1.10)* $A \supseteq \mathbb{N}_G(Q^\dagger) \supseteq \mathrm{Stab}_G(Q)$ and so Theorem 1.2 implies that there exists a unique prime ideal T of $K[A]$ with $T \cap K[\mathrm{nio}(G)] = \cap_{a\in A} Q^a$ and $T^G = P$. Then we have :

(1.11) <u>The sources of</u> P <u>(for the given</u> H) <u>are precisely the minimal covering primes of</u> $T \cap K[\nabla_G(H)]$.

(See [4 , Theorem 2.4 (ii), proof] .) The relations between P and its vertex H and source L are of course quite interesting. For example, if Q(.) denotes the classical ring of quotients, then the centers $\mathfrak{Z}(Q(K[G]/P))$ and $\mathfrak{Z}(Q(K[\nabla_G(H)]/L))$ have the same transcendence degree over K , in short

(1.12) $$\mathrm{c.r.}(P) = \mathrm{c.r.}(L)$$

(c.r. = central rank ; see [3].). It follows that, for K nonabsolute, P is primitive if and only if L has finite codimension in $K[\nabla_G(H)]$. Finally, one can give an expression for the height ht(P) of P involving a certain group theoretic invariant, depending upon $H = \mathrm{vx}(P)$, and the central rank of L . For details we refer to [4].

§ 2. - <u>Vertices of Primes in</u> K[G]

The vertex vx(P) of any prime P in K[G] is, by definition, an isolated orbital subgroup of G . The following lemma shows that all isolated orbital subgroup of G do in fact occur this way.

(2.1) <u>Lemma</u> <u>Let</u> $H \leqslant G$ <u>be an isolated orbital subgroup of</u> G . <u>Then there exists a prime ideal</u> P <u>in</u> K[G] <u>with</u> $\mathrm{vx}(P) =_G H$.

<u>Proof</u> If $H \leqslant G$ is isolated orbital, then $\nabla_G(H)/H \cong \mathbf{Z}^n$ for some n and so the augmentation ideal $L = (\omega H)\, K[\nabla_G(H)]$ satisfies $K[\nabla_G(H)]/L \cong K[X_1^{\pm 1}, X_2^{\pm 1}, \ldots, X_n^{\pm 1}]$. Thus L is prime and is clearly almost faithful sub H . By Theorem 1.8(iii) we

conclude that $P = L^G$ is a prime ideal of $K[G]$ with $vx(P) =_G H$.

By definition of $nio(G)$, we have for any prime ideal P of $K[G]$:

$$\mathbb{N}_G(vx(P)) \supseteq nio(G) . \tag{2.2}$$

The extreme case of a normal vertex is certainly of interest.

(2.3) Lemma Let P be a prime ideal of $K[G]$. Then $vx(P)$ is normal in G if and only if P is virtually standard.

Proof First assume that $H = vx(P)$ is normal in G and set $\nabla = \nabla_G(H)$. Then $\nabla \trianglelefteq G$ and, by Theorem 1.8, $P = L^G$ for some prime ideal L of $K[\nabla]$ with $[H : L^\dagger] < \infty$. Since $\nabla \trianglelefteq G$, we have $P = (P \cap K[\nabla]) K[G]$ and $P \cap K[\nabla] = \cap_{g\in G} L^g$, by (1.1). In particular, it follows that $P^\dagger = P^\dagger \cap \nabla = \cap_{g\in G} (L^\dagger)^g \subseteq H$. Note that $[H : \cap_{g\in G}(L^\dagger)^g] < \infty$, since each $(L^\dagger)^g$ is a subgroup of H of index $[H : L^\dagger]$ and there are only finitely many such subgroups. Thus $[H : P^\dagger] < \infty$ and, in particular, $[L^\dagger : P^\dagger] < \infty$. Moreover, the definition of ∇ easily implies that $\Delta(G/P^\dagger) = \nabla/P^\dagger$. Thus, if $\bar{\ }: K[G] \to K[G/P^\dagger]$ denotes the natural map then $\bar{\nabla} = \Delta(\bar{G})$, $\bar{L}$ is almost faithful in $K[\bar{\nabla}]$ and $\bar{P} = \bar{L}^G$. This proves that P is virtually standard.

Conversely, assume that P is virtually standard and let D denote the complete inverse image of $\Delta(G/P^\dagger)$ in G . Then $P = L^G$ for some prime L of $K[D]$ with $[L^\dagger : P^\dagger] < \infty$. Let $H/P^\dagger$ be the torsion subgroup of $D/P^\dagger$. Then $[H : P^\dagger] < \infty$, and H is easily seen to be isolated orbital and normal in G . In particular, L is almost faithful sub H , and since $P = L^G$ and $D = \nabla_G(H)$ we deduce from the Uniqueness Theorem (Theorem 1.8(ii)) that $H = vx(P)$. Thus $vx(P)$ is normal in G , and the lemma is proved.

Recall that, by definition, G is orbitally sound if and only if all isolated orbital subgroups of G are normal. Thus Lemmas 2.1 and 2.3 immediately give the following result :

(2.4) Corollary G is orbitally sound if and only if all primes in K[G] are virtually standard.

Note that this contains Roseblade's Theorem C1 in [8] as the "only if" -direction.

In the other extrema case, namely $\mathbb{N}_G(vx(P)) = nio(G)$, we have $P = (P \cap K[nio(G)])K[G]$. This follows from the following slightly more general observation, together with (1.1).

(2.5) Lemma Let P be a prime ideal of K[G] and let A be a subgroup of G with $A \supseteq \nabla_G(vx(P))$. Then $P = I^G$ for some prime ideal I of K[A].

Proof By Theroem 1.8(i), P is induced from $\nabla_G(vx(P))$ and, since induction is transitive, P is also induced from A. Thus there exists an ideal I of K[A] with $I^G = P$. Choosing I maximal with this property we can get I to be prime. Indeed, if J_1 and J_2 are ideals of K[A] containing I such that $J_1.J_2 \subseteq I$, then $J_1^G . J_2^G \subseteq (J_1.J_2)^G \subseteq I^G = P$, and hence $J_1^G \subseteq P$ or $J_2^G \subseteq P$. The maximality of I now yields $J_1 = I$ or $J_2 = I$, as required.

Now consider prime ideals P_1, and P_2, of K[G] with $P_1 \subseteq P_2$. Then it follows from (1.10) that we have :

$$vx(P_1) \subseteq vx(P_2) \quad (2.6)$$

(up to G-conjugation, of course). For, if we write $P_i \cap K[nio(G)] = \cap_{g\in G} Q_i^g$ for suitable primes Q_i of K[nio(G)], we see that $Q_2 \supseteq \cap_{g\in G} Q_1^g$ and hence $Q_2 \supseteq Q_1^g$ for some $g \in G$. Replacing Q_1 by a G-conjugate if necessary, we may assume that $Q_2 \supseteq Q_1$. Thus $Q_2^\dagger \supseteq Q_1^\dagger$ and so $i_G(Q_2^\dagger) \supseteq i_G(Q_1^\dagger)$, since $i_G(.)$ is monotonic ([8, § 3.1]). (2.6) now follows from (1.10). Note that (2.6) and (1.7) imply that :

$$\nabla_G(vx(P_1)) \subseteq \nabla_G(vx(P_2)). \quad (2.7)$$

We now consider the case $vx(P_1) = vx(P_2)$. For any isolated orbital subgroup H of G we set :

$$\mathrm{Spec}_H (K[G]) = \{ P \in \mathrm{Spec}(K[G]) \mid vx(P) =_G H \},$$

a nonempty subset of $\mathrm{Spec}(K[G])$, by Lemma 2.1. We have

$$\mathrm{Spec}(K[G]) = \dot{\bigcup}_H \mathrm{Spec}_H(K[G]),$$

a disjoint union with H ranging over a complete set of non-conjugate isolated orbital subgroups of G . Our goal is to describe $\mathrm{Spec}_H(K[G])$ for H a fixed isolated orbital subgroup of G . Set $A = \mathbb{N}_G(H)$ and $\nabla = \nabla_G(H)$ so that $H \trianglelefteq \nabla \trianglelefteq A$ and $\nabla/H = \Delta(A/H)$. Note also that H is normal and isolated orbital in ∇ , the latter since ∇/H is torsion-free abelian. Thus $\mathrm{Spec}_H(K[\nabla])$ is defined and is in fact easily seen to be identical with the set of all primes of $K[\nabla]$ which are almost faithful sub H . Now A acts on $\mathrm{Spec}_H(K[\nabla])$ by conjugation, and we let

$$\mathscr{S}_H = \mathrm{Spec}_H(K[\nabla]) / A$$

denote the set orbits under this action. The A-orbit of $L \in \mathrm{Spec}_H(K[\nabla])$ will be written as $[L]$. We remark that <u>each such orbit is finite</u>. To see this, choose a normal subgroup N of A with $N \subseteq L^{\dagger}$ and $[H : N] < \infty$. Then we have $\Delta(A/N) = \nabla/N$ and thus there exists a subgroup X of A of finite index which centralizes ∇/N . Clearly, $X \subseteq \mathrm{Stab}_A(L)$ and so the latter has finite index in A. For $L_1, L_2 \in \mathrm{Spec}_H(K[\nabla])$ define :

$$[L_1] \leqslant [L_2] \quad \text{if and only if} \quad L_1 \subseteq L_2{}^a \quad \text{for some } a \in A .$$

If $[L_1] \leqslant [L_2] \leqslant [L_1]$, then $L_1 \subseteq L_2{}^a \subseteq L_1{}^b$ for suitable $a,b \in A$, and the fact that $[L_1]$ is finite implies that we have equality throughout so that $[L_1] = [L_2]$. Thus $\leqslant$ defines a partial order on $\mathscr{S}_H$. Note that, surely, $L^G = (L^a)^G$ for any $L \in \mathrm{Spec}_H(K[\nabla])$ and $a \in A$. Hence the induction map $(.)^G$ can be defined on $\mathscr{S}_H$, and Theorem 1.8 says that $(.)^G$ is a one-to-one map of $\mathcal{S}_H$ onto $\mathrm{Spec}_H(K[G])$. The preimage of $P \in \mathrm{Spec}_H(K[G])$ is the set of all sources of P corresponding to H.

(2.8) Proposition. The map :

$$(.)^G : \mathcal{S}_H \longrightarrow \mathrm{Spec}_H([G])$$

induces a 1-1 correspondence between these two sets such that for any $[L_1], [L_2] \in \mathcal{S}_H$ we have :

$$[L_1]^G \subseteq [L_2]^G \text{ if and only if } [L_1] \leqslant [L_2] .$$

Moreover, in this case :

$$\mathrm{ht}\,([L_2]^G / [L_1]^G) = \mathrm{c.r.}\,(L_1) - \mathrm{c.r.}\,(L_2) .$$

Here, of course, $\mathrm{ht}([L_2]^G / [L_1]^G)$ denotes the height of the prime ideal $[L_2]^G / [L_1]^G$ of $K[G]/[L_1]^G$. Note also that $\mathrm{c.r.}(L_i)$ is surely on invariant of $[L_i]$, since the factor rings corresponding to the elements of $[L_i]$ are pairwise isomorphic. In addition, we know by (1.12) that :

$$\mathrm{c.r.}\,(L_i) = \mathrm{c.r.}\,([L_i]^G) .$$

Proof of (2.8) The fact that $(.)^G$ is one-to-one and onto has been noted above, and $(.)^G$ is clearly order preserving, i.e. $[L_1] \leqslant [L_2]$ implies $[L_1]^G \subseteq [L_2]^G$. It remains to show that, conversely, $[L_1]^G \subseteq [L_2]^G$ implies $[L_1] \leqslant [L_2]$ and to verify the height formula.

Write $P_i = [L_i]^G$ $(i = 1,2)$ and assume that $P_1 \subseteq P_2$. We reconstruct $[L_1]$ and $[L_2]$ by using (1.10), (1.11). Thus write $P_i \cap K[\mathrm{nio}(G)] = \cap_{g \in G} Q_i^{\,g}$ for suitables primes Q_i of $K[\mathrm{nio}(G)]$ and, as we have remarked earlier, we may assume that $Q_1 \subseteq Q_2$. By (1.10), we have $H = i_G(Q_1^{\dagger}) = i_G(Q_2^{\dagger})$. As above, let $A = \mathbb{N}_G(H)$. Then (1.4) yields $A \supseteq \mathbb{N}_G(Q_i^{\dagger}) \supseteq \mathrm{Stab}_G(Q_i) \supseteq \mathrm{nio}(G)$ for $i = 1,2$, and hence it follows from Theorem 1.2 that $P_i = T_i^{\,G}$ for certain uniquely determined prime ideals T_i of $K[A]$ with $T_i \cap K[\mathrm{nio}(G)] = \cap_{a \in A} Q_i^{\,a} (i = 1,2)$. In fact, we know that T_i is the unique minimal covering prime of $P_i \cap K[A]$ which does not contain $\cap_{g \in G \setminus A} Q_i^{\,G}$. We claim that $T_2 \supseteq T_1$. If not, then T_2 contains

contains $\cap_{g\in G\setminus A} Q_1^{\,g}$ and we obtain that $Q_2 \supseteq \cap_{a\in A} Q_2^{\,a} = T_2 \cap K[\mathrm{nio}(G)] \supseteq \cap_{g\in G\setminus A} Q_1^{\,g}$. Therefore, $Q_2 \supseteq Q_1^{\,g}$ for some $g \in G\setminus A$ so that $H = i_G(Q_2\dagger) \supseteq i_G((Q_1^{\,g})\dagger) = (i_G(Q_1\dagger))^g = H^g$. Since H is orbital, it follows that $H = H^g$, contradicting the fact that $g \notin A = \mathbb{N}_G(H)$. Thus we must have $T_2 \supseteq T_1$, and hence $T_2 \cap K[\nabla] \supseteq T_1 \cap K[\nabla]$. (Here, $\nabla = \nabla_G(H)$, as above). By (1.11), the sources of P_i , for the given H , are precisely the minimal corering primes of $T_i \cap K[\nabla]$ $(i = 1,2)$. Since any minimal covering prime of $T_2 \cap K[\nabla]$ contains a minimal covering prime of $T_1 \cap K[\nabla]$, we conclude that $[L_2] \geqslant [L_1]$. This proves the first assertion.

As to the height formula, first note that the members of any chain of primes in $K[G]$ leading from $[L_1]^G$ to $[L_2]^G$ belong to $\mathrm{Spec}_H(K[G])$, by (2.6) . Thus, by the foregoing, we conclude that :

$$\mathrm{ht}\ ([L_2]^G / [L_1]^G) = \mathrm{ht}\ ([L_2] / [L_1]) ,$$

where the latter of course denotes the maximal length, n , of a saturated chain $[L_1] = S_0 \lneqq S_1 \lneqq \dots \lneqq S_n = [L_2]$ with $S_i \in \mathcal{S}_H$. Now each such chain yields a chain $L_1 = I_0 \subsetneqq I_1 \subsetneqq \dots \subsetneqq I_n$ with $I_j \in \mathrm{Spec}_H(K[\nabla])$ and $I_n \in [L_2]$, and conversely. Thus in order to complete the proof of the proposition, it suffices to establish the following sublemma :

<u>Sublemma</u> <u>Let</u> H <u>and</u> ∇ <u>be as above and let</u> $I \subseteq J$ <u>be prime ideals of</u> $K[\nabla]$ <u>which are almost faithful</u> sub H . <u>Then</u> $\mathrm{ht}(J/I) = \mathrm{c.r.}(I) - \mathrm{c.r.}(J)$.

<u>Proof</u> We have $I\dagger \lhd \nabla$ and $\nabla/I^\dagger$ is finite-by-abelian, since $[H : I\dagger] < \infty$ and ∇/H is abelian. Let $^-: K[\nabla] \to K[\nabla/I^\dagger]$ denote the natural map. Then $\bar{I}$ and $\bar{J}$ are primes of $K[\nabla]$ with $\bar{I} \subseteq \bar{J}$ and $\mathrm{c.r.}(\bar{I}) = \mathrm{c.r.}(I)$, $\mathrm{c.r.}(\bar{J}) = \mathrm{c.r.}(J)$. Moreover, $\mathrm{ht}(\bar{J}/\bar{I}) = \mathrm{ht}(J/I)$, and so we may assume that ∇ is finite-by-abelian. In particular, ∇ contains a torsion-free central subgroup Z of finite index. Now $X = I \cap K[Z] \subseteq Y = J \cap K[Z]$ are prime in $K[Z]$, and $\mathrm{c.r.}(X) = \mathrm{c.r.}(I)$, $\mathrm{c.r.}(Y) = \mathrm{c.r.}(J)$ (see [3, Lemma 4.3]). It is not hard to show that $\mathrm{ht}(J/I) = \mathrm{ht}(Y/X)$. Indeed, $\leqslant$ follows from Incomparability ([4, Lemma 1.3(ii)] , for example), and $\geqslant$ is a consequence of the more general Proposition 3.3 . Thus we

may assume that $\nabla = Z$, and hence $K[\nabla] \cong K[X_1^{\pm 1}, X_2^{\pm 1}, \ldots, X_m^{\pm 1}]$ for some m. Since the assertion is classical in this case (cf. [7 , p. 84/85]), the sublemma, and hence the proposition, are proved.

The above proposition shows that $\mathrm{Spec}_H(K[G])$ and $\mathscr{S}_H$ may be identified for our purposes. Moreover, as we have seen in the above sublemma, the situation in $\mathscr{S}_H$ is almost classical. So, for example, if n denotes the rank of the free abelian group ∇/H , then we know that any prime ideal L of $K[\nabla]$ which is almost faithful sub H has central rank at most n . Thus, by (1.12), we have for any $P \in \mathrm{Spec}_H(K[G])$.

$$\text{(2.9)} \qquad \text{c.r.}(P) \leq \mathrm{rank}(\nabla/H) = n .$$

Proposition 2.8 further implies that any chain of primes in $\mathrm{Spec}_H(K[G])$ has length at most n . On the other hand, if K is nonabsolute then there always exists a chain $L_0 \subsetneqq L_1 \subsetneqq \ldots \subsetneqq L_n$ of primes L_i in $K[\nabla]$ with $L_i^{\dagger} = H$ for all i . Indeed, if we let $\bar{\ }: K[\nabla] \twoheadrightarrow K[\nabla/H]$ denote the canonical map, then $\bar{\nabla} = \prod_{i=i}^{n} \langle x_i \rangle \cong \mathbb{Z}^n$ can be embedded in K° . For $i = 1,2,\ldots,n$ let $\zeta_i \in K^{\circ}$ denote the image of x_i under this embedding, and for each $\ell = 0,1,\ldots,n$ let $\varphi_\ell\ K[\bar{\nabla}] \longrightarrow K[\bar{\nabla}]$ be the K-algebra map given by $\varphi_\ell(x_i) = \zeta_i = \zeta$ for $i \leq \ell$ and $\varphi_\ell(x_i) = x_i$ for $i > \ell$. Then each $\bar{L}_\ell = \mathrm{Ker}\,\varphi_\ell$ is a faithful prime in $K[\bar{\nabla}]$, and $\bar{L}_0 \subsetneqq \bar{L}_1 \subsetneqq \ldots \subsetneqq \bar{L}_n$. Thus for nonabsolute K we have :

$$\text{(2.10)} \qquad \dim \mathrm{Spec}_H(K[G]) = \mathrm{rank}(\nabla/H) .$$

This is however no longer true if K is absolute. For, in this case the image of ∇ in every simple homomorphic image of $K[\nabla]$ is finite. Another consequence of Proposition (2.8) is that for any two given primes $P_1 \subseteq P_2$ in $\mathrm{Spec}_H(K[G])$ all saturated chains $P_1 = Q_0 \subsetneqq Q_1 \subsetneqq \ldots \subsetneqq Q_r = P_2$ of primes Q_i in $K[G]$ have length $r = \text{c.r.}(P_1) - \text{c.r.}(P_2)$.

§ 3 - Catenarity

It is an interesting question, whether or not the fact described in the last paragraph of Section 2 holds quite generally in $\mathrm{Spec}(K[G])$. That is, given any two primes $P_1 \subseteq P_2$ in $K[G]$, do all saturated chains of primes

$P_1 = Q_o \subsetneq Q_1 \subsetneq \dots \subsetneq Q_r = P_2$ have the same length $r = ht(P_2) - ht(P_1)$? In short, are group algebras of polycyclic-by-finite groups G catenary ? Roseblade has proved a positive answer to this for G orbitally sound or, slightly more generally, for G a $\mathfrak{Z}$-group (see [8], [9]). In this section we will show that our methods do at least quite easily yield an extension of Roseblade's result from orbitally sound to orbitally sound-by-finite nilpotent groups, that is polycyclic-by-finite groups G with a normal subgroup N such that N is orbitally sound and G/N is finite nilpotent.

Recall that two primes $P_1 \subseteq P_2$ of K[G] are called <u>adjacent</u> (or <u>neighbors</u>) if there exists no prime in K[G] lying strictly between P_1 and P_2. K[G] catenary if and only if for any two adjacent primes $P_1 \subseteq P_2$ of K[G] we have $ht(P_2) = ht(P_1) + 1$.

(3.1) <u>Lemma</u> <u>Let</u> H <u>be a normal subgroup of</u> G <u>of finite index and assume that</u> K[H] <u>is catenary</u>. <u>Let</u> $P_1 \subseteq P_2$ <u>be adjacent primes of</u> K[G] <u>such that</u> P_1 <u>is induced from</u> H (i.e. $P_1 = (P_1 \cap K[H]).K[G]$). <u>Then</u> $ht(P_2) = ht(P_1) + 1$.

<u>Proof</u> For $i = 1,2$ write $P_i \cap K[H] = \cap_{g\in G} Q_i^g$ for suitable primes Q_i of K[H] such that $Q_1 \subseteq Q_2$. It is well-known (see for example [8, § 8.1]) that $ht(Q_i) = ht(P_i)$. Thus it suffices to show that $ht(Q_2) = ht(Q_1) + 1$ or, since K[H] is catenary, that Q_1 and Q_2 are adjacent. Assume otherwise so that $Q_1 \subsetneq Q \subsetneq Q_2$ for some prime Q of K[H]. Then we have

$$P_1 \cap K[H] = \cap_{g\in G} Q_1^g \subsetneq I = \cap_{g\in G} Q^g \subsetneq P_2 \cap K[H] = \cap_{g\in G} Q_2^g ,$$

where the inclusions are strict since all occuring intersections are finite. It follows that $P_1 = (P_1 \cap K[H])K[G] \subsetneq IK[G] \subsetneq (P_2 \cap K[H])K[G] \subsetneq P_2$. Note that IK[G] is an ideal of K[G] and, moreover, every minimal covering prime of IK[G] intersects K[H] in I (see [5, Lemma 4.1] or [8,Lemma 8]). Since P_2 contains such a minimal covering prime, P, we obtain that $P_1 \subsetneq P \subsetneq P_2$, contradicting the fact that P_1 and P_2 are adjacent. Thus Q_1 and Q_2 have to be adjacent, and the lemma is proved.

(3.2) <u>Proposition</u> <u>Assume</u> G <u>is orbitally sound-by-finite nilpotent and let</u> K <u>be any field</u>. <u>Then</u> K[G] <u>is catenary</u>.

<u>Proof</u> Let N be an orbitally sound normal subgroup of G such that G/N is finite nilpotent. We argue by induction on |G/N| . The case G = N is due to Roseblade so we assume that $G \supsetneq N$. Let $P_1 \subseteq P_2$ be adjacent primes in K[G]. We have to

show that $ht(P_2) = ht(P_1) + 1$. For this, we may assume P_1 to be faithful. Indeed, writing $P_i \cap K[N] = \cap_{g\in G} Q_i^g$ for suitable primes Q_i of $K[N]$ with $Q_1 \subseteq Q_2$, we have $ht(Q_i) = ht(P_i)$, and $ht(Q_2) = ht(Q_1) + 1$ holds if and only if Q_1 and Q_2 are adjacent, by Roseblade's result. Thus $ht(P_2) = ht(P_1) + 1$ if and only if Q_1 and Q_2 are adjacent, and the latter surely holds if and only if the images of Q_1 and Q_2 under $K[G] \twoheadrightarrow K[G/P_1^{\dagger}]$ are adjacent. Thus P_1 will be faithful in the following.

First assume that the vertex of P_1, $vx(P_1)$, is normal in G. Then, by Lemma 2.3, P_1 is virtually standard. In particular, since P_1 is faithful, we have $P_1 = (P_1 \cap K[\Delta])K[G]$, where $\Delta = \Delta(G)$ is contained in nio(G), by (1.5). Thus P_1 is induced from nio(G), and since $K[nio(G)]$ is catenary, by Roseblade's result, we may apply Lemma 3.1 to conclude that $ht(P_2) = ht(P_1) + 1$.

Now assume that $\mathbb{N}_G(vx(P_1))$ is a proper subgroup of G. By (2.2), $\mathbb{N}_G(vx(P_1))$ contains nio(G) and hence N. Since G/N is nilpotent, there exists a proper normal subgroup H of G with $H \supseteq \mathbb{N}_G(vx(P_1))$. By Lemma 2.5, P_1 is induced from H and, by induction, $K[H]$ is catenary. Thus Lemma 3.1 again yields $ht(P_2) = ht(P_1) + 1$, and we are done.

We remark that group algebras of finitely generated abelian-by-finite groups are catenary. This follows from work of Schelter on affine PI-rings ([10]). We close with a related result on certain crossed products. For the definition and basic facts concerning crossed products we refer to [5]. Here we just note that if $S = R \star G$ is a crossed product of the finite group G over the commutative ring R, then G acts on R, and if P is a prime ideal of S, then $P \cap R = \cap_{g\in G} Q^g$ for some prime ideal Q_i of R which is unique up to G-conjugacy and satisfies $ht(Q) = ht(P)$ (See [5, § 4]).

(3.3) Proposition Let R be a finitely generated commutative K-algebra and and let $S = R \star G$ be a crossed product with G a finite group. Let $P_1 \subseteq P_2$ be primes in S and write $P_i \cap R = \cap_{g\in G} Q_i^g$ for suitable primes Q_i of R with $Q_1 \subseteq Q_2$. If P_1 and P_2 are adjacent then so are Q_1 and Q_2.

Proof Upon dividing out by $(P_1 \cap R)S$ we may assume that $P_1 \cap R = 0$. Set $T = S/P_1$ and let P denote the image of P_2 in T. Then $R \subseteq T$ and P has height 1 in T. Since G is finite, the fixed subring $R^G = \{r \in R \mid r^g = r$ for all $g \in G\}$ is a finitely generated K-algebra, as R is, and R is a finitely generated module over R^G. Moreover, R^G is central in T. By the Noether normalization theorem ([7, p.91]), R^G contains a subring V such that

$V \cong K[X_1,X_2,\ldots,X_\ell]$ for some ℓ and R^G is a finitely generated module over V. Therefore, T is a finitely generated module over the central subring $V \cong K[X_1,X_2,\ldots,X_\ell]$.

Now assume, by way of contradiction, that $Q_1 \subsetneq Q \subsetneq Q_2$ for some prime Q of R. Then the Incomparability theorem ([1, p.61]) implies that $0 \neq X = Q \cap V \subsetneq X_2 = Q_2 \cap V$. Note that $X_2 = (\bigcap_{g \in G} Q_2^g) \cap V = P \cap V$, since $V \subset R^G$. Note further that V is integrally closed and T is a prime PI-algebra, being a finitely generated module over a commutative K-algebra. Hence, by Schelter [10, Theorem 3], the Going Down theorem holds for the extension $V \subseteq T$. Thus there exists a prime ideal P' in T with $P' \subseteq P$ and $P' \cap V = X$. In particular, $0 \neq P' \cap V \subsetneq P \cap V$ and so $0 \neq P' \subsetneq P$, contradicting the fact that P has height 1. Therefore, we conclude that Q_1 and Q_2 are adjacent, and the proposition is proved.

REFERENCES

[1] M.F. ATIYAH and I.G. MACDONALD - Introduction to Commutative Algebra, Addison-Wesley, Reading, Mass. (1969).

[2] D. FARKAS and R. SNIDER - Induced representations of polycyclic groups, Proc. LMS (3) 39 (1979) 193-207.

[3] M. LORENZ and D.S. PASSMAN - Centers and prime ideals in group algebras of polycyclic-by-finite groups, J. Algebra 57 (1979) 355-386.

[4] M. LORENZ and D.S. PASSMAN - Prime ideals in group algebras of polycyclic-by-finite groups, Proc. LMS (to appear).

[5] M. LORENZ and D.S. PASSMAN - Prime ideals in crossed products of finite groups, Israel J. Math. 33 (1979) 89-132.

[6] M. LORENZ and D.S. PASSMAN - Addendum-Prime ideals in crossed products of finite groups, Israel J. Math. (to appear).

[7] H. MATSUMURA - Commutative Algebra, Benjamin, New York (1970).

[8] J.E. ROSEBLADE - Prime ideals in group rings of polycyclic groups, Proc. LMS (3) 36 (1978) 385-447.

[9] J.E. ROSEBLADE - Corrigendum-Prime ideals in group rings of polycyclic groups, Proc LMS (3) 38 (1979) 216-218.

[10] W. SCHELTER - Non-commutative affine PI-rings are catenary, J. Algebra 51(1978) 12-18.

Martin LORENZ
Fachbereich Mathematik
Universität Essen
4300 ESSEN 1

STRUCTURE OF INTEGRAL GROUP RINGS

K.W.Roggenkamp (Stuttgart)

In his important paper on the structure of blocks of defect one R.Brauer in 1941 investigated as a main tool integral group rings in order to pass from ordinary character theory to modular representation theory - this is also nowadays one of the main applications of integral representation theory.

To state Brauer's result we need the following

Notation: Let G be a finite group, p a rational prime divisor of $|G|$ with $|G|=p^a n$, $p \nmid n$ and T an irreducible complex representation of degree $m \equiv 0(p^{a-1})$. A field K is said to be a normal minimally ramified splitting field for T, if

(i) K is a normal extension of $\mathbb{Q}$ such that the ramification index of p in K is the same as the ramification index of p in $\mathbb{Q}(\chi_T(g))$ - adjoining the character values of T to $\mathbb{Q}$.

In the same paper Brauer proved the existence of such a K.

(ii) Let now $\mathfrak{p}$ lie above p in K and let $S_\mathfrak{p}$ be the localisation of the ring of integers S of K at $\mathfrak{p}$. Moreover, it is no restriction if we assume that T is realized in $S_\mathfrak{p}$.

(iii) Let t denote the number of non-isomorphic modular constituents of T i.e. different composition factors of T/pT.

Theorem O (Brauer 1941, Theorem 11): The $S_\mathfrak{p}G$-submodules of L, the $S_\mathfrak{p}G$-lattice on which T acts, are linearly ordered

$$L = L_o \supset L_1 \supset \ldots \supset L_{t-1} \supset L_t = \mathfrak{p}L \supset \mathfrak{p}L_1 \supset \ldots$$

and the (absolutely) irreducible $S_\mathfrak{p}/\mathfrak{p}G$-modules L_{i-1}/L_i $(1 \leq i \leq t)$ are pairwise non-isomorphic.

Supported by DFG grant.

where $\tilde{B}$ is a separable order (i.e. $p\tilde{B} = \mathrm{rad}\,\tilde{B}$) and $\mathrm{rad}\,B = p\uparrow^G \eta + pB$. (This uses heavily ($\alpha$)).

(iv) Let ε be a primitive idempotent in KP, then

(i) $(\mathrm{rad}(RP\varepsilon))\uparrow^G \eta = [\mathrm{rad}(RG\varepsilon)]\eta$,

(ii) $[\mathrm{rad}(RG\varepsilon)]\eta \simeq (RG\varepsilon)\eta$.

Proof: If S is the simple module, we have the exact sequence

$$0 \to \mathrm{rad}(RP\varepsilon) \to RP\varepsilon \to S \to 0 .$$

Since B is a block, it is a flat RP-module, and so we get

$$0 \to B \otimes_{RP} \mathrm{rad}\,RP\varepsilon \to B\varepsilon \to B \otimes_{RP} S \to 0 ,$$

and $B \otimes_{RP} S$ is a $\tilde{B}$-module (cf. (**)), and so it is semi-simple; i.e. $B \otimes_{RP} \mathrm{rad}\,RP\varepsilon \supset \mathrm{rad}\,B\varepsilon$.
On the other hand, P is normal and ε is central, thus $B \otimes_{RP} \mathrm{rad}\,RP\varepsilon$ is nilpotent modulo pRG; i.e.

$B \otimes_{RP} \mathrm{rad}\,RP\varepsilon = \mathrm{rad}\,B\varepsilon$, this proves (i).

But because of (ii) $\mathrm{rad}\,RP\varepsilon \simeq RP\varepsilon$ and so

$B\varepsilon \simeq \mathrm{rad}\,B\varepsilon$ proving (ii).

(v) If now e is a central primitive idempotent in KB, then either $\tilde{B}e \neq 0$, in which case Be is even separable, or $\tilde{B}e = 0$, in which case there exists a primitive idempotent ε of KP with $\varepsilon e = e$, and the statement follows from (iii).#

Remark: In order to describe the group ring RG one has to know how the various projections RGe are linked together. For this we make the following

Definition: Let Λ be an R-order in the semi-simple K-algebra A. Λ is said to be a Bäckström-order if there exists a hereditary R-order Γ in A with $\mathrm{rad}\,\Lambda = \mathrm{rad}\,\Gamma$.

Remarks: (i) The good thing about Bäckström-orders is that one can write them down explicitely, if one knows Γ and the embedding ring of integers over R in D .

An R-order Γ in A is said to be hereditary if every submodule of a projective Γ-lattice is itself projective.

Theorem 1 (Auslander-Goldman 61, Harada 63, Brumer 63, Jacobinski 71): For an R-order Γ in A the following are equivalent:

(i) Γ is hereditary

(ii) $\mathrm{rad}\,\Gamma$ is projective

(iii) $\{x \in A:\ x\,\mathrm{rad}\,\Gamma \subset \mathrm{rad}\,\Gamma\} = \Gamma$

(iv)

$$\Gamma \ \widetilde{\text{Morita}} \ \begin{pmatrix} \Omega & \Pi\Omega & \cdots & \Pi\Omega \\ \vdots & \ddots & \ddots & \vdots \\ \vdots & & \ddots & \Pi\Omega \\ \Omega & \cdots & \cdots & \Omega \end{pmatrix}_m$$

Proposition 1: Let B be a block of RG with defect group P and block idempotent η such that

α.) $P \lhd G$

β.) P is abelian

γ.) every subgroup of P is G-invariant.

(i) If e is a central primitive idempotent in KB, then Be is a hereditary order.

(ii) If $B/\mathfrak{p}B$ contains an absolutely simple module, then the Schur-indices of all simple modules in KB are one.

Sketch of the proof:

(i) The hypotheses on P imply that every primitive idempotent ε in KP is central (β) and G-invariant (γ).

(ii) Because of (β), $RP\varepsilon$ is a maximal order - in particular hereditary - observe that P operates on $RP\varepsilon$ via a primitive p^s-th root of unity.

(iii) (Michler 1975) If $\mathfrak{p}$ is the augmentation ideal of P ; i.e. $\mathfrak{p}$ is the kernel of the homomorphism $RP \to R$, then we have an exact sequence of twosided B-modules

(*) $$0 \to \mathfrak{p}{\uparrow}^G\eta \to B \to \widetilde{B} \to 0 \quad , \quad \mathfrak{p}{\uparrow}^G = \mathrm{Ker}(RG \to RG/P)$$

$\Lambda/\mathrm{rad}\,\Lambda \to \Gamma/\mathrm{rad}\,\Lambda$ i.e. an embedding of one semi-simple $\mathfrak{k}$-algebra into another.

(ii) With the embedding $\Lambda/\mathrm{rad}\,\Lambda \to \Gamma/\mathrm{rad}\,\Lambda$ one can associate a valued graph γ, in the sense of Dlab/Ringel and the representation theory of Λ is the same as that of the non-simple γ-modules (Ringel/Roggenkamp 1979).

<u>Corollary:</u> Let e be a primitive central idempotent, in $K\mathfrak{p}\uparrow^G\eta$ and write $\eta = \eta_1 + \eta_2$ according to the exact sequence (*) with $\eta_2\widetilde{B} = \widetilde{B}$. Then $RG(e + \eta_2)$ is a Bäckström-order.

<u>Proof:</u> One uses (iv) and the sequence (*) to show

$$\mathrm{rad}\,RG(e+\eta_2) = \mathfrak{p}\uparrow^G e \oplus \mathfrak{p}\widetilde{B} \quad . \qquad \#$$

<u>Examples:</u> 1.) Let $G = C_p \rtimes C_q$ be the Frobenius-group of order pq, p a rational prime and q a divisor of $p-1$. One should imagine G as the semi-direct product of the group of primitive p-th roots of unity with the subgroup of order q of the Galois-group of $K(\zeta)$ over K, where ζ is a primitive p-th root of unity, acting. If we put $S = \mathrm{Fix}_{C_q}(R(\zeta))$, then RG has the following structure with $\mathfrak{p} = \mathrm{rad}\,S$:

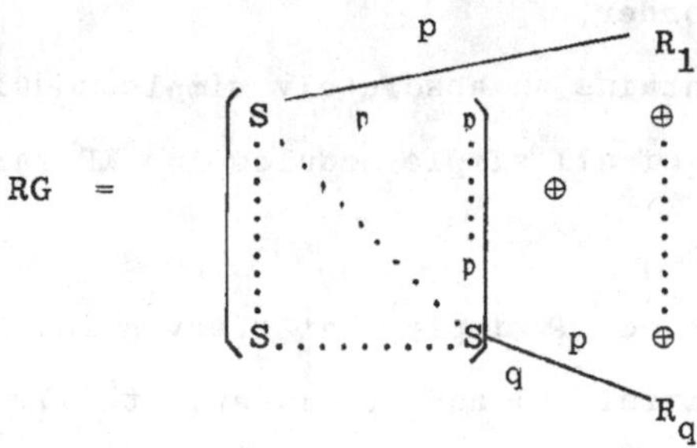

where $R_i = R$ for $1 \leq i \leq q$ and $S \underset{p}{-} R$ is the pullback of

$$\begin{array}{ccc} S \underset{\mathfrak{p}}{-} R & \to & R \\ \downarrow & & \downarrow \\ S & \to & S/\mathfrak{p}S \simeq R/pR \end{array} \quad .$$

One can use - in case q is also a prime - the local information to write down the group ring $\mathbb{Z}G$ explicitely. Let ξ be a primitive

<u>This result can be phrased as follows</u>: Let e_T be the central idempotent of KG corresponding to T , then

$$(**) \qquad S_p G e_T \quad \widetilde{\text{Morita}} \quad \begin{pmatrix} S_p & p & \cdots & p \\ \vdots & \ddots & & \vdots \\ \vdots & & \ddots & p \\ S_p & \cdots & \cdots & S_p \end{pmatrix}_t .$$

<u>Remarks:</u> (i) The condition $\deg(T) \equiv 0(p^{a-1})$ implies that T belongs to a block of defect one.

(ii) If one works with an arbitrary splitting field $L \supset K$ then p in (**) must be replaced by p^s where s is the ramification index.

(iii) E.C.Dade in 1966 has generalized Brauers results for blocks of defect one to blocks with cyclic defect group, and he remarks that he has not found an analogue to Brauer's Theorem (11). In his Habilitationsschrift W.Plesken (1980) has - using the description of Peacock (1977) for blocks with cyclic defect group - found the structure of $S_p G e$ for a central idempotent e in KG , in case KG is split .

(iv) In applications - crystallographic groups, descriptions of units in $\mathbb{Z}G$, automorphisms of $\mathbb{Z}G$ - it is often necessary to find $\mathbb{Z}Ge$ resp. $\hat{\mathbb{Z}}_p Ge$; and the difficulty is that there can occur skewfields.

So the aim of this talk is to give a description of blocks of $\hat{\mathbb{Z}}_p G$ of defect one, $\hat{\mathbb{Z}}_p$ being the p-adic complete integers.

We first fix the notation for the sequel and recall some results about hereditary orders.

<u>Notation:</u> p is a rational prime and $\hat{\mathbb{Z}}_p$ with field of quotients $\hat{\mathbb{Q}}_p$ denotes the p-adic completion of $\mathbb{Z}$, and R with fractions K is a finite unramified extension of $\hat{\mathbb{Z}}_p$. We put $\ell = R/pR$. Let $A = (D)_n$ be a simple K-algebra, D a skewfield and Ω with $\Pi\Omega = \text{rad}\,\Omega$ the

q-th root of unity, then

$$\mathbb{Z}[\xi]/p\mathbb{Z}[\xi] \simeq \mathbb{Z}/p\mathbb{Z}^{(q-1)} \quad ,$$

and if $S = \mathrm{Fix}_{C_q}\ \mathbb{Z}[\zeta]$, we can form the pullback

$$\begin{array}{ccc} \mathbb{Z}[\xi] \underset{p}{-} (\underbrace{S \oplus \ldots \oplus S}_{q-1}) & \rightarrow & S^{(q-1)} \\ \downarrow & & \downarrow \\ \mathbb{Z}[\xi] & \rightarrow & \mathbb{Z}/p\mathbb{Z}^{(q-1)} \end{array} \quad ,$$

and also the pullback

$$\begin{array}{ccc} \mathbb{Z}[\xi] \underset{q}{-} \mathbb{Z} & \rightarrow & \mathbb{Z} \\ \downarrow & & \downarrow \\ \mathbb{Z}[\xi] & \rightarrow & \mathbb{Z}/q\mathbb{Z} \end{array} \quad .$$

The group ring then has the following form

$\mathbb{Z}G =$

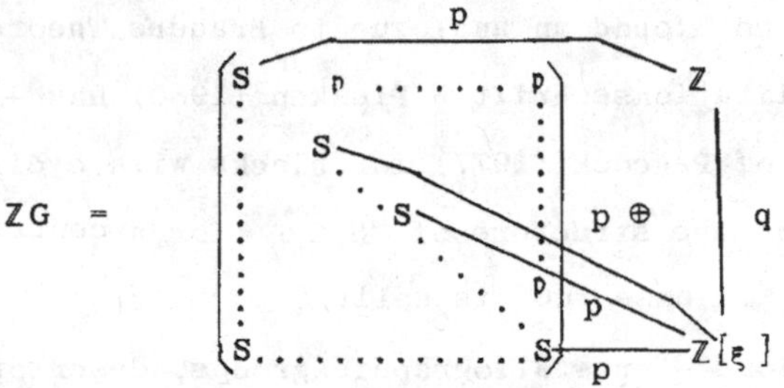

2.) Let G_1 be the Frobenius-group of order 21, and form the pullback

$$\begin{array}{ccccccccc} 1 & \rightarrow & C_7 & \rightarrow & G_1 & \rightarrow & C_3 & \rightarrow & 1 \\ & & \| & & \uparrow & & \uparrow & & \\ 1 & \rightarrow & C_7 & \rightarrow & G & \rightarrow & C_9 & \rightarrow & 1 \end{array} \quad ,$$

C_i the cyclic group of order i .

Let $R = \hat{\mathbb{Z}}_7$, then the principal block of RG is isomorphic to RG_1 and is described as above. There are two more algebraically cojugate blocks B_1 and B_2 with

$$KB_1 = L \oplus D \quad ,$$

where $L = K(\zeta)$, ζ a primitive 9th root of unity and D the unique skewfield over $F = \mathrm{Fix}_{C_3}(K(\sqrt[7]{1}))$ of index 3 .

If S is the ring of integers in L and Ω the maximal order in D, then $S/\mathrm{rad}\,S \simeq \Omega/\mathrm{rad}\,\Omega = \mathbb{F}_{7^3}$ and B_1 is the pullback

$$\begin{array}{ccc} S - \Omega & \to & S \\ \downarrow & & \downarrow \\ \Omega & \to & \mathbb{F}_{7^3} \end{array} ,$$

where $\mathbb{F}_{7^3}$ is the field with 7^3 elements.

Remark: If it is possible to write down in such an explicit way the integral group rings, one ought to be able, to describe the group of units in $\mathbb{Z}G$.

So let us detour for a moment for some remarks on units of integral group rings.

We denote by $U(\mathbb{Z}G)$ the units in $\mathbb{Z}G$ and by $V(\mathbb{Z}G)$ the group of normalized units; i.e. the units which have augmentation one. Since G is a subgroup of $V(\mathbb{Z}G)$ there are some questions connected with this embedding (cf. Dennis (1976)):

(1) When does there exist an exact sequence

$$1 \to V_o \to V(\mathbb{Z}G) \to G \to 1 ,$$

which is split by the embedding $G \to V(\mathbb{Z}G)$.

(2) When is V_o torsion free. In all the examples which are known, (1) and (2) have positive answers. (The symmetric group on three letters - Hughes-Pearson 1972, the dihedral groups of odd order - Miyata 1979, the alternating group on 4 letters - Allan and Hobby 1979)

Zassenhaus showed in 1975 that the order of every unit of finite order in $V(\mathbb{Z}G)$ divides the exponent of G. This-together with the known examples - lead Zassenhaus to conjecture.

(3a) If U is a finite subgroup of $V(\mathbb{Z}G)$, then U is isomorphic to a subgroup of G.

(3b) Let u be a unit of finite order in $V(\mathbb{Z}G)$ then there exist a unit $x \in \mathbb{Q}G$ such that $xux^{-1} \in G$.

It should be noted that if G has the property that V_o is torsion free and $\mathbb{Z}H \simeq \mathbb{Z}G$, then $H \simeq G$; i.e. (2) is stronger than the isomorphism problem.

In my opinion, the isomorphism problem has a general positive answer only for metabelian groups (Whitcomb 1968), and I think that (3a) and (3b) are true for metabelian groups, but not in general.

In connection with (1) and (2) we have the following result.

<u>Theorem 2:</u> Let $G = A \rtimes \overline{G}$ be a metabelian Frobeniusgroup, $\overline{G}$ abelian of exponent 2,3,4 or 6, then V_o is torsionfree.

In order to sketch the proof we first show

<u>Lemma 1:</u> Let $G = A \rtimes \overline{G}$ be a semi-direct product with abelian kernel A, and assume there exists an isomorphism

$$\bar{\varphi} : V(\mathbb{Z}\overline{G}) \to \overline{G} \quad , \quad \text{e.g. } \overline{G} \text{ of exponent } 2,3,4,6 ,$$

then $V(\mathbb{Z}G)$ is described by the split exact sequence

$$1 \to V_o \to V(\mathbb{Z}G) \to G \to 1 \quad ,$$

where V_o is given by the split exact sequence

$$1 \to U_o \to V_o \to V_1 \to 1$$

with $V_1 \simeq \operatorname{Ker}\bar{\varphi}$ and

$$U_o = \{1 + x : x \in \mathfrak{a}\mathfrak{g}\} \cap V(\mathbb{Z}G) \quad ,$$

where $\mathfrak{a}$ and $\mathfrak{g}$ are the respective augmentation ideals of A and G.

<u>Proof:</u> We put

$$U_1 = \{1 + x : x \in \mathfrak{a}\uparrow^G\} \cap V(\mathbb{Z}G) \quad .$$

The pullback $\mathcal{E}_1$ of the extension

$$\begin{array}{lccccccccc} & 1 & \to & A & \to & G & \to & \overline{G} & \to & 1 \\ & & & \| & & \uparrow & & \uparrow \overline{\varphi} & & \\ \mathfrak{E}_1 : & 1 & \to & A & \to & Y & \to & V(\mathbb{Z}\overline{G}) & \to & 1 \end{array}$$

is split.

On the other hand we have the split exact sequence

$$1 \to U_1 \to V(\mathbb{Z}G) \to V(\mathbb{Z}\overline{G}) \to 1 \quad ;$$

note that the right hand map is surjective, since $G \to \overline{G}$ is split. Moreover, we have an epimorphism

$$\psi : U_1 \to A \ ,$$

defined as $(1 + x)\psi = x\psi_o$, where

$$\psi_o : \mathfrak{a}\uparrow^G \to A$$

is induced by $(a - 1)\psi_o = a$. Since $\operatorname{Ker}\psi_o = \mathfrak{a}\,\mathfrak{g}$, ψ is a group homomorphism, and we can form the pushout via ψ :

$$\begin{array}{lccccccccc} \mathfrak{E}_2 : & 1 & \to & A & \to & Y' & \to & V(\mathbb{Z}\overline{G}) & \to & 1 \\ & & & \psi \uparrow & & \uparrow & & \| & & \\ & 1 & \to & U_1 & \to & V(\mathbb{Z}G) & \to & V(\mathbb{Z}\overline{G}) & \to & 1 \ . \end{array}$$

Then $\mathfrak{E}_2$ is a split exact sequence and $V(\mathbb{Z}\overline{G})$ acts on A as $\overline{G}$. Thus $\mathfrak{E}_1$ and $\mathfrak{E}_2$ are equivalent, and we obtain a commutative diagram with exact rows and columns

$$\begin{array}{lccccccccc} & & & 1 & & 1 & & 1 & & \\ & & & \uparrow & & \uparrow & & \uparrow & & \\ & 1 & \to & A & \to & G & \to & \overline{G} & \to & 1 \\ & & & \uparrow \psi & & \uparrow \varphi & & \uparrow \overline{\varphi} & & \\ \mathbb{D} & 1 & \to & U_1 & \to & V(\mathbb{Z}G) & \to & V(\mathbb{Z}\overline{G}) & \to & 1 \\ & & & \uparrow & & \uparrow & & \uparrow & & \\ & 1 & \to & U_o & \to & \operatorname{Ker}\varphi & \to & V_1 & \to & 1 \\ & & & \uparrow & & \uparrow & & \uparrow & & \\ & & & 1 & & 1 & & 1 & & . \end{array}$$

This proves the lemma.

<u>Remark:</u> It should be noted that for abelian $\overline{G}$ by Higman's theorem (1940)

$$V(\mathbb{Z}\overline{G}) \simeq \overline{G} \oplus V_1$$

with V_1 free abelian.

The theorem is now proved by calculating U_o and showing that it is torsion free. Since that is rather technical, we shall here only prove the simplest case: The following argument is due to A. Wiedemann.

Let $G = C_p \rtimes C_q$ be the Frobenius-group from Example 1 and $R = \hat{\mathbb{Z}}_p$. Then it is easily seen that with $A = C_p$

$$\mathfrak{a}\mathfrak{g} = \begin{pmatrix} p & R & \cdots & R \\ \vdots & \ddots & & \vdots \\ p & & \ddots & R \\ p^2 & p & \cdots & p \end{pmatrix}_q$$

and

$$\mathfrak{a}\uparrow^G = \begin{pmatrix} p & R & \cdots & R \\ \vdots & \ddots & p \ \ddots & \vdots \\ \vdots & & \ddots & R \\ p & \cdots & \cdots & p \end{pmatrix}_q .$$

By the usual localizing argument one finds that

$$\{1 + x : x \in \mathfrak{a}\uparrow^G\}$$

has only units of p'-order. To show that U_o has no p-torsion, let ζ be a primitive p-th root of unity, and let $\mu(X)$ be the minimum polynomial of $\theta - 1$, then $\mu(X)$ has degree q and

$$\mu(X) \equiv X^q \bmod(p) ,$$

and the coefficient of X^o lies in $p \setminus p^2$.

In particular if A is the compagnion matrix of $\mu(X)$ in $(S)_q$, then $(1 + A)^p = 1$ and $\det A \in p \setminus p^2$. If now $1 + B$ is a unit of order p with $B \in \mathfrak{a}\mathfrak{g}$, $B \neq 0$. Then the minimum polynomial and the characteristic polynomial of B coincide with $\mu(X)$; in particular

A and B are conjugate and so $\det B \in \mathfrak{p}\backslash\mathfrak{p}^2$. On the other hand every element in $\mathfrak{a}\mathfrak{g}$ has determinant in $\mathfrak{p}^2$, a contradiction. #

E.C.Dade in 1971 showed that there are two non-isomorphic metabelian groups G,H such that $KG \simeq KH$ for all fields K. Analyzing the proof, one sees that $\hat{\mathbb{Z}}_pG \simeq \hat{\mathbb{Z}}_pH$ for all primes $p \in \max(\mathbb{Z})$. We shall now exhibit a large class of metabelian groups G for which (3a) and (3b) are false for $V(\mathbb{Z}_pG)$ for every prime p, where $\mathbb{Z}_p$ is the localization of $\mathbb{Z}$ at p, but (1) and (2) are valid.

We shall now consider the following groups:

Let G_1 be the Frobeniusgroup of order pq, where p and q are rational primes and $q \mid (p-1)$. We then form the pullpback

$$\begin{array}{ccccccccc} 1 & \to & C_p & \to & G_1 & \to & C_q & \to & 1 \\ & & \| & & \uparrow & & \uparrow & & \\ 1 & \to & C_p & \to & G & \to & C_{q^2} & \to & 1 \end{array}.$$

We shall first describe the groupring $\hat{\mathbb{Z}}_pG$. For this we put $\hat{R} = \mathrm{Fix}_{C_q}(\hat{\mathbb{Z}}_p\sqrt[p]{T})$, and denote by $\hat{\mathfrak{p}}$ the maximal ideal of $\hat{R}$.

<u>Case 1:</u> $q^2 \nmid (p-1)$.

Let $\hat{S} = \hat{\mathbb{Z}}_p[\sqrt[q^2]{T}]$, then $|\hat{S} : \hat{\mathbb{Z}}_p| = q$ and $\hat{S}$ is unramified; i.e. $p\hat{S} = \mathrm{rad}\,\hat{S}$. Let now $\hat{D}_i$ with ring of integers $\hat{\Omega}_i$ and $\hat{\mathfrak{P}}_i = \mathrm{rad}\,\hat{\Omega}_i$ be the unique skewfield with centre $\mathbb{Q}_p \otimes_{\hat{\mathbb{Z}}_p} \hat{R} = \hat{K}$ of index q and with invariant i, $0 < i < q$ (cf. Hasse 1931). Then $\hat{\Omega}_i/\hat{\mathfrak{P}}_i \simeq F_{p^q} \simeq \hat{S}/p\hat{S}$. We form the pullback

$$\begin{array}{ccc} \hat{\Omega}_i \underset{p}{-} \hat{S} & \to & \hat{S} \\ \downarrow & & \downarrow \\ \hat{\Omega}_i & \to & F_{p^q} \end{array}.$$

<u>Claim 1:</u> $\hat{\mathbb{Z}}_p G$ has q blocks of defect one; $\hat{B}_o \simeq \hat{\mathbb{Z}}_p G_1$ and $q-1$ algebraically conjugate blocks $\hat{B}_i = (\hat{\Omega}_i \; \overline{p} \; \hat{S})$, $1 \le i \le q-1$.

<u>Proof:</u> Let ϵ_i, $0 \le i \le q-1$, be the central primitive idempotents in $\hat{\mathbb{Z}}_p C_q$; since C_q is the centre of G, the ϵ_i are also orthogonal central idempotents in $\hat{\mathbb{Z}}_p G$ and so $\hat{\mathbb{Z}}_p G = \hat{B}_o \amalg (\prod_{i=1}^{q-1} \hat{B}_i)$ with $\hat{B}_i = \hat{\mathbb{Z}}_p G \epsilon_i$ and $\hat{B}_1, \dots, \hat{B}_{q-1}$ are algebraically conjugate. We have to show that the $\hat{B}_i$ are blocks. Now, $\mathrm{rad}\, \hat{\mathbb{Z}}_p G = p \hat{\mathbb{Z}}_p G + \hat{\mathbb{Z}}_p \otimes_{\mathbb{Z}} c_p{\uparrow}^G$ and $\hat{\mathbb{Z}}_p G / \mathrm{rad}\, \hat{\mathbb{Z}}_p G - \mathbb{F}_p C_{q^2} \simeq \mathbb{F}_p^q \oplus \mathbb{F}_{p^q}{}^{(q-1)}$. Since $\hat{B}_o / \mathrm{rad}\, \hat{B}_o \simeq \hat{\mathbb{Z}}_p G_1 / \mathrm{rad} \hat{\mathbb{Z}}_p G_1 \simeq \mathbb{F}_p^q$ (cf. above), there are at most q blocks. Hence except $\hat{B}_o$ there are $q-1$ blocks which have each one isomorphism type of an indecomposable projective module. Moreover, $\hat{B}_i$ is a Bäckström-order with graph •——• $1 \le i \le q-1$. Since $\hat{\mathbb{Q}}_p \hat{B}_i \simeq \hat{\mathbb{Q}}_p \otimes_{\hat{\mathbb{Z}}_p} \hat{S} \oplus$ (simple algebra), the simple algebra must be a skewfield of index q and the centre is $\hat{K}$, $1 \le i \le q-1$. Hence the block structure is as claimed.

<u>Case 2:</u> $q^2 \mid (p-1)$.

In this case one has still q blocks for $\hat{\mathbb{Z}}_p G$, $\hat{B}_o, \dots, \hat{B}_{q-1}$, but all blocks are algebraically conjugate to $\hat{B}_o = \hat{\mathbb{Z}}_p G_1$.

We have

$$\mathbb{Q}G \simeq \mathbb{Q} \amalg \mathbb{Q}(\sqrt[q]{1}) \amalg \mathbb{Q}(\sqrt[q^2]{1}) \amalg (\mathrm{Fix}_{C_q} \mathbb{Q}(\sqrt[p]{1}))_q \amalg A ,$$

where for $q^2 \nmid (p-1)$, A is a skewfield D with centre $\mathbb{Q} \otimes_{\mathbb{Z}} R[\sqrt[q]{1}]$, where $R = \mathrm{Fix}_{C_q} \mathbb{Z}[\sqrt[p]{1}]$. For $q^2 \mid (p-1)$ $A = (\mathbb{Q} \otimes_{\mathbb{Z}} R)_q$.

$\mathbb{Z}_p G = B_o \amalg B_1$, where $B_o \simeq \mathbb{Z}_p G_1$ and $\mathbb{Q} \otimes_{\mathbb{Z}_p} B_1 \simeq \mathbb{Q}(\sqrt[q^2]{1}) \amalg A$.

This is clear, since the sequence

$$0 \to \mathbb{Z}_p \otimes_{\mathbb{Z}} c_q{\uparrow}^G \to \mathbb{Z}_p G \to \mathbb{Z}_p G_1 \to 0$$

is two-sided split.

<u>Claim 3:</u> Let

$$U_1(p) = \{1 + x ,\ x \in \mathbb{Z}_p \otimes_{\mathbb{Z}} c_p{\uparrow}^G\} \cap V(\mathbb{Z}_p G) ,$$

then $U_1(p)$ contains an elementary abelian subgroup $C_p \times C_p$, and

prime ideal in S lying above q_i and $\hat{S}_{\tilde{q}_i}$ is the corresponding completion. Moreover, $\hat{B}_o$ has defect 2 and $\hat{B}_i$, $1 \le i \le r$ has defect 1 .

<u>Proof:</u> The sequence

$$0 \to \hat{\mathbb{Z}}_q \otimes_{\mathbb{Z}} c_p\!\uparrow^G \to \hat{\mathbb{Z}}_q G \to \hat{\mathbb{Z}}_q C_{q^2} \to 0$$

is two-sided split and $B_o = \hat{\mathbb{Z}}_q C_{q^2}$ is the principal block of defect 2. We now consider the exact sequence

$$0 \to \hat{\mathbb{Z}}_q \otimes_{\mathbb{Z}} c_q\!\uparrow^G \to \hat{\mathbb{Z}}_q G \to \hat{\mathbb{Z}}_q G_1 \to 0$$

and restrict it to $\hat{\mathbb{Z}}_q C_p$, which is a separable order. Then $\hat{\mathbb{Z}}_q c_p$ is projective over $\hat{\mathbb{Z}}_q C_p$ and we thus obtain the exact sequence

$$0 \to \hat{\mathbb{Z}}_q \otimes_{\mathbb{Z}} c_q\!\uparrow^G \otimes_{\hat{\mathbb{Z}}_q C_p} c_p \to \hat{\mathbb{Z}}_q \otimes_{\mathbb{Z}} c_p\!\uparrow^G \to \hat{\mathbb{Z}}_q G_1 \otimes_{\hat{\mathbb{Z}}_q C_p} c_p \to 0 .$$

However, from the above remarks it follows that

$$\hat{\mathbb{Z}}_q G_1 \otimes_{\hat{\mathbb{Z}}_q C_p} c_p = \hat{\mathbb{Z}}_q c_p\!\uparrow^{G_1} = \prod_{i=1}^{r} (\hat{R}_{q_i})_q .$$

Now this is a separable order, and hence it follows that $\hat{\mathbb{Z}}_q \otimes_{\mathbb{Z}} c_q\!\uparrow^G$ decomposes into blocks with defect one. From the structure of the rational group algebra it then follows that $\hat{\mathbb{Z}}_q \otimes_{\mathbb{Z}} c_q\!\uparrow^G$ decomposes into r blocks $\hat{B}_i$ which are again Bäckström-orders and have hence the form described in the claim 4 .

From these results it follows readily that $\mathbb{Z}_q G$ decomposes into $B_o = \mathbb{Z}_q C_{q^2}$ and $B_1 = \mathbb{Z}_q \otimes_{\mathbb{Z}} c_q$, where the latter is 2-sided indecomposable in

$$(\mathbb{Q} \otimes_{\mathbb{Z}} R)_q \quad \Pi \quad A \qquad (A \text{ as above}) ,$$

and the only congruences are congruences modulo q .

<u>Claim 5:</u> B_1 contains two primitive q^{th} roots of unity, which are not conjugate in $\mathbb{Q}G$, in case $q^2 | (p-1)$.

<u>Proof:</u> We write the elements in B_1 as pair (x_1, x_2) with $x_1 \in (\mathbb{Q} \otimes_{\mathbb{Z}} R)_q$ and $x_2 \in A$. Then B_1 contains a primitive q^2-th root of

$$U_o(p) = \{1 + x \ , \ x \in \mathbb{Z}_p \otimes_{\mathbb{Z}} c_p g\} \cap V(\mathbb{Z}_p G)$$

has torsion.

<u>Proof:</u> From the above remark it follows that $\mathbb{Z}_p \otimes_{\mathbb{Z}} c_p\uparrow^G \simeq I_1 \oplus I_2$ with $\mathbb{Q} \otimes_{\mathbb{Z}_p} I_1 = (\mathbb{Q} \otimes_{\mathbb{Z}} R)_q$ and $\mathbb{Q} \otimes_{\mathbb{Z}_p} I_2 = A$. Now $c_p\uparrow^G$ is a cyclic $\mathbb{Z}_p G$-module generated by $c-1$ with $C_p = \langle c \rangle$. Hence $(c-1) = (x_1, x_2)$ with $x_i \in I_i$. Let e_1 and e_2 be the central idempotents corresponding to $\mathbb{Q} \otimes_{\mathbb{Z}_p} I_1$ and $\mathbb{Q} \otimes_{\mathbb{Z}_p} I_2$. Since $e_i \notin I_i$, $x_i \neq 0, \pm e_i$ (note $p \neq 2$), we conclude $c = (e_1 + x_1 \ , \ e_2 + x_2)$ with $e_i + x_i \in \mathbb{Z}_p G$. Hence $C_p \times C_p \simeq \langle e_1 + x_1 \ , \ e_2 + x_2 \rangle \leq U_1(p)$. Since $U_1(p)/U_o(p) \simeq C_p$, $U_o(p)$ can not be torsionfree.

We <u>remark</u> that for the pullback H

$$\begin{array}{ccccccccc} 1 & \to & C_p & \to & G_1 & \to & C_q & \to & 1 \\ & & \| & & \uparrow & & \uparrow & & \\ 1 & \to & C_p & \to & H & \to & C_{q^n} & \to & 1 \end{array} ,$$

the corresponding subgroup $U_1(p) < V(\mathbb{Z}_p H)$ contains a subgroup isomorphic to $C_p^{(n)}$.

We now consider our groupring $\mathbb{Z}G$ at the place q. (Here we do not have to make the distinction $q^2 | (p-1)$ or $q^2 \nmid (p-1)$. Let $R = \mathrm{Fix}_{C_q} \mathbb{Z}[\sqrt[p]{1}]$ then q is unramified in R and so $qR = \mathfrak{q}_1, \ldots, \mathfrak{q}_r$ where $\mathfrak{q}_i$ are prime ideals in R - moreover, the residue class degree of $\mathfrak{q}_i$ is the smallest integer f such that $q^f \equiv 1(p)$. We put $S = R[\sqrt[q]{1}]$.

<u>Claim 4:</u> $\hat{\mathbb{Z}}_q G$ decomposes into $r + 1$ blocks $\hat{B}_o, \hat{B}_1, \ldots, \hat{B}_r$, where $\hat{B}_o = \hat{\mathbb{Z}}_q C_{q^2}$ and $\hat{B}_i$, $1 \leq i \leq r$, is the pullback

$$\begin{array}{ccc} \hat{B}_i & \to & (\hat{R}_{\mathfrak{q}_i})_q \\ \downarrow & & \downarrow \\ (\hat{S}_{\tilde{\mathfrak{q}}_i})_q & \to & (F_{q^f})_q \end{array} ;$$

here $R_{\mathfrak{q}_i}$ denotes the completion of R at $\mathfrak{q}_i$, and $\tilde{\mathfrak{q}}_i$ is the unity corresponding to the generator of C_{q^2}, say (u_1,u_2) then $u_1{}^q = 1$ and so $(1,u_2{}^q)$ is a primitive q-th root of unity in B_1, lying in the centre of A.

If $q^2|(p-1)$, then $A = (\mathbb{Q} \otimes_{\mathbb{Z}} S)_q$ and $R_q \subset S_{\tilde{q}_1 \cdots \tilde{q}_2}$ -recall $S_{\tilde{q}_1 \cdots \tilde{q}_2} = R_q[\sqrt[q]{1}]$.

Then B_1 is the pullback of

$$\begin{array}{ccc} B_1 & \to & (R_q)_q \\ \downarrow & & \downarrow \\ (S_{\tilde{q}_1 \cdots \tilde{q}_r})_q & \to & (\prod_{i=1}^{r} R/q_i)_q \simeq (\Pi\, S/\tilde{q}_i)_q \end{array} .$$

Then $(R_q)_q$ contains a primitive q-th root of unity, say $\tilde{u}_1$ corresponding to the element of order q in G_1. Via the injection $(Rq)_q \to (S_{\tilde{q}_1 \cdots q_r})_q$ we obtain a root of unity $\tilde{u}_2$ such that the pair $(\tilde{u}_1,\tilde{u}_2)$ is a primitive q-th root of unity in B_1. Obviously $(\tilde{u}_1,\tilde{u}_2)$ and $(1,u_2{}^q)$ are not conjugate in $\mathbb{Q}G$.

Let us summarize our results:

<u>Proposition 2:</u> Let G be the above semi-direct product $C_p \rtimes C_{q^2}$, $q^2|(p-1)$. Then both $V(\mathbb{Z}_pG)$ and $V(\mathbb{Z}_qG)$ contain units of finite order, which are not conjugate in $\mathbb{Q}G$ to a group element. Moreover, these units even lie in $U_1(p)$ and $U_1(q)$ resp. $U_1(p)$ contains a subgroup $C_p \times C_p$.

<u>Proposition 3:</u> Let G be the semi-direct product $C_p \rtimes C_{q^2}$, $q|(p-1)$ then there exists a split exact sequence

$$* \qquad 1 \to V_o \to V(\mathbb{Z}G) \to G \to 1$$

and V_o is torsion free.

<u>Proof:</u> In view of L1 we know that we have the exact sequence (*) and because of that lemma, it suffices to show that

$$U_o = \{1 + x : x \in c_p g\} \cap V(\mathbb{Z}G)$$

has no torsion.

Let e be the identity element in $\mathbb{Q} \otimes_{\mathbb{Z}} c_p g$. Then $\hat{\mathbb{Z}}_p e + \hat{\mathbb{Z}}_p \otimes_{\mathbb{Z}} c_p g$ is a local order, and so U_o can only have p-torsion. Let now $u_o \in U_o$ be an element of order p. If $\varphi: \mathbb{Z}G \to \mathbb{Z}G_1$ is the canonical homomorphism. Then $u_o \varphi = 1$, and hence $u_o \in \{1 + y : y \in c_q \uparrow^G\} \cap U_o = \overline{U}$. But C_q is the centre of G and so $\overline{U}$ contains only trivial units of finite order; i.e. units of order q, thus $u_o = 1$.

We have now finished our detour on units in integral group rings and return to the structure of blocks, and we keep the local notation as introduced earlier. The handicap of Proposition 1 is that it assumes "normal defect groups". In general one surpasses this by using Brauer- and Green-correspondence. However, if D is the defect group of the block B, and $N = N_G(D)$, then in general - under Green-correspondence - a central primitive idempotent in KN will not correspond to a central primitive idempotent. Moreover, W.Plesken has told me that Prop.1 is not valid without the assumption of normality (e.g. Sl(2,8) at p = 3). Nevertheless the analogue of Prop.1 is valid for blocks of defect one.

<u>Theorem 3:</u> Let B be a block of defect one in RG. Then

(i) If e is a central primitive idempotent in KB, then Be is a hereditary order.

(ii) B is a Bäckström-order; i.e. rad B = rad Γ, where $\Gamma = \bigoplus_{i=1}^{\tau} \Gamma_i$ is a hereditary order in $KB = \bigoplus_{i=1}^{\tau} (D_i)_{n_i}$.

(iii) If P is an indecomposable projective B-module, then $KP = U \oplus U'$ where U and U' are non-isomorphic simple KB-modules. Hence one can form the rational p-adic Brauertree T_p of B, whose vertices are the non-isomorphic simple KB-modules U_i, $1 \le i \le \tau$, and there is an edge between U_i and U_j if there exists an indecomposable projective B-module P with $KP \simeq U_i \oplus U_j$. (This is a tree because of the injectivity of the Cartan-map.)

(iv) The vertices of T_p can also be identified with the hereditary orders $\{\Gamma_i\}$ and a vertex i_o has order n in T_p - i.e. there are n edges meeting in i_o - if and only if Γ_{i_o} has exactly n non-isomorphic indecomposable projective Γ_{i_o}-modules.

(v) B has $\tau - 1$ non-isomorphic indecomposable projective modules and $2(\tau - 1)$ non-isomorphic indecomposable non-projective lattices; these are the projective Γ_i-modules $1 \le i \le \tau$.

(vi) If $S_1, \ldots, S_{\tau-1}$ are the non-isomorphic simple B-modules, then $End_B(S_i) = \mathfrak{k}_o$ is the same for $1 \le i \le \tau - 1$, and the Schur-indices of the simple KB-module U_j , $1 \le j \le \tau$, are bounded by $d = |\mathfrak{k}_o : \mathfrak{k}|$.

(vii) If B has an absolutely simple module then T_p coincides with the ordinary Brauer-tree T_o. In this case the skew-fields D_i , $1 \le i \le \tau$ coincide with K except for the exceptional vertex i_o ; here D_{j_o} is a totally ramified extension of K , and the ramification index of D_{j_o} is the multiplicity of j_o .

<u>Remarks:</u> (i) extends the result of Brauer, mentioned in the beginning. (ii) and (iii) were independently obtained by H. Jacobinski (1979). (v) was for metacyclic groups obtained by Pu (1965) and for the principal block by J.A. Green (1974).
(vi) For the principal block D.H. Gluck (1979) has shown that the Schur-indices are one.

The ordinary Brauertree T_o of B is a d-fold covering for our Brauertree T_p. E.q. The group of order 63 (Ex.2) has T_p : •——• and T_o : •——•< for the non-principal blocks - here $d = 3$.
For explicit calculations it should be noted that the ordinary Brauertree can be read off - modulo ε - from the charactertable of G , and then the above results allow to write down the principal block.

It also should be noted that the results of Theorem 3 contain all the known informations on the modular representation theory of blocks of defect one - including results about liftability of modules.

<u>Example 3:</u> Let $G = Su(8)$ be the simple group of order $2^6 \cdot 5 \cdot 7 \cdot 13 = 29\,120$. Then the Brauertree of the principal block has the form

14

1 64 ③ 35

14

If $p = 13$ and $R = \hat{Z}_{13}$, $S = \mathrm{Fix}_{C_4}(R[\sqrt[13]{1}\,])$, then the principal block has the following form

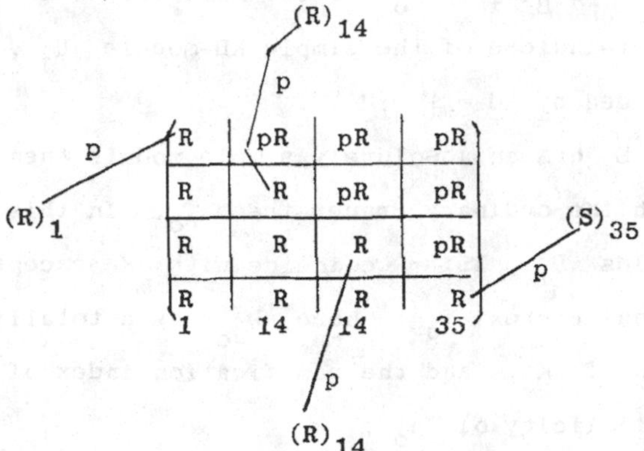

Some words to the proof, which may be found in detail in Roggenkamp (1979):

Let $N = N_G(D)$ where D is the defect group of B. It is then fairly easy to prove the results for the blocks of RN (cf. Green (1974)). Green-correspondence is then used to prove (v). Now, B is contained in a Bäckström-order Λ which has the same number of non-isomorphic indecomposable modules as Λ, namely $2(\tau-1)$. A careful analysis of the module structure of Λ shows that Λ has altogether $3(\tau-1)$ indecomposable lattices; by (v) this is also the number of non-isomorphic indecomposable B-lattices; hence $\Lambda = B$. The remaining statements follow from combinatorical considerations.

R E F E R E N C E S

Auslander, M. - O.Goldman: Maximal orders.
Trans.Am.Math.Soc. 97 (1960), 1-24.

Bäckström, K.J.: Orders with finitely many indecomposable lattices.
Ph.D.Thesis 1972, Göteborg.

Brauer, R.: Investigations on group characters.
Ann.of Math. 42 (1941), 936-958.

Brumer, A.: Structure of hereditary orders.
Bull.Am.Math.Soc. 69 (1963), 721-729.

Dade, E.C.: Blocks with cyclic defect groups.
Ann.of Math. 84 (1966), 20-48.

Dennis, R.K.: The structure of the unit group of group rings.
Proc.Conf.Ring Theory, Oklahoma 1976. Marcel Dekker.

Dlab, V. - C.M.Ringel: Indecomposable represenstations of graphs and algebras.
Memoirs of the AMS No.173 (1976).

Gluck, D.H.: A character table bound for the Schur index.
(1979) to appear.

Green, J.A.: Walking around the Brauertree.
J.of the Australian Math.Soc. XVII (1974), 197-213.

Harada, M.: Structure of hereditary orders over local rings.
J.of Mathematics, Osaka City Univ.14 (1963), 1-22.

Higman, G.: The units of group rings.
Proc.London Math.Soc. (2), 46 (1940), 231-248.

Hughes, I. - K.R.Pearson: The group of units of the integral group ring $\mathbb{Z} S_3$.
Can.Math.Bull. 15 (1972), 529-534.

Jacobinski, H.: Two remarks about hereditary orders.
Proc.AMS 28 (1971), 1-8.

Michler, G.: Green correspondence between blocks with cyclic defect groups II.
Proc.ICRA, Springer Lecture Notes Nr.488, 210-235.

Miyata, T.: On the units of the integral group ring of a dihedral group.
(1979) to appear J.Math.Soc.Japan.

Peacock, R.M.: Blocks with a cyclic defect group.
J.Algebra 34 (1975), 232-259.

Plesken, W.: Gruppenringe über lokalen Dedekindbereichen.
Habilitationsschrift, Aachen (1980).

Pu, L.C.: Integral representations of non-abelian groups of order p q.
Mich.Math.J. 12 (1965), 231-246.

Ringel, C.M. - K.W.Roggenkamp: Diagrammatic methods in the representation theory of orders.
J.Algebra 60 (1979), 11-42.

Roggenkamp, K.W.: Representation theory of finite groups.
(1979) to appear Presses de l'Université de Montréal.

Whitcomb, A.: The group ring problem.
Ph.D.Thesis, Univ.of Chicago, Illinois (1968).

Zassenhaus, H.: On the torsion units of finite group rings.
Studies in Math. (in honor of A.Almeida Costa), Instituto de alta Cultura, Lisboa (1974), 119-126.

Hasse, H.: Über p-adische Schiefkörper und ihre Bedeutung für die Arithmetik hyperkomplexer Zahlsysteme.
Math.Ann. 104 (19), 495-534.

Allan P.J. - C.Hobby: A characterization of units in $\mathbb{Z}[A\ ]$.
Abstracts of papers presented to the Am.Math.Soc. 1 (1980), 48, 773-20-10.

K.W.Roggenkamp
Mathematisches Institut B
Universität Stuttgart
Pfaffenwaldring 57
D 7000 Stuttgart-80

February 1980

Standard Monomial Theory*

By C. Musili and C.S. Seshadri

We give here a survey of Geometry of G/P-I,...,IV (cf. [15], [16], [17], [22]). The main aim of these papers is to extend the classical Hodge-Young standard monomial theory (cf. [10], [11]) of $SL(n)$ to the case of an arbitrary semi-simple linear algebraic group G. Roughly speaking, the problem is to give an explicit basis for the space $H^o(G/B,L)$ of sections of a line bundle L (in the dominant chamber) on the flag variety G/B of G (or more generally for $H^o(X,L)$, where X is a Schubert subvariety of G/B) in terms of some nicely chosen bases of $H^o(G/B,L_i)$ (or $H^o(X,L_i)$), where L_i are the line bundles on G/B associated to the fundamental weights. Because of the Borel-Weil theorem, which states that when the base-field is of characteristic zero, any irreducible representation of G is of the form $H^o(G/B,L)$, a particular case of our problem can be viewed as finding explicit bases of any irreducible representation of G (when the base field is of characteristic zero), in terms of some nicely chosen bases of the fundamental representations ($= H^o(G/B,L_i)$). Our results provide a complete solution to this problem when G is a classical group, as well as partial answers when G is exceptional.

Our original motivation to get at a standard monomial theory for any semi-simple algebraic group G, which is an extension of the Hodge-Young theory for $SL(n)$ was to prove statements of the type

---

* This is an expanded version of a talk given by the second author in Séminaire d'algèbre Paul Dubreil et Marie Paule Malliavin, 1980.

that $H^i(X,L) = 0$, $i > 0$, L being in the dominant chamber and that the singularities of X (X Schubert variety in $G/_B$) are Cohen-Macaulay. When the base field is of characteristic zero, these results have been proved by Demasure (cf. [4]) but when the base field is of arbitrary characteristic, they are still not proved in complete generality. When X is the Grassmannian, it was realized that these results are consequences of the Hodge-Young theory, the important technical point being that this theory provided a good hold of the ideal theory of Schubert varieties in the Grassmannian (cf. [8], [12], [18], [20]). It was therefore natural to search for a generalisation of the Hodge-Young standard monomial theory to the case of an arbitrary semi-simple algebraic group. As a consequence of our standard monomial theory, we show that if X is any Schubert variety in $G/_B$, G a <u>classical group</u>, $H^i(X,L) = 0$, $i > 0$, L being a line bundle in the dominant chamber; however even in this case (i.e. G is classical), it has not been shown that the singularities of <u>any</u> Schubert variety X in $G/_B$ are Cohen-Macaulary (and normal), though there are many partial results in this direction (see Theorem 2.3 and Remark 2.4 later). The vanishing theorem of Kempf, namely that $H^i(G/_B, L) = 0$, $i > 0$, G any semi-simple group and L any line in the dominant chamber, can be deduced as a consequence of standard monomial theory (cf. [13], [15]). Recall that a beautiful and short proof of this result has been found independently by Andersen (cf. [1]) and Haboush (cf. [7]).

The formulation of our standard monomial theory was arrived at by suitably interpreting the work of De Concini and Procesi on classical invariant theory (cf. [3]; G/p-II, [17]).

We take this occasion to point out some corrections and modifications to be made in G/p-IV, (see Remarks 4.1, 5.1 and 5.2).

§ 1. Hodge-Young theory.

Let $G = SL(n)$, B = the Borel subgroup of G consisting of the upper triangular matrices and T = the maximal torus of G consisting of the diagonal matrices. Let P be a maximal parabolic subgroup of G with respect to B (i.e. an algebraic subgroup of G containing B, $\neq G$ and maximal with these properties) so that G/p is the Grassmannian of r-dimensional subspaces of an n-dimensional vector space for some r, $1 \leqslant r \leqslant n$. One knows that Pic $G/p \approx \mathbb{Z}$ and that a generator of this group can be taken to be ample; in fact this generator L is a very ample line bundle on G/p and its sections give an imbedding of G/p into a projective space called the Plucker imbedding. Now $H^0(G/p,L)$ is a G-module and the T-stable vectors* in $H^0(G/p,L)$ are called weight vectors and one has the notion of highest and lowest weight vectors (the highest weight vector being B-stable and the lowest weight being B^- stable, where B^- = the Borel subgroup opposite to B i.e. consisting of the lower triangular matrices). Let f be a lowest weight vector in $H^0(G/p,L)$ (f is uniquely determined upto a constant). Let $W = W(G)$ denote the Weyl group of G. One sees that the subgroup of W fixing the one dimensional linear

(*) i.e. the 1-dimensional space spanned by the vector is T-stable

space spanned by f, is precisely the Weyl group W(P) of P. The translates of f by W can therefore by indexed by $W/W(P)$ and we set

$$p(\tau) = \tau \cdot f, \ \tau \in W/W(P)$$

It is known that $\{p(\tau)\}$ are precisely the weight vectors and they form a basis of $H^0(G/p, L)$. The translates of the highest or lowest weight vector by W are called <u>extremal weight vectors</u> (in the general context). Now $H^0(G/p, L)$ is a <u>fundamental representation</u> of $G = SL(n)$ and a basis of this representation consists of extremal weight vectors. Now this property is not always true in the general context (i.e. when $G \neq SL(n)$, G being a semi-simple algebraic group) and if this property is true we say that this fundamental representation is <u>minuscule</u> (and that the corresponding maximal parabolic P as well as the fundamental weight are minuscule). The B-stable <u>closed subvarieties</u> of G/p are called <u>Schubert varieties</u> (in G/p); it is known that the number of Schubert varieties in G/p is equal to # W/W(P) and hence the Schubert subvarieties can be indexed as $X(\tau)$, $\tau \in W/W(P)$. One has a canonical <u>partial order</u> and the notion of <u>length</u> in W and W/W(P). Besides one has the property:

$$\tau_1 \geq \tau_2 \Longleftrightarrow X(\tau_1) \supseteq X(\tau_2), \ \tau_i \in W/W(P).$$

There is a unique maximal element in W/W(P) and the associated Schubert variety is G/p. One can also choose the indexing of Schubert varieties (by $W/W(P)$) so that

$$X(\tau) = \text{common zero set of all } p(\lambda) \text{ such that } \tau \not\geq \lambda,\ \lambda \in W/W(P)$$

i.e.

$$p(\lambda) \mid X(\tau) \not\equiv 0 \Leftrightarrow \tau \geq \lambda .$$

Let us call a standard young diagram of length m on G/p <u>(resp. $X(\tau)$)</u> a subset of m elements of $W/W(P)$ of the form

$$\tau_1 \geq \tau_2 \geq \ldots \geq \tau_m \text{ (resp. } \tau \geq \tau_1 \geq \tau_2 \geq \ldots \geq \tau_m \text{)}$$

We call a <u>standard monomial of length m on G/p (resp. $X(\tau)$)</u> an expression of the form

$$p(\tau_1)\, p(\tau_2) \ldots p(\tau_m) \text{ such that}$$
$$\tau_1 \geq \tau_2 \geq \ldots \geq \tau_m \text{ (resp. } \tau \geq \tau_1 \geq \ldots \geq \tau_m \text{)}$$

Now a standard monomial of length m on G/p (resp. $X(\tau)$) represents an element of $H^0(G/p, L^m)$ (resp. $H^0(X(\tau), L^m)$). By the Hodge-Young standard monomial theory, we mean essentially the following

<u>Theorem 1.1</u> : Standard monomials of length m on $X(\tau)$ form a basis of $H^0(X(\tau), L^m)$. In particular

$$\dim H^0(X(\tau), L^m) = \# \{\text{standard Young diagrams of length } m \text{ on } X(\tau)\}$$
$$\dim H^0(G/p, L^m) = \# \{\text{standard Young diagrams of length } m \text{ on } G/p\}$$

<u>Remark 1.1</u> : It can be shown that we have a canonical identification

of $W/W(P)$ with the following set of indices :

$$I_n(r)=\{(i_1,\dots,i_r)\mid 1\leq i_1<i_2<\dots<i_r\leq n\}$$

Further if $\tau_1=(i_1,\dots,i_r)$ and $\tau_2=(j_1,\dots,j_r)$, then

$$\tau_1\geq\tau_2 \text{ in } W/W(P)\Leftrightarrow i_k\geq j_k,\ 1\leq k\leq r.$$

With this identification, the $p(\underline{\tau})$ are the usual Plucker coordinates.

Remark 1.2 : Theorem 1.1 is proved in the book of Hodge and Pedoe [11] (see also [10]) when the base field is of characteristic zero. In arbitrary characteristic, it has been proved by several authors (cf. [8], [12], [18], [20]). The above theorem solves the problem of giving canonical bases for irreducible representations of $SL(n)$ whose highest weights are multiples of fundamental weights, the base field being of characteristic zero. The case of an arbitrary irreducible representation is dealt with in the next theorem.

Let $W_1,\dots,W_{n-1}$ be the Weyl groups of the distinct $(n-1)$ parabolic subgroups $P_1,\dots,P_{n-1}$ containing B such that

$$W/W_k\approx I_n(k),\quad 1\leq k\leq n-1;$$

$$\text{i.e. } W/W_1\approx\{(i_1)\mid 1\leq i_1\leq n-1\},$$

$$W/W_2\approx\{(i_1,i_2)\mid 1\leq i_1<i_2\leq n\},$$

etc.

Let $\tau=(i_1,\dots,i_k)\in W/W_k$ and $\varphi=(j_1,\dots,j_\ell)\in W/W_\ell$. Then we say that $\tau \geqslant \varphi$ (or $\varphi \leqslant \tau$) if

(i) $k \leqslant \ell$

(ii) $i_1 \geqslant j_1$; $i_2 \geqslant j_2, \ldots, i_k \geqslant j_k$.

Let us call a <u>standard Young diagram</u> in W (or the <u>flag variety</u> G/B) of <u>type</u> or (<u>multi degree</u>) $(m_1, \ldots, m_{n-1})$ a set of $\underline{\delta} = m_1 + m_2 + \ldots + m_{n-1}$ elements from $W/W_1 \cup \ldots \cup W/W_{n-1}$ of the form

(1) $\tau_1 \geqslant \tau_2 \geqslant \ldots \geqslant \tau_\delta$ (in the sense defined above)

with the further condition that the first m_1 elements i.e. $\tau_1, \ldots, \tau_{m_1}$ are in W/W_1, the next m_2 elements are in W/W_2, the next m_3 elements are in W/W_3 etc. By a <u>standard monomial of type</u> $(m_1, \ldots, m_{n-1})$ on G/B, we mean an expression

$$p(\tau_1) \cdot p(\tau_2) \ldots p(\tau_\delta), \; \delta = m_1 + \ldots + m_{n-1}$$

such that $(\tau_1, \ldots, \tau_\delta)$ is a standard Young diagram of type $(m_1, \ldots, m_{n-1})$ in W. Let $L_i \in \operatorname{Pic} G/P_i$, $1 \leqslant i \leqslant (n-1)$, denote the canonical ample generators of $\operatorname{Pic} G/P_i$. Then a standard monomial of type $(m_1, \ldots, m_{n-1})$ can be identified with an element of

$$H^0(G/B, L_1^{m_1} \otimes L_2^{m_2} \otimes \ldots \otimes L_{n-1}^{m_{n-1}}).$$

Then we have the following result

<u>Theorem 1.2</u> : Standard monomials of type $(m_1, \ldots, m_{n-1})$ form a basis

of $H^0(G/_B, L_1^{m_1} \otimes L_2^{m_2} \otimes \ldots \otimes L_{n-1}^{m_{n-1}})$, $m_i \geqslant 0$.

<u>Remark 1.3</u> : This theorem is proved by Hodge in [10] assuming that the base field is of characteristic zero. Thanks to the Borel-Weil theorem by which every irreducible G-module is of the form $H^0(G/_B, L_1^{m_1} \otimes .. \otimes L_{n-1}^{m_{n-1}})$, $m_i \geqslant 0$ for some choice of m_i, the above theorem solves the problem of giving canonical bases of any irreducible G-module in terms of some nices bases of the fundamental representations when $G = SL(n)$ and the base field is of characteristic zero. Unlike Theorem 1.1, Hodge does <u>not</u> give an analogue of Theorem 1.2 for <u>any</u> Schubert variety in $G/_B$; his proof gives it for a very special class of Schubert varieties in $G/_B$. There is some difficulty in giving an analogue for <u>any</u> Schubert variety in $G/_B$. The definition of standard monomials on an arbitrary Schubert variety in $G/_B$ has to be done with greater care and this will be done in § 3.

§ 2. <u>Main results for the case of one maximal parabolic group.</u>

Let $G_{\mathbb{Z}}$ denote a semi-simple, simply-connected, Chevalley group scheme over the ring of integers $\mathbb{Z}$ and let G_k(or simply G) = $G_{\mathbb{Z}} \times_{\mathbb{Z}} k$ denote the base change of $G_{\mathbb{Z}}$ by Spec $k \longrightarrow$ Spec $\mathbb{Z}$, k being an algebraically closed <u>ground field</u>. We fix a maximal torus subgroup scheme $T_{\mathbb{Z}}$ in $G_{\mathbb{Z}}$ and a Borel subgroup scheme $B_{\mathbb{Z}}$ of $G_{\mathbb{Z}}$ such that $T_{\mathbb{Z}} \hookrightarrow B_{\mathbb{Z}}$. We set

$$T_k \text{ (or } T) = T_{\mathbb{Z}} \times k \text{ , } B = B_{\mathbb{Z}} \times k$$

Relative to T and B, we talk of roots, weights etc. and we denote by W the Weyl group. We write

$$\Delta\,(\text{resp}\,\Delta^+) = \text{system of roots (resp. system of positive roots)}$$

$$S = \text{set of simple roots}\,\{\alpha_1,\ldots,\alpha_\ell\},\ \ell = \text{rank of } G.$$

We denote by (,) a W-invariant scalar product on $\mathrm{Hom}(T,\mathbb{G}_m)$ ($\mathbb{G}_m$ = multiplicative group scheme of dimension one) and we write

$$(\lambda,\alpha^*) = \frac{2(\lambda,\alpha)}{(\alpha,\alpha)},\ \alpha\in\Delta \quad .$$

We denote by $G_{\mathbb{Q}}$, $B_{\mathbb{Q}}$... etc. the base changes $G_{\mathbb{Q}} = G_{\mathbb{Z}} \times \mathbb{Q}$, $B_{\mathbb{Q}} = B_{\mathbb{Z}} \times \mathbb{Q}$ etc.

The $B_{\mathbb{Z}}$ stable (resp B stable) closed subschemes (resp. subvarieties) of $G_{\mathbb{Z}}/B_{\mathbb{Z}}$ (resp. $G/_B$) are known as <u>Schubert subschemes</u> (resp. <u>subvarieties</u>) of $G_{\mathbb{Z}}/B_{\mathbb{Z}}$ (resp. $G/_B$). It is known that the number of Schubert subschemes (resp. subvarieties) in $G_{\mathbb{Z}}/B_{\mathbb{Z}}$ (resp. $G/_B$) is equal to $\# W$ and that they can be indexed as $X_{\mathbb{Z}}(\tau)$ (resp. $X(\tau)$, $\tau\in W$ such that :

$$X_{\mathbb{Z}}(\tau) \text{ is } \mathbb{Z}\text{-flat and } X(\tau) = X_{\mathbb{Z}}(\tau) \underset{\mathbb{Z}}{\times} k.$$

More generally, if $P_{\mathbb{Z}}$ is a parabolic subgroup scheme of $G_{\mathbb{Z}}$ such that $P_{\mathbb{Z}} \supset B_{\mathbb{Z}}$ so that the parabolic subgroup $P = P_{\mathbb{Z}} \times k$ contains B, we call $B_{\mathbb{Z}}$(resp. B) stable closed subschemes (resp. subvarieties) of $G_{\mathbb{Z}}/P_{\mathbb{Z}}$ (resp. $G/_P$) Schubert subschemes (resp. subvarieties) of $G_{\mathbb{Z}}/P_{\mathbb{Z}}$ (resp. $G/_P$). These can be indexed by $W/W(P)$ ($W(P)$ = Weyl group of P) such that one has :

$$X_{\mathbb{Z}}(\tau) \text{ is } \mathbb{Z}\text{-flat and } X(\tau) = X_{\mathbb{Z}}(\tau) \times k\ .$$

Suppose that P is now a <u>maximal parabolic</u> subgroup of G. We denote by L the ample generator of Pic G/P and by $L_{\mathbb{Z}}$ the ample (relative to $\mathbb{Z}$) line bundle on $G_{\mathbb{Z}}/P_{\mathbb{Z}}$ such that $L = L_{\mathbb{Z}} \times k$. One knows that P is associated to a <u>fundamental weight</u> ϖ ; this association can be done in such a manner that the $G_{\mathbb{Q}}$ irreducible module $H^0(G_{\mathbb{Q}}/P_{\mathbb{Q}}, L_{\mathbb{Q}})$ has highest weight $i(\varpi)$ (where i is the Weyl involution $-w_0$, w_0 being the element of maximal <u>length</u> in W) and consequently of lowest weight $-\varpi$. We say that P (resp. $P_{\mathbb{Z}}$, resp. ϖ, resp. the fundamental representation $H^0(G_{\mathbb{Q}}/P_{\mathbb{Q}}, L))$ is of <u>classical type</u> if

$$|(\varpi,\alpha^*)| = \frac{2|(\varpi,\alpha)|}{(\alpha,\alpha)} \leq 2 \ \forall \ \alpha \in \Delta \ .$$

If G is a <u>classical group</u>, every maximal parabolic subgroup P of G is of classical type; in fact G is classical if and only if every maximal parabolic subgroup of G is of classical type. Further if G is <u>any</u> semi-simple algebraic group, it is easily seen that there is always a maximal parabolic subgroup of G which is of classical type. For $\tau \in W/W(P)$, let us denote by $[X(\tau)]$ the element of the <u>Chow ring</u> Ch(G/P) of G/P determined by the Schubert variety $X(\tau)$ in G/P. Let H denote the unique codimension one Schubert subvariety of G/P (the line bundle defined by H is L). It can be shown that

$$[X(\tau)] \cdot [H] = \sum_i d_i [X(\tau_i)], \ d_i > 0$$

where $\cdot$ denotes multiplication in Ch(G/P) and τ_i runs over the set of all $\lambda \in W/W(P)$ such that $X(\lambda)$ is of codimension one in $X(\tau)$. We call

d_i the multiplicity of $X(\tau_i)$ in $X(\tau)$. We see that $d_i \leq 2$ since P is of classical type (using a formula of Chevalley, cf.[4], [15]). It can also be seen easily that $d_i = 1$ if Θ or P is minuscule.

A pair of elements (τ, φ) in $W/W(P)$ is called an admissible pair if either $\tau = \varphi$ (in which case it is called a trivial admissible pair) or $\tau \neq \varphi$ and there exist $\{\tau_i\}$, $1 \leq i \leq s$, $\tau_i \in W/W(P)$ such that

(i) $\tau = \tau_1 \geq \tau_2 \geq \ldots \geq \tau_s = \varphi$

(ii) $X(\tau_i)$ is of codimension one in $X(\tau_{i-1})$ and the multiplicity of $X(\tau_i)$ in $X(\tau_{i-1})$ is 2.

Note that if P is minuscule, every admissible pair is trivial. Then we have

Theorem 2.1 : There is a basis $\{P(\tau,\varphi)\}$ of $H^0(G_{\mathbb{Z}}/P_{\mathbb{Z}}, L_{\mathbb{Z}})$ indexed by admissible pairs (τ,φ) in $W/W(P)$ such that

(i) $P(\tau,\varphi)$ is a weight vector (under $T_{\mathbb{Z}}$) of weight $= -\frac{1}{2}(\tau(\Theta)+\varphi(\Theta))$

(ii) the canonical rational morphism

$$G_{\mathbb{Z}}/P_{\mathbb{Z}} \longrightarrow \mathbb{P}(H^0(G_{\mathbb{Z}}/P_{\mathbb{Z}}, L_{\mathbb{Z}}))$$

is in fact a closed immersion

(iii) Set $p(\tau,\varphi) = P(\tau,\varphi) \otimes 1$, $p(\tau,\varphi)$ being the canonical image of $P(\tau,\varphi)$ in $H^0(G/P, L)$. Then the restriction of $p(\tau,\varphi)$ to the Schubert variety $X(\theta)$ is not identically equal to zero if and only if $\theta \geq \tau$.

<u>Remark 2.1</u> : It is <u>not</u> true that the basis $\{P(\tau,\varphi)\}$ of $H^{o}(G_{\mathbb{Z}}/P_{\mathbb{Z}},L_{\mathbb{Z}})$ is uniquely determined (upto ± 1) by the properties (i) and (iii). However, when P is <u>minuscule</u>, the basis $\{P(\tau,\varphi)\}$ is uniquely determined (upto ± 1) and it is precisely the set of <u>extremal</u> weight vectors in $H^{o}(G_{\mathbb{Z}}/P_{\mathbb{Z}},L_{\mathbb{Z}})$.

<u>Remark 2.2</u> : It is true that $\{p(\tau,\varphi)\}$, $\theta \geq \tau$, form a basis of $H^{o}(X(\theta),L)$ (see Theorem 2.2 below).

Let us call an expression of the type

$$P(\tau_1,\varphi_1)P(\tau_2,\varphi_2)\ldots P(\tau_m,\varphi_m)$$

$$(\text{resp } p(\tau_1,\varphi_1)\ldots\ldots p(\tau_m,\varphi_m))$$

a <u>standard monomial of length m on a Schubert</u> scheme (resp. <u>variety</u>) $X_{\mathbb{Z}}(\theta)$ (resp. $X(\theta)$) if we have

(i) $\theta \geq \tau_1$ and

(ii) $\tau_1 \geq \varphi_1 \geq \tau_2 \geq \varphi_2 \geq \ldots \geq \tau_m \geq \varphi_m$

We then say that

$$(\tau_1,\varphi_1),(\tau_2,\varphi_2),\ldots,(\tau_m,\varphi_m)$$

is a <u>standard Young diagram</u> on $X_{\mathbb{Z}}(\theta)$ (or $X(\theta)$). We see that a standard monomial of length m on $X_{\mathbb{Z}}(\theta)$ (resp. $X(\theta)$) defines an element of $H^{o}(X_{\mathbb{Z}}(\theta),L^{m}_{\mathbb{Z}})$ (resp. $H^{o}(X(\theta),L^{m})$). Then we have

<u>Theorem 2.2</u> : (i) Standard monomials of length m on $X_{\mathbb{Z}}(\theta)$ (resp. $X(\theta)$)

form a basis of $H^0(X_{\mathbb{Z}}(\theta), L_{\mathbb{Z}}^m)$ (resp. $H^0(X(\theta),L^m)$).

(ii) $H^i(X(\theta),L^m) = 0$, $i > 0$ and $m \geqslant 0$.

(this implies by the semi-continuity theorem that $H^i(X_{\mathbb{Q}}(\theta),L_{\mathbb{Q}}^m) = 0$, $i > 0$, $m \geqslant 0$ and this has been proved by Demazure [4]).

<u>Remark 2.3</u> : Denote by (τ,φ) a standard Young diagram on $X(\theta)$ of length m of the form

$$(\tau,\varphi) = (\tau_1,\varphi_1), (\tau_2,\varphi_2), \ldots, (\tau_m,\varphi_m).$$

Set

$$\chi(\tau,\varphi) = -\sum_i \frac{1}{2}(\tau_i(\varpi) + \varphi_i(\varpi))$$

Then since standard monomials are weight vectors (under $T_{\mathbb{Z}}$ or T), we have the following character formula as an immediate consequence of Theorem 2.2 :

Character of $H^0(X(\theta),L^m) = \Sigma \exp\chi(\tau,\varphi)$, where (τ,φ) runs over all standard Young diagrams on $X(\theta)$ of length m.

<u>Theorem 2.3</u> : The cone $\widehat{X(\theta)}$ over the Schubert variety (for the projective imbedding defined by (ii) of Theorem 2.1) is Cohen-Macaulay and normal (in particular $X(\theta)$ is Cohen Macaulay and normal).

<u>Remark 2.4</u> : Theorem 2.3 has been proved by C. DeConcini and V. Lakshmibai [2]. The essential point to prove is that $\widehat{X(\theta)}$ is Cohen-Macaulay since one

knows that $\widehat{X(\theta)}$ is non-singular in codimension one. This theorem uses Theorem 2.2 and also the ideas introduced by Eisenbud, De Concini and Procesi (cf [6]) and constitutes a significant application of standard monomial theory. For particular cases of this theorem, see Kempf [13] and G/P-III,[15].

§ 3. Main results for the mixed case

Let G be now a classical group of rank ℓ. Let $P_1,\ldots,P_\ell$ denote the set of maximal parabolic subgroups of $G(\supset B)$ taken in some order which we fix in the equal. By a Young diagram of type (or multidegree) $m = (m_1,\ldots,m_\ell)$, $m_i \geqslant 0$, we mean a pair (τ,φ) defined as follows :

$$\begin{cases} \tau = (\tau_{ij}), \ \varphi = (\varphi_{ij}), \ (\tau_{ij},\varphi_{ij}) \text{ admissible pair in } W/W(P_i) \\ 1 \leqslant j \leqslant m_i, \ 1 \leqslant i \leqslant \ell . \end{cases}$$

We say that a Young diagram $(\tau.\varphi)$ is standard on $X(\Theta)$ (or on $X_{\mathbb{Z}}(\Theta)$) for $\Theta \in W$ (and written as $\Theta \geqslant (\tau,\varphi)$), if there is a pair (α,β) called a defining pair for (τ,φ), defined as follows :

(i) $\alpha = (\alpha_{ij})$, $\beta = (\beta_{ij})$, α_{ij} and β_{ij} are in W

(ii) Each α_{ij} (resp. β_{ij}) is a lift in W of τ_{ij}(resp. φ_{ij})

(iii) $\Theta \geqslant \alpha_{11} \geqslant \beta_{11} \cdots \geqslant \alpha_{ij} \geqslant \beta_{ij} \geqslant \alpha_{i,j+1} \geqslant \beta_{i,j+1} \geqslant \cdots$

$\cdots \geqslant \alpha_{i,m} \geqslant \beta_{i,m} \geqslant \alpha_{i+1,1} \geqslant \beta_{i+1,1} \cdots \geqslant \alpha_{\ell,m_\ell} \geqslant \beta_{\ell,m_\ell}$

(with respect to the partial order in the Weyl group W).

To a Young diagram (τ,φ) of multi degree m, we define the monomials $P(\tau,\varphi)$ (resp. $p(\tau,\varphi)$) as follows :

$$P(\tau,\varphi) = \prod_{1\leq i\leq \ell}\ \prod_{1\leq j\leq m} P(\tau_{ij},\varphi_{ij})$$

(resp. a similar definition for $p(\tau,\varphi)$)

where $P(\tau_{ij},\varphi_{ij})$ (resp. $p(\tau_{ij},\varphi_{ij})$) are the elements defined as in Theorem 2.1. We say that $P(\tau,\varphi)$ (resp. $p(\tau,\varphi)$) is <u>a standard monomial of type</u> or <u>multi degree</u> $m = (m_1,\dots m_\ell)$ on $X_{\mathbb{Z}}(\theta)$ (resp. $X(\theta)$) if (τ,φ) is standard on $X(\theta)$) of type m. Let $L^m_{\mathbb{Z}}$ (resp. L^m) denote the line bundle on $G_{\mathbb{Z}}/B_{\mathbb{Z}}$ (resp. G/B) defined by

$$L^m_{\mathbb{Z}} = L^{m_1}_{1,\mathbb{Z}} \otimes \dots \otimes L^{m_\ell}_{\ell,\mathbb{Z}}$$

$$(\text{resp. } L^m = L_1^{m_1} \otimes \dots \otimes L_\ell^{m_\ell})$$

where $L_{i,\mathbb{Z}}$ (resp. L_i) are the ample generators of Pic $G_{\mathbb{Z}}/P_{i,\mathbb{Z}}$ (resp. G/P_i). We see that a standard monomial $P(\tau,\varphi)$ (resp. $p(\tau,\varphi)$) of type m on $X_{\mathbb{Z}}(\theta)$ (resp. $X(\theta)$) defines an element of $H^0(X_{\mathbb{Z}}(\theta), L^m_{\mathbb{Z}})$ (resp. $H^0(X(\theta),L^m)$). Note that the elements of $H^0(X_{\mathbb{Z}}(\theta),L^m_{\mathbb{Z}})$ (resp. $H^0(X(\theta),L^m)$) which are given by standard monomials, are restrictions to $X_{\mathbb{Z}}(\theta)$ (resp. $X(\theta)$) of elements of $H^0(G_{\mathbb{Z}}/B_{\mathbb{Z}},L^m_{\mathbb{Z}})$ (resp. $H^0(G/B,L^m)$). We have then

<u>Theorem 3.1</u>. (i) Standard monomials of type m on $X_{\mathbb{Z}}(\theta)$ (resp. $X(\theta)$) form a basis of $H^0(X_{\mathbb{Z}}(\theta), L^m_{\mathbb{Z}})$ (resp. $H^0(X(\theta),L^m)$).

(ii) $H^i(X(\theta),L^m) = 0$, $i > 0$, $m \geq 0$ (i.e. $m_i \geq 0$, $1 \leq i \leq \ell$).

i.e. all the higher cohomologies of the restrictions to $X(\theta)$ of any line bundle in the dominant chamber, are zero. (By the semi-continuity theorem this implies the corresponding statement in characteristic zero, a fact which has been proved by Demazure (cf.[4]) for an arbitrary semi-simple G).

Remark 3.1 : If (τ,φ) is a Young diagram in W with $\tau = (\tau_{ij})$, $\varphi = (\varphi_{ij})$, we set

$$\chi(\tau,\varphi) = \sum_{1 \leqslant i \leqslant \ell,\, 1 \leq j \leqslant m_i} -\tfrac{1}{2}(\tau_{ij}(\varpi_i) + \varphi_{ij}(\varpi_i))$$

Then we have the following character formula as an immediate consequence of Theorem 3.1.

$$\text{Character of } H^0(X(\theta), L^m) = \Sigma \exp \chi(\tau,\varphi)$$

where (τ,φ) runs over all standard Young diagrams on $X(\theta)$ of type m.

Remark 3.2 : The basis of $H^0(X(\theta),L^m)$ given in Theorem 3.1 depends upon the order in which we have taken the maximal parabolic groups $P_1,\dots,P_\ell$. Hence, for example, the basis of $H^0(G/_B,L^m)$ depends upon two things, first on the choice of a basis of $H^0(G/_{P_i},L_i)$ (cf. Remark 2.1; note that when $G = SL(n)$ this basis is however uniquely determined) and secondly on the choice of an ordering of the maximal parabolic subgroups of G (containing B).

Let U denote the universal enveloping algebra of the Lie algebra Lie G of G . Let $U_{\mathbb{Z}}$ (resp. $U^+_{\mathbb{Z}}, U^-_{\mathbb{Z}}$) denote the canonical $\mathbb{Z}$-form in U i.e. the $\mathbb{Z}$-subalgebra of U spanned by

$$\frac{X^n_\alpha}{n!}, \alpha \in \Delta \quad (\text{resp. } \alpha \in \Delta^+, \Delta^-)$$

where $\{X_\beta, H_\alpha\}$, $\beta \in \Delta$, $\alpha \in S$, denote a Chevalley basis of Lie G in the usual notation (cf. [23]). Let λ be a dominant weight and let the $\mathbb{Q}$ vector space V_λ denote the finite dimensional irreducible $G_{\mathbb{Q}}$ module with highest weight λ. Let e_λ denote a highest weight vector (determined uniquely upto a constant factor) in V_λ. For $\tau \in W$, set

$$V_{\lambda,\mathbb{Z}}(\tau) = U^+_{\mathbb{Z}}\, e_\tau,\ e_\tau = \tau \cdot e_\lambda,\ \tau \in W$$

(we see that $\tau \in W$ can be represented by a $\mathbb{Z}$-valued point of $G_{\mathbb{Z}}$ since the "Weyl group scheme" $W_{\mathbb{Z}} = N(T_{\mathbb{Z}})/T_{\mathbb{Z}}$ ($N(T_{\mathbb{Z}})$ = normaliser of $T_{\mathbb{Z}}$) is a constant group scheme, W being the underlying group. We see that e_τ is well-determined upto ± 1). We write

$$V_{\lambda,\mathbb{Z}}(w_o) = V_{\lambda,\mathbb{Z}},\ w_o = \text{the element of } W \text{ of maximal length.}$$
$$(\text{one has } V_\lambda = V_{\lambda,\mathbb{Z}} \otimes_{\mathbb{Z}} \mathbb{Q}).$$

One knows that $V_{\lambda,\mathbb{Z}}$ is a $U_{\mathbb{Z}}$ stable $\mathbb{Z}$-module or equivalently a "$G_{\mathbb{Z}}$ - $\mathbb{Z}$ module". We have then the following (for classical groups) which was conjectured by Demazure for all semi-simple G (cf. [4]).

<u>Theorem 3.2</u> : The $\mathbb{Z}$-submodule $V_{\lambda,\mathbb{Z}}(\tau)$ of $V_{\lambda,\mathbb{Z}}$ is a direct summand.

§ 4. <u>Outline of proof of Theorems 2.1 and 2.2.</u>

Let us first outline the proof of Theorem 2.1. First one proves the character formula for $H^o(X(\theta),L)$ (i.e. the formula of Remark 2.3 for $m = 1$) when the ground field is of characteristic zero. This is done by induction on the dimension of the Schubert variety $X(\theta)$. Let $X(\delta)$ be a Schubert subvariety of $X(\theta)$ of codimension one of the form

$$\delta = s_\alpha \theta, \alpha \text{ a simple root}$$

Such an $X(\delta)$ is called a <u>moving divisor in $X(\theta)$</u> (with respect to α) following Kempf (cf. [13]). To prove the required character formula, we suppose that this is true on $X(\delta)$. One then makes use of the character formula of Demazure [4] which generalizes the character formula of Weyl (when the ground field is of characteristic zero) to that of $H^0(X(\tau),M)$, where $X(\tau)$ is an arbitrary Schubert variety $X(\tau) \subset G/B$, G being an arbitrary semi-simple group and M is any line bundle in the dominant chamber. We are then required to prove that this character formula of Demazure for $H^0(X(\theta),L)$ is equal to the character formula of Remark 2.3 (for the case $m = 1$) by supposing that these two are the same for $X(\delta)$. This involves a counting argument (cf. p. 300-301, G/P-IV, [16]) and one has to know how admissible pairs on $X(\theta)$ could be obtained by knowing them on $X(\delta)$; the basic simple observation in this connection being that <u>any</u> Schubert subvariety $X(\lambda)$ of $X(\theta)$ is of the form (cf. Lemma 1.9, p. 104, [15])

$$\text{either } X(\lambda) \subset X(\delta) \text{ or } X(\lambda) = X(s_\alpha \mu) \text{ with } X(\mu) \subset X(\delta).$$

One in fact shows that an admissible pair on $X(\theta)$ which is not admissible on $X(\delta)$ is of the form (cf. Lemma 3.11, p. 297, G/P-IV, [16])

$(s_\alpha \varphi_1, \varphi_2)$ or $(s_\alpha \varphi_1, s_\alpha \varphi_2)$ for an admissible pair (φ_1, φ_2) on $X(\delta)$.

The next step is the construction of the basis elements $P(\tau,\varphi)$ of Theorem 2.1. Let us set

$$V_{\mathbb{Q},\mathbb{Z}}(\tau) = V_{\mathbb{Z}}(\tau) \quad \text{and} \quad V_{\mathbb{Q},\mathbb{Z}}(\tau) \otimes \mathbb{Q} = V_{\mathbb{Q}}(\tau)$$

$$V_{\mathbb{Z}}(w_o) = V_{\mathbb{Z}} \quad \text{and} \quad V(w_o) = V_{\mathbb{Q}}$$

By the work of Demazure [4], one knows that

$$H^0(X_{\mathbb{Q}}(\tau), L) = (\text{dual})\, V_{\mathbb{Q}}(\tau)^{\vee} \text{ of } V_{\mathbb{Q}}(\tau)$$

The elements $\{P(\tau,\varphi)\}$ constitute the basis dual to a basis $\{Q(\tau,\varphi)\}$ of $V_{\mathbb{Z}}(\theta)$ which will be chosen inductively; i.e. the required basis $\{Q(\tau,\varphi)\}$ is first assumed chosen for $V_{\mathbb{Z}}(\delta)$ and then new basis elements are chosen for $V_{\mathbb{Z}}(\theta)$. They are of the form

$$\begin{cases} X_{-\alpha} Q(\tau,\varphi) \text{ or } \dfrac{X^2_{-\alpha}}{2} Q(\tau,\varphi), \\ Q(\tau,\varphi) \text{ basis element of } V_{\mathbb{Z}}(\delta) \end{cases}$$

We make use of the relation (cf. Lemma 5.2, p. 309, [16])

$$U_{-\alpha} V_{\mathbb{Q}}(\delta) = V_{\mathbb{Q}}(\theta),$$

$$U_{-\alpha,\mathbb{Z}} V_{\mathbb{Z}}(\delta) = V_{\mathbb{Z}}(\theta)$$

where $U_{-\alpha,\mathbb{Z}}$ (resp. $U_{-\alpha}$) denotes the $\mathbb{Z}$-subalgebra (resp. $\mathbb{Q}$-subalgebra) of $U_{\mathbb{Z}}$ (resp. U) generated by $\dfrac{X^n_{-\alpha}}{n!}$ (resp. $X^n_{-\alpha}$). The property (iii) of Theorem 2.1 can be translated into a property for the basis $\{Q(\tau,\varphi)\}$ of

$V_{\mathbb{Z}}(\theta)$ and this property is again checked inductively in its construction (cf. Lemma 5.5, p. 312, G/p-IV, [16]). By these considerations we would get elements $\{P(\tau,\varphi)\}$ of $H^0(X_{\mathbb{Z}}(\theta),L_{\mathbb{Z}})$ such that they are linearly independent over $\mathbb{Q}$ and generate $V_{\mathbb{Q}}(\theta)$ and their "reductions mod p" namely $p(\tau,\varphi)$ also remain linearly independent as elements of $H^0(X(\theta),L)$ (the property (iii) of Theorem 2.1 is crucial for this purpose). Then one gets that $\{P(\tau,\varphi)\}$ is a basis of $H^0(X_{\mathbb{Z}}(\theta),L_{\mathbb{Z}})$; however the fact that $\{p(\tau,\varphi)\}$ is a basis of $H^0(X(\theta),L)$ is not immediate and follows only after the property (ii) of Theorem 2.2 is proved. It is to be remarked that the property (ii) of Theorem 2.1 is valid for any semi-simple group G and uses a lemma of Deodhar (cf. Prop. 5.7 and Lemma 5.8, G/p-IV, [16]).

We shall now take up the proof of Theorem 2.2. One shows that standard monomials in $p(\tau,\varphi)$ on $X(\theta)$ of length m are linearly independent. The proof of this is formal and the idea is the same as for the case of the Grassmannian (cf. [20] also G/p-I, [22]); one has to use in addition the "special quadratic" relations, namely

$$P(\tau,\varphi)^2 = \pm P(\tau)P(\varphi) \text{ on } X_{\mathbb{Z}}(\tau)$$

$$(\text{resp. } p(\tau,\varphi)^2 = \pm p(\tau)p(\varphi) \text{ on } X(\tau))$$

On account of this; to show that standard monomials of length m on $X_{\mathbb{Z}}(\theta)$ form a basis of $H^0(X_{\mathbb{Z}}(\theta),L^m_{\mathbb{Z}})$, it suffices to prove the corresponding statement when the <u>base field is $\mathbb{Q}$</u> i.e. it suffices to show that the character of $H^0(X(\theta),L^m)$ is as stated in Remark 2.3.

One easily reduces this question to the case $m = 2$. The proof of the character formula of Remark 2.3 for the case $m = 2$ is treated in the same way as far $m = 1$ given above, except that the details for this case get a bit involved.

We will now indicate a proof of the statement that $H^i(X(\theta),L^m) = 0$, $i > 0$, $m \geqslant 0$. This would also imply that standard monomials of length m on $X(\theta)$ form a basis of $H^0(X(\theta),L^m)$. Because of the standard monomial theory already established, one gets a good hold of the ideal theory of Schubert varieties (unions, intersections etc.) and their hyperplane sections. The proof of the vanishing theorem is again by induction on $\dim X(\theta)$ and uses standard arguments using exact sequences (cf. § 9, G/P-IV, [16]) and they are very similar to the ones in 20 and G/P-III, [15].

Remark 4.1 : The argument given in Theorem 8.3, G/P-IV, [16] that $X(\tau)$ is projectively normal is not complete* (the proof only shows that the depth at the vertex of the cone over $X(\tau)$ is $\geqslant 2$); however this follows from the recent results of C. De Concini and V. Lakshmibai [2] (see Theorem 2.3). Note that in the proof of Theorem 2.2 the considerations of normality or depth do not enter (cf. § 7, [15] and § 9, [16]).

§ 5. Outline of proof of Theorems 3.1 and 3.2

The proof of the linear independence of standard monomials of multi degree m on $X(\theta)$ in G/B uses the existence of a unique

---

* This was pointed out by Hochster (as we learn from Lakshmibai).

maximal (resp. minimal) defining pair (α,β) for a given standard Young diagram (α,β) on $X(\theta)$ (cf. Lemmas 11.1 and 11.1', G/p-IV, [16]); otherwise the arguments for the proof of (i) of Theorem 3.1 run on the same lines as for the case of a maximal parabolic subgroup given in § 4. As in § 4, one proves first that standard monomials of multi degree m on $X_{\mathbb{Z}}(\theta)$ form a basis of $H^0(X_{\mathbb{Z}}(\theta), L^m_{\mathbb{Z}})$. This is done by proving the corresponding statement when the base field is $\mathbb{Q}$. This question is easily reduced to the case of type (1,1) i.e.

$$m = (m_1, \ldots, m_\ell),\ \Sigma m_i = 2,\ m_i \geq 0$$

The proof of the character formula of Remark 3.1 for type (1,1) is almost the same as for the proof of the character formula of Remark 2.3 for $m = 2$ indicated above (a complete proof of this is not given in G/p-IV, [16]).

One then gets that the canonical map

$$(*)\quad H^0(G_{\mathbb{Z}}/B_{\mathbb{Z}}, L^m_{\mathbb{Z}}) \longrightarrow H^0(X_{\mathbb{Z}}(\theta), L^m_{\mathbb{Z}})$$

is surjective. Now it can be shown that (*) implies Theorem 3.2 (Demazure's conjecture), the proof being the same (see Remark 5.1 below) as in Remark 9.6, p. 337, G/p-IV, [16]. Now by the work of Demazure [4], Theorem 3.2 implies the "vanishing theorem" i.e. the assertion (ii) of Theorem 3.1. Then as a consequence we see that standard monomials on $X(\theta)$ of multidegree m form a basis of $H^0(X(\theta), L^m)$ which is one of the assertions of Theorem 3.1. Consequently we get also that

$$(**)\quad H^0(G/B, L^m) \longrightarrow H^0(X(\theta), L^m) \rightarrow 0 \quad \text{is exact.}$$

Remark 5.1 : Note that in the proof of Remark 9.6 in G/P-IV, [16] (i.e. of Demazure's conjecture) one has made use of the hypothesis (**) above, whereas in the above outline of proof of Theorems 3.1 and Theorem 3.2, we deduced (**) after proving Theorem 3.2. One notes however that the (weaker) hypothesis (*) instead of (**) would suffice in the proof of Remark 9.6 in G/P-IV, [16].

Now by the work of Demazure [4], Theorem 3.2 implies the "vanishing theorem" i.e. the assertion (ii) of Theorem 3.1. (Note that (**) follows a posteriori).

Remark 5.2 : In Remark 16.3, p. 361, G/P-IV, [16], it has been stated that the vanishing theorem i.e. the assertion (ii) of Theorem 3.1 can also be proved on the lines of the proof of the assertion (ii) of Theorem 2.2 indicated above in § 4. (note that the assertion in Remark 16.3, [16] is a little more general). This appears possible, at least for the present case, though we have not worked out all the details. By standard monomial theory one would get a hold of the ideal theory of Schubert varieties and then to get at the required vanishing theorem, one proceeds in the same way as in § 7.2, G/P-III, [15] to get the Kempf vanishing theorem.

§ 6. Classical invariant theory.

Consider the flag variety G/B (or more generally G/P, P being any parabolic subgroup of G), G being an arbitrary semi-simple algebraic group. Recall that a Schubert cell (or Bruhat cell) is by definition a B-orbit in G/P and that it is isomorphic to an affine space. A Schubert variety in G/P is just the Zariski closure (in G/P) of a Schubert cell in G/P. The big cell C in G/P is just the B-orbit in G/P of the biggest dimension and the opposite big cell C^{o} in G/P is by definition $C^{o} = w_o C$, w_o an element of G representing the element of the Weyl group of largest length. If X is any Schubert variety in G/P, the opposite big cell in X (resp. the big cell in X) is by definition $X \cap C^{o}$ (resp. $X \cap C$). Note that the opposite big cell in G/P is again a cell isomorphic to the big cell. However the opposite big cell in a Schubert variety X is not necessarily isomorphic to the big cell in X which is again isomorphic to an affine space. In fact the opposite big cell in X may have singularities. The opposite big cell C^{o} in G/P meets every Schubert variety and the affine piece $X \cap C^{o}$ of X reflects much of the singularity and information about X.

Consider now the Grassmannian $G_{m,2m}$ of m dimensional linear subspaces of a $2m$ dimensional linear space. We have then $G_{m,2m} = SL(2m)/P$ where P is the parabolic subgroup, $P = \begin{pmatrix} * & * \\ 0 & * \end{pmatrix}$, 0 represents the zero $(m \times m)$ matrix. The big (or opposite big) cell in $G_{m,2m}$ is an affine space of dimension m^2. Let us then identify

Opposite big cell in $G_{m,2m} \approx$ Space M_m of $(m \times m)$ matrices. In fact if Z is the subgroup

$$Z = \begin{pmatrix} Id & 0 \\ Y & Id \end{pmatrix}, \quad Y \in M_m$$

of $SL(2m)$ and Ψ is the canonical map $SL(2m) \to SL(2m)/P$, Ψ maps Z

isomorphically onto the opposite big cell in $G_{m,2m}$. It can be shown (cf.[17],[21]) that we have a canonical bijective map

$$W/W(P) \xrightarrow{\sim} \bigcup_{0 \leq k \leq m} (I_m(k) \times I_m(k))$$

($k = 0$ represents the element of smallest length in $W/W(P)$)

where we recall (cf. § 1)

$$I_m(k) = \left\{(i_1,\ldots,i_k) \mid 1 \leq i_1 < \ldots < i_k \leq m\right\}$$

We simply write for $\tau \in W/W(P)$

$$\tau = ((i),(j)) \; ; \; (i),(j) \in I_m(k)$$

where $((i), (j))$ denotes the image of τ under the above canonical map. We call $((i), (j))$ the <u>canonical dual pair</u> associated to $\tau \in W/W(P)$. The dual pair <u>reverses the order</u> i.e. if

$$\tau = ((i),(j)), \; \tau' = ((i)',(j)') ; \tau, \tau' \in W/W(P)$$

then

$$\tau \geq \tau' \Leftrightarrow \left.\begin{array}{l} (i) \leq (i)' \\ (j) \leq (j)' \end{array}\right\} \text{ in the sense defined in § 1.}$$

Besides if $p_{(i),(j)}$ denotes the polynomial function

$$p_{(i),(j)} : M_m \longrightarrow k$$

which associates to every $(m \times m)$ matrix the determinant of its minor corresponding to $(i), (j)$; then we can canonically identify $p_{(i),(j)}$

with the restriction of $p(\tau)$ (the element of $H^0(G_{m,2m},L)$ associated to $\tau \in W/W(P)$, cf. § 1) to the opposite big cell M_m.

Let now D_k denote the closed subset of M_m defined by the vanishing of all $(n \times n)$ minors such that $n \geqslant (k+1)$; D_k is called a <u>determinantal variety</u>. Let $\tau \in W/W(P)$ be the element defined by a dual pair as follows :

$$\tau = ((1,\ldots,k),(1,\ldots,k))$$

It can be shown that

$$D_k = X(\tau) \cap M_m$$

i.e. D_k is the opposite big cell of the Schubert variety $X(\tau)$. Then as a consequence of standard monomial theory (cf. Theorem 1.1) we get (cf. [3], [17]) :

<u>Theorem 6.1</u> : The coordinate ring R_k of D_k has a basis consisting of standard monomials as follows :

$$\left\{\begin{array}{l} p_{(i),(j)}\, p_{(i)',(j)'}\, p_{(i)'',(j)''} \ldots, \\ \#\text{ of any } (i),(i)'\ldots, \text{ or } (j),(j)',\ldots \text{ is } \leqslant k, \\ \quad (i) \leqslant (i)' \leqslant (i)'' < \ldots , \\ \quad (j) \leqslant (j)' \leqslant (j)'' < \ldots \end{array}\right.$$

<u>Remark 6.1</u> : A particular case of the above theorem gives a basis of the algebra of polynomials on M_m indexed by <u>double standard tableaux</u> (cf. Doubilet, Rota, Stein [5]).

Let Sp(2m) denote the <u>sympletic</u> subgroup of SL(2m) of rank m, which we take as the <u>fixed point set</u> of the <u>involution</u> σ on SL(2m) :

$$\sigma(A) = \begin{pmatrix} 0 & J \\ -J & 0 \end{pmatrix} ({}^{t}A)^{-1} \begin{pmatrix} 0 & J \\ -J & 0 \end{pmatrix}^{-1}, \quad J = \begin{pmatrix} 0 & & 1 \\ & \cdot^{\cdot^{\cdot}} & \\ 1 & & 0 \end{pmatrix}$$

Then $Q = Sp(2m) \cap P$ is also a maximal parabolic subgroup of Sp(2m) and Sp(2m)/Q is the symplectic Grassmannian formed by maximal (i.e. of dimension m) isotropic subspaces of a 2m dimensional vector space. We see that the subgroup Z of SL(2m) is σ-stable and the fixed point subset Z^{σ} of Z under σ can be identified with the subset

$$\begin{pmatrix} Id & 0 \\ Y & Id \end{pmatrix}, \quad Y = J\,{}^{t}Y\,J \Longleftrightarrow (Id, Y) \begin{pmatrix} 0 & J \\ -J & 0 \end{pmatrix} \begin{pmatrix} Id \\ Y \end{pmatrix} = 0$$

and that Z^{σ} can be identified with the opposite big cell of Sp(2m)/Q. Setting Y = JX, we have

$$Y = J\,{}^{t}Y\,J \Longleftrightarrow {}^{t}X = X$$

i.e. the σ fixed subset of Z can be identified with the space Sym M_m of all <u>symmetric</u> (m x m) matrices. Thus we can identify <u>Sym M_m with the opposite big cell</u> in <u>Sp(2m)/Q</u>. It can also be checked easily (cf.[17]) that

$$(*) \quad W(Sp(2m))/W(Q) = \left\{ \tau \in W(SL(2m))/W(P) \;\middle|\; \tau = ((i),(j)) \text{ such that } (i) = (j) \right\}$$

and that in this identification of W(Sp(2m))/W(Q) as a subset of W(SL(2m))/W(P), the corresponding partial orders are preserved.

Because of (*) we see that

$$W(SL(2m))/W(P) \hookrightarrow W(Sp(2m))/W(Q) \times W(Sp(2m))/W(Q)$$

in fact, if

$$\lambda \in W(SL(2m))/W(P), \quad \lambda = ((i),(j)), \text{ then}$$

$$\left.\begin{array}{l} \lambda \longmapsto (\mu,\nu); \quad \mu = ((i),(i)) \\ \qquad\qquad\qquad \nu = ((j),(j)) \end{array}\right\} \text{ are in } W(Sp(2m))/W(Q).$$

Now the maximal parabolic group Q is <u>not minuscule</u> and hence there are admissible pairs in $W(Sp(2m))/W(Q)$ which are not trivial. In fact, we have the following (cf. Lemma 5.1, G/P-II, [17]).

<u>Lemma 6.1</u> : Let $\tau = ((i),(i))$ and $\varphi = ((j),(j))$ be in $W(Sp(2m))/W(Q)$. Then (τ,φ) is an <u>admissible pair</u> if and only if $(j) \leqslant (i)$.

<u>Remark 6.2</u> : Note that an admissible pair (τ,φ) in $W(Sp(2m))/W(Q)$ can be identified with an element λ in $W(SL(2m))/W(P)$; in fact if $\tau = ((j),(j))$, $\varphi = ((i),(i))$, then the associated element λ of $W(SL(2m))/W(P)$ is precisely the element given by the dual pair $((j),(j))$. Let us now define the element $p(\tau,\varphi)$ in $H^0(Sp(2m)/Q, L)$ associated to the admissible pair (τ,φ) as the restriction of $p(\lambda) \in H^0(SL(2m)/P, L)$ to the subvariety $Sp(2m)/Q$ of $SL(2m)/P$. It is not difficult to see that $\{p(\tau,\varphi)\}$ satisfy the properties (ii) and (iii) of Theorem 2.1. The $\{p(\tau,\varphi)\}$ in fact provide a basis satisfying the properties of Theorem 2.1. Now the restriction of $p(\lambda)$ to M_m can be identified with

the function $p_{(i),(j)}$ as we saw above. We write by the same $p_{(i),(j)}$ the restriction of $p_{(i),(j)}$ to $\mathrm{Sym}\, M_m$.

Let $\mathrm{Sym}\, D_k = D_k \cap \mathrm{Sym}\, M_m$; we call this a <u>symmetric determinantal variety</u>. The element

$$\tau = ((1,\ldots,k),(1,\ldots,k))$$

in fact lies in $W(Sp(2m))/W(Q)$ and if $X(\tau)$ denotes the Schubert variety in $Sp(2m)/_Q$ defined by τ, it can be seen that $X(\tau) \cap \mathrm{Sym}\, M_m = \mathrm{Sym}\, D_k$ i.e. the symmetric determinantal variety $\mathrm{Sym}\, D_k$ is the opposite big cell of the Schubert variety $X(\tau)$ of $Sp(2m)/_Q$. The preceding discussion together with Theorem 2.2 then yield the following (cf. [3], [17]).

<u>Theorem 6.2</u> : The coordinate ring of the symmetric determinantal variety $\mathrm{Sym}\, D_k$ has a basis consisting of <u>standard monomials</u> as follows :

$$\begin{cases} p_{(i),(j)}\, p_{(i)',(j)'}\, p_{(i)'',(j)''} \cdots , \\ (i) \leq (j) \leq (i)' \leq (j)' \leq (i)'' \leq (j)'' < \ldots \end{cases}$$

Let now X stand for the affine space (represented by its geometric points).

$$X = \underbrace{V \times \ldots \times V}_{m \text{ times}} \times \underbrace{V^{\vee} \times \ldots \times V^{\vee}}_{m \text{ times}}, \quad V \text{ vector space of dimension } n.$$

Let us take the canonical diagonal action of $GL(n)$ on X and let R denote the coordinate ring of X. Write

$$x \in X,\ x = (x_1,\ldots,x_m,\ y_1,\ldots,y_m).$$

Consider then the canonical morphism

$$\Psi : X \longrightarrow M_m$$

defined by

$$x \longmapsto \text{the matrix } |\langle x_i, y_j \rangle|$$

where $\langle\ ,\ \rangle$ denotes the canonical bilinear form on $V \times V^{\vee}$. Then we have (cf. [3], [8], [9], [17]):

<u>Theorem 6.3</u> : The morphism Ψ factors through the determinantal variety D_n i.e. we have a commutative diagram

$$\begin{array}{ccc} X & \xrightarrow{\Psi} & M_m \\ & \varphi\searrow & \nearrow \\ & & D_n \end{array}$$

Besides the morphism $\varphi : X \longrightarrow D_n$ is <u>surjective</u> and we have a canonical identification (through φ) of the coordinate ring of D_n with the sub-ring $R^{GL(n)}$ of R formed by $GL(n)$ invariants, so that by Theorem 6.1, we get a basis of $R^{GL(n)}$ by <u>standard monomials</u> as in that theorem.

Let now X stand for the affine space (represented by its geometric points)

$$X = \underbrace{V \times \ldots \times V}_{m \text{ times}}, \ V \text{ vector space of dimension } 2n.$$

Consider the diagonal action of the orthogonal group $O(2n)$ on X

(base field of characteristic $\neq 2$). Let R = coordinate ring of X. Consider the canonical morphism

$$\Upsilon : X \longrightarrow \mathrm{Sym}\, M_m \text{ , defined by}$$

$$x = (x_1, \ldots, x_m) \longmapsto |\langle x_i, x_j \rangle|$$

where $\langle\ ,\ \rangle$ denotes a scalar product on V such that $O(2n)$ is the group preserving this scalar product. Then we have (cf. [3], [14], [17]).

<u>Theorem 6.4</u> : The morphism Υ factors through the symmetric determinantal variety $\mathrm{Sym}\, D_{2n}$ i.e. we have a commutative diagram

$$\begin{array}{ccc} X & \xrightarrow{\ \Upsilon\ } & \mathrm{Sym}\, M_m \\ & {\scriptstyle\varphi}\searrow \quad \nearrow & \\ & \mathrm{Sym}\, D_{2n} & \end{array}$$

Besides the morphism φ is <u>surjective</u> and we have a canonical identification of the subring $R^{O(2n)}$ of R formed by $O(2n)$ invariants, so that by Theorem 6.2 we have a basis of $R^{O(2n)}$ by standard monomials as in that theorem.

<u>Remark 6.3</u> : A similar theorem holds for the canonical diagonal action of $Sp(2n)$ on $V \times \ldots \times V$ (m times), $\dim V = 2n$ (cf. [3], [17]).

Let us very briefly outline a proof of Theorems 6.3 and 6.4. It is easy to see that Υ factors through D_n (resp. $\mathrm{Sym}\, D_{2n}$) and that Υ is a $GL(n)$ (resp. $O(2n)$) <u>invariant morphism</u> (i.e. for the trivial action of $GL(n)$ (resp. $O(2n)$) on D_{2n} (resp. $\mathrm{Sym}\, D_{2n}$), Υ is a $GL(n)$

(resp. $O(2n)$ morphism). Further, it is not difficult to see that over a non-empty Zariski open subset of D_n (resp. Sym D_{2n}), φ is a $GL(n)$ (resp. $O(2n)$) principal fibration (for this one takes m sufficiently large and this would suffice for the proof of the theorem). These considerations show that φ induces a canonical __birational morphism__

$$j : \text{Spec } R^G \longrightarrow D_n \text{ (resp. Sym } D_{2n})$$

We are required to show that j is an __isomorphism__. It is not difficult to show that j is a __finite morphism__; this is an easy consequence of the fact if $x \in X$ and is __semi-stable__ in the sense of Mumford, $\Upsilon(x) \neq 0$ (cf. Mumford [19]). Obviously R^G is a normal ring. Hence to show that j is an isomorphism it suffices to know that D_n (resp. Sym D_{2n}) is a normal variety. This is a consequence of Theorem 2.3.

__Remark 6.4__ : The most important point in the above proof is the __normality__ of D_n (resp. Sym D_{2n}) and in practice, in similar questions, this appears to be the hardest to achieve.

__Remark 6.5__ : Theorems 6.1, 6.2, 6.3 and 6.4 (esp. 6.3 and 6.4, the case of the linear group is perhaps more classical, cf. [8], [9]) were proved by DeConcini and Procesi, cf. [3]. However, there was no connection with Schubert varieties in [3]. The connection with Schubert varieties, especially for the case of envariants under the orthogonal group was noted in G/p-II, [17]. The basis written by De Concini and Procesi for

this case provided the clue to index a basis of a non-minuscule fundamental representation by means of admissible pairs (see Lemma 6.1 above and Lemma 5.1, [17]) and the main results on standard monomial theory (esp in § 2) were then conjectured in G/p-II, [17].

References

1. H.H. Andersen : The Frobenius morphism on the cohomology of homogeneous vector bundles on G/B- Ann. Math., vol.112, 1980, p. 113-121.

2. C. DeConcini and V. Lakshmibai : Arithmetic Cohen-Macanlayness and Arithmetic normality of Schubert varieties, To appear in Amer. J. Maths.

3. C. De Concini and C. Procesi : A characteristic free approach to invariant theory, Advances in Math., vol. 21, p. 330-354.

4. M. Demazure : Désingularisation des variétés de Schubert generalisées Ann. Sc. Ec. Norm. Sup, t. 7, 1974, p. 53 - 88.

5. P. Doubilet, Gian-Carlo Rota and J. Stein : On the foundations of Combinatorial theory : IX (Combinatorial methods in invariant theory), Studies in applied mathematics, vol.L III, 1974, p. 185-216.

6. D. Eisenbud : Introduction to algebras with straightening laws, Ring theory and algebra III, Proceedings of the third Oklahoma conference, edited by B.R. Mc Donald, Marcel Dekker, Inc.,1980.

7. W.J. Haboush : A short proof of the Kempf vanishing theorem, Inventiones Math., vol. 56, 1980, p. 109-112.

8. M. Hochster : Grassmannians and their Schubert varieties are Cohen-Macaulay, J. Algebra, vol. 25, 1973, p. 40-57.

9. M. Hochster and J.A. Eagon : Cohen Macaulay rings, invariant theory and the generic perfection of the determinal loci, Amer. J. Math ., Vol. XC III, 1971, p. 1020 - 1058.

10. W.V.D. Hodge : Some enumerative results in the theory of forms, Proc. Camb. Phil. Soc., Vol. 39, p. 22-30.

11. W.V.D. Hodge and D. Pedue : Methods of algebraic geometry, Cambridge Univ. Press, vol.II

12. G. Kempf : Schubert methods with an application to algebraic curves, Stichting mathematisch centrum, Amsterdam, 1971.

13. G. Kempf : Linear systems on homogeneous spaces, Ann. of Math. (Second Series), vol. 103, p. 557-591.

14. R.E. Kutz : Cohen-Macaulay rings and ideal theory in rings of invariants of algebraic groups, Trans.Amer. Math. Soc., vol.194, 1974, p. 115-129.

15. V. Lakshmibai, C. Musili and C.S. Seshadri : Geometry of G/p-III (Standard monomial theory for a quasi-minuscule P), Proc. Indian Acad. Sci, vol. 87, 1978, p. 93-177.

16. V. Lakshmibai, C. Musili and C.S. Seshadri : Geometry of G/p-IV, Proc. Indian Acad. Sci., vol. 88 A, 1979, p. 280-362.

17. V. Lakshmibai and C.S. Seshadri : Geometry of G/p-II (the work of De Concini and Procesi and the basic conjectures), Proc. Indian Acad. Sci. vol. 87, 1978, p. 1-54.

18. D. Laksov : The arithmetic Cohen-Macaulay character of Schubert schemes, Acta Mathematica, vol. 129, p. 1-9.

19. D. Mumford : Geometric Invariant theory, Ergebnisse der Math. und Ihrer Grenzgebiete, 34, Springer-Verlag, Berlin, 1965.

20. C. Musili : Postulation formula for Schubert varieties, J. Indian Math. Soc., vol. 36, 1972, p. 143-171.

21. C. Musili : Some properties of Schubert varieties, J. Indian Math. Soc., vol. 38, 1974, p. 131-145.

22. C.S. Seshadri : Geometry of G/p-I (Standard monomial theory for a minuscule P), C.P. Ramanujam : A Tribute, 207 (Springer-Verlag), Published for the Tata Institute of Fundamental Research, Bombay, 1978, p. 207-239.

23. R.Steinberg: Lectures on Chevalley groups, Lecture Notes in Mathematics, (New Haven, Conn, Yale Univ.), 1967, p.243-268.

School of Mathematics
Tata Institute of Fundamental Research
Homi Bhabha Road
Bombay 400 005 (India)

29.1.81

Vol. 700: Module Theory, Proceedings, 1977. Edited by C. Faith and S. Wiegand. X, 239 pages. 1979.

Vol. 701: Functional Analysis Methods in Numerical Analysis, Proceedings, 1977. Edited by M. Zuhair Nashed. VII, 333 pages. 1979.

Vol. 702: Yuri N. Bibikov, Local Theory of Nonlinear Analytic Ordinary Differential Equations. IX, 147 pages. 1979.

Vol. 703: Equadiff IV, Proceedings, 1977. Edited by J. Fábera. XIX, 441 pages. 1979.

Vol. 704: Computing Methods in Applied Sciences and Engineering, 1977, I. Proceedings, 1977. Edited by R. Glowinski and J. L. Lions. VI, 391 pages. 1979.

Vol. 705: O. Forster und K. Knorr, Konstruktion verseller Familien kompakter komplexer Räume. VII, 141 Seiten. 1979.

Vol. 706: Probability Measures on Groups, Proceedings, 1978. Edited by H. Heyer. XIII, 348 pages. 1979.

Vol. 707: R. Zielke, Discontinuous Čebyšev Systems. VI, 111 pages. 1979.

Vol. 708: J. P. Jouanolou, Equations de Pfaff algébriques. V, 255 pages. 1979.

Vol. 709: Probability in Banach Spaces II. Proceedings, 1978. Edited by A. Beck. V, 205 pages. 1979.

Vol. 710: Séminaire Bourbaki vol. 1977/78, Exposés 507–524. IV, 328 pages. 1979.

Vol. 711: Asymptotic Analysis. Edited by F. Verhulst. V, 240 pages. 1979.

Vol. 712: Equations Différentielles et Systèmes de Pfaff dans le Champ Complexe. Edité par R. Gérard et J.-P. Ramis. V, 364 pages. 1979.

Vol. 713: Séminaire de Théorie du Potentiel, Paris No. 4. Edité par F. Hirsch et G. Mokobodzki. VII, 281 pages. 1979.

Vol. 714: J. Jacod, Calcul Stochastique et Problèmes de Martingales. X, 539 pages. 1979.

Vol. 715: Inder Bir S. Passi, Group Rings and Their Augmentation Ideals. VI, 137 pages. 1979.

Vol. 716: M. A. Scheunert, The Theory of Lie Superalgebras. X, 271 pages. 1979.

Vol. 717: Grosser, Bidualräume und Vervollständigungen von Banachmoduln. III, 209 pages. 1979.

Vol. 718: J. Ferrante and C. W. Rackoff, The Computational Complexity of Logical Theories. X, 243 pages. 1979.

Vol. 719: Categorial Topology, Proceedings, 1978. Edited by H. Herrlich and G. Preuß. XII, 420 pages. 1979.

Vol. 720: E. Dubinsky, The Structure of Nuclear Fréchet Spaces. V, 187 pages. 1979.

Vol. 721: Séminaire de Probabilités XIII. Proceedings, Strasbourg, 1977/78. Edité par C. Dellacherie, P. A. Meyer et M. Weil. VII, 647 pages. 1979.

Vol. 722: Topology of Low-Dimensional Manifolds. Proceedings, 1977. Edited by R. Fenn. VI, 154 pages. 1979.

Vol. 723: W. Brandal, Commutative Rings whose Finitely Generated Modules Decompose. II, 116 pages. 1979.

Vol. 724: D. Griffeath, Additive and Cancellative Interacting Particle Systems. V, 108 pages. 1979.

Vol. 725: Algèbres d'Opérateurs. Proceedings, 1978. Edité par P. de la Harpe. VII, 309 pages. 1979.

Vol. 726: Y.-C. Wong, Schwartz Spaces, Nuclear Spaces and Tensor Products. VI, 418 pages. 1979.

Vol. 727: Y. Saito, Spectral Representations for Schrödinger Operators With Long-Range Potentials. V, 149 pages. 1979.

Vol. 728: Non-Commutative Harmonic Analysis. Proceedings, 1978. Edited by J. Carmona and M. Vergne. V, 244 pages. 1979.

Vol. 729: Ergodic Theory. Proceedings, 1978. Edited by M. Denker and K. Jacobs. XII, 209 pages. 1979.

Vol. 730: Functional Differential Equations and Approximation of Fixed Points. Proceedings, 1978. Edited by H.-O. Peitgen and H.-O. Walther. XV, 503 pages. 1979.

Vol. 731: Y. Nakagami and M. Takesaki, Duality for Crossed Products of von Neumann Algebras. IX, 139 pages. 1979.

Vol. 732: Algebraic Geometry. Proceedings, 1978. Edited by K. Lønsted. IV, 658 pages. 1979.

Vol. 733: F. Bloom, Modern Differential Geometric Techniques in the Theory of Continuous Distributions of Dislocations. XII, 206 pages. 1979.

Vol. 734: Ring Theory, Waterloo, 1978. Proceedings, 1978. Edited by D. Handelman and J. Lawrence. XI, 352 pages. 1979.

Vol. 735: B. Aupetit, Propriétés Spectrales des Algèbres de Banach. XII, 192 pages. 1979.

Vol. 736: E. Behrends, M-Structure and the Banach-Stone Theorem. X, 217 pages. 1979.

Vol. 737: Volterra Equations. Proceedings 1978. Edited by S.-O. Londen and O. J. Staffans. VIII, 314 pages. 1979.

Vol. 738: P. E. Conner, Differentiable Periodic Maps. 2nd edition, IV, 181 pages. 1979.

Vol. 739: Analyse Harmonique sur les Groupes de Lie II. Proceedings, 1976–78. Edited by P. Eymard et al. VI, 646 pages. 1979.

Vol. 740: Séminaire d'Algèbre Paul Dubreil. Proceedings, 1977–78. Edited by M.-P. Malliavin. V, 456 pages. 1979.

Vol. 741: Algebraic Topology, Waterloo 1978. Proceedings. Edited by P. Hoffman and V. Snaith. XI, 655 pages. 1979.

Vol. 742: K. Clancey, Seminormal Operators. VII, 125 pages. 1979.

Vol. 743: Romanian-Finnish Seminar on Complex Analysis. Proceedings, 1976. Edited by C. Andreian Cazacu et al. XVI, 713 pages. 1979.

Vol. 744: I. Reiner and K. W. Roggenkamp, Integral Representations. VIII, 275 pages. 1979.

Vol. 745: D. K. Haley, Equational Compactness in Rings. III, 167 pages. 1979.

Vol. 746: P. Hoffman, τ-Rings and Wreath Product Representations. V, 148 pages. 1979.

Vol. 747: Complex Analysis, Joensuu 1978. Proceedings, 1978. Edited by I. Laine, O. Lehto and T. Sorvali. XV, 450 pages. 1979.

Vol. 748: Combinatorial Mathematics VI. Proceedings, 1978. Edited by A. F. Horadam and W. D. Wallis. IX, 206 pages. 1979.

Vol. 749: V. Girault and P.-A. Raviart, Finite Element Approximation of the Navier-Stokes Equations. VII, 200 pages. 1979.

Vol. 750: J. C. Jantzen, Moduln mit einem höchsten Gewicht. III, 195 Seiten. 1979.

Vol. 751: Number Theory, Carbondale 1979. Proceedings. Edited by M. B. Nathanson. V, 342 pages. 1979.

Vol. 752: M. Barr, *-Autonomous Categories. VI, 140 pages. 1979.

Vol. 753: Applications of Sheaves. Proceedings, 1977. Edited by M. Fourman, C. Mulvey and D. Scott. XIV, 779 pages. 1979.

Vol. 754: O. A. Laudal, Formal Moduli of Algebraic Structures. III, 161 pages. 1979.

Vol. 755: Global Analysis. Proceedings, 1978. Edited by M. Grmela and J. E. Marsden. VII, 377 pages. 1979.

Vol. 756: H. O. Cordes, Elliptic Pseudo-Differential Operators – An Abstract Theory. IX, 331 pages. 1979.

Vol. 757: Smoothing Techniques for Curve Estimation. Proceedings, 1979. Edited by Th. Gasser and M. Rosenblatt. V, 245 pages. 1979.

Vol. 758: C. Năstăsescu and F. Van Oystaeyen; Graded and Filtered Rings and Modules. X, 148 pages. 1979.

Vol. 759: R. L. Epstein, Degrees of Unsolvability: Structure and Theory. XIV, 216 pages. 1979.

Vol. 760: H.-O. Georgii, Canonical Gibbs Measures. VIII, 190 pages. 1979.

Vol. 761: K. Johannson, Homotopy Equivalences of 3-Manifolds with Boundaries. 2, 303 pages. 1979.

Vol. 762: D. H. Sattinger, Group Theoretic Methods in Bifurcation Theory. V, 241 pages. 1979.

Vol. 763: Algebraic Topology, Aarhus 1978. Proceedings, 1978. Edited by J. L. Dupont and H. Madsen. VI, 695 pages. 1979.

Vol. 764: B. Srinivasan, Representations of Finite Chevalley Groups. XI, 177 pages. 1979.

Vol. 765: Padé Approximation and its Applications. Proceedings, 1979. Edited by L. Wuytack. VI, 392 pages. 1979.

Vol. 766: T. tom Dieck, Transformation Groups and Representation Theory. VIII, 309 pages. 1979.

Vol. 767: M. Namba, Families of Meromorphic Functions on Compact Riemann Surfaces. XII, 284 pages. 1979.

Vol. 768: R. S. Doran and J. Wichmann, Approximate Identities and Factorization in Banach Modules. X. 305 pages. 1979.

Vol. 769: J. Flum, M. Ziegler, Topological Model Theory. X, 151 pages. 1980.

Vol. 770: Séminaire Bourbaki vol. 1978/79 Exposés 525–542. IV, 341 pages. 1980.

Vol. 771: Approximation Methods for Navier-Stokes Problems. Proceedings, 1979. Edited by R. Rautmann. XVI, 581 pages. 1980.

Vol. 772: J. P. Levine, Algebraic Structure of Knot Modules. XI, 104 pages. 1980.

Vol. 773: Numerical Analysis. Proceedings, 1979. Edited by G. A. Watson. X, 184 pages. 1980.

Vol. 774: R. Azencott, Y. Guivarc'h, R. F. Gundy, Ecole d'Eté de Probabilités de Saint-Flour VIII-1978. Edited by P. L. Hennequin. XIII, 334 pages. 1980.

Vol. 775: Geometric Methods in Mathematical Physics. Proceedings, 1979. Edited by G. Kaiser and J. E. Marsden. VII, 257 pages. 1980.

Vol. 776: B. Gross, Arithmetic on Elliptic Curves with Complex Multiplication. V, 95 pages. 1980.

Vol. 777: Séminaire sur les Singularités des Surfaces. Proceedings, 1976-1977. Edited by M. Demazure, H. Pinkham and B. Teissier. IX, 339 pages. 1980.

Vol. 778: SK_1 von Schiefkörpern. Proceedings, 1976. Edited by P. Draxl and M. Kneser. II, 124 pages. 1980.

Vol. 779: Euclidean Harmonic Analysis. Proceedings, 1979. Edited by J. J. Benedetto. III, 177 pages. 1980.

Vol. 780: L. Schwartz, Semi-Martingales sur des Variétés, et Martingales Conformes sur des Variétés Analytiques Complexes. XV, 132 pages. 1980.

Vol. 781: Harmonic Analysis Iraklion 1978. Proceedings 1978. Edited by N. Petridis, S. K. Pichorides and N. Varopoulos. V, 213 pages. 1980.

Vol. 782: Bifurcation and Nonlinear Eigenvalue Problems. Proceedings, 1978. Edited by C. Bardos, J. M. Lasry and M. Schatzman. VIII, 296 pages. 1980.

Vol. 783: A. Dinghas, Wertverteilung meromorpher Funktionen in ein- und mehrfach zusammenhängenden Gebieten. Edited by R. Nevanlinna and C. Andreian Cazacu. XIII, 145 pages. 1980.

Vol. 784: Séminaire de Probabilités XIV. Proceedings, 1978/79. Edited by J. Azéma and M. Yor. VIII, 546 pages. 1980.

Vol. 785: W. M. Schmidt, Diophantine Approximation. X, 299 pages. 1980.

Vol. 786: I. J. Maddox, Infinite Matrices of Operators. V, 122 pages. 1980.

Vol. 787: Potential Theory, Copenhagen 1979. Proceedings, 1979. Edited by C. Berg, G. Forst and B. Fuglede. VIII, 319 pages. 1980.

Vol. 788: Topology Symposium, Siegen 1979. Proceedings, 1979. Edited by U. Koschorke and W. D. Neumann. VIII, 495 pages. 1980.

Vol. 789: J. E. Humphreys, Arithmetic Groups. VII, 158 pages. 1980.

Vol. 790: W. Dicks, Groups, Trees and Projective Modules. IX, 127 pages. 1980.

Vol. 791: K. W. Bauer and S. Ruscheweyh, Differential Operators for Partial Differential Equations and Function Theoretic Applications. V, 258 pages. 1980.

Vol. 792: Geometry and Differential Geometry. Proceedings, 1979. Edited by R. Artzy and I. Vaisman. VI, 443 pages. 1980.

Vol. 793: J. Renault, A Groupoid Approach to C*-Algebras. III, 160 pages. 1980.

Vol. 794: Measure Theory, Oberwolfach 1979. Proceedings 1979. Edited by D. Kölzow. XV, 573 pages. 1980.

Vol. 795: Séminaire d'Algèbre Paul Dubreil et Marie-Paule Malliavin. Proceedings 1979. Edited by M. P. Malliavin. V, 433 pages. 1980.

Vol. 796: C. Constantinescu, Duality in Measure Theory. IV, 197 pages. 1980.

Vol. 797: S. Mäki, The Determination of Units in Real Cyclic Sextic Fields. III, 198 pages. 1980.

Vol. 798: Analytic Functions, Kozubnik 1979. Proceedings. Edited by J. Ławrynowicz. X, 476 pages. 1980.

Vol. 799: Functional Differential Equations and Bifurcation. Proceedings 1979. Edited by A. F. Izé. XXII, 409 pages. 1980.

Vol. 800: M.-F. Vignéras, Arithmétique des Algèbres de Quaternions. VII, 169 pages. 1980.

Vol. 801: K. Floret, Weakly Compact Sets. VII, 123 pages. 1980.

Vol. 802: J. Bair, R. Fourneau, Etude Géometrique des Espaces Vectoriels II. VII, 283 pages. 1980.

Vol. 803: F.-Y. Maeda, Dirichlet Integrals on Harmonic Spaces. X, 180 pages. 1980.

Vol. 804: M. Matsuda, First Order Algebraic Differential Equations. VII, 111 pages. 1980.

Vol. 805: O. Kowalski, Generalized Symmetric Spaces. XII, 187 pages. 1980.

Vol. 806: Burnside Groups. Proceedings, 1977. Edited by J. L. Mennicke. V, 274 pages. 1980.

Vol. 807: Fonctions de Plusieurs Variables Complexes IV. Proceedings, 1979. Edited by F. Norguet. IX, 198 pages. 1980.

Vol. 808: G. Maury et J. Raynaud, Ordres Maximaux au Sens de K. Asano. VIII, 192 pages. 1980.

Vol. 809: I. Gumowski and Ch. Mira, Recurences and Discrete Dynamic Systems. VI, 272 pages. 1980.

Vol. 810: Geometrical Approaches to Differential Equations. Proceedings 1979. Edited by R. Martini. VII, 339 pages. 1980.

Vol. 811: D. Normann, Recursion on the Countable Functionals. VIII, 191 pages. 1980.

Vol. 812: Y. Namikawa, Toroidal Compactification of Siegel Spaces. VIII, 162 pages. 1980.

Vol. 813: A. Campillo, Algebroid Curves in Positive Characteristic. V, 168 pages. 1980.

Vol. 814: Séminaire de Théorie du Potentiel, Paris, No. 5. Proceedings. Edited by F. Hirsch et G. Mokobodzki. IV, 239 pages. 1980.

Vol. 815: P. J. Slodowy, Simple Singularities and Simple Algebraic Groups. XI, 175 pages. 1980.

Vol. 816: L. Stoica, Local Operators and Markov Processes. VIII, 104 pages. 1980.